Palladium Reagents and Catalysts

Palladium Reagents and Catalysts

Innovations in Organic Synthesis

JIRO TSUJI

Okayama University of Science, Okayama, Japan

JOHN WILEY & SONS

Chichester · New York · Brisbane · Toronto · Singapore

Reprinted with corrections, December 1996
Reprinted August 1997, March 1998, March 1999

Other Wiley Editorial Offices

John Wiley & Sons, Inc., 605 Third Avenue,
New York, NY 10158-0012, USA

WILEY-VCH Verlag GmbH, Pappelallee 3,
D-69469 Weinheim, Germany

Jacaranda Wiley Ltd, 33 Park Road, Milton,
Queensland 4064, Australia

John Wiley & Sons (Asia) Pte Ltd, 2 Clementi Loop #02-01,
Jin Xing Distripark, Singapore 129809

John Wiley & Sons (Canada) Ltd, 22 Worcester Road,
Rexdale, Ontario M9W 1L1, Canada

British Library Cataloguing in Publication Data

A catalogue record for this book is available from the British Library

ISBN   0 471 95483 7 (HB)   0 471 97202 9 (PB)

Typeset in 10/12pt Times by Alden Multimedia
Printed and bound in Great Britain by Biddles Ltd, Guildford and King's Lynn

This book is printed on acid-free paper responsibly manufactured from sustainable forestry,
in which at least two trees are planted for each one used for paper production.

Contents

Preface

Modern palladium chemistry started in 1960 with the ingenious invention of an industrial process for acetaldehyde production by the air oxidation of ethylene, catalyzed by PdCl$_2$ and CuCl$_2$, which is called the Wacker process. In 1965 I was fortunate enough to discover that carbon–carbon bond formation can be achieved by reacting π-allylpalladium and π-PdCl$_2$ complexes of 1,5-cyclooctadiene with carbon nucleophiles, typically active methylene compounds. This discovery was significant because all hitherto known organometallic compounds react with electrophiles. Since then, remarkable progress has been made in organic synthesis using Pd compounds both as stoichiometric reagents and catalysts.

In 1980, I wrote a book entitled *Organic Synthesis with Palladium Compounds* to summarize significant reactions which had been shown to be catalyzed or promoted by Pd(0) and Pd(II) compounds. The book covered the literature up to 1978. Some 15 years later, even more remarkable progress has been made in organic synthesis using palladium compounds. It is true that many transition metals are now used in organic synthesis, but it is widely recognized that palladium is the most versatile in promoting or catalyzing reactions, particularly those involving carbon–carbon bond formation, which is not always easy to achieve with other transition metals. Indeed, it is difficult these days to find a single issue of a major journal of organic chemistry which does not contain a paper involving the use of palladium in synthesis. I feel that another comprehensive book is now needed to cover the explosive growth of the chemistry of palladium over the last decade. Mention should be made of a book entitled *Palladium Reagents in Organic Synthesis* by R. F. Heck, which appeared in 1985 with emphasis on experimental procedures.

I undertook the present task to give a birds-eye view of the broad field of palladium in organic synthesis. I have tried to accomplish this task by citing many references; these were selected from a much larger number which I have collected over the years. I tried to be as comprehensive as possible by selecting those references which reported original ideas and new reactions, or evident synthetic utility. Synthetic utility is clearly biased towards catalytic rather than stoichiometric reactions and this emphasis is apparent in the selection of the

xi

references. In this book, references published before 1993 and a few published in 1994 are covered. The overall task of selecting references to include was very difficult, and I imagine that many researchers will complain that their important papers were not cited. I apologize for significant omissions. I also apologize for the errors and incorrect citations which must inevitably be present: these are my responsibility alone. I believed that it was better to finish an imperfect book than never to finish a perfect one. I have devoted most of my research life to the organic chemistry of palladium and I will be more than happy if this book stimulates in any way further developments in this field.

I wish to acknowledge valuable suggestions and corrections given by Professor H. Okumoto, Kurasiki University of Science and the Arts), who read the whole manuscript. I also thank to my wife Yoshiko, who helped enormously, and in particular produced all the chemical structures in the book.

Jiro Tsuji
July 1994, Okayama

Abbreviations

Ac	acetyl
acac	acetylacetonato
Ar	aryl
atm	atmospheric pressure
BBEDA	*N, N′*-bis(benzylidene)ethylenediamine
9-BBN	9-borabicyclo[3.3.1]nonanyl
9-BBNH	9-borabicyclo[3.3.1]nonane
Bn	benzyl
BINAP	2,2′-bis(diphenylphosphino)-1,1′binaphthyl
BPPFA	1-[1,2-bis(diphenylphosphino)ferrocenyl]ethyldimethylamine
(*R*)-(*S*)-BPPFOH	(*R*)-2[(*S*)-1′,2-bis(diphenylphosphino)ferrocenyl]ethyl alcohol
BPPM	(2*S*, 4*S*)-*N*-t-butoxycarbonyl-2,4-bis(diphenylphosphino)methyl-pyrrolidine
bpy	2,2′-bipyridyl or bipyridine
Boc	*t*-butoxycarbonyl
BQ	1,4-benzoquinone
Bu	butyl
BSA	*N,O*-bis(trimethylsilyl)acetamide
CDT	1,5,9-cyclododecatriene
COD	1,5-cyclooctadiene
Cp	cyclopentadienyl
dba	dibenzylideneacetone
DBU	1,8-diazabicyclo[5.4.0]undec-7-ene
dchpe	bis(dicyclohexylphosphino)ethane
DEAD	diethyl azodicarboxylate
dippp	bis(diisopropylphosphino)propane
DIOP	2,3-*O*-isopropylidene-2,3-dihydroxy-1,4-bis(diphenylphosphino)-butane
DMAD	dimethyl acetylenedicarboxylate
DMBA	*N,N*-dimethylbarbituric acid

DPMSPP	diphenyl(*m*-sulfophenyl)phosphine
dppb	bis(diphenylphosphino)butane
dppe	bis(diphenylphosphino)ethane
dppf	1,1′-bis(diphenylphosphino)ferrocene
dppp	bis(diphenylphosphino)propane
EWG	electron-withdrawing group
MA	maleic anhydride
MOM	methoxymethyl
MOP	monodentate optically active phosphine
NORPHOS	(*R,R*)-5,6-bis(diphenylphosphino)-2-norbornene
Nu	nucleophile
phen	1,10-phenanthroline
PhH	benzene
PhMe	toluene
PHMS	poly(hydromethylsiloxane)
PPFA	*N,N*-dimethyl-1,2-(diphenylphosphino)ferrocenylethylamine
py	pyridine
TASF	tris(diethylamino)sulfonium difluoro(trimethyl)silicate
TBAC	tetrabutylammonium chloride
TBAF	tetrabutylammonium fluoride
TBDMS	*t*-butyldimethylsilyl
TCPC	2,3,4,5-tetrakis(methoxycarbonyl)palladacyclopentadiene
TDMPP	tri(2,6-dimethoxyphenyl)phosphine
Tf	trifluoromethylsulfonyl (triflyl)
TFP	tri(2-furyl)phosphine
TMEDA	*N,N,N′,N′*-tetramethylethylenediamine
TMM	trimethylenemethane
TMPP	trimethylolpropane phosphite or 4-ethyl-2,6,7-trioxa-1-phospho-bicyclo[2.2.2]octane
TMS	trimethylsilyl
TMSPP	tri(*m*-sulfophenyl)phosphine
Tol	tolyl
Ts	tosyl
TsOH	*p*-toluenesulfonic acid
TTMPP	tri(2,4,6-trimethoxyphenyl)phosphine
Tu	thiourea

Chapter 1

The Basic Chemistry of Organopalladium Compounds

1 Characteristic Features of Pd-Promoted or -Catalyzed Reactions

There are several features which make reactions involving Pd particularly useful and versatile among many transition metals used for organic synthesis. Most importantly, Pd offers many possibilities of carbon–carbon bond formation. The importance of carbon–carbon bond formation in organic synthesis needs no explanation, and no other transition metals can offer such versatile methods for carbon–carbon bond formations as Pd. The tolerance of Pd reagents to many functional groups such as carbonyl and hydroxy groups is the second important feature. Pd-catalyzed reactions can be carried out without protection of these functional groups. Although reactions involving Pd should be carried out carefully, Pd reagents and catalysts are not very sensitive to oxygen and moisture, or even to acid. Ni(0) complexes are extremely sensitive to oxygen. On the other hand, in many reactions catalyzed by Pd–phosphine complexes, it is enough to apply precautions to avoid oxidation of the phosphine, and this can be done easily.

Of course, Pd is a noble metal and expensive, but it is much less expensive than Rh, Pt, and Os. Also, the toxicity of Pd has posed no problem so far. The fact that a number of industrial processes (more than ten at least) based on Pd-catalyzed reactions have been developed and are now operated reflects these advantages of using Pd catalysts commercially[1].

2 Palladium Compounds, Complexes, and Ligands Widely Used in Organic Synthesis

In organic synthesis, two kinds of Pd compounds, namely Pd(II) salts and Pd(0) complexes, are used. Pd(II) compounds are used either as stoichiometric reagents or as catalysts and Pd(0) complexes as catalysts. Pd(II) compounds such as $PdCl_2$ and $Pd(OAc)_2$ are commercially available and widely used as

1

unique oxidants. Pd(acac)$_2$ is also used. These reagents are stable. They can be used in two important ways: as unique stoichiometric oxidizing agents, and as precursors of Pd(0) complexes.

PdCl$_2$ is stable, but it has low solubility in water and organic solvents. It is soluble in dilute HCl and becomes soluble in organic solvents by forming a PdCl$_2$(PhCN)$_2$ complex[2]. M$_2$PdCl$_4$ (M = Li, Na, K) are soluble in water, lower alcohols, and some organic solvents. Pd(OAc)$_2$ is commercially available and is stable and soluble in organic solvents. It can be prepared from metallic Pd by dissolving it in AcOH containing nitric acid[3]. Sometimes the quality of Pd(OAc)$_2$ is variable[4] and it may contain nitrate anion. It becomes partially insoluble in organic solvents by forming a polymer. In such a case, Pd(OAc)$_2$ is purified by dissolving it in hot benzene and concentrating the benzene solution after removing the insoluble part. Pure Pd(acac)$_2$ can be obtained as needle-like crystals by recrystallization.

Pd(II) salts can be used as sources of Pd(0). For example, stable PdCl$_2$(Ph$_3$P)$_2$ is reduced to Pd(Ph$_3$P)$_n$ with i-Bu$_2$AlH, BuLi[5], and aqueous KOH[6]. Of particular importance, Pd(OAc)$_2$ is easily reduced to Pd(0) complexes *in situ* in the presence of phosphine ligands with several reducing agents, such as metal hydrides (NaBH$_4$, LiAlH$_4$), alkenes, CO[7], alcohols, organometallic compounds to form Pd(0)(R$_3$P)$_n$. Phosphines can also be used as reducing agents. For example, when Pd(OAc)$_2$ is treated with Ph$_3$P, Pd(0) species and phosphine oxide are formed slowly (eq. 1)[8,9]. An especially active Pd(0) catalyst can be prepared by a rapid reaction of Pd(OAc)$_2$ or Pd(acac)$_2$ with n-Bu$_3$P in a 1 : 1 ratio in THF or benzene[10]. Bu$_3$P is oxidized rapidly to butylphosphine oxide and a phosphine-free Pd(0) species is formed besides Ac$_2$O (eq. 2). This catalyst is very active, but not stable, and must be used immediately; black Pd metal begins to precipitate after 30 min if no substrate is added. The *in situ* generation of Pd(0) species with n-Bu$_3$P is a very convenient preparative method for catalytic species.

$$\text{Pd(OAc)}_2 + \text{Ph}_3\text{P} + \text{H}_2\text{O} \longrightarrow \text{Pd(0)} + \text{Ph}_3\text{PO} + 2\text{ AcOH} \quad (1)$$

$$\text{Pd(OAc)}_2 + \text{Bu}_3\text{P} \longrightarrow \text{Pd(0)}\text{......} \text{O=PBu}_3 + \text{Ac}_2\text{O} \quad (2)$$

Pd(0) is d^{10} and has four coordination sites. Two Pd(0) complexes are commercially available. Pd(Ph$_3$P)$_4$ is light-sensitive, unstable in air, yellowish green crystals and a coordinatively saturated Pd(0) complex. The complex is prepared from PdCl$_2$(Ph$_3$P)$_2$ with various reducing agents such as hydrazine[11] or alkali metal alkoxides[12] in the presence of Ph$_3$P[13]. In solution, two Ph$_3$P dissociate to form a coordinatively unsaturated Pd(0) species. Sometimes Pd(Ph$_3$P)$_4$ is less active as a catalyst because it has too many ligands to allow the coordination of some reactants.

$Pd_2(dba)_3$–$CHCl_3$ (dba = dibenzylideneacetone) is another commercially available Pd(0) complex in the form of purple needles which contain one molecule of $CHCl_3$ when $Pd(dba)_2$, initially formed in the process of preparation, is recrystallized from $CHCl_3$. $Pd(dba)_2$ corresponds to $Pd_2(dba)_3$–dba. Both $Pd_2(dba)_3$ and $Pd(dba)_2$ are used in this book as a complex of the same nature. One of the dba molecules in $Pd_2(dba)_3$-dba does not coordinate to Pd and is displaced by $CHCl_3$ to form $Pd_2(dba)_3$–$CHCl_3$ when it is recrystallized from $CHCl_3$. In $Pd_2(dba)_3$, dba behaves as two monodentate ligands, and not one bidentate ligand, and each Pd is coordinated with three double bonds of three molecules of dba, forming a 16-electron complex **1**. It is an air-stable complex, prepared by the reaction of $PdCl_2$ and dba and recrystallization from $CHCl_3$[14,15]. $Pd_2(dba)_3$, in which Pd is coordinated by olefinic bonds of dba, is converted into PdL_n (L is mainly a phosphine ligand) by a ligand-exchange reaction in solution, as shown in eq. 3. $Pd_2(dba)_3$ itself without phosphine is an active catalyst in some reactions.

1

$$Pd_2(dba)_3 + n\ R_3P \longrightarrow 2\ Pd(R_3P)_n + 3\ dba \qquad (3)$$

Attention should be paid to the fact that the ratio of Pd and phosphine ligand in active catalysts is crucial for determining the reaction paths. It is believed that dba is displaced completely with phosphines when $Pd_2(dba)_3$ is mixed with phosphines in solution. However the displacement is not complete[16]. Also, it should be considered that dba itself is a monodentate alkene ligand, and it may inhibit the coordination of a sterically hindered olefinic bond in substrates. In such a case, no reaction takes place, and it is recommended to prepare Pd(0) catalysts by the reaction of $Pd(OAc)_2$ with a definite amount of phosphines[10]. In this way a coordinatively unsaturated Pd(0) catalyst can be generated. Preparation of $Pd_3(tbaa)_3$ (tbaa = tribenzylidene-acetylacetone) was reported[17], but the complex actually obtained was $Pd(dba)_2$[18].

Highly reactive Pd(0) powder is prepared by the reduction of Pd(II) salts with Li or K and used for catalytic reactions[19,20]. Pd on carbon in the presence of Ph_3P is used as an active catalyst similar to $Pd(Ph_3P)_n$[21].

Several phosphines are used as ligands of Pd, Ph_3P being by far the most commonly used. Any contaminating phosphine oxide is readily removed by recrystallization from ethanol. However, in some catalytic reactions, more electron-donating alkylphosphines such as n-Bu_3P and tricyclohexylphosphine or aryl phosphines such as tri(2,4,6-trimethoxyphenyl)phosphine (TTMPP) and tri(2,6-dimethoxyphenyl)phosphine (TDMPP) are used, because these phosphines accelerate the 'oxidative' addition step. Sulfonated triphenylphosphine [tri(m-sulfophenyl)phosphine (TMSPP)] (2) and monosulfonated triphenylphosphine [diphenyl(m-sulfophenyl)phosphine (DPMSPP)] (3) are special phosphines as water-soluble ligands, with which Pd goes into the aqueous phase and catalytic reactions proceed in water[22–25]. Another water-soluble phosphine is 2-(diphenylphosphinoethyl)trimethylammonium halide[26]. Pd complexes coordinated by these phosphines are soluble in water, and Pd-catalyzed reactions can be carried out in water, which is said to have an accelerating effect in some catalytic reactions. Bidentate phosphines such as dppe (4), dppp (5), and dppb (6) play important roles in some reactions. Another bidentate phosphine is dppf (7), which is different from other bidentate phosphines, showing its own characteristic activity.

Phosphites, such as triisopropyl and triphenyl phosphite, are weaker electron donors than the corresponding phosphines, but they are used in some reactions because of their greater π-accepting ability. The cyclic phosphite trimethylolpropane phosphite (TMPP) or 4-ethyl-2,6,7-trioxa-1-phosphabicyclo[2.2.2]octane (8), which has a small cone angle and small steric hindrance, shows high catalytic activity in some reactions. It is not commercially available, but can be prepared easily[27].

$$Ph_2PCH_2CH_2PPh_2$$
4. dppe

2. TMSPP

$$Ph_2PCH_2CH_2CH_2PPh_2$$
5. dppp

7. dppf

3. DPMSPP $$Ph_2PCH_2CH_2CH_2CH_2PPh_2$$
6. dppb

8. TMPP

The roles of phosphines are not clearly understood and are unpredictable. Therefore, in surveying optimum conditions of catalytic reactions, it is advisable to test the activity of all these important types of phosphines and phosphites, which have different steric effects and electron-donating properties.

Although Pd is cheaper than Rh and Pt, it is still expensive. In Pd(0)- or Pd(II)-catalyzed reactions, particularly in commercial processes, repeated use of Pd catalysts is required. When the products are low-boiling, they can be separated from the catalyst by distillation. The Wacker process for the production of acetaldehyde is an example. For less volatile products, there are several approaches to the economical uses of Pd catalysts. As one method, an alkyldiphenylphosphine 9, in which the alkyl group is a polyethylene chain, is prepared as shown. The Pd complex of this phosphine has low solubility in some organic solvents such as toluene at room temperature, and is soluble at higher temperature[28]. Pd(0)-catalyzed reactions such as an allylation reaction of nucleophiles using this complex as a catalyst proceed smoothly at higher temperatures. After the reaction, the Pd complex precipitates and is recovered when the reaction mixture is cooled.

$$n \ CH_2{=}CH_2 \ + \ BuLi \xrightarrow{\text{TMEDA}} Bu(CH_2CH_2)_nCH_2CH_2Li \xrightarrow{\text{ClPPh}_2}$$

$$Bu(CH_2CH_2)_nCH_2CH_2PPh_2 \xrightarrow{\text{Pd(Ph}_3\text{P)}_4} [Bu(CH_2CH_2)_nCH_2CH_2PPh_2]_4Pd$$

$$\mathbf{9}$$

Pd can also be recovered as insoluble complexes such as the dimethylglyoxime complex, or $PdCl_2(Ph_3P)_2$ by treatment with HCl and Ph_3P. When water-soluble phosphines are used, the catalyst always remains in the aqueous phase and can be separated from a product in the organic phase, and is used repeatedly.

3 Fundamental Reactions of Pd Compounds

Fundamental reactions of Pd are briefly explained in order to understand how reactions either promoted or catalyzed by Pd proceed. In schemes written for the explanation, phosphine ligands are omitted for simplicity. First, a brief explanation of chemical terms specific to organopalladium chemistry is given.

3.1 'Oxidative' Addition Reaction

The term 'oxidative' might sound strange to organic chemists who are not familiar with organometallic chemistry. The term 'oxidative' used in organometallic chemistry has different meaning to 'oxidation' used in organic chemistry such as oxidation of secondary alcohols to ketones. An 'oxidative' addition is the addition of a molecule X—Y to Pd(0) with cleavage of its covalent bond, forming two new bonds (eq. 4)[29]. Since the two previously nonbonding electrons of Pd are involved in bonding, the Pd increases its formal oxidation state by two, namely Pd(0) is oxidized to Pd(II). This process is similar to the formation of Grignard reagents from alkyl halides and Mg(0) (eq. 5). In the preparation of Grignard reagents, Mg(0) is oxidized to Mg(II) by the 'oxidative' addition of alkyl halides to form two covalent bonds. Another example, which shows the clear difference between 'oxidation' in organic chemistry and 'oxidative' addition in organometallic chemistry, is the 'oxidative' addition of H_2 to Pd(0) to form Pd(II) hydride. In other words, Pd(0) is 'oxidized' to Pd(II) by H_2. This sounds strange to organic chemists, because H_2 is a reducing agent in organic chemistry.

$$\text{Pd(0)} + \text{X-Y} \xrightarrow{\text{Oxidative addition}} \text{X-Pd(II)-Y} \tag{4}$$

$$\text{Mg(0)} + \text{CH}_3\text{-I} \longrightarrow \text{CH}_3\text{-Mg-I} \tag{5}$$

A number of different covalent bonds are capable of undergoing the oxidative addition to Pd(0). The most widely known are C—X (X = halogen and pseudo-halogen), C—O, H—H, C—H, Si—H, M—H, M—M (M = main group metals), and H—X bonds. Also N—H, X—X, O—H, and even some C—C bonds undergo oxidative addition. Most frequently observed is the oxidative addition of organic halides of sp^2 carbons (eq. 6), and the rate of the addition decreases in the following order: C—I > C—Br ≫ C—Cl ⋙ C—F. Typical bonds which undergo the oxidative addition are following: R—X (R = alkenyl and aryl, X = halogen and pseudo-halogen), acyl halides (RCO—X) (eq. 7), aldehydes (RCO—H), allylic compounds (RCH=CHCH$_2$—X, X = halogen, esters, NO$_2$, SO$_2$R, etc.), sulfonyl halides (RSO$_2$—X), H—H, H—SnR$_3$, H—SiR$_3$, Ar—H, etc. The oxidative addition takes place with coordinatively unsaturated Pd(0) complexes. The saturated (four-coordinate, 18 electrons) Pd(0) complex, Pd(Ph$_3$P)$_4$ undergoes reversible dissociation *in situ* in solution to give the unsaturated 14-electron species Pd(Ph$_3$P)$_2$ (10), which is capable of undergoing the oxidative addition. Various σ-bonded Pd complexes are formed by the oxidative addition. In many cases, this is the first step of catalytic reactions.

$$\text{Pd(Ph}_3\text{P)}_4 \xrightarrow{\qquad} \text{Pd(Ph}_3\text{P)}_2 \xrightarrow{\text{Ph-I}} \text{Ph-Pd-I(Ph}_3\text{P)}_2 \tag{6}$$

$$\downarrow 2\ \text{Ph}_3\text{P}$$

$$\text{RCO-Cl} + \text{Pd(0)} \longrightarrow \text{RCO-Pd-Cl} \tag{7}$$

3.2 Insertion Reaction

The reaction of Grignard reagents with a carbonyl group can be understood as an insertion reaction of an unsaturated C=O bond of the carbonyl group into the Mg—C bond to form the Mg alkoxide **11**. Similarly, various unsaturated bonds insert to Pd—C σ-bonds. The insertion is understood as the migration of a one-electron ligand from Pd to an unsaturated ligand. Insertion has the same meaning as palladation of alkenes and alkynes. The insertion takes place in two ways: α,β- (or 1,2-) and α, α- (or 1,1-) insertions. The first and widely observed one is the α, β-insertion of unsaturated bonds as expressed by eq. 8.

$$X\text{-}Pd\text{-}Y + A\text{=}B \longrightarrow X\text{-}A\text{-}B\text{-}Pd\text{-}Y \tag{8}$$

Migration of a hydride ligand from Pd to a coordinated alkene (insertion of alkene) to form an alkyl ligand (alkylpalladium complex) (**12**) is a typical example of the α, β-insertion of alkenes. In addition, many other unsaturated bonds such as in conjugated dienes, alkynes, CO_2, and carbonyl groups, undergo the α, β-insertion to Pd-X σ-bonds. The insertion of an internal alkyne to the Pd—C bond to form **13** can be understood as the *cis*-carbopalladation of the alkyne. The insertion of butadiene into a Ph—Pd bond leads to the π-allylpalladium complex **14**. The insertion is usually highly stereospecific.

$$Ph\text{-}Pd\text{-}I + R\text{-}C\text{≡}C\text{-}R \xrightarrow{\text{Carbopalladation}} \mathbf{13}$$

CO is a representative species for α,α-insertion; its insertion into C—Pd bonds affords acylpalladium complexes such as **15**. Mechanistically, the CO insertion is 1,2-alkyl migration to coordinated CO. This is an important step in carbonylation. SO_2, isonitriles, and carbenes are other species which undergo α,α-insertion.

$$\text{Ph-Pd-I} + \text{CO} \longrightarrow \underset{\textbf{15}}{\text{Ph}} \overset{\overset{\displaystyle O}{\|}}{\underset{}{\diagup}} \text{Pd-I}$$

It should be emphasized that in some cases the insertions take place several times sequentially. For example, in carbonylation, the insertion of alkene is followed by CO insertion. Sometimes, further insertions of another alkene and CO take place. Particularly useful is the formation of polycyclic compounds by intramolecular sequential insertions of alkenyl and alkynyl bonds.

3.3 Transmetallation Reaction

Organic compounds M—R and hydrides M—H of main group metals such as Mg, Zn, B, Al, Sn, Si, and Hg react with A—Pd—X complexes formed by oxidative addition, and an organic group or hydride is transferred to Pd by exchange reaction of X with R or H. In other words, the alkylation of Pd takes place (eq. 9). A driving force of the reaction, which is called transmetallation, is ascribed to the difference in the electronegativities of two metals. A typical example is the phenylation of phenylpalladium iodide with phenyltributyltin to form diphenylpalladium (**16**).

$$\text{A-Pd-X} + \text{M-R} \;\rightleftarrows\; \text{A-Pd} \overset{\displaystyle R}{\underset{\displaystyle X}{\diagdown M}} \;\rightleftarrows\; \text{A-Pd-R} + \text{MX} \qquad (9)$$

$$\text{M = main group metal}$$

$$\text{Ph-Pd-I} + \text{Ph-SnBu}_3 \longrightarrow \underset{\textbf{16}}{\text{Ph-Pd-Ph}} + \text{Bu}_3\text{SnI}$$

3.4 The Final Step of Pd-Promoted or -Catalyzed Reactions

3.4.1 Reductive Elimination Reaction

Similarly to 'oxidative,' the term 'reductive' used in organometallic chemistry has a different meaning to 'reduction' in organic chemistry. This unimolecular decomposition pathway is the reverse of the 'oxidative' addition, and involves the loss of two one-electron ligands of *cis* configuration from the Pd center, combining them to form a single elimination product. In other words, coupling of two groups coordinated to Pd liberates a product in the last step of catalytic reactions (eq. 10). By the 'reductive' elimination, both the coordination number and the formal oxidation state of Pd are reduced by two to generate Pd(0), as shown in eq. 11. In other words, Pd(II) is reduced to Pd(0). This is why the step is called 'reductive' elimination. Pd(0) species, thus regenerated, can start the oxidative addition again. Thus a catalytic cycle becomes possible by the 'reductive' elimination. Without the 'reductive' elimination, the reaction ends up as a stoichiometric one. For example, Grignard reactions are stoichiometric, because the 'reductive' elimination to generate Mg(0) is not possible.

$$X\text{-}Pd\text{-}R \longrightarrow R\text{-}X \; + \; Pd(0) \tag{10}$$

$$Ph\text{-}Pd\text{-}Ph \longrightarrow Ph\text{-}Ph \; + \; Pd(0) \tag{11}$$

3.4.2 Elimination of β-Hydrogen

Another reaction in the last step is the *syn* elimination of β-hydrogen with Pd as H—Pd—X, which takes place with alkyl Pd complexes, and the Pd hydride and an alkene are formed. The insertion of an alkene into Pd hydride and the elimination of β-hydrogen are reversible steps. The elimination of β-hydrogen generates the alkene, and both the hydrogen and the alkene coordinate to Pd, increasing the coordination number of Pd by one. Therefore, the β-elimination requires coordinative unsaturation on Pd complexes. The β-hydrogen eliminated should be *syn* to Pd.

The elimination of β-hydrogen of Pd alkoxide (**17**) to afford a carbonyl compound is a similar reaction.

3.5 How are Catalytic Reactions Possible?

The most useful reaction of Pd is a catalytic reaction, which can be carried out with only a small amount of expensive Pd compounds. The catalytic cycle for the Pd(0) catalyst, which is understood by the combination of the afore-mentioned reactions, is possible by reductive elimination to generate Pd(0). The Pd(0) thus generated undergoes oxidative addition and starts another catalytic cycle. A Pd(0) catalytic species is also regenerated by β-elimination to form Pd—H which is followed by the insertion of the alkene to start the new catalytic cycle. These relationships can be expressed as shown.

As a typical example, the catalytic reaction of iodobenzene with methyl acrylate to afford methyl cinnamate (**18**) is explained by the sequences illustrated for the oxidative addition, insertion, and β-elimination reactions.

4 References

1. Review of industrial applications of Pd; J. Tsuji, *Synthesis*, 739 (1990).
2. M. S. Kharasch, R. C. Seyler, and R. R. Mayo, *J. Am. Chem. Soc.*, **60**, 882 (1938).
3. T. A. Stephenson, S. M. Morehouse, A. R. Powell, J. P. Heffer, and G. Wilkinson, *J. Chem. Soc.*, 3632 (1965): T. Hosokawa, S. Miyagi, S. Murahashi, and A. Sonoda, *J. Org. Chem.*, **43**, 2752 (1978).
4. D. H. R. Barton, J. Khamsi, N. Ozbalik, M. Ramesh, and J. C. Sarma, *Tetrahedron Lett.*, **30**, 4661 (1989).
5. E. Negishi, T. Takahashi, and K. Akiyoshi, *Chem. Commun.*, 1138 (1986).
6. V. V. Grushin and H. Alper, *Organometallics*, **12**, 1890 (1993)
7. T. A. Stromnova, M. N. Vargaftik, and I. I. Moiseev, *J. Organomet. Chem.*, **252**, 113 (1983): *Pure Appl. Chem.*, **61**, 1755 (1989).
8. C. Amatore, A. Jutand, and M. A. M'Barki, *Organometallics*, **11**, 3009 (1992).
9. F. Ozawa, A. Kubo, and T. Hayashi, *Chem. Lett.*, 2177 (1992); T. Hayashi, A. Kubo, and F. Ozawa, *Pure Appl. Chem.*, **64**, 421 (1992).
10. T. Mandai, T. Matsumoto, J. Tsuji, and S. Saito, *Tetrahedron Lett.*, **34**, 2513 (1993).
11. D. R. Coulson, *Inorg. Synth.*, **13**, 121 (1972); **28**, 107 (1990).
12. P. Roffia, G. Gregorio, F. Conti, and G. F. Pregaglia, *J. Mol. Catal.*, **2**, 191 (1977).
13. M. Ioele, G. Ortaggi, M. Scarsella, and G. Sletter, *Polyhedron*, **10**, 2475 (1991).
14. Y. Takahashi, T. Ito, S. Sakai, and Y. Ishii, *Chem. Commun.*, 1065 (1970); T. Ukai, H. Kawazura, Y. Ishii, J. J. Bennet, and J. A. Ibers, *J. Organomet. Chem.*, **65**, 253 (1974).
15. M. F. Retting and P. M. Maitlis, *Inorg. Synth.*, **17**, 135 (1977), 28, 110 (1990).

16. C. Amatore, A. Jutand, F. Khalil, M. A. M'Barkl, and L. Mottier, *Organometallics*, **12**, 3168 (1993).
17. Y. Ishii, S. Hasegawa, S. Kimura, and K. Itoh, *J. Organomet. Chem.*, **73**, 411 (1974).
18. A. M. Eschavarren and J. K. Stille, *J. Organomet. Chem.*, **356**, C35 (1988).
19. R. D. Riecke, A. V. Kavaliunas, L. D. Rhyne, and D. J. J. Fraser, *J. Am. Chem. Soc.*, **101**, 246 (1979); *J. Org. Chem.*, **44**, 3069 (1979).
20. D. Savoia, C. Trombini, A. Umani-Ronchi, and G. Verardo, *Chem. Commun.*, 540, 541 (1981).
21. D. E. Bergbreiter, B. Chen, and D. Weatherford, *J. Mol. Catal.*, **74**, 409 (1992).
22. Synthesis: S. Ahrland, J. Chatt, N. R. Davies, and A. A. Williams, *J. Chem. Soc.*, 276 (1958).
23. Review: E. G. Kuntz, *Chemtech*, 570 (1987).
24. Review: W. A. Herrmann and C. W. Kohlpaintner, *Angew. Chem., Int. Ed. Engl.*, **32**, 1524 (1993).
25. E.g.: A. L. Casanuovo and J. C. Calabrese, *J. Am. Chem. Soc.*, **112**, 4324 (1990).
26. G. Peiffer, S. Chan, A. Bendayan, B. Waegell, and J. P. Zahra, *J. Mol. Catal.*, **59**, 1 (1990).
27. Preparative method: T. J. Hutleman, *J. Inorg. Chem.*, **4**, 950 (1965).
28. D. E. Bergbreiter and D. A. Weatherford, *J. Org. Chem.*, **54**, 2726 (1989).
29. Review: J. K. Stille and K. S. Y. Lau, *Acc. Chem. Res.*, **10**, 434 (1977).

Books†

On Palladium Chemistry

1. P. M. Maitlis, *The Organic Chemistry of Palladium*, Vols 1 and 2, Academic Press, New York, 1971.
2. J. Tsuji, *Organic Synthesis with Palladium Compounds*, Springer, Berlin, 1980.
3. P. M. Henry, *Palladium Catalyzed Oxidation of Hydrocarbons*, D. Reidel Pub. Co., Dordrecht, 1980.
4. B. M. Trost and T. R. Verhoeven, Organopalladium Compounds in Organic Synthesis and in Catalysis, in *Comprehensive Organometallic Chemistry*, Vol 8, Pergamon Press, Oxford, 1982, p. 799.
5. R. F. Heck, *Palladium Reagents in Organic Syntheses*, Academic Press, New York, 1985.

On Organometallic Chemistry

1. S. G. Davies, *Organotransition Metal Chemistry: Applications to Organic Synthesis*, Pergamon Press, Oxford, 1982.
2. A. J. Pearson, *Metallo-organic Chemistry*, Wiley, Chichester, 1985.
3. A. Yamamoto, *Organotransition Metal Chemistry*, Academic Press, New York, 1986.

---

† These book references are not cited in the text.

4. J. P. Collman, L. S. Hegedus, J. R. Norton, and R. G. Finke, *Principles and Applications of Organotransition Metal Chemistry*, University Science Books, Mill Valley, CA, 1987.
5. P. J. Harrington, *Transition Metals in Total Synthesis*, Wiley, New York, 1990
6. F. J. McQuillin, *Transition Metals Organometallics for Organic Synthesis*, Cambridge University Press, Cambridge, 1991.
7. H. M. Colquhoun, D. J. Thompson, and M. V. Twigg, *Carbonylation*, Plenum Press, New York, 1991.

Chapter 2

Classification of the Reactions Involving Pd(II) Compounds and Pd(0) Complexes in This Book

1 Stoichiometric and Catalytic Reactions

A rational classification of reactions based on mechanistic considerations is essential for the better understanding of such a broad research field as that of the organic chemistry of Pd. Therefore, as was done in my previous book, the organic reactions of Pd are classified into stoichiometric and catalytic reactions. It is essential to form a Pd—C σ-bond for a synthetic reaction. The Pd—C σ-bond is formed in two ways depending on the substrates. σ-Bond formation from 'unoxidized' forms [1] of alkenes and arenes (simple alkenes and arenes) leads to stoichiometric reactions, and that from 'oxidized' forms of alkenes and arenes (typically halides) leads to catalytic reactions. We first consider how these two reactions differ.

2 Stoichiometric Oxidative Reactions with Pd(II) Compounds in which Pd(II) is Reduced to Pd(0)

The Pd—C σ-bond can be prepared from simple, unoxidized alkenes and aromatic compounds by the reaction of Pd(II) compounds. The following are typical examples. The first step of the reaction of a simple alkene with Pd(II) and a nucleophile X^- or Y^- to form **19** is called palladation. Depending on the nucleophile, it is called oxypalladation, aminopalladation, carbopalladation, etc. The subsequent elimination of β-hydrogen produces the nucleophilic substitution product **20**. The displacement of Pd with another nucleophile (X) affords the nucleophilic addition product **21** (see Chapter 3, Section 2). As an example, the oxypalladation of 4-pentenol with PdX_2 to afford furan **22** or **23** is shown.

π-Allylpalladium complex formation from alkenes takes place by the displacement of an allylic hydrogen of alkene with Pd(II) (see Chapter 3, Section

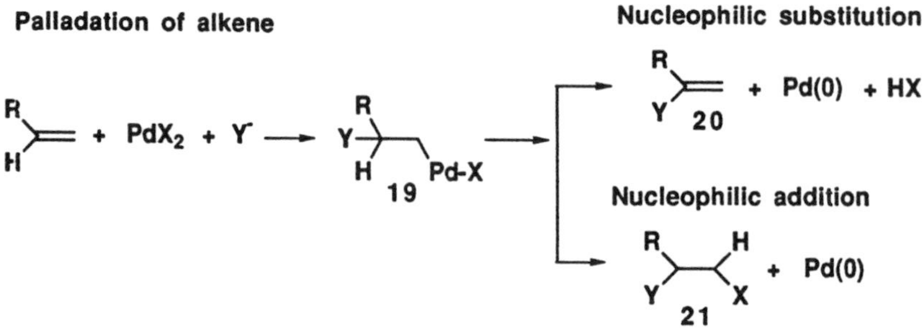

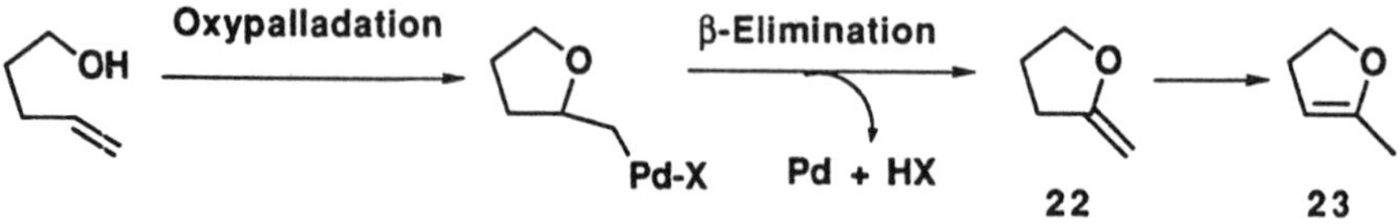

Palladation of alkene

Nucleophilic substitution

Nucleophilic addition

Oxypalladation   **β-Elimination**

3). Insertion of one of two double bonds of butadiene into Pd—X forms substituted a π-allylpalladium complex **24** (see Chapter 3, Section 4).

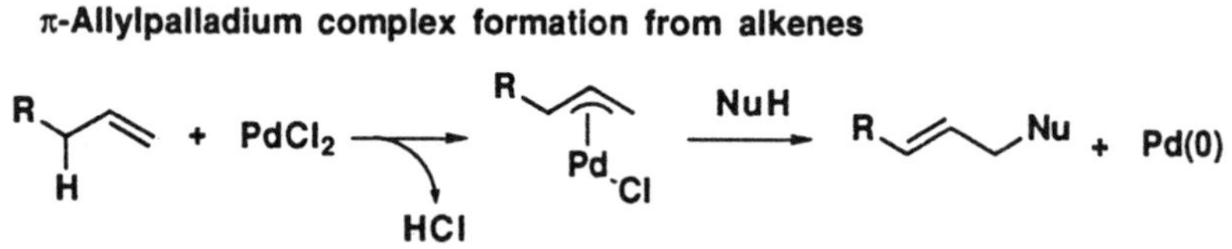

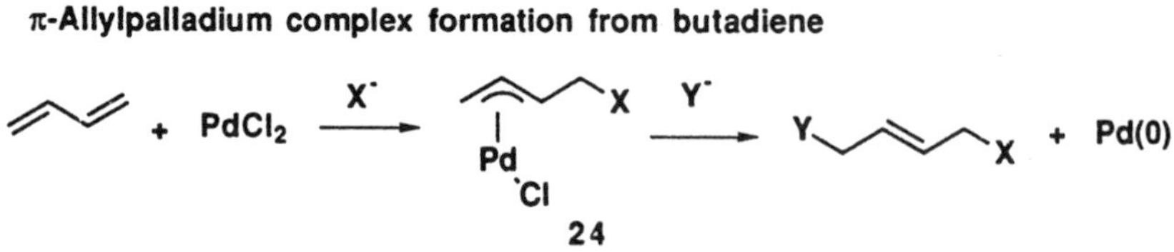

π-Allylpalladium complex formation from alkenes

π-Allylpalladium complex formation from butadiene

Palladation of aromatic compounds with Pd(OAc)$_2$ gives the arylpalladium acetate **25** as an unstable intermediate (see Chapter 3, Section 5). A similar complex **26** is formed by the transmetallation of PdX$_2$ with arylmetal compounds of main group metals such as Hg.

Those intermediates which have the Pd—C σ-bonds react with nucleophiles or undergo alkene insertion to give oxidized products and Pd(0) as shown below. Hence, these reactions proceed by consuming stoichiometric amounts of Pd(II) compounds, which are reduced to the Pd(0) state. Sometimes, but not always, the reduced Pd(0) is reoxidized *in situ* to the Pd(II) state. In such a case, the whole oxidation process becomes a catalytic cycle with regard to the Pd(II) compounds. This 'catalytic' reaction is different mechanistically, however, from the Pd(0)-catalyzed reactions described in the next section. These stoichiometric and 'catalytic' reactions are treated in Chapter 3.

Palladation of aromatics

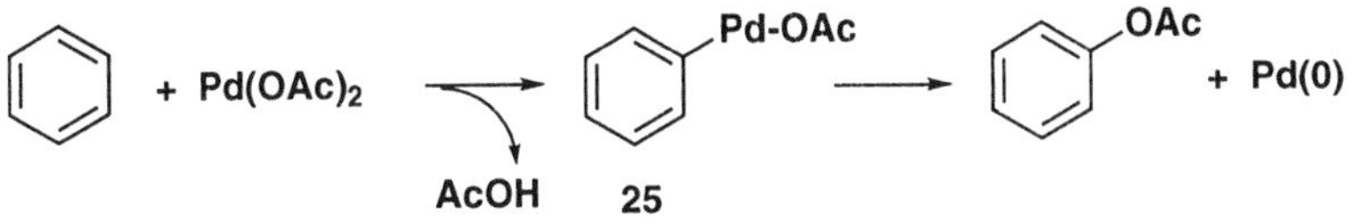

Transmetallation

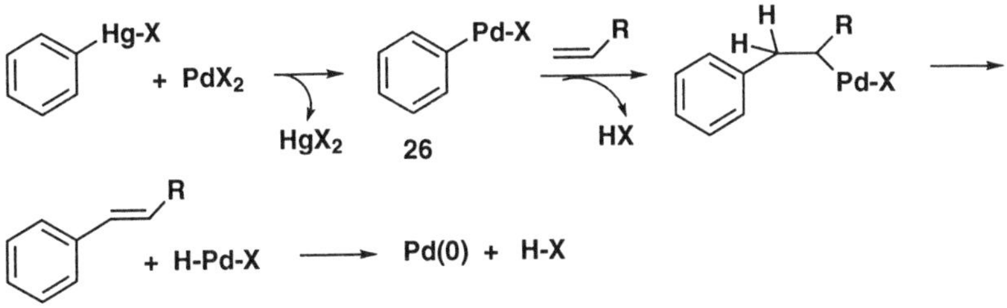

3 Catalytic Reactions Mainly with Pd(0) Complexes

The reactions of the second class are carried out by the reaction of 'oxidized' forms[1] of alkenes and aromatic compounds (typically their halides) with Pd(0) complexes, and the reactions proceed catalytically. The 'oxidative' addition of alkenyl and aryl halides to Pd(0) generates Pd(II)—C σ-bonds (**27** and **28**), which undergo several further transformations.

Oxidative addition and insertion

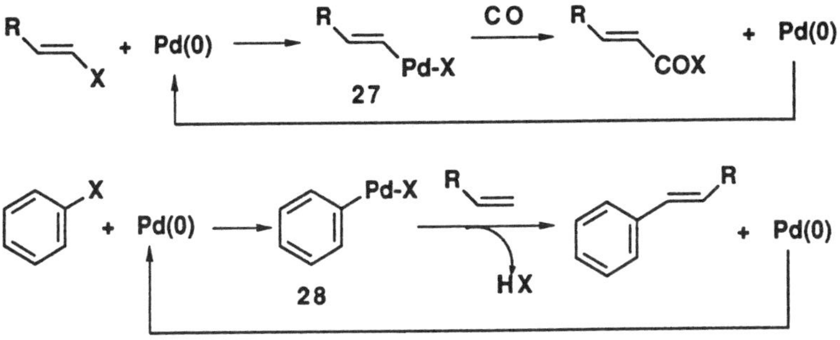

Formation of a π-allylpalladium complex **29** takes place by the oxidative addition of allylic compounds, typically allylic esters, to Pd(0). The π-allylpalladium complex is a resonance form of σ-allylpalladium and a coordinated π-bond. π-Allylpalladium complex formation involves inversion of stereochemistry, and the attack of the soft carbon nucleophile on the π-allylpalladium complex is also inversion, resulting in overall retention of the stereochemistry. On the other hand, the attack of hard carbon nucleophiles is retention, and hence overall inversion takes place by the reaction of the hard carbon nucleophiles.

π-Allyl complex formation and its reaction with a nucleophile

The 'oxidative' coupling of two molecules of butadiene with Pd(0) forms the bis-π-allylpalladium complex **31**, which is the resonance form of 2,5-divinyl-palladacyclopentane (**30**) formed by 'oxidative' cyclization.

Oxidative cyclization of butadiene and trapping with a nucleophile

All these intermediate complexes undergo various transformations such as insertion, transmetallation, and trapping with nucleophiles, and Pd(0) is regenerated at the end in every case. The regenerated Pd(0) starts the catalytic cycle again, making the whole process catalytic. These reactions catalyzed by Pd(0) are treated in Chapter 4.

Another type of catalytic reaction involving simple alkenes and alkynes proceeds with Pd(0) in the presence of acids (H—X). The oxidative addition of HCl or HI and even AcOH generates H—Pd—X **32**. Carbonylation of alkenes is explained by the insertion of an alkene and CO to form an acylpalladium complex **33** which reacts with alcohol to form an ester and regenerates the catalytic species **32**. Alkyne insertion into **32**, followed by alkene insertion generates an alkylpalladium intermediate **34**, which undergoes elimination of β-hydrogen to give a diene **35** and regenerates **32**. In this way, these reactions can be carried out catalytically in the presence of acids.

4 The Characteristic Features of Pd—C Bonds

The most characteristic feature of the Pd—C bonds in these intermediates of both the stoichiometric and catalytic reactions is their reaction with nucleophiles, and Pd(0) is generated by accepting two electrons from the nucleophiles as exemplified for the first time by the reactions of π-allylpalladium chloride[2] or PdCl$_2$-COD[3] complex with malonate and acetoacetate. It should be noted

H-X ; H-Cl, AcO-H etc

H–X + Pd(0) ⟶ H-Pd-X

32

that metal–carbon bonds in many organometallic compounds of other metals such as Mg, Al, Zn, and even Ni are attacked by electrophiles, and two-electron oxidation of these metals takes place, generating Mg(II), Ni(II), etc. In other words, the Grignard reaction starts with Mg(0) and ends as Mg(II) by reacting with electrophiles. Mg, Zn, etc., are base metals and Mg(II) is more stable than Mg(0)

R-Pd-X + Nu-H ⟶ R-Nu + HX + Pd(0)

R-M-X + El-Y ⟶ R-El + Y-M(II)Y

M = Mg, Zn, Ni, Fe, etc

Nu = nucleophile, El = electrophile

π-Allylpalladium chloride (**36**) reacts with the nucleophiles, generating Pd(0), whereas π-allylnickel chloride (**37**) and allylmagnesium bromide (**38**) reacts with electrophiles (carbonyl), generating Ni(II) and Mg(II). Therefore, it is understandable that the Grignard reaction cannot be carried out with a catalytic amount of Mg, whereas the catalytic reaction is possible with the regeneration of an active Pd(0) catalyst. Pd is a noble metal and Pd(0) is more stable than Pd(II). The carbon–metal bonds of some transition metals such as Ni and Co react with nucleophiles and their reactions can be carried out catalytically, but not always. In this respect, Pd is very unique.

Another feature of the Pd—C bonds is the excellent functional group tolerance. They are inert to many functional groups, except alkenes and alkynes and iodides and bromides attached to sp^2 carbons, and not sensitive to H_2O, ROH, and even RCO_2H. In this sense, they are very different from Grignard reagents, which react with carbonyl groups and are easily protonated.

Reactions of another class are catalyzed by Pd(II) compounds which act as Lewis acids, and are treated in Chapter 5 and partly in Chapter 4. From the above-mentioned explanation, the reactions catalyzed by Pd(0) and Pd(II) are clearly different mechanistically. In this book the stoichiometric and catalytic reactions are classified further according to reacting substrates. However, this classification has some problems, viz. it leads to separate treatment of some unit reactions in different chapters. The carbonylation of alkenes is an example. Oxidative carbonylation of alkenes is treated in Chapter 3 and hydrocarbonylation in Chapter 4.

5 References

1. For the definition of oxidation state of carbon, see J. B. Hendrickson, D. J. Cram, and G. S. Hammond, *Organic Chemistry*, 3rd Ed., McGraw-Hill, New York, 1970, p. 739.
2. J. Tsuji, H. Takahashi, and M. Morikawa, *Tetrahedron Lett.*, 4387 (1965).
3. J. Tsuji and H. Takahashi, *J. Am. Chem. Soc.*, **87**, 3275 (1965); **90**, 2387 (1968).

Chapter 3

Oxidative Reactions with Pd(II) Compounds

1 Introduction

As with Pb(IV) or Hg(II) compounds, two-electron oxidation is possible with Pd(II) compounds, but there are many unique oxidation reactions (dehydrogenation) specific to Pd(II). After the oxidation is completed, Pd(II) is reduced to Pd(0). If a stoichiometric amount of expensive Pd(II) compounds is consumed, the reaction cannot be a truly useful synthetic method. The Wacker process, in which acetaldehyde is produced from ethylene, is the first example of the oxidation of an organic compound with Pd(II). The essence of the great success of the Wacker process is the invention of an ingenious catalytic cycle, in which the reduced Pd(0) is reoxidized *in situ* to Pd(II) with $CuCl_2$, and in turn CuCl is easily reoxidized to $CuCl_2$ with oxygen. Consequently, ethylene (or other organic compounds) is oxidized (dehydrogenated) indirectly with oxygen without consuming $PdCl_2$ and $CuCl_2$ by the combination of these redox reactions. The catalytic oxidation of organic compounds with $PdCl_2$ and $CuCl_2$ can be summarized as shown[1].

$$PdCl_2 + AH + BH \longrightarrow A{-}B + Pd(0) + 2\,HCl$$
$$Pd(0) + 2\,CuCl_2 \longrightarrow PdCl_2 + 2\,CuCl$$
$$2\,CuCl + 2\,HCl + 1/2\,O_2 \longrightarrow 2\,CuCl_2 + H_2O$$
$$\overline{\quad AH + BH + 1/2\,O_2 \longrightarrow A{-}B + H_2O \quad}$$

In addition to $CuCl_2$, some other compounds such as $Cu(OAc)_2$, $Cu(NO_3)_2$, $FeCl_3$, dichromate, HNO_3, potassium peroxodisulfate, and MnO_2 are used as oxidants of Pd(0). Also heteropoly acid salts comtaining P, Mo, V, Si, and Ge are used with $PdSO_4$ as the redox system[2]. Organic oxidants such as benzoquinone (BQ), hydrogen peroxide and some organic peroxides are used for oxidation. Alkyl nitrites are unique oxidants which are used in some industrial

19

processes[3]. Efficient reoxidation of Pd(0) in some oxidation reactions with O_2 alone without other reoxidants is possible in DMSO[4,5]. It should be pointed out that the *in situ* reoxidation of Pd(0) is not always easy. The Pd(0) state is the more stable state than Pd(II). Pd metal is called a noble metal because it is not easily oxidized, whereas Cu is a base metal, because it is easily oxidized. Therefore, the oxidation of Pd(0) with base metal salts such as $CuCl_2$ and $FeCl_3$ seems to be rather abnormal and unexpected[6]. The very small equilibrium constant calculated for the oxidation reaction of metallic Pd with free Cu(II) ion suggests the difficulty of oxidizing Pd(0) with Cu(II) salts. The reaction becomes easier in the presence of chloride ion, which stabilizes Pd(II) and Cu(I) states by the complex formation [7].

$$Pd_{metal} + 2\ Cu(II) \underset{}{\overset{K}{\rightleftharpoons}} Pd(II) + 2\ Cu(I)$$

$$K = 10^{-28.2}$$

$$Pd_{metal} + 2\ Cu(II) + 8\ Cl^- \underset{}{\overset{K'}{\rightleftharpoons}} [PdCl_4]^{-2} + 2\ [CuCl_2]^{-1}$$

$$K'_{25^\circ} = 7.9 \times 10^{-6}, \quad K'_{100^\circ} = 7.9 \times 10^{-5}$$

In the Wacker process, the reaction is actually carried out in dilute HCl at a high concentration of chloride ion and an elevated temperature. The high concentration of $CuCl_2$ shifts the equilibrium further to the right.

The oxidation of Pd(0) should not be treated as an easy process, and the proper solvents, reaction conditions and reoxidants should be selected to carry out smooth catalytic reactions. In many cases, the catalytic cycle cannot be achieved. In this case, the reduced Pd must be recovered and then reused after oxidation with a suitable strong oxidizing agent such as HNO_3. The procedure for the recovery and reoxidation of Pd(0) is tedious and time consuming. Hence, the stoichiometric process is tolerable only for the synthesis of rather expensive organic compounds in limited quantities. This is a serious limitation in the application of oxidation reactions involving Pd(II). One important method for the oxidation reaction with Pd(II) is a gas-phase reaction using a solid Pd catalyst supported on active carbon or alumina. The supported Pd catalyst behaves similarly to the Pd(II)–Cu(II) redox system in the oxidation reactions[8,9]. Actually industrial production of vinyl acetate is carried out in the gas-phase by using a supported Pd catalyst[10]. Much expertise in connection with the preparation of supported Pd catalysts has been published, especially in patents. The patent literature is difficult to summarize. Also, the gas-phase reaction is a rather different technique. Therefore, mainly liquid-phase reactions by Pd(II) compounds are treated in this chapter. Oxidation of various organic compounds is possible with Pd(II), and it is surveyed by further subdivisions based on substrates.

2 Reactions of Alkenes

2.1 Introduction

Pd(II) compounds coordinate to alkenes to form π-complexes. Roughly, a decrease in the electron density of alkenes by coordination to electrophilic Pd(II) permits attack by various nucleophiles on the coordinated alkenes. In contrast, electrophilic attack is commonly observed with uncomplexed alkenes. The attack of nucleophiles with concomitant formation of a carbon–palladium σ-bond **1** is called the palladation of alkenes. This reaction is similar to the mercuration reaction. However, unlike the mercuration products, which are stable and isolable, the product **1** of the palladation is usually unstable and undergoes rapid decomposition. The palladation reaction is followed by two reactions. The elimination of H—Pd—Cl from **1** to form vinyl compounds **2** is one reaction path, resulting in nucleophilic substitution of the olefinic proton. When the displacement of the Pd in **1** with another nucleophile takes place, the nucleophilic addition of alkenes occurs to give **3**. Depending on the reactants and conditions, either nucleophilic substitution of alkenes or nucleophilic addition to alkenes takes place.

XH, YH = nucleophiles, H_2O, ROH, RCO_2H, RNH_2, CH_2E_2

Typical nucleophiles known to react with coordinated alkenes are water, alcohols, carboxylic acids, ammonia, amines, enamines, and active methylene compounds[11,12]. The intramolecular version is particularly useful for syntheses of various heterocyclic compounds[13,14]. CO and aromatics also react with alkenes. The oxidation reactions of alkenes can be classified further based on these attacking species. Under certain conditions, especially in the presence of bases, the π-alkene complex **4** is converted into the π-allylic complex **5**. Various stoichiometric reactions of alkenes via π-allylic complex **5** are treated in Section 4.

$$R\diagup\diagdown\diagup\!\!\!\diagdown \;+\; PdCl_2 \longrightarrow R\diagdown\diagup\diagdown\!\!\!\diagup \underset{\substack{PdCl_2 \\ 4}}{} \longrightarrow \underset{\substack{Cl \swarrow \overset{Pd}{|}\diagdown \\ 5}}{R\diagdown\diagup\diagup} \;+\; HCl$$

2.2 Reaction with Water

Formation of acetaldehyde and metallic Pd by passing ethylene into an aqueous solution of $PdCl_2$ was reported by Phillips in 1894[15] and used for the quantitative analysis of Pd(II)[16]. The reaction was highlighted after the industrial process for acetaldehyde production from ethylene based on this reaction had been developed[1,17,18]. The Wacker process (or reaction) involves the three unit reactions shown. The unique feature in the Wacker process is the invention of the *in situ* redox system of $PdCl_2$–$CuCl_2$.

$$CH_2{=}CH_2 \;+\; H_2O \;+\; PdCl_2 \longrightarrow CH_3CHO \;+\; 2\,HCl \;+\; Pd(0)$$

$$Pd(0) \;+\; 2\,CuCl_2 \longrightarrow PdCl_2 \;+\; 2\,CuCl$$

$$2\,CuCl \;+\; 2\,HCl \;+\; 1/2\,O_2 \longrightarrow 2\,CuCl_2 \;+\; H_2O$$

$$CH_2{=}CH_2 \;+\; 1/2\,O_2 \longrightarrow CH_3CHO$$

Extensive studies on the Wacker process have been carried out in industrial laboratories. Also, many papers on mechanistic and kinetic studies have been published[17–22]. Several interesting observations have been made in the oxidation of ethylene. Most important, it has been established that no incorporation of deuterium takes place by the reaction carried out in D_2O, indicating that the hydride shift takes place and vinyl alcohol is not an intermediate[1,17]. The reaction is explained by oxypalladation of ethylene, β-elimination to give the vinyl alcohol **6**, which complexes to H–PdCl, reinsertion of the coordinated vinyl alcohol with opposite regiochemistry to give **7**, and aldehyde formation by the elimination of Pd—H.

$$CH_2{=}CH_2 \;+\; D_2O \;+\; PdCl_2 \longrightarrow CH_3CHO \;+\; Pd(0) \;+\; 2\,DCl$$

$$CH_2{=}CH_2 \;+\; H_2O \;+\; PdCl_2 \longrightarrow$$

The attack of OH obeys the Markovnikov rule. Higher alkenes are oxidized to ketones and this unique oxidation of alkenes has extensive synthetic applications[23]. The oxidation of propylene affords acetone. Propionaldehyde is not obtained. This means that the oxidation of terminal alkenes affords methyl ketones, which have widespread uses in organic synthesis. Based on this reaction, the terminal alkenes can be regarded as masked methyl ketones[24,25]. The reaction is sometimes called the Wacker–Tsuji reaction.

The Wacker process is carried out in dilute HCl solution and low-boiling acetaldehyde is removed continuously by distillation. On the other hand, the oxidation of higher alkenes is carried out in organic solvents which can mix both alkenes and water. DMF is widely used for this purpose[26]. 3-Methylsulfolane and NMP give better results than DMF in the oxidation of 3,3-dimethyl-1-butene[27]. The oxidation proceeds faster in alcohols, but double bond isomerization also occurs extensively in alcohols[28]. Polyethylene glycol (PEG 400) is a good solvent for the oxidation of terminal and internal alkenes[29]. 1-Octene, nonene, and decene are oxidized selectively without attacking lower or higher alkenes in the presence of α-cyclodextrin[30,31]. The oxidation of ethylene proceeds even in aqueous ammonia to afford two pyridine derivatives (**8** and **9**) selectively via acetaldehyde[32].

In addition to $CuCl_2$, several other oxidizing agents are used in the oxidation of alkenes. Sometimes chlorination of carbonyl compounds is observed with the use of $CuCl_2$. The use of CuCl, after facile preoxidation to Cu(II) with oxygen, is recommended because no chlorination takes place with CuCl[24,25,33]. A typical oxidation procedure of 1-decene gives a 68–73% yield[34]. Also, catalytic amounts of $Cu(NO_3)_2$ and $Cu(OAc)_2$ are used[28,35]. $FeCl_3$, HNO_3, and MnO_2 can oxidize Pd(0), but stoichiometric amounts are necessary. Oxidation of cyclopentene is carried out with a catalytic system of $PdCl_2$–$Fe(ClO_4)_3$ combined with electrochemical oxidation[36]. Nitro cobalt[37] and nitroso and nitro complexes such as [PdClNO][MeCN][38-41] and $[PdCl(NO_2)(MeCN)]_2$ in the presence of amides[42] oxidize alkenes without reoxidants under oxygen, and the reaction is mechanistically different from the Wacker-type redox system[43]. Water-soluble heteropoly acids are used as co-catalysts[44-46]. Some organic compounds are used as oxidants in stoichiometric amounts. Benzoquinone is most widely used[47]. The oxidation can be carried out using a catalytic amount of benzoquinone by the combination of electrochemical oxidation as shown [48], or with iron phthalocyanin[49,50]. Electrochemical Wacker oxidation was carried out using Pd(II) and tri(4-bromophenyl)amine[51]. Peroxides such as H_2O_2[52,53] and *t*-butyl hydroperoxid [54,55] are other oxidants.

The oxidation of higher alkenes in organic solvents proceeds under almost neutral conditions, and hence many functional groups such as ester or lactone[26,56–59], sulfonate[60], aldehyde[61–63], acetal[60], MOM ether[64], carbobenzoxy[65], *t*-allylic alcohol[66], bromide[67,68], tertiary amine[69], and phenylselenide[70] can be tolerated. Partial hydrolysis of THP ether[71] and silyl ethers under certain conditions was reported. Alcohols are oxidized with Pd(II)[72–74] but the oxidation is slower than the oxidation of terminal alkenes and gives no problem when alcohols are used as solvents[75,76].

Higher terminal alkenes are oxidized to methyl ketones and this unique oxidation of alkenes has extensive synthetic applications[23]. The terminal alkenes can be regarded as masked methyl ketones, which are stable to acids, bases, and nucleophiles[24]. The oxidation of terminal alkenes to methyl ketones has extensively applied to syntheses of many natural products[77].

Several 1,4-dicarbonyl compounds are prepared based on this oxidation. Typically, the 1,4-diketone **10** or the 1,4-keto aldehyde **12** can be prepared by the allylation of a ketone[24] or aldehyde[61,62], followed by oxidation. The reaction is a good annulation method for cyclopentenones (**11** and **13**). Syntheses of pentalenene[78], laurenene[67], descarboxyquadrone[79], muscone (**14** R = Me)[80] and the coriolin intermediate **15**[71] have been carried out by using allyl group as the masked methyl ketone (facing page).

Conjugate addition of vinyllithium or a vinyl Grignard reagent to enones and subsequent oxidation afford the 1,4-diketone **16**[25]. 4-Oxopentanals are synthesized from allylic alcohols by [3,3]sigmatropic rearrangement of their vinyl ethers and subsequent oxidation of the terminal double bond. Dihydrojasmone (**18**) was synthesized from allyl 2-octenyl ether (**17**) based on Claisen rearrangement and oxidation[25] (page 26).

1,5-diketone **19** is prepared by 3-butenylation of a ketone, followed by Pd-catalyzed oxidation, and is used for annulation to form the cyclohexenone **20**[24]. In this method, the 3-butenyl group is a masked methyl vinyl ketone. Lewis acid-promoted Michael addition of allylsilane to the α, β-unsaturated ketone **21** followed by Pd-catalyzed oxidation affords the 1,5-diketone **22**[81]. The 1,6-diketone **24** was prepared from 1,7-diene **23** and used for cyclization[82].

The 5-oxohexanal **27** is prepared by the following three-step procedure: (1) 1,2-addition of allylmagnesium bromide to an α, β-unsaturated aldehyde to give the 3-hydroxy-1,5-diene **25**, (2) oxy-Cope rearrangement of **25** to give **26**, and (3) palladium catalyzed oxidation to afford **27**. The method was applied to the synthesis of $\Delta^{3,4}$-2-octalone (**28**), which is difficult to prepare by the Robinson annulation[25].

In some cases where there is a neighboring group participation effect, aldehydes are formed. The α-vinyl group in the β-lactam **29** is mainly oxidized to aldehyde **30**[83].

Oxidative rearrangement takes place in the oxidation of the 1-vinyl-1-cyclobutanol **31**, yielding the cyclopentenone derivative **32**[84]. Ring contraction to cyclopropyl methyl ketone (**34**) is observed by the oxidation of 1-methylcyclobutene (**33**)[85], and ring expansion to cyclopentanone takes place by the reaction of the methylenecyclobutane **35**. [86,87]

The methyl enol ether **37** is oxidized to the α,β-unsaturated aldehyde **39** via hemiacetal **38**. Unsaturated aldehyde **39**, elongated one carbon from the aldehyde **36**, is prepared by the Wittig reaction of **36** to give **37**, and application of this reaction[88].

The oxidation of simple internal alkenes is very slow. The clean selective oxidation of a terminal double bond in **40**, even in the presence of an internal double bond, is possible under normal conditions[89,90]. The oxidation of cyclic alkenes is difficult, but can be carried out under selected conditions. Addition of strong mineral acids such as $HClO_4$, H_2SO_4, and HBF_4 accelerates the oxidation of cyclohexene and cyclopentene[48,91]. A catalyst system of $PdSO_4-H_3PMo_6W_6O_{40}$ [92] or $PdCl_2-CuCl_2$ in EtOH is used for the oxidation of cyclopentene and cyclohexene[93].

In the presence of some functional groups at an appropriate position, internal alkenes can be oxidized by their participation. In addition, the oxidation is highly regioselective. The α,β-unsaturated ester **41** and ketone **43** are hardly oxidized in DMF, but they can be oxidized regioselectively in 50% AcOH, *i*-PrOH or NMP to give β-keto ester **42** or 1,3-diketone **44** by use of Na_2PdCl_4 and *t*-butyl hydroperoxide[54]. α,β-Unsaturated ester **45** is oxidized more easily when a 6- or 7-hydroxy group is present[94]. β,γ-Unsaturated esters and ketones such as **46** are converted into the π-allylpalladium complex **47** in DMF without being oxidized[95], but they are oxidized in aqueous dioxane or THF to give the γ-keto ester **48** or a 1,4-diketone[96]. Cyclic γ,δ-unsaturated ketones **49** and **51** are oxidized regioselectively to 1,4-diketones **50** and **52**, which are used for annulation[97,98].

Allylic ether **53** is oxidized regioselectively to the β-alkoxy ketone **54**, which is converted into the α,β-unsaturated ketone **55** and used for annulation[99]. The ester of homoallylic alcohol **56** is oxidized mainly to the γ-acetoxy ketone **57**[99].

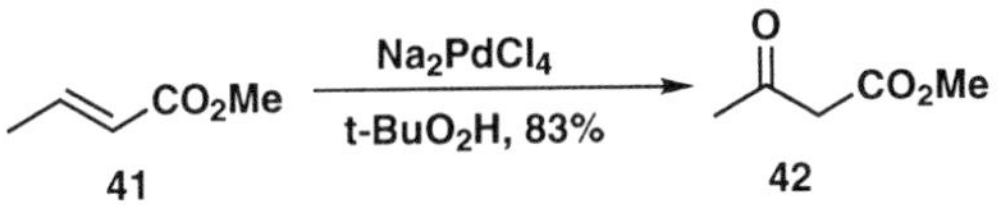
CO2Me
Na2PdCl4
t-BuO2H, 83%
CO2Me
41
42

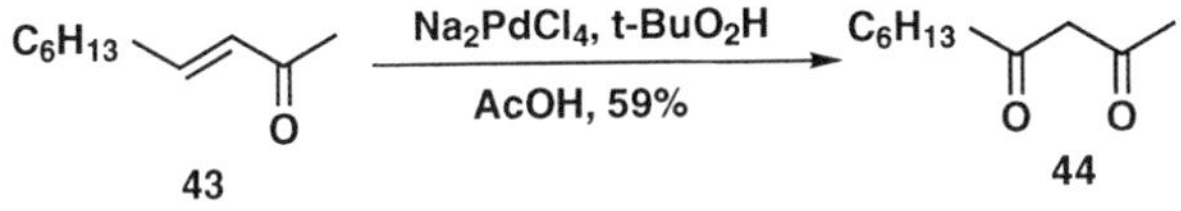
C6H13
Na2PdCl4, t-BuO2H
AcOH, 59%
C6H13
43
44

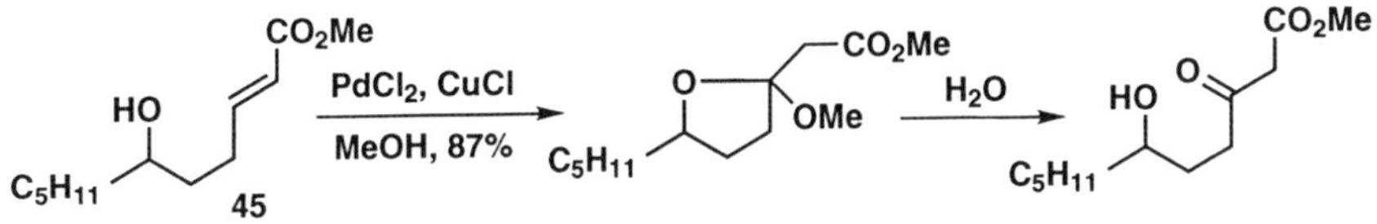
CO2Me
HO
C5H11
45
PdCl2, CuCl
MeOH, 87%
CO2Me
OMe
C5H11
H2O
CO2Me
HO
C5H11

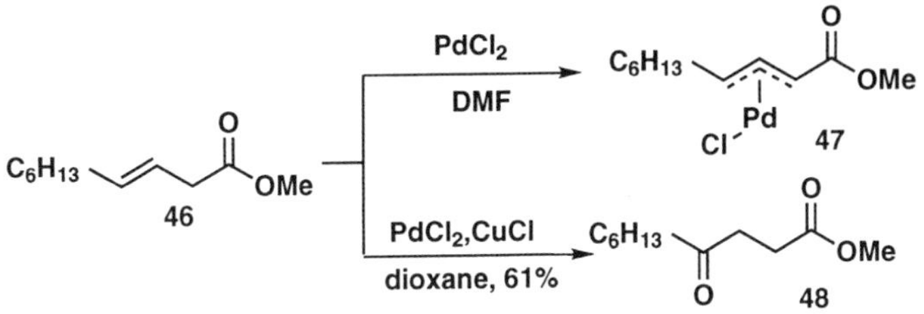
C6H13
OMe
46
PdCl2
DMF
C6H13
OMe
Pd
Cl
47
PdCl2,CuCl
dioxane, 61%
C6H13
OMe
48

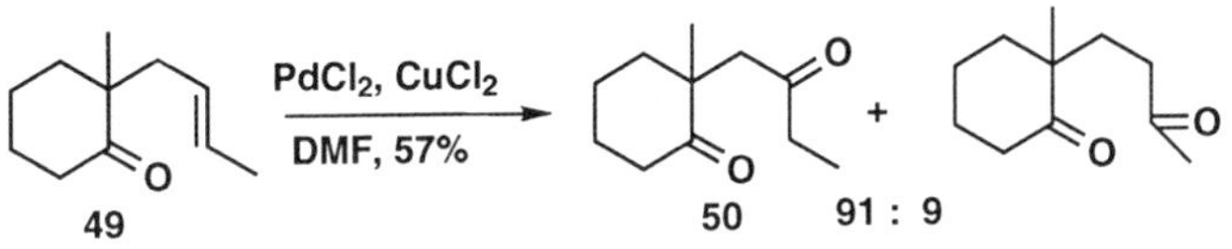
49
PdCl2, CuCl2
DMF, 57%
50
91 : 9

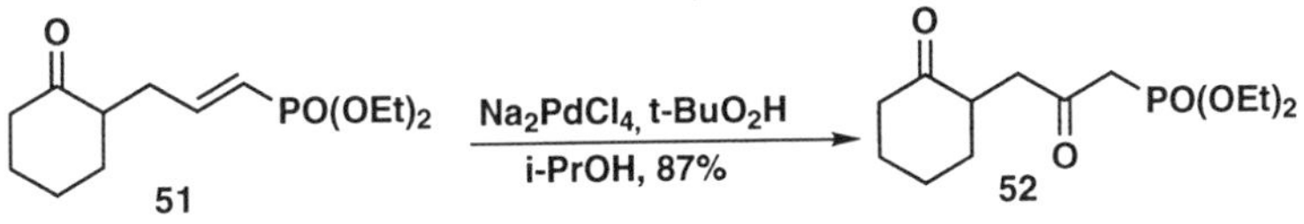
PO(OEt)2
51
Na2PdCl4, t-BuO2H
i-PrOH, 87%
PO(OEt)2
52

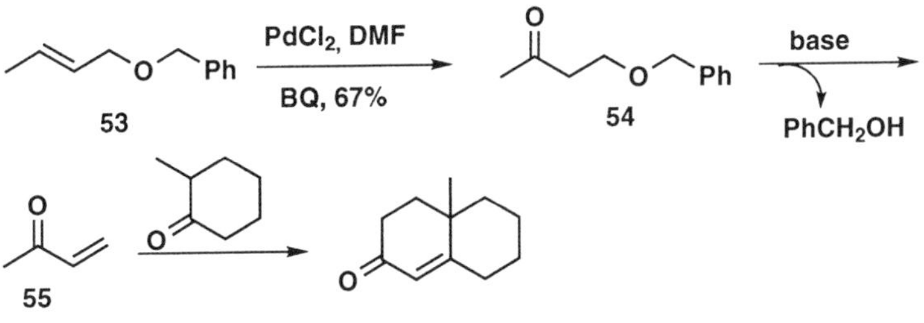
Ph
53
PdCl2, DMF
BQ, 67%
Ph
54
base
PhCH2OH
55

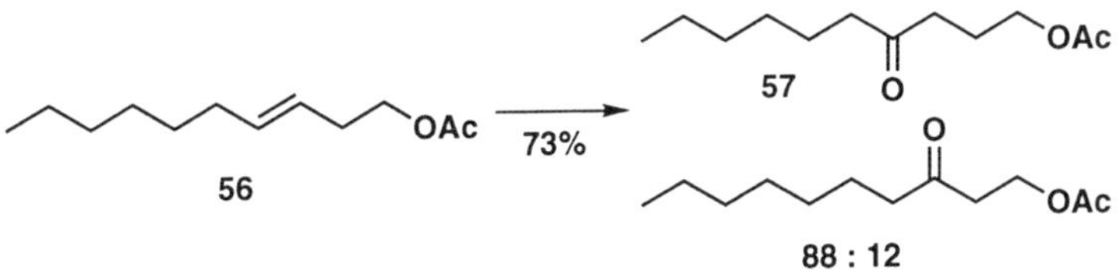
OAc
57
OAc
56
73%
OAc
88 : 12

In contrast to oxidation in water, it has been found that 1-alkenes are directly oxidized with molecular oxygen in anhydrous, aprotic solvents, when a catalyst system of PdCl$_2$(MeCN)$_2$ and CuCl is used together with HMPA. In the absence of HMPA, no reaction takes place[100]. In the oxidation of 1-decene, the O$_2$ uptake correlates with the amount of 2-decanone formed, and up to 0.5 mol of O$_2$ is consumed for the production of 1 mol of the ketone. This result shows that both O atoms of molecular oxygen are incorporated into the product, and a bimetallic Pd(II) hydroperoxide coupled with a Cu salt is involved in oxidation of this type, and that the well known redox catalysis of PdX$_2$ and CuX$_2$ is not always operative[101]. The oxidation under anhydrous conditions is unique in terms of the regioselective formation of aldehyde **59** from *N*-allyl-*N*-methylbenzamide (**58**), whereas the use of aqueous DME results in the predominant formation of the methyl ketone **60**. Similar results are obtained with allylic acetates and allylic carbonates[102]. The complete reversal of the regioselectivity in PdCl$_2$-catalyzed oxidation of alkenes is remarkable.

Chlorohydrin **61** is formed by the nucleophilic addition to ethylene with PdCl$_2$ and CuCl$_2$[103,104]. Regioselective chlorohydroxylation of the allylic amine **62** is possible by the participation of the heteroatom to give chlorohydrin **63**. Allylic sulfides behave similarly[105].

2.3 Reactions with Alcohols, Phenols, and Other Hydroxy Compounds

Oxidation of ethylene in alcohol with $PdCl_2$ in the presence of a base gives an acetal and vinyl ether[106,107]. The reaction of alkenes with alcohols mediated by $PdCl_2$ affords acetals **64** as major products and vinyl ethers **65** as minor products. No deuterium incorporation was observed in the acetal formed from ethylene and MeOD, indicating that hydride shift takes place and the acetal is not formed by the addition of methanol to methyl vinyl ether[108]. The reaction can be carried out catalytically using $CuCl_2$ under oxygen[28].

$$R \diagup\!\!\!\diagup + 2\,R'OH + PdCl_2 \longrightarrow R\!-\!\!\!\underset{OR'}{\overset{OR'}{|}}\!\!\!- + \underset{R}{\overset{R'O}{\diagdown}}\!\!=\!\!= + Pd(0) + 2HCl$$

$$\quad\quad\quad\quad\quad\quad\quad\quad\quad\quad\quad\quad\quad\quad\quad \textbf{64} \quad\quad\quad \textbf{65}$$

$$CH_2{=}CH_2 + PdCl_2 + 2\,MeOD \longrightarrow CH_3CH(OMe)_2 + 2\,DCl + Pd$$

The syntheses of brevicomin (**67**)[109,110] and frontalin[111] have been achieved as an elegant application of the intramolecular acetal formation with the diol **66** in dry DME. Optically active frontalin (**68**) has been synthesized by this cyclization in triglyme[112].

The oxidation of terminal alkenes with an EWG in alcohols or ethylene glycol affords acetals of aldehydes chemoselectively. Acrylonitrile is converted into 1,3-dioxolan-2-ylacetonitrile (**69**) in ethylene glycol and to 3,3-dimethoxy-propionitrile (**70**) in methanol[28]. 3,3-Dimethoxypropionitrile (**70**) is produced commercially in MeOH from acrylonitrile by use of methyl nitrite (**71**) as a unique reoxidant of Pd(0). Methyl nitrite (**71**) is regenerated by the oxidation of NO with oxygen in MeOH. Methyl nitrite is a gas, which can be separated easily from water formed in the oxidation[3].

Efficient acetalization of alkenes bearing various EWG with an optically active 1,3-diol **72** proceeds smoothly utilizing $PdCl_2$, CuCl, and O_2 in DME to give the 1,3-dioxane **73**[113]. Methacrylamide bearing 4-t-butyloxazolidin-2-one **74** as a chiral auxiliary reacts with MeOH in the presence of $PdCl_2$ catalyst to give the corresponding acetal **75** in a high yield and with a high chiral induction[114]. Selective terminal acetalization takes place with α-cyanoallyl acetate in MeOH and HMPA[115]. Both acetalization to give **77** and Michael addition to give **78** take place with the vinyl ketone **76**. Regioselective acetalization takes place in the presence of Na_2HPO_4[113]. The Pd-catalyzed diastereoselective acetalization of methyl (2*S*,3*S*)-2-allyl-3-hydroxybutyrate (**79**) in MeOH gives (2*R*,4*S*,5*S*)-2,5-dimethyl-2-methoxy-4-methoxycarbonyltetrahydrofuran (**80**) in 88% *de*[116]. The methoxy group in **80** can be displaced with allylsilane or hydrosilane with high stereoselectivity.

Na₂HPO₄	75%	0
none	41%	33%

The 4-hydroxy-1-alkene (homoallylic alcohol) **81** is oxidized to the hemiacetal **82** of the aldehyde by the participation of the OH group when there is a substituent at C3. In the absence of the substituent, a ketone is obtained. The hemiacetal is converted into butyrolactone **83**[117]. When Pd nitro complex is used as a catalyst in *t*-BuOH under oxygen, acetals are obtained from homoallylic alcohols even in the absence of a substituent at C-3[118]. *t*-Allylamine is oxidized to the acetal **84** of the aldehyde selectively by participation of the amino group[119].

Oxypalladation of a vinyl ether followed by alkene insertion is an interesting synthetic method for functionalized cyclic ethers. The attack of the allylic alcohol **85** to vinyl ethers (oxypalladation of vinyl ether) to give **86** is followed by an intramolecular alkene insertion, forming the tetrahydrofuran **87** in a one-pot reaction[120]. Similarly, in the following prostaglandin synthesis, the oxypalladation of ethyl vinyl ether with monoprotected cyclopentenediol **88**, the intramolecular and intermolecular alkene insertions and β-elimination take place in a one-pot reaction at room temperature, giving the final product **90** in 72% yield[121]. The stereochemistry of the product shows that *cis*-alkene insertion takes place. It should be noted that the elimination of β-hydrogen from the intermediate **89** is not possible, because there is no β-hydrogen *cis* to the Pd. Tandem intramolecular oxypalladation of the 5-hydroxyalkene **91** and insertion of acrylate proceed to give the tetrahydrofuran **92** with a catalytic amount of Pd(OAc)₂ and a stoichiometric amount of CuCl under oxygen[122].

The oxidation of the cyclic enol ether **93** in MeOH affords the methyl ester **95** by hydrolysis of the ketene acetal **94** formed initially by regioselective attack of the methoxy group at the anomeric carbon, rather than the α-alkoxy ketone[35]. Similarly, the double bond of the furan part in khellin (**96**) is converted ino the ester **98** via the ketene acetal **97**[123].

The γ, δ-unsaturated alcohol **99** is cyclized to 2-vinyl-5-phenyltetrahydro-furan (**100**) by *exo* cyclization in aqueous alcohol[124]. On the other hand, the dihydropyran **101** is formed by *endo* cyclization from a γ, δ-unsaturated alcohol substituted by two methyl groups at the δ-position. The direction of elimination of β-hydrogen to give either enol ethers or allylic ethers can be controlled by using DMSO as a solvent and utilized in the synthesis of the tetronomycin precursor **102**[125]. The oxidation of the optically active 3-alkene-1,2-diol **103** affords the 2,5-dihydrofuran **104** in high *ee*. It should be noted that β-OH is eliminated rather than β-H at the end of the reaction[126].

Phenolic oxygen participates in facile oxypalladation. The intramolecular reaction of 2-hydroxychalcone (**105**) produces the flavone **106**[127]. The benzofuran **107** is formed from 2-allylphenol by *exo* cyclization with Pd(OAc)$_2$, but benzopyran **108** is obtained by *endo* cyclization with PdCl$_2$[128]. Normal cyclization takes place to form the furan **109** from 2-(1-phenylethenyl)phenol[129]. Benzofuran formation by this method has been utilized in the synthesis of aklavinione (**110**)[130].

Unexpectedly, a completely different reaction took place in the oxidation of 2-(1-propenyl)phenol (**111**) with PdCl$_2$. Carpanone (**112**) was obtained in one step in 62% crude yield. This remarkable reaction is explained by the formation of *o*-quinone, followed by the radical coupling of the side-chain. Then the intramolecular cycloaddition takes place to form carpanone[131].

The furo- and pyranobenzopyranones **114** and **115** are prepared by the reaction of *O*-enolate of β-keto lactone **113**[132]. The isoxazole **117** is obtained by the oxidation of the oxime **116** of α, β- or β, γ-unsaturated ketones with PdCl$_2$ and Na$_2$CO$_3$ in dichloromethane[133], but the pyridine **118** is formed with PdCl$_2$(Ph$_3$P)$_2$ and sodium phenoxide[134].

2.4 Reactions with Carboxylic Acids

Soon after the invention of the Wacker process, the formation of vinyl acetate by the reaction of ethylene with $PdCl_2$ in AcOH in the presence of sodium acetate was reported[106,107]. No reaction takes place in the absence of base. The reaction of $Pd(OAc)_2$ with ethylene forms vinyl acetate.

Extensive studies on both the stoichiometric and catalytic formation of vinyl acetate using the Pd(II)–Cu(II) redox system under oxygen have been carried out[135–137]. In the catalytic reaction, considerable amounts of acetaldehyde and ethylidene diacetate are formed in addition to vinyl acetate. Also, acetic anhydride is formed in an amount equimolar with aldehyde. No deuterium incorporation in ethylidene diacetate was observed in the reaction in AcOD[108]. It was found that $PdCl_2$ catalyzes the reaction of AcOH and vinyl acetate to give acetic anhydride and acetaldehyde[138]. Industrial production of vinyl acetate from ethylene and AcOH has been developed by Imperial Chemical Industries, initially in the liquid phase[139]. However, owing to operational problems, mainly due to corrosion, the liquid-phase process was abandoned and a gas-phase process using a supported Pd catalyst was developed[140,141]. At present, vinyl acetate is produced commercially based on this reaction in the gas phase using the Pd supported on carbon or silica as a catalyst[10].

$$CH_2=CH_2 \ + \ AcOH \ + \ O_2 \ \xrightarrow{\text{Pd(II), Cu(II)}}$$

$$\overset{=}{\diagdown}_{OAc} \ + \ CH_3CHO \ + \ (Ac)_2O \ + \ CH_3CH(OAc)_2$$

$$AcOH \ + \ \overset{=}{\diagdown}_{OAc} \ \xrightarrow{\text{PdCl}_2} \ (Ac)_2O \ + \ CH_3CHO$$

$$CH_2=CH_2 \ + \ AcOH \ + \ 1/2 \ O_2 \ \xrightarrow{\text{Pd / Cu}} \ \overset{=}{\diagdown}_{OAc} \ + \ H_2O$$

With higher alkenes, three kinds of products, namely alkenyl acetates, allylic acetates and dioxygenated products are obtained[142]. The reaction of propylene gives two propenyl acetates (**119** and **120**) and allyl acetate (**121**) by the nucleophilic substitution and allylic oxidation. The chemoselective formation of allyl acetate takes place by the gas-phase reaction with the supported Pd(II) and Cu(II) catalyst. Allyl acetate (**121**) is produced commercially by this method[143]. Methallyl acetate (**122**) and 2-methylene-1,3-diacetoxypropane (**123**) are obtained in good yields by the gas-phase oxidation of isobutylene with the supported Pd catalyst[144].

$$\overset{=}{\diagdown} \ + \ AcOH \ \xrightarrow{\text{Pd(OAc)}_2} \ \underset{AcO}{\diagup}\overset{=}{} \ + \ \overset{=}{\diagdown}_{OAc} \ + \ \overset{=}{\diagup}\diagdown_{OAc}$$

$$\qquad\qquad\qquad\qquad\quad \textbf{119} \qquad\qquad\quad \textbf{120} \qquad\qquad \textbf{121}$$

$$\overset{=}{\diagdown} \ + \ AcOH \ \xrightarrow[\text{Cu(OAc)}_2]{\text{Pd(OAc)}_2} \ \overset{=}{\diagup}\diagdown_{OAc}$$

$$\qquad\qquad\qquad\qquad\qquad\qquad\qquad \textbf{121}$$

$$\overset{=}{\diagdown}\diagup \ + \ AcOH \ \xrightarrow{\text{Pd(OAc)}_2} \ \overset{=}{\diagdown}\diagup-OAc \ + \ \overset{=}{\diagdown}\overset{-OAc}{\underset{-OAc}{}}$$

$$\qquad\qquad\qquad\qquad\qquad \textbf{122} \qquad\qquad \textbf{123}$$

The oxidation of cycloalkenes to cyclic ketones with PdCl$_2$ is difficult, but their allylic oxidation (acetoxylation) proceeds smoothly. Reaction of cyclohexene with Pd(OAc)$_2$ gives 3-acetoxycyclohexene[145–150]. Allylic acetoxylation of cyclohexene to give 3-acetoxycyclohexene (**124**) can be carried out with Pd(OAc)$_2$ (5 mol%) and Fe(NO$_3$)$_3$ (5 mol%) using oxygen as final oxidant[151]. In the oxidation of cyclohexene, 1-acetoxycyclohexene is not formed because no β-hydrogen *syn* to Pd is available on the acetoxy-bearing carbon, and *syn*-elimination of β-hydrogen yields allylic acetate (3-acetoxycyclohexene) (**124**). The allylic oxidation of various alkenes is carried out smoothly with palladium trifluoroacetate and benzoquinone. Cycloheptene is oxidized to 3-acetoxycycloheptene (**125**) with Pd(OAc)$_2$, MnO$_2$ and benzoquinone[152]. One of the two double bonds of geranylacetone (**126**) was oxidized to give **127** and **128**. *o*-Methoxyacetophenone is used as a ligand[153]. Regioselective acetoxylation of β,γ-unsaturated ketone and esters such as **129** proceeds to give an γ-acetoxy-(E)-α,β-unsaturated ketone and ester, e.g. **131**, using alkyl nitrite as a reoxidant via the π-allylpalladium complex **130**. Yields are lower with other reoxidants[154]. The oxidative cyclization of the 1,5-diene **132** with Pd(OAc)$_2$ affords the 1-methylene-3-acetoxycyclopentane **133**[155,156]. 2,6-Diacetoxybicyclo[3.3.0]octane (**134**) is obtained by the transannular reaction of 1,5-COD with PdCl$_2$ and Pb(OAc)$_4$ in AcOH[157].

1,2-Dioxygenation by nucleophilic addition to alkenes takes place in the presence of nitrate anion. The reaction of ethylene with Pd(OAc)$_2$ in the presence of LiNO$_3$ affords ethylene glycol monoacetate (**135**)[158–160]. Propylene glycol monoacetates **136** and **137** are formed from propylene in the presence of LiNO$_3$[161,162]. These reactions attract attention as potentially useful for the commercial production of ethylene glycol and propylene glycol.

Acetoxylchlorination of norbornene (**138**) proceeds with skeletal rearrangement in the presence of an excess of CuCl$_2$ to give *exo*-2-chloro-*syn*-7-acetoxy-norbornane (**139**). This is a good synthetic method for *syn*-7-norbornenol[163]. Similarly, a brendane derivative (tricyclo[4.2.1.0^{3,7}]nonane) **141** was prepared in one step by the oxidative acetoxychlorination of commercially available vinyl-norbornene (**140**) using PdCl$_2$–CuCl$_2$ as catalyst[164]. Bicyclo[3.2.1.]octa-2,7-diene was converted into *exo*-6-acetoxytricyclo [3.2.1.0^{2,7}]-3-octene (**142**) in a good yield[165].

The intramolecular reaction of alkenes with various O and N functional groups offers useful synthetic methods for heterocycles[13,14,166]. The reaction of unsaturated carboxylic acids affords lactones by either *exo-* or *endo-* cyclization depending on the positions of the double bond. The reaction of sodium salts of the 3-alkenoic acid **143** and 4-alkenoic acid **144** with Li$_2$PdCl$_4$ affords mostly five-membered lactones in 30–40% yields[167]. Both 5-hexenoic acid (**145**) and 4-hexenoic acid (**146**) are converted to five- or six-membered lactones depending on the solvents and bases[168]. Conjugated 2,4-pentadienoic acid (**147**) is cyclized with Li$_2$PdCl$_4$ to give 2-pyrone (**148**) in water[169].

The isocoumarin **151** is prepared by the intramolecular reaction of 2-(2-propenyl)benzoic acid (**149**) with one equivalent of PdCl$_2$(MeCN)$_2$. However, the (Z)-phthalide **150** is obtained from the same acid with a catalytic amount of Pd(OAc)$_2$ under 1 atm of O$_2$ in DMSO. O$_2$ alone is remarkably efficient in reoxidizing Pd(0) in DMSO. The isocoumarin **151** is obtained by the reaction of 2-(1-propenyl)benzoic acid (**152**) under the same conditions[4]. 2-Vinylbenzoic acid (**153**) is also converted into the isocoumarin **154,** but not to the five-membered lactone[167,170].

Although it is not a reaction of alkenes, oxidation of some alkanes with Pd(II) is cited here. 1-Adamantyl trifluoroacetate (**155**) was obtained in above 50% yield by the reaction of adamantane with Pd(OAc)$_2$ in trifluoroacetic acid at 80 °C[171].

2.5 Reactions with Amines

Formation of enamines is expected by the oxidative reaction of amines with alkenes using Pd(II) salts via the aminopalladation, but it is not easy to achieve the enamine synthesis. Aliphatic amines are strong ligands to electrophilic Pd(II), and the aminopalladation is possible only in a special case. Amide nitrogen reacts more easily than amines, because amidation of amines reduces the complexing ability of the amines. Tertiary amines **156** are obtained by the reaction of alkene–PdCl$_2$ complexes with secondary amines at –50 °C, followed by the reduction of the aminopalladation products with hydrogen or hydride reagents. Terminal alkenes are converted into amines in high yields[172,173]. It has been confirmed that the aminopalladation is a *trans* addition. The (β-acetoxyalkyl)amines **157** are obtained by the oxidation of the aminopalladation product with Pb(OAc)$_4$[174]. For example, 2-dimethylamino-1-butanol was obtained from 1-butene and dimethylamine in 84% yield after hydrolysis. The stereospecific *cis* diamination of alkenes takes place to give the 1,2-diaminoalkanes **158** by the *in situ* oxidation of the aminopalladation products in the presence of amines. MCPBA is suitable for this oxidation. Bromine and NBS are also used. From 1-hexene, 1,2-bis(dimethylamino)hexane (**159**) was obtained in 56% yield[175].

Furthermore, treatment of the aminopalladation product with bromine affords aziridines[176]. The aziridine **160** was obtained stereoselectively from methylamine and 1-decene in 43% yield. The aminopalladation of $PdCl_2$ complexes of ethylene, propylene, and 1-butene with diethylamine affords the unstable σ-alkylpalladium complex **161**, which is converted into the stable chelated acylpalladium complex **162** by treatment with CO[177].

Unlike the intermolecular reaction, the intramolecular aminopalladation proceeds more easily[13,14,166]. Methylindole (**164**) is obtained by the intramolecular *exo* amination of 2-allylaniline (**163**). If there is another olefinic bond in the same molecule, the aminopalladation product **165** undergoes intramolecular alkene insertion to give the tricyclic compound **166**[178]. 2,2-Dimethyl-1,2-dihydroquinoline (**168**) is obtained by *endo* cyclization of 2-(3,3-dimethylallyl)aniline (**167**). The oxidative amination proceeds smoothly with aromatic amines which are less basic than aliphatic amines, whereas it is difficult to use aliphatic amines for the cyclization under similar conditions. The successful oxidative amination is possible with tosylated aliphatic amines **169** to give **170**. The tosylation of amines also reduces the strong complexing ability of aliphatic amines. This is the reason why amides or tosylamides are used for smooth reaction with alkenes[179].

The usefulness of the Pd-catalyzed reactions was amply demonstrated in the total synthesis of clavicipitic acid[180]. The first step is the intramolecular aminopalladation of the 2-vinyltosylamide **171** with Pd(II) to give the indole **172**. Then stepwise Heck reactions of the iodide and bromide of **173** with different alkenes in the presence and absence of a phosphine ligand are carried out to give **174**. In the last step of the synthesis, the intramolecular aminopalladation of **174** is carried out with a catalytic amount of Pd(II) to give the cyclized product **176**. It should be noted that aminopalladation is a stoichiometric reaction by nature. However, in this case, instead of the elimination of β-hydrogen, HO—PdCl, a Pd(II) species, is the elimination product, as shown by **175**, and reoxidation of Pd(0) to Pd(II) is not necessary, hence the reaction is catalytic without a reoxidant. As another example of the catalytic version, the optically active 1-amino(or hydroxy)-3-en-2-ol **177** undergoes diastereoselective cyclization by aminopalladation and elimination of the hydroxy group as HO—PdCl to give **178** and Pd(II), and the reaction proceeds catalytically with 30 mol% of Pd(II)[181].

Intramolecular aminopalladation has been applied to the total synthesis of the complex skeleton of bukittinggine (**179**). For this reaction, Pd(CF$_3$CO$_2$)$_2$ (10 mol%) and benzoquinone (1.1 equiv.) are used. It is important to use freshly recrystallized benzoquinone for successful cyclization. Formation of a π-allylpalladium species as an intermediate in this amination reaction has been suggested[182]. The amino group in *o*-bromoaniline reacts first with acrylate in the presence of PdCl$_2$ to give **180**, and then intramolecular Heck reaction of the resulting alkene **180** with Pd(0) catalyst affords indolecarboxylate **181**[183].

In contrast to the difficult intermolecular reaction of amines with alkenes, amides react more easily. Ethylene–PdCl$_2$ complex reacts with pyrrolidone or caprolactam to give *N*-vinylpyrrolidone and *N*-vinylcaprolactam[184]. The catalytic oxidative amidation of acrylates using lactams proceeds smoothly under oxygen in the presence of PdCl$_2$(MeCN)$_2$ (5 mol%) and CuCl (5 mol%) in DME at 60 °C to give **182**[185]. Cyclic carbamates are more reactive. As an intramolecular version, the 2-pyridone **184** is obtained by the oxidative cyclization of the 2,4-alkadienamide **183** with PdCl$_2$[186]. Similarly, the hydrazide of α, β-unsaturated acid **185** was cyclized to form the 3-pyrazolone **186** [187]. Even the acylurea derivative of an unsaturated acid can be cyclized with PdCl$_2$ to pyrimidinedione. An interesting application is the synthesis of uracil (**188**) in 42% yield by the reaction of acryloylurea (**187**)[188]. Furthermore, 3-methyl-*N*-methylisoquinolone (**189**) is prepared from 2-allyl-*N*-methylbenzamide with PdCl$_2$ and NaH[170,189].

The intermediate **190** of the intramolecular aminopalladation of an allenic bond with *N*-tosylcarbamate undergoes insertion of allylic chloride. Subsequent elimination of PdCl$_2$ occurs to afford the 1, 4-diene system **191**. The regeneration of Pd(II) species makes the reaction catalytic without using a reoxidant[190].

2.6 Reactions with Carbon Nucleophiles

Alkenes coordinated by Pd(II) are attacked by carbon nucleophiles, and carbon–carbon bond formation takes place. The reaction of alkenes with carbon nucleophiles via π-allylpalladium complexes is treated in Section 3.1.

Facile reaction of a carbon nucleophile with an olefinic bond of COD is the first example of carbon–carbon bond formation by means of Pd. COD forms a stable complex with PdCl$_2$. When this complex **192** is treated with malonate or acetoacetate in ether under heterogeneous conditions at room temperature in the presence of Na$_2$CO$_3$, a facile carbopalladation takes place to give the new complex **193**, formed by the introduction of malonate to COD. The complex has π-olefin and σ-Pd bonds. By the treatment of the new complex **193** with a base, the malonate carbanion attacks the σ-Pd—C bond, affording the bicyclo[6.1.0]-nonane **194**. The complex also reacts with another molecule of malonate which attacks the π-olefin bond to give the bicyclo[3.3.0]octane **195** by a transannulation reaction[12,191]. The formation of **194** involves the novel cyclopropanation reaction of alkenes by nucleophilic attack of two carbanions.

In contrast to the facile reaction of COD, the reaction of malonate with simple alkenes is not facile and optimum yields were obtained by using 2 equivalents of Et$_3$N and the carbanion in THF at $-50\,^{\circ}$C to give **196**. In some cases, the products are isolated after the carbonylation[192]. The carbopalladation of the double bond of the benzyl *N*-vinylcarbamate **197** with benzyl acetoacetate to generate **198,** followed by the carbonylation, affords the amino ester **199**, which is converted into the β-lactam **200**[193]. In a similar fashion, negamycin (**203**) has been prepared. The carbopalladation of the optically active *N*-vinyl cyclic carbamate **201** with malonate at $-78\,^{\circ}$C, followed by the carbonylation and trapping the acylpalladium with a vinyltin reagent afforded the ketone **202** in 95% *de*[194]. 2,3-Dihydrofuran reacts with a carbanion to afford the 5-alkylated 4,5-dihydrofuran **204** regioselectively. The double bond isomerization takes place after the alkylation by the *syn* elimination of hydrogen from C-3, followed by readdition and elimination[195]. The reaction of acetylacetone with styrene promoted by PdCl$_2$ and CuCl$_2$ in AcOH affords the nucleophilc addition products **205** and **206**[196].

The phenylation of styrene with phenyl Grignard reagents as a hard carbon nucleophile proceeds in 75% yield in the presence of $PdCl_2$, LiCl, and K_2CO_3 at room temperature to give stilbene (**207**). Selection of the solvent is crucial and the best results are obtained in MeCN. The reaction can be made catalytic by the use of $CuCl_2$[197]. Methyllithium reacts with styrene in the presence of $Pd(acac)_2$ or $Pd(OAc)_2$ to give β-methylstyrene (**208**) in 90% yield[198].

The silyl enol ethers **209** and **212** are considered to be sources of carbanions, and their transmetallation with $Pd(OAc)_2$ forms the Pd enolate **210**, or oxy-π-allylpalladium, which undergoes the intramolecular alkene insertion and β-elimination to give 3-methylcyclopentenone (**211**) and a bicyclic system **213**[199]. Five- and six-membered rings can be prepared by this reaction[200]. Use of benzoquinone makes the reaction catalytic. The reaction has been used for syntheses of skeletons of natural products, such as the phyllocladine intermediate **214**[201], capnellene[202], the stemodin intermediate **215**[203] and hirsutene [204].

In the prostaglandin synthesis shown, silyl enol ether **216**, after transmetallation with Pd(II), undergoes tandem intramolecular and intermolecular alkene insertions to yield **217**[205]. It should be noted that a different mechanism (palladation of the alkene, rather than palladium enolate formation) has been proposed for this reaction, because the corresponding alkyl enol ethers, instead of the silyl ethers, undergo a similar cyclization[201].

2.7 Oxidative Carbonylation

As a unique reaction of Pd(II), the oxidative carbonylation of alkenes is possible with Pd(II) salts. Oxidative carbonylation is mechanistically different from the hydrocarboxylation of alkenes catalyzed by Pd(0), which is treated in Chapter 4, Section 7.1. The oxidative carbonylation in alcohol can be understood in the following way. The reaction starts by the formation of the alkoxycarbonylpalladium **218**. Carbopalladation of alkene (alkene insertion) with **218** gives **219**. Then elimination of β-hydrogen of this intermediate **219** proceeds to yield the α, β-unsaturated ester **220**. The formation of **220** is regarded as the nucleophilic substitution of alkenes with CO_2R. Further CO insertion in **219** gives the acylpalladium intermediate **221** and its alcoholysis yields the succinate derivative **223**. The β-alkoxy ester **222** is formed by nucleophilic substitution of **219** with alkoxide. Formation of the esters **222** and **223** can be regarded as nucleophilic addition to alkenes promoted by Pd(II).

The first report of oxidative carbonylation is the reaction of alkenes with CO in benzene in the presence of $PdCl_2$ to afford the β-chloroacyl chloride **224**[12,206]. The oxidative carbonylation of alkene in alcohol gives the α, β-unsaturated ester **225** and β-alkoxy ester **226** by monocarbonylation, and succinate **227** by dicarbonylation depending on the reaction conditions[207–209]. The scope of the reaction has been studied[210]. Succinate formation takes place at room temperature and 1 atm of CO using Pd on carbon as a catalyst in the presence of an excess of CuCl$_2$, although the reaction is slow (100% conversion after 9 days)[211].

The synthesis of acrylic acid or its ester (**228**) from ethylene has been investigated in AcOH from the standpoint of its practical production[212]. The carbonylation of styrene is a promising commercial process for cinnamate (**229**) production[207,213,214]. Asymmetric carbonylation of styrene with Pd(acac)$_2$ and benzoquinone in the presence of TsOH using 2,2'-dimethoxy-6, 6'-bis(diphenylphosphino)biphenyl (**231**) as a chiral ligand gave dimethyl phenylsuccinate **230** in 93% *ee*, although the yield was not satisfactory, showing that phosphine coordination influences the stereochemical course of the oxidative carbonylation with Pd(II) salt[215].

The dicarboxylation of cyclic alkenes is a useful reaction. All-*exo*-methyl-7-oxabicyclo[2.2.1]heptane-2,3,5,6-tetracarboxylate (**233**) was prepared from the cyclic alkene **232** using Pd on carbon and CuCl$_2$ in MeOH at room temperature with high diastereoselectivity[216]. The dicarbonylation of cyclopentene takes place to form *cis*-1,2- and -1,3-diesters **234** and **236** in a ratio of 68 : 27. The latter is formed by the elimination of Pd—H to form **235** and its read-dition[217]. The dicarbonylation of isobutylene in the presence of a base affords the diester **237** via rearrangement. The alkyl nitrite **238** is a unique reoxidant of Pd(0) and is used for efficient oxidative carbonylation of alkenes to produce succinate derivatives **239**[218]. The oxidative carbonylation with nitrites is different mechanistically from $PdCl_2$–$CuCl_2$-catalyzed carbonylation[219].

The carbonylation of COD–$PdCl_2$ complex in aqueous sodium acetate produces *trans*-2-hydroxy-5-cyclooctenecarboxylic acid β-lactone (**240**). The lactone is obtained in 79% yield directly by the carbonylation of the COD complex in aqueous sodium acetate solution[220]. β-Propiolactone (**241**) is obtained in 72% yield by the reaction of the $PdCl_2$ complex of ethylene with CO and water in MeCN at −20 °C. β-Propiolactone synthesis can be carried out with a catalytic amount of $PdCl_2$ and a stoichiometric amount of $CuCl_2$[221].

The carbonylation of alkene in AcOH–acetic anhydride in the presence of NaCl affords the β-acetoxycarboxylic anhydride **242** in good yields and the method offers a good synthetic method for β-hydroxycarboxylic acid **243**[222].

The intramolecular oxidative carbonylation has wide synthetic application. The γ-lactone **247** is prepared by intramolecular oxycarbonylation of the alkenediol **244** with a stoichiometric amount of Pd(OAc)$_2$ under atmospheric pressure[223]. The intermediate **245** is formed by oxypalladation, and subsequent CO insertion gives the acylpalladium **246**. The oxycarbonylation of alkenols and alkanediols can be carried out with a catalytic amount of PdCl$_2$ and a stoichiometric amount of CuCl$_2$, and has been applied to the synthesis of frenolicin[224] and frendicin B (**249**) from **248**[225]. The carbonylation of the 4-penten-1,3-diol **250**, catalyzed by PdCl$_2$ and CuCl$_2$, afforded in the *cis*-3-hydroxytetrahydrofuran-2-acetic acid lactone **251**[226]. The cyclic acetal **253** is prepared from the dienone **252** in the presence of trimethyl orthoformate as an accepter of water formed by the oxidative reaction[227].

The structures 248, 249, 250, 251, 252, 253 with reaction schemes:

248 + CO → (Pd(II), 70%) → 249   trans / cis = 3 / 1

250 + CO → (PdCl₂, CuCl₂, AcONa, 79%) → 251

252 + CO + MeOH → (PdCl₂, CuCl₂, HC(OMe)₃, 85%) → 253

Aminopalladation and subsequent carbonylation are also facile reactions. The carbonylation of substituted 3-hydroxy-4-pentenylamine as a carbamate (254) proceeds smoothly via the aminopalladation product 255 in AcOH to give 256[228]. The protection of the amino group of the carbamate as tosyl amide is important in the carbonylation of 257 to give 258[229].

254 + CO → (PdCl₂, CuCl₂, AcOH, AcONa, 95%) → [255] → (CO) → 256

257 + CO + MeOH → (PdCl₂, CuCl₂, AcONa, 100%) → 258

2.8 Reactions with Aromatic Compounds

Similarly to mercuration reactions, Pd(OAc)₂ undergoes facile palladation of aromatic compounds. On the other hand, no reaction of aromatic compounds takes place with PdCl₂. PdCl₂ reacts only in the presence of bases. The aromatic palladation product is an unstable intermediate. It can be isolated only when stabilized by chelation. The palladation product of aromatics as the intermediate undergoes three reactions. The first is the homocoupling to form biaryls and the second is acetoxylation of aromatic rings. These reactions are treated in Section 5. As the third reaction, it reacts with alkenes to form styrene derivatives. This reaction is treated in this section.

The Pd(II)-mediated reaction of benzene with alkenes affords styrene derivatives 259[230,231].

In addition to benzene and naphthalene derivatives, heteroaromatic compounds such as ferrocene[232], furan, thiophene, selenophene[233,234], and cyclobutadiene iron carbonyl complex[235] react with alkenes to give vinyl heterocycles. The ease of the reaction of styrene with substituted benzenes to give stilbene derivatives 260 increases in the order benzene < naphthalene < ferrocene < furan. The effect of substituents in this reaction is similar to that in the electrophilic aromatic substitution reactions[236].

R	ortho %	meta %	para %
Me	17	24	33
Et	11	23	31
OMe	30	5	48
NO$_2$	4	29	4
Cl	12	20	30

Mechanistic studies show that the arylation of alkenes proceeds via the palladation of aromatic compounds to form a σ-aryl–Pd bond (261), into which insertion of alkene takes place to form 262. The final step is β-elimination to form the arylated alkenes 259 and Pd(0).

Indene derivatives 264a and 264b are formed by the intramolecular reaction of 3-methyl-3-phenyl-1-butene (263a) and 3,3,3-triphenylpropylene (263b) [237]. Two phenyl groups are introduced into the β-substituted β-methylstyrene 265 to form the β-substituted β-diphenylmethylstyrene 267 via 266 in one step[238]. Allyl acetate reacts with benzene to give 3-phenylcinnamaldehyde (269) by acyl—O bond fission. The primary product 268 was obtained in a trace amount[239].

Pd-catalyzed intramolecular allylation of the amine **270** gives the indole **271**. Further intramolecular reaction of **271** with a double bond has been applied to ibogamine synthesis using $PdCl_2(MeCN)_2$ and $AgBF_4$ in MeCN in the presence of Et_3N. Palladium tetrafluoroborate is formed and seems to be more reactive than $PdCl_2$. The σ-alkyl–Pd species as an intermediate is reduced with $NaBH_4$ to give the saturated cyclized product **272** in 45% yield[240]. In the total synthesis of (+)-paraherquamide B, the heptacycle **274** was prepared in 80% yield by the cyclization of **273** using $PdCl_2$ and $AgBF_4$ and hydrogenolysis with $NaBH_4$[241].

The regioselective arylation of enol ethers in glycals is possible. A phenyl group was introduced into the enone **275** derived from glycal. In addition to the β-phenylenone **276**, which is formed by *anti* elimination of hydrogen, the β-phenyl ketone **277** a Michael-type addition (or hydrophenylation) product, was obtained by insertion and hydrogenolysis, rather than *anti* elimination[242]. Phenylation of the glycal **278** followed by *syn* elimination of β-hydrogen afforded the expected allylic ether **279**[243].

Heteroaromatics such as furan, thiophene, and even the 2-pyridone **280** react with acrylate to form **281**[244–246]. Benzene and heteroaromatic rings are introduced into naphthoquinone (**282**) as an alkene component[247]. The pyrrole ring is more reactive than the benzene ring in indole.

Although several attempts have been made to achieve the catalytic arylation of alkenes under an oxygen atmosphere or in the presence of other reoxidants, the catalytic turnovers observed so far are not high[248,249]. An efficient reoxidant is *t*-butyl perbenzoate. The reaction of acrylate with 2-methyl-furan (**283**) using Pd(OCOPh)$_2$ (1 mol%) and *t*-butyl perbenzoate (100 mol%) in AcOH gave the coupled product **284** in 67% yield[250]. A catalytic reaction of benzene with acrylate was carried out with montmorillonite ethylsilyldiphenylphosphine–palladium chloride as a heterogenized homogeneous catalyst[251].

2.9 Reactions with Hydrogen Cyanide and Chloride

Unsaturated nitriles are formed by the reaction of ethylene or propylene with Pd(CN)$_2$[252]. The synthesis of unsaturated nitriles by a gas-phase reaction of alkenes, HCN, and oxygen was carried out by use of a Pd catalyst supported on active carbon. Acrylonitrile is formed from ethylene. Methacrylonitrile and crotononitrile are obtained from propylene[253]. Vinyl chloride is obtained in a high yield from ethylene and PdCl$_2$ using highly polar solvents such as DMF. The reaction can be made catalytic by the use of chloranil[254].

2.10 Oxidative Coupling Reactions

The oxidative coupling of alkenes which have two substituents at the 2-position, such as isobutylene, styrene, 2-phenylpropene, 1,1-diphenylethylene, and methyl methacrylate, takes place to give the 1,1,4,4-tetrasubstituted butadienes **285** by the action of Pd(OAc)$_2$ or PdCl$_2$ in the presence of sodium acetate[255–257]. Oxidation of styrene with Pd(OAc)$_2$ produces 1,4-diphenylbutadiene (**285, R = H**) as a main product and α- and β-acetoxystyrenes as minor products[258]. Prolonged oxidation of the primary coupling product **285** (R = Me) of 2-phenylpropene with an excess of Pd(OAc)$_2$ leads slowly to *p*-

terphenyl (**286**)[256]. The oxidative coupling of vinyl acetate using Pd(OAc)₂
affords 1,4-diacetoxy-1,3-butadiene (**287**). In addition, considerable amounts
of 1,1,4-triacetoxy-2-butene (**288**) and 1,1,4,4-tetracetoxybutane (**289**) are
formed[259].

The cross-coupling of two alkenes also takes place. Alkenes such as acrylate
react regioselectively with 1,3-dimethyluracil (**290**) to afford 5-(1-alkenyl)ura-
cils such as **291** in a high yield[260].

2.11 Reactions with Alkynes

The co-trimerization of diphenylacetylene with ethylene promoted by
PdCl₂(PhCN)₂ afforded *trans,trans*-3,4-diphenylhexa-2,4-diene (**292**) in 2–6 h,
but 3-ethyl-1-methyl-2-phenylindene (**293**) and 1-ethyl-3-methyl-2-phenylin-
dene (**294**) were formed in 24 h[261]. Other terminal alkenes react similarly[262].
On the other hand, mono- and disubstituted alkynes react with styrene by use of
PdCl₂ and LiCl in AcOH at room temperature to give the chlorodiene **295**.
Under oxygen, the reaction proceeds catalytically. The reaction can be
explained by the chloropalladation of the triple bond, followed by the alkene
insertion. The final step is the elimination of Pd—H to give the diene[263].
When allyl chloride is used as an alkene component, PdCl₂ is regenerated in
the final step and the reaction proceeds catalytically without oxygen. The intramolecular reaction of *o*-(2-phenylethynyl)(2-phenylvinyl)benzene (**296**) with
PdCl₂(PhCN)₂ affords the indenone **297**[264].

3 Stoichiometric Reactions of π-Allyl Complexes

3.1 Stoichiometric and Catalytic π-Allylpalladium Chemistry

π-Allylpalladium chloride reacts with a soft carbon nucleophile such as malonate and acetoacetate in DMSO as a coordinating solvent, and facile carbon–carbon bond formation takes place[12,265]. This reaction constitutes the basis of both stoichiometric and catalytic π-allylpalladium chemistry. Depending on the way in which π-allylpalladium complexes are prepared, the reaction becomes stoichiometric or catalytic. Preparation of the π-allylpalladium complexes **298** by the oxidative addition of Pd(0) to various allylic compounds (esters, carbonates etc.), and their reactions with nucleophiles, are catalytic, because Pd(0) is regenerated after the reaction with the nucleophile, and reacts again with allylic compounds. These catalytic reactions are treated in Chapter 4, Section 2. On the other hand, the preparation of the π-allyl complexes **299** from alkenes requires Pd(II) salts. The subsequent reaction with the nucleophile forms Pd(0). The whole process consumes Pd(II), and ends as a stoichiometric process, because the *in situ* reoxidation of Pd(0) is hardly attainable. These stoichiometric reactions are treated in this section.

3.2 Preparation of π-Allylpalladium Complexes from Alkenes and Their Reactions with Carbon Nucleophiles

π-Allylpalladium complexes are prepared by the reaction of alkenes with $PdCl_2$ under basic conditions. Efficient complex formation takes place in DMF in the presence of bases[266–268]. Efficient complex formation takes place in AcOH in the presence of $CuCl_2$ and sodium acetate. It is well known that π-allylpalladium complexes react with soft carbon nucleophiles[265]. Combination of these two reactions permits the allylic alkylation of alkenes with carbon nucleophiles via π-allylpalladium complexes as a stoichiometric reaction. This reaction offers a method for the oxidative functionalization of alkenes. The reaction of π-allylpalladium complexes prepared from alkenes has been applied to the syntheses of a number of natural products[269–271]. Functionalization of pinene was carried out by the reaction of the π-allylpalladium 300 derived from pinene with a phenylsulfinyl group, followed by oxidative elimination of the phenylsulfinyl group to give the pinene derivative 301[272]. In the synthesis of vitamin A (305), the π-allylpalladium complex 303 was prepared using $CuCl_2$ from prenyl acetate (302) without attacking the allylic acetate moiety[273]. The reaction of the carbon nucleophile 304 is carried out in DMSO in the presence of an excess of Ph_3P. Functionalization of geranylacetone[274] and carbon chain extension of farnesoate (306) via regioselective formation of the π-allylpalladium complex 307 to give geranylgeraniol (309) using methyl 4-methyl-2-phenylsulfonyl-3-pentenoate (308) as a nucleophile are other examples[275].

π-Allyl complex formation takes place particularly easily from the α, β- or β, γ-unsaturated carbonyl compound 310, because the elimination of α- or γ-hydrogen is facilitated by their high acidity[276]. The reaction of the complex 311 with a carbon nucleophile to lead to 312 constitutes γ-alkylation of α, β-unsaturated ketones or esters[277]. The facile formation of the π-allylpalladium complex 313 from α, β-unsaturated keto steroids (cholest-4-en-3-one, testosterone, and progesterone) takes place[278], and malonate is introduced regioselectively at the 6β-position (γ-alkylation of the ketone) to give 314 stereoselectively by the reaction of the complex 313 in DMSO in the absence of Ph_3P. The 6β-product 314 initially formed is isomerized to the more stable 6α-epimer[279].

The enamine **315** as a carbon nucleophile reacts with π-allylpalladium complexes to give allyl ketones after hydrolysis[265].

Interestingly, some nucleophiles attack the central carbon of the π-allyl system to form a palladacyclobutane **316** and its reductive elimination gives cyclopropanes. Li enolates of carboxylic acids are such nucleophiles for the cyclopropanation[280,281]. Li enolate of ketone **317**, isobutyryloxazolidinone, and isopropyl phenyl sulfones react with π-allylpalladium to afford the α-cyclopropyl ketone **318** and the corresponding cyclopropanes in the presence of TMEDA under a CO atmosphere[282].

Hard carbon nucleophiles of organometallic compounds react with π-allyl-palladium complexes. A steroidal side-chain is introduced regio- and stereo-selectively by the reaction of the steroidal π-allylpalladium complex **319** with the alkenylzirconium compound **320**[283].

3.3 Carbonylation, Elimination, and Oxidation Reactions

Treatment of π-allylpalladium chloride with CO in EtOH affords ethyl 3-butenoate (**321**)[284]. β, γ-Unsaturated esters, obtained by the carbonylation of π-allylpalladium complexes, are reactive compounds for π-allyl complex formation and undergo further facile transformation via π-allylpalladium complex formation. For example, ethyl 3-butenoate (**321**) is easily converted into 1-carboethoxy-π-allylpalladium chloride (**322**) by the treatment with Na$_2$PdCl$_4$ in ethanol. Then the repeated carbonylation of the complex **322** gives ethyl 2-

pentenedioate (**323**), which can be converted further into the 1,3-dicarboethoxy–π-allylpalladium complex (**324**)[285].

Elimination of H—Pd—Cl from π-allylpalladium generates conjugated dienes[286–288]. The conjugated 4,6-dienone **326** can be prepared from a steroidal enone via the π-allylpalladium complex **325** with KCN[289]. Catalytic oxidative dehydrogenation of ethyl 3-butenedicarboxylate (**327**) to butadiene-dicarboxylate (muconate) (**329**) proceeds in a high yield using PdCl$_2$ and CuCl$_2$ under an oxygen atmosphere in AcOH containing sodium acetate[290]. The reaction proceeds via the π-allylpalladium complex **328** and the subsequent elimination of H—Pd—Cl.

Oxidation of π-allylpalladium complexes with hydroperoxide or peracids produces allylic alcohols regioselectively with retention of the configuration. A steroidal α-oriented π-allylpalladium complex is converted into the 4α-allylic alcohol **330**, and the corresponding β-oriented complex is oxidized to the 6β-allylic alcohol regioselectively[291,292]. The π-allylpalladium complex **331** substituted with an EWG is oxidized regioselectively with molecular oxygen under irradiation to give the α, β-unsaturated ketone **332**[293]. The α, β-unsaturated ketone **332** is also obtained by the oxidation of the allylsilane **333** substituted by the EWG under irradiation using a catalytic amount of Pd(OCO$_2$CF$_3$) [294].

Isomerization of double bonds in vitamin D analogs such as calciferol by oxidation and reduction has been carried out via the formation of the π-allylpalladium complex **334** with PdCl$_2$(PhCN)$_2$ in 70% yield, followed by hydride reduction to afford **335**[295].

4 Reactions of Conjugated Dienes

When butadiene is treated with PdCl$_2$, the 1-chloromethyl-π-allylpalladium complex **336** (X = Cl) is formed by the chloropalladation. In the presence of nucleophiles, the substituted π-methallylpalladium complex **336** (X = nucleophile) is formed[296–299]. In this way, the nucleophile can be introduced at the terminal carbon of conjugated diene systems. For example, a methoxy group is introduced at the terminal carbon of 3,7-dimethyl-1,3,6-octatriene to give **337** as expected, whereas myrcene (**338**) is converted into the π-allyl complex **339** after the cyclization[288].

The π-allylpalladium complexes formed from conjugated dienes are reactive and react further with a nucleophile to give the 1,4-difunctionalized products **340**. Based on this reaction, various nucleophiles are introduced into conjugated dienes to form 1,4-difunctionalized 2-alkenes. Acetoxy, alkoxy, halo, and

X = Cl, OR, OAc

amino groups, carbon nucleophiles, and CO are introduced into dienes. This is the basis of the synthetic application of the Pd(II)-promoted oxidative difunctionalization of conjugated dienes[300]. The reaction itself is stoichiometric with respect to Pd(II) salts, but it can be made catalytic by the use of the reoxidants of Pd(0). The first example of the 1,4-difunctionalization is carbonylation. Treatment of the isoprene complex **341** with CO in ethanol at room temperature affords ethyl 5-ethoxy-3-methyl-3-pentenoate (**342**) as a major product and 4-methyl-3-hexenedioate (**343**) at 100 °C[301].

1,4-Difunctionalization with similar or different nucleophiles has wide synthetic applications. The oxidative diacetoxylation of butadiene with Pd(OAc)$_2$ affords 1,4-diacetoxy-2-butene (**344**) and 1,2-diacetoxy-3-butene (**345**). The latter can be isomerized to the former. An industrial process has been developed based on this reaction. The commercial process for 1,4-diacetoxy-2-butene (**344**) has been developed using the supported Pd catalyst containing Te in AcOH. 1,4-Butanediol and THF are produced commercially from 1,4-diacetoxy-2-butene (**344**)[302].

1, 4-Diacetoxylation of various conjugated dienes including cyclic dienes has been extensively studied. 1,3-Cyclohexadiene was converted into a mixture of isomeric 1,4-diacetoxy-2-cyclohexenes of unknown stereochemistry[303]. The stereoselective Pd-catalyzed 1,4-diacetoxylation of dienes is carried out in AcOH in the presence of LiOAc and /or LiCl and benzoquinone[304,305]. In the presence of acetate ion and in the absence of chloride ion, *trans*-diacetoxylation occurs, whereas addition of a catalytic amount of LiCl changes the stereochemistry to *cis* addition. The coordination of a chloride ion to Pd makes the *cis* migration of the acetate from Pd impossible. From 1,3-cyclohexadiene, *trans*- and *cis*-1,4-diacetoxy-2-cyclohexenes (**346** and **347**) can be prepared stereoselectively. For the 6-substituted 1,3-cycloheptadiene **348**, a high diastereoselectivity is observed. The stereoselective *cis*-diacetoxylation of 5-carbomethoxy-1,3-cyclohexadiene (**349**) has been applied to the synthesis of *dl*-shikimic acid (**350**).

The diacetoxylation of (*E,E*)- and (*E,Z*)-2,4-hexadiene (**351** and **353**) is stereospecific, and 2,5-dimethylfurans (**352** and **354**) of different stereochemistry have been prepared from the isomers. Two different carboxylates are introduced with high *cis* selectivity by the reaction of 1,3-cyclohexadiene and

1,3-cycloheptadiene (**355**) in AcOH in the presence of CF_3CO_2H, CF_3CO_2Li, MnO_2, and benzoquinone[306]. The product **356** is synthetically useful, because the trifluoroacetate group can be selectively replaced in the Pd-catalyzed nucleophilic substitution without affecting the acetoxy group, or hydrolyzed selectively to give the monoacetate **357**.

It is possible to prepare 1-acetoxy-4-chloro-2-alkenes from conjugated dienes with high selectivity. In the presence of stoichiometric amounts of LiOAc and LiCl, 1-acetoxy-4-chloro-2-butene (**358**) is obtained from butadiene[307], and *cis*-1-acetoxy-4-chloro-2-cyclohexene (**360**) is obtained from 1,3-cyclohexadiene with 99% selectivity[308]. Neither the 1,4-dichloride nor 1,4-diacetate is formed. Good stereocontrol is also observed with acyclic dienes[309]. The chloride and acetoxy groups have different reactivities. The Pd-catalyzed selective displacement of the chloride in **358** with diethylamine gives **359** without attacking allylic acetate, and the chloride in **360** is displaced with malonate with retention of the stereochemistry to give **361**, while the uncatalyzed reaction affords the inversion product **362**.

The *meso*-diacetate **363** can be transformed into either enantiomer of the 4-substituted 2-cyclohexen-1-ol **364** via the enzymatic hydrolysis. By changing the relative reactivity of the allylic leaving groups (acetate and the more reactive carbonate), either enantiomer of 4-substituted cyclohexenyl acetate is accessible by choice. Then the enantioselective synthesis of (*R*)- and (*S*)-5-substituted 1,3-cyclohexadienes **365** and **367** can be achieved. The Pd(II)-catalyzed acetoxylactonization of the diene acids affords the lactones **366** and **368** of different stereochemistry[310]. The tropane alkaloid skeletons **370** and **371** have been constructed based on this chemoselective Pd-catalyzed reactions of 6-benzyloxy-1,3-cycloheptadiene (**369**)[311].

An intramolecular version offers useful synthetic methods for heterocycles. The total syntheses of α- and γ-lycoranes (**373** and **374**) have been carried out by applying the intramolecular aminochlorination of the carbamate of 5-(2-aminoethyl)-1,3-cyclohexadiene (**372**) as a key reaction[312,313]. Interestingly, the 4,6- and 5,7-diene amides **375** and **377** undergo the intramolecular amination twice via π-allylpalladium to form alkaloid skeletons of pyrrolizidine (**376**) and indolizidine (**378**), showing that amide group is reactive[314].

In MeOH, 1,4-dimethoxy-2-cyclohexene (**379**) is obtained from 1,3-cyclohexadiene[315]. Acetoxylation and the intramolecular alkoxylation took place in the synthesis of the naturally occurring tetrahydrofuran derivative **380** and is another example of the selective introduction of different nucleophiles[316]. In intramolecular 1,4-oxyacetoxylation to form the fused tetrahydrofurans and tetrahydropyrans **381**, *cis* addition takes place in the presence of a catalytic amount of LiCl, whereas the *trans* product is obtained in its absence[317]. The stereocontrolled oxaspirocyclization proceeds to afford the *trans* product **382** in the presence of Li₂CO₃ and the *cis* product in the presence of LiCl[318,319].

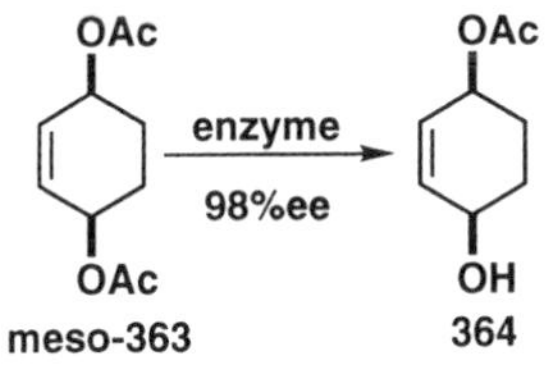
OAc
enzyme
98%ee
OAc
OAc
OH
meso-363
364

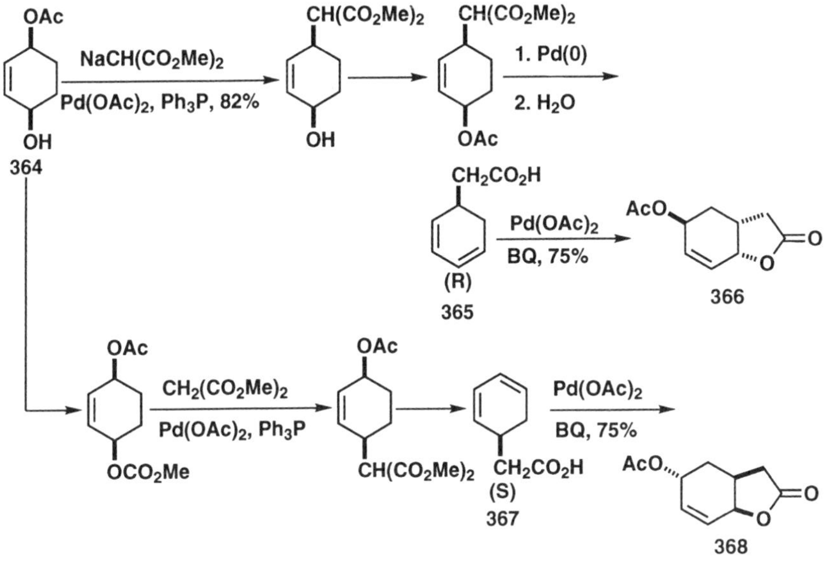
OAc
NaCH(CO2Me)2
Pd(OAc)2, Ph3P, 82%
OH
364
CH(CO2Me)2
OH
CH(CO2Me)2
OAc
1. Pd(0)
2. H2O
CH2CO2H
(R)
365
Pd(OAc)2
BQ, 75%
AcO
O
O
366
OAc
CH2(CO2Me)2
Pd(OAc)2, Ph3P
OCO2Me
OAc
CH(CO2Me)2
CH2CO2H
(S)
367
Pd(OAc)2
BQ, 75%
AcO
O
O
368

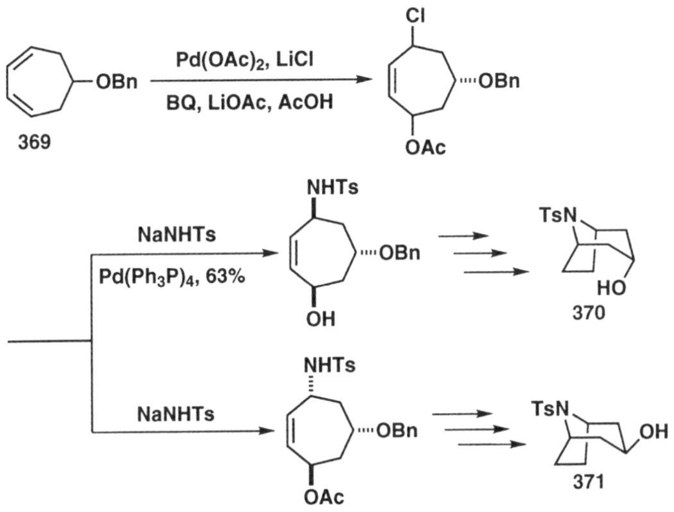
OBn
369
Pd(OAc)2, LiCl
BQ, LiOAc, AcOH
Cl
OBn
OAc
NaNHTs
Pd(Ph3P)4, 63%
NHTs
OBn
OH
TsN
HO
370
NaNHTs
NHTs
OBn
OAc
TsN
OH
371

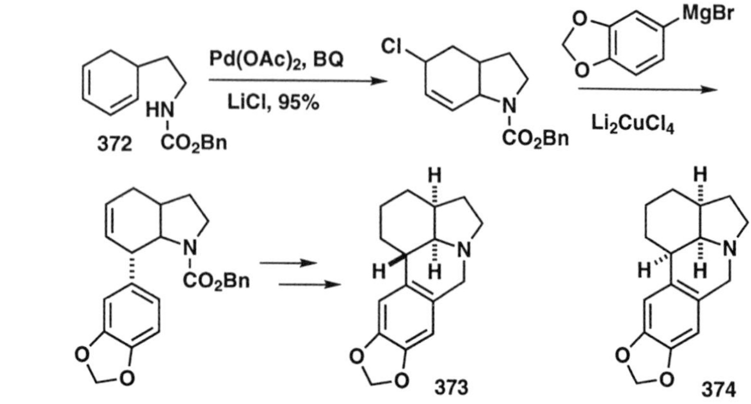

375

376

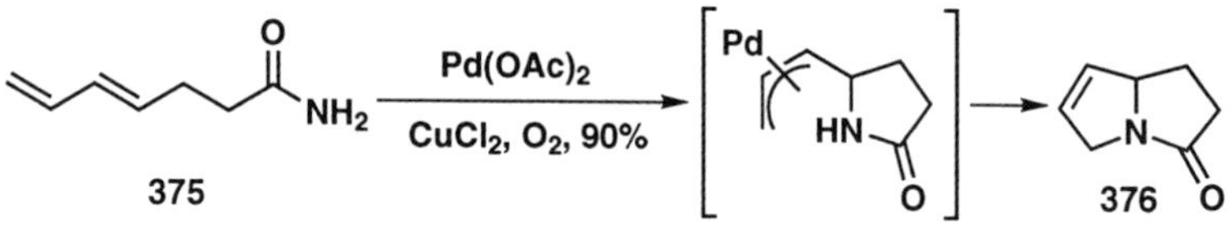

377

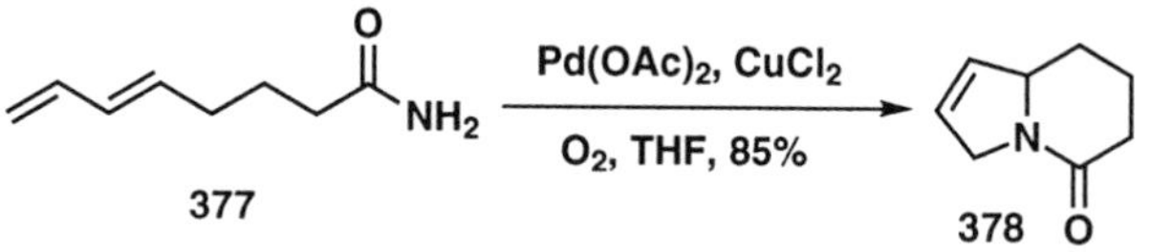

378

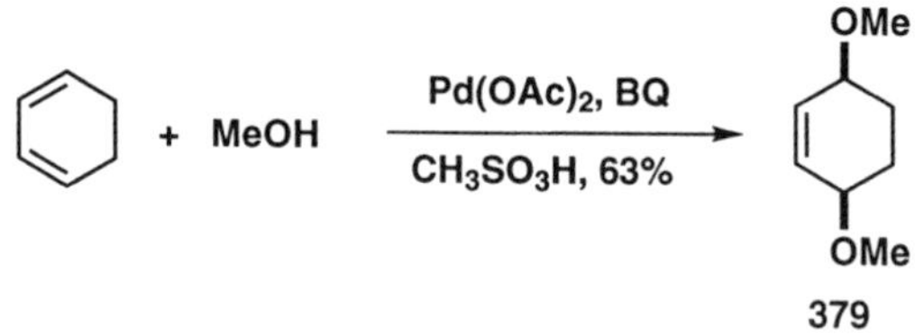

379

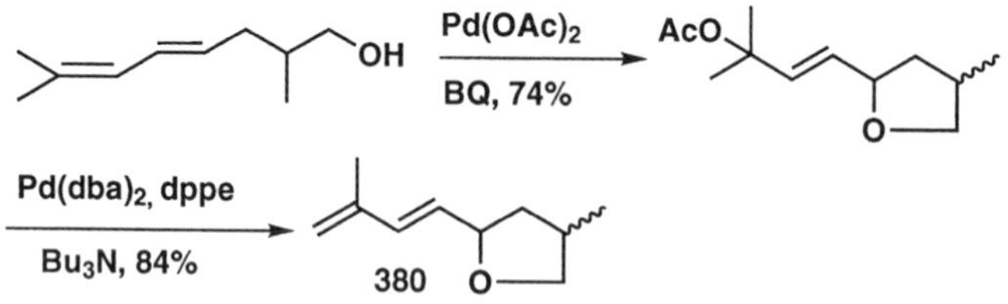

380

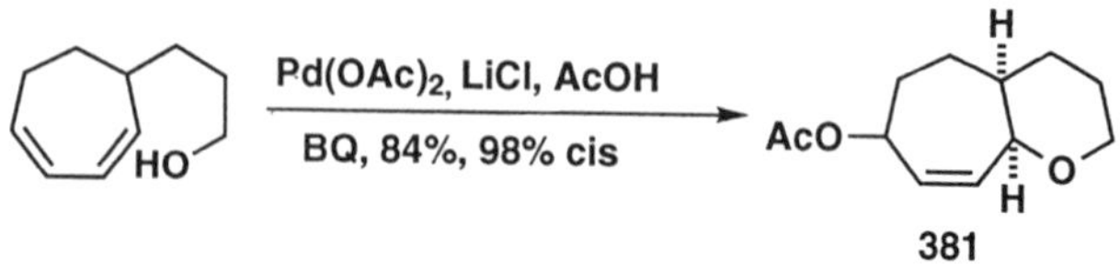

381

382

In order to make these oxidative reactions of 1,3-dienes catalytic, several reoxidants are used. In general, a stoichiometric amount of benzoquinone is used. Furthermore, Fe–phthalocyanine complex or Co–salen complex is used to reoxidize hydroquinone to benzoquinone. Also, it was found that the reaction is faster and stereoselectivity is higher when (phenylsulfinyl)benzoquinone (383) is used owing to coordination of the sulfinyl group to Pd. Thus the reaction can be carried out using catalytic amounts of Pd(OAc)$_2$ and (arylsulfinyl)benzoquinone in the presence of the Fe or Co complex under an oxygen atmosphere[320]. Oxidative dicyanation of butadiene takes place to give 1,4-dicyano-2-butene(384) (40%) and 1,2-dicyano-3-butene (385)[321].

Aryl- or alkenylpalladium complexes can be generated *in situ* by the transmetallation of the aryl- or alkenylmercury compounds 386 or 389 with Pd(II) (see Section 6). These species react with 1,3-cyclohexadiene via the formation of the π-allylpalladium intermediate 387, which is attacked intramolecularly by the amide or carboxylate group, and the 1,2-difunctionalization takes place to give 388 and 390[322]. Similarly, the *ortho*-thallation of benzoic acid followed by transmetallation with Pd(II) forms the arylpalladium complex, which reacts with butadiene to afford the isocoumarin 391, achieving the 1,2-difunctionalization of butadiene[323].

5 Reactions of Aromatic Compounds

Three oxidative reactions of benzene with Pd(OAc)$_2$ via reactive σ-aryl–Pd complexes are known. The insertion of alkenes and elimination afford arylalkenes. The oxidative functionalization of alkenes with aromatics is treated in Section 2.8. Two other reactions, oxidative homocoupling[324,325] and the acetoxylation[326], are treated in this section. The palladation of aromatic compounds is possible only with Pd(OAc)$_2$. No reaction takes place with PdCl$_2$.

5.1 Homocoupling Reaction

The oxidative homocoupling of benzene with Pd(OAc)$_2$, generated *in situ* from PdCl$_2$ and AcONa, affords biphenyl in 81% yield. In the absence of AcONa, no reaction took place. Pd(OAc)$_2$ itself is a good reagent for the coupling[324–326]. The scope of the reaction has been studied[327,328].

$$2 \; \text{(benzene)} \; + \; Pd(OAc)_2 \longrightarrow \text{(biphenyl)} \; + \; Pd \; + \; 2 \, AcOH$$

In the reaction of aromatic compounds with $Pd(OAc)_2$ in AcOH, addition of $HClO_4$ shows a profound effect, shortening the reaction time at 100 °C from 16 h to 5 min, and the reaction at lower temperature eliminates the acetoxylation[325]. Electron-donating groups increase the rate, whereas the opposite holds for EWGs[324]. A mixture of isomers is obtained with substituted benzenes. For example, the following isomer distribution was observed in the reaction of toluene with $Pd(OAc)_2$ in AcOH containing $HClO_4$[325]:

	2,2'	2,3'	2,4'	3,3'	3,4'	4,4'
50 °C, 90 min	0.7	4.6	11.4	6.9	34.2	42.2
100 °C, 60 min	0.5	4.8	9.4	9.8	38.2	37.3

Under oxygen pressure, the reaction proceeds catalytically to some extent. The oxygen pressure favors homocoupling and inhibits ring acetoxylation[325]. The reaction is made catalytic by using various oxidants such as Cu(II) salts, HNO_3, peroxides[329], sodium peroxodisulfate[330], and thallium trifluoroacetate[331]. The reaction is applied to the industrial production of the biphenyltetracarboxylates **392** by oxidative coupling of dimethyl phthalate using $Pd(OAc)_2$ and $Cu(OAc)_2$. Addition of phenanthroline is important for the regioselective formation of the most important 3,4,3'4'-isomer[332].

$$\text{(o-phthalate, } CO_2R, CO_2R) \xrightarrow[Cu(OAc)_2]{Pd(OAc)_2} \text{(biphenyltetracarboxylate } RO_2C, RO_2C, CO_2R, CO_2R)$$

392

The cyclized products **393** can be prepared by the intramolecular coupling of diphenyl ether or diphenylamine[333,334]. The reaction has been applied to the synthesis of an alkaloid **394**[335]. The intramolecular coupling of benzoyl-*N*-methylindole affords 5-methyl-5,10-dihydroindenol[1,2-b]indol-10-one (**395**) in 60% yield in AcOH[336]. Staurosporine aglycone (**396**) was prepared by the intramolecular coupling of an indole ring[337].

The oxidative coupling of thiophene, furan[338] and pyrrole[339,340] is also possible. The following order of reactivity was observed in the coupling of substituted furans[338]: R = H > Me > CHO > CO_2Me > $CH(OAc)_2$ > CO_2H. The cross-coupling of furans and thiophenes with arene is possible, and 4-phenylfurfural (**397**) is the main product of the cross-coupling of furfural and benzene[341].

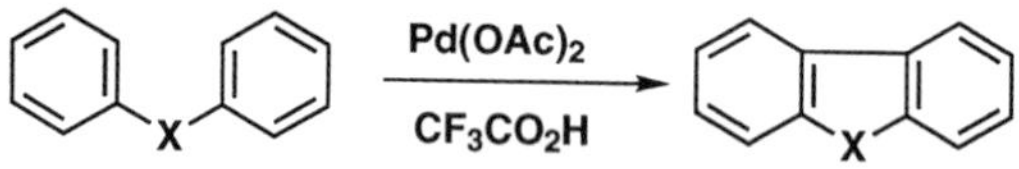

X = O, NH   393

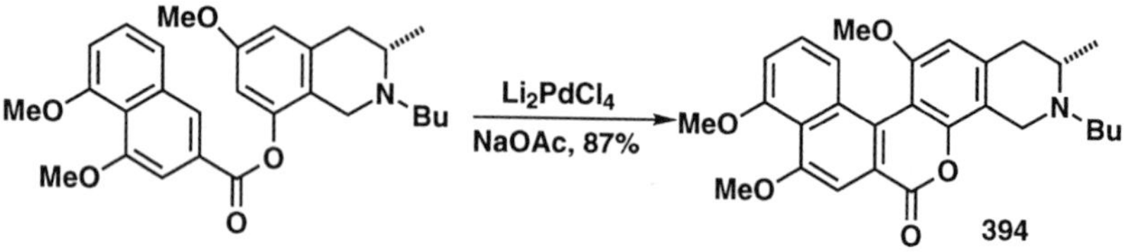

394

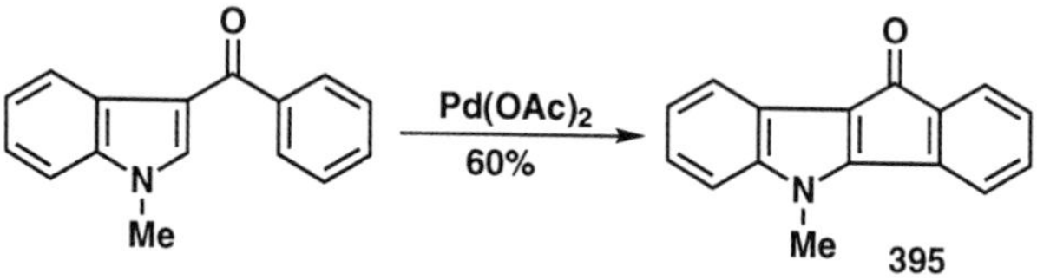

395

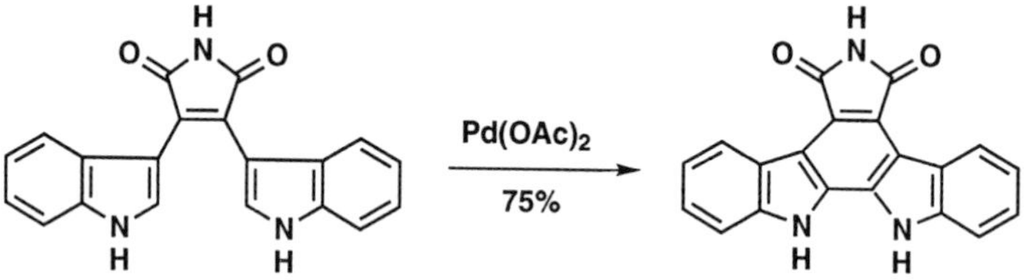

396

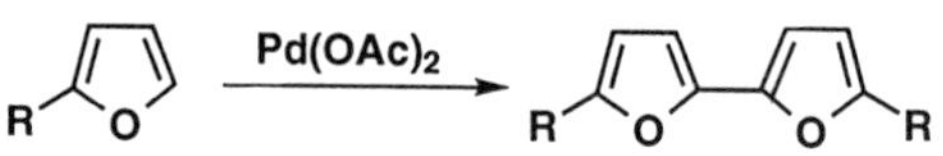

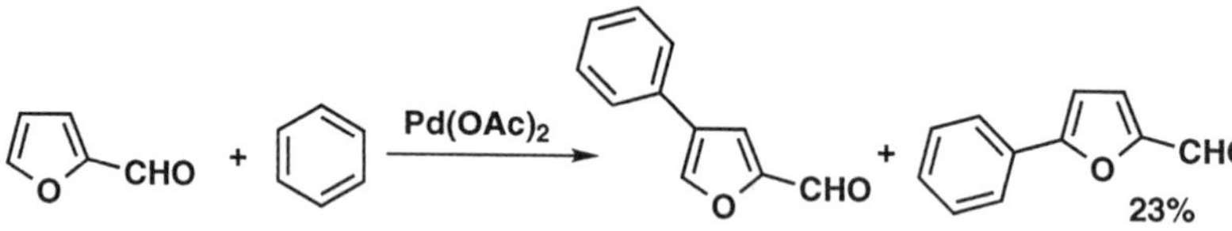

48%   397   23%

5.2 Oxidative Substitution Reactions

Acetoxybenzene is prepared by the reaction of benzene with Pd(OAc)$_2$[325,342–345]. This reaction is regarded as a potentially useful method for phenol production from benzene, if carried out with only a catalytic amount of Pd(OAc)$_2$. Extensive studies have been carried out on this reaction in order to achieve a high catalytic turnover. In addition to oxygen and Cu(II) salts, other oxidants, such as HNO$_3$, nitrate[346,347], potassium peroxodisulfate[348], and heteropoly acids[349,350], are used. HNO$_3$ is said to be a good reoxidant of Pd(0) for the efficient catalytic reaction. Further reaction of acetoxybenzene produces diacetoxybenzene. When HNO_3 is used as the reoxidant, the reaction is accompanied by the formation of explosive polynitrated products as byproducts. The direct oxidation of benzene to phenol with $Pd(OAc)_2$ is carried out under oxygen and CO in the presence of phenanthroline (phen)[351].

Meta-Acetoxylation takes place mainly with aromatic compounds substituted by an electron-donating group, whereas *ortho*- and *para*-acetoxylations are observed with aromatics substituted with an EWG. The isomer distribution in the acetoxylation shows an opposite trend to the ordinary electrophilic aromatic substitution. The *ortho : meta : para* ratio of the acetoxylation of chlorobenzene with $Pd(OAc)_2$ in AcOH under oxygen atmosphere has been found to be 3 : 88 : 9[352]. With toluene, the ratio was 20 : 50 : 30. In addition, toluene undergoes two other reactions with $Pd(OAc)_2$, namely the homocoupling and the acetoxylation of the methyl group. Selective acetoxylation of the methyl group in toluene is a promising preparative method for benzyl acetate (**398**)[325,326,343,352,356] and addition of excess AcONa, AcOK, or $Sn(OAc)_2$ and active carbon has profound effects on the selective formation of benzyl acetate[353]. The heterogeneous catalyst prepared by supporting $Pd(OAc)_2$, $Sn(AcO)_2$, and AcOK on carbon is used for the smooth preparation of benzyl acetate[354]. Naphthalene is acetoxylated equally at the α- and β-positions with $Pd(OAc)_2$[355].

Nitration of aromatic rings is possible by use of Pd(NO₃)₂[356], Pd(OAc)₂–NaNO₂[357], Pd(OAc)₂–NO₂[358], and Pd(0)–NO₂[359]. The nitration can be carried out fully catalytically by Pd(OAc)₂–NO₂ and oxygen. This reaction offers a promising new method of nitration without using mixed acids of HNO₃ and H₂SO₄.

Benzoic acid and naphthoic acid are formed by the oxidative carbonylation by use of Pd(OAc)₂ in AcOH. *t*-BuO₂H and allyl chloride are used as reoxidants. Addition of phenanthroline gives a favorable effect[360]. Furan and thiophene are also carbonylated selectively at the 2-position[361,362]. Indole-3-carboxylic acid is prepared by the carboxylation of 1-acetylindole using Pd(OAc)₂ and peroxodisulfate (Na₂S₂O₈)[362a]. Benzoic acid derivatives are obtained by the reaction of benzene derivatives with sodium palladium malonate in refluxing AcOH[363].

$$\text{Benzene} + CO \xrightarrow[\text{Cl, 1 atm., 75°}]{Pd(OAc)_2, \text{ t-BuO}_2H} \text{benzene-CO}_2H$$

1300% [based on Pd(II)]

6 Synthetic Reactions by Transmetallation of Organometallic Compounds

The transmetallation of various organometallic compounds (Hg, Tl, Sn, B, Si, etc.) with Pd(II) generates the reactive σ-aryl, alkenyl, and alkyl Pd compounds. These carbopalladation products can be used without isolation for further reactions. Pd(II) and Hg(II) salts have similar reactivity toward alkenes and aromatic compounds, but Hg(II) salts form stable mercuration products with alkenes and aromatic rings. The mercuration products are isolated and handled easily. On the other hand, the corresponding palladation products are too reactive to be isolated. The stable mercuration products can be used for various reactions based on facile transmetallation with Pd(II) salts to generate the very reactive palladation products **399** and **400** *in situ*[364,365].

In connection with mechanistic studies on the Wacker reaction, the transmetallation of β-ethoxy- and β-hydroxyethylmercury(II) chloride with PdCl$_2$ has been carried out, giving ethyl vinyl ether and acetaldehyde[366]. The reaction proceeds by the formation of β-ethoxy- and β-hydroxyethylpalladium chlorides (**401**), which decompose as soon as they are formed.

Palladation products formed from arylmercurials, carboalkoxymercurials, and alkylmercurials, which have no β-hydrogen, are used *in situ* for the reaction of alkenes[367]. Particularly, the arylation of alkenes is synthetically useful. Styrene derivatives **402** and **403** are formed by the reaction of a phenylmercury compound, $Pd(OAc)_2$, and alkenes. Extensive studies have been carried out with various alkenes and mercury compounds[368–370]. The reaction of arylmercury(II) chloride, alkenes, $CuCl_2$, and a catalytic amount of $PdCl_2$ affords the β-arylalkyl chlorides 404[367].

$$PhHgX + PdX_2 \longrightarrow [Ph\text{-}Pd\text{-}X] \longrightarrow \left[\begin{array}{c} R \\ \diagdown \diagup \\ \text{Ph} \quad \text{Pd-X} \end{array} + \begin{array}{c} R \\ \diagdown \diagup \\ \text{X-Pd} \quad \text{Ph} \end{array} \right]$$

$$\longrightarrow \begin{array}{c} R \\ \diagdown \diagup \\ \text{Ph} \\ 402 \end{array} + \begin{array}{c} R \\ \diagdown \diagup \\ \text{Ph} \\ 403 \end{array} + Pd + X^-$$

$$PhHgCl + \begin{array}{c} R \\ \diagdown \diagup \end{array} + CuCl_2 \xrightarrow{PdCl_2} \begin{array}{c} Ph \diagdown \diagup R \\ | \\ 404 \quad Cl \end{array}$$

The isoflavone 406 is prepared by the indirect α-phenylation of a ketone by reaction of phenylmercury(II) chloride with the enol acetate 405, prepared from 4-chromanone[371]. A simple synthesis of pterocarpin (409) has been achieved based on the oxypalladation of the *ortho*-mercurated phenol derivative 408 with the cyclic alkene 407[372,373].

Carbopalladation products of norbornene have no possibility of *syn* elimination of β-hydrogen, and hence they undergo further transformations, such as insertions of alkene, alkyne, and CO, and anion trapping. The prostaglandin endoperoxide analog 411 was prepared by the transmetallation of ethyl (acetoxymercurio)acetate (410) with $PdCl_2$, followed by carbopalladation of norbornene and insertion of 1-octen-3-one[374]. Trapping the carbopalladation product 412 with an alkenyltin reagent to give 413 is another example[375].

The Pd-mediated coupling involving organomercury and alkenes is useful for the synthesis of C-5 substituted 2'-deoxyribonucleosides[376]. The ethyluridine 415 is prepared by the reaction of the 5-chloromercuriuridine 414 with

$AcOHgCH_2CO_2Et$ + Li_2PdCl_4 +
410

ethylene and $PdCl_2$, followed by hydrogenolysis[377]. (1,3-Dimethyl-2,4-pyrimidinedion-5-yl)mercury(II) acetate (**416**) reacts with dihyrofuran or 3,4-dihydropyran to give the coupled product **417**[378].

The reaction of alkenyl mercurials with alkenes forms π-allylpalladium intermediates by the rearrangement of Pd via the elimination of H—Pd—Cl and its reverse readdition. Further transformations such as trapping with nucleophiles or elimination form conjugated dienes[379]. The π-allylpalladium intermediate **418** formed from 3-butenoic acid reacts intramolecularly with carboxylic acid to yield the γ-vinyl-γ-lactone **419**[380]. The β,γ-unsaturated amide **421** is obtained by the reaction of 4-vinyl-2-azetidinone (**420**) with an organomercurial. Similarly homoallylic alcohols are obtained from vinylic oxetanes[381].

Addition of several organomercury compounds (methyl, aryl, and benzyl) to conjugated dienes in the presence of Pd(II) salts generates the π-allylpalladium complex **422**, which is subjected to further transformations. A secondary amine reacts to give the tertiary allylic amine **423** in a modest yield along with diene **424** and reduced product **425**[382,383]. Even the unconjugated diene **426** is converted into the π-allyllic palladium complex **427** by the reaction of PhHgCl via the elimination and reverse readdition of H—Pd—Cl[383].

The oxidative coupling of toluene using Pd(OAc)$_2$ via *p*-tolylmercury(II) acetate (**428**) forms bitolyl[384]. The aryl–aryl coupling proceeds with copper and a catalytic amount of PdCl$_2$ in pyridine[385]. Conjugated dienes are obtained by the coupling of alkenylmercury(II) chlorides[386].

Me—⟨◯⟩—HgOAc + Pd(OAc)₂ → Me—⟨◯⟩—⟨◯⟩—Me + Hg(OAc)₂ + Pd

428

Pd(II) salts promote the carbonylation of organomercury compounds. Reaction of phenylmercury chloride and PdCl₂ under CO pressure affords benzophenone (**429**)[387]. Both esters and ketones are obtained by the carbonylation of furylmercury(II) chloride in alcohol[388]. Although the yields are not satisfactory, esters are obtained by the carbonylation of aryl- and alkylmercury(II) chlorides[389,390]. One-pot catalytic carbonylation of thiophene, furan, and pyrrole (**430**) takes place at the 2-position via mercuration and transmetallation by the use of PdCl₂, Hg(NO₃), and CuCl₂[391].

2 PhHgCl + PdCl₂ + CO ⟶

Ph⟍C⟋Ph
      ‖
      O

429

EtHgCl + CO + ROH $\xrightarrow{PdCl_2}$ EtCO₂R

⟨furan⟩ + CO + ROH $\xrightarrow[CuCl_2]{PdCl_2, Hg(NO_3)_2}$ ⟨furan⟩—CO₂R

430   Z = O, S, NH

The stereo-defined enol ester **432** is prepared by the reaction of the vinyl-mercurial **431**, obtained by acetoxymercuration of 2-butyne, with mercury(II) carboxylates using a catalytic amount of Pd(OAc)₂[392].

Me≡Me + Hg(OAc)₂ $\xrightarrow{NaCl}$

AcO⟍   Me
    C=C
Me⟋   HgCl

431

$\xrightarrow[Pd(OAc)_2]{Hg(OCOPh)_2}$

AcO⟍   Me
    C=C
Me⟋   O—C(Ph)(O)

432

Thallation of aromatic compounds with thallium tris(trifluoroacetate) proceeds more easily than mercuration. Transmetallation of organothallium compounds with Pd(II) is used for synthetic purposes. The reaction of alkenes with arylthallium compounds in the presence of Pd(II) salt gives styrene derivatives (**433**). The reaction can be made catalytic by use of CuCl₂[393,394]. The arylation of methyl vinyl ketone was carried out with the arylthallium compound **434**[395]. The β-alkoxythallium compound **435**, obtained by oxythallation of styrene, is converted into acetophenone by the treatment with PdCl₂[396].

ortho-Thallation takes place mainly when coordinating functional groups are present. The *ortho*-thallation of benzoic acid, followed by the transmetallation with Pd(II), forms an arylpalladium complex, which reacts with allyl chloride to afford the isocoumarin **436**[323]. A similar reaction of arylacetic acids and benzyl alcohols produces seven-membered lactones and six-membered ethers. Esters and lactones are prepared by the carbonylation of organothallium compounds. Arylthallium compounds are carbonylated to give esters in alcohol[397]. The anhydride **437** is obtained by the carbonylation of phenylacetic acid in the presence of Tl(III). The carbonylation proceeds with a catalytic amount of PdCl$_2$; Tl(III) behaves as the reoxidant of Pd(0)[398].

Organoboranes undergo transmetallation. 1-Hexenylboronic acid (**438**) reacts with methyl acrylate via the transmetallation with Pd(OAc)$_2$, giving methyl 2,4-nonadienoate (**439**)[399]. The (*E*)-alkenylboranes **440**, prepared by the hydroboration of terminal alkynes, are converted into the alkylated (*E*)-alkenes **441** by treatment with an equivalent amount of Pd(OAc)$_2$ and triethylamine[400]. The (*E*)-octenylborane **442** reacts with CO in MeOH in the presence of benzoquinone and a catalytic amount of $PdCl_2$ to give methyl 2-nonenoate (**443**)[401]. However, it was reported that the carbonylation is slightly catalytic with $Pd(Ph_3P)_4$[402]. The coupling of alkenylboronic acid to give the conjugated dienes also proceeds with a catalytic amount of $PdCl_2$ and LiCl[403].

Arylstannanes are coupled with the olefinic bond of the dihydrofuran **444** or

2,3-dihydropyran[404]. The alkenylstannane of benzoquinone **445** is coupled using $PdCl_2$ and benzoquinone[405]. The homocoupling of the alkenylstannane **446** using $Pd(OAc)_2$ and $t\text{-}BuO_2H$ as a reoxidant gives a 1,3-diene[406]. However, the coupling of the alkenylstannane **447** using a catalytic amount of $PdCl_2(MeCN)_2$ in HMPA without the reoxidant has been reported[407]. Homocoupling of the alkenylstannane **448** was carried out with $PdCl_2(MeCN)_2$ and $CuCl_2$ after conversion to the alkenyllithium **449**[408].

Alkenylmercury compounds are coupled to give conjugated dienes by the transmetallation with Pd(II). A mixture of (*E*)- and (*Z*)-dibenzylidenesuccinic acids (**451** and **452**) was obtained by the transmetallation of 2-chloromercurio-3-phenylacrylic acid (**450**) with Li_2PdCl_4 in the presence of $CuCl_2$[409,410].

III6-6

$$+ \ HgCl_2 \ + \ Pd \ + \ 2 \ LiCl$$

Conjugated dienes and diaryls are formed by the coupling of alkenylsilanes **453**[411] and alkenyl and arylstannanes. The reaction of styrene with tetraphenylsilane and Li_2PdCl_4 produces stilbene (**454**)[412]. Alkenyl- and phenylpentafluorosilicates **455** undergo facile transmetallation with Pd(II) salts. The organopalladium compounds **456** thus formed *in situ* are used for homocoupling, alkene insertion, and cross-coupling with alkenyl halides. Esters are formed by the carbonylation[413]. For example, the (*E*)-α,β-unsaturated ester **458** is obtained in a high yield by the carbonylation of the (*E*)-alkenylpentafluorosilicate **457** at room temperature and 1 atm of CO using $PdCl_2$.

$$K_2[RSiF_5] + PdCl_2 \longrightarrow [R-PdCl]$$

455   456

With branches to: R-R ; $\diagup\diagdown CO_2Me$; R. $\diagup\diagdown CO_2Me$

CO | MeOH ↓

RCO$_2$Me

$$K_2\begin{bmatrix} C_6H_{13}\diagdown\diagup\diagdown_{SiF_5} \end{bmatrix} + CO + MeOH \xrightarrow[91\%]{PdCl_2} C_6H_{13}\diagdown\diagup\diagdown CO_2Me$$

457   458

Organotellurium(II and IV) compounds undergo transmetallation with Pd(II)[414]. The carbonylation of the alkenylphenyltellurium(II) **459** gives the α,β-unsaturated ester **460** and benzoate, **460** being the main product[415]. Reductive coupling of diaryl, dialkyl, and aryl alkyltellurides **461** to give **462** proceeds by treatment with Pd(OAc)$_2$[416,417].

$$Ph\diagdown\diagup TePh + CO + MeOH \xrightarrow[CuCl_2]{PdCl_2} Ph\diagdown\diagup CO_2Me + PhCO_2Me$$

459   460

$$PhCH_2CH_2TeCH_2Ph \xrightarrow{Pd(OAc)_2} Ph(CH_2)_3Ph$$

461   462

Biphenyl formation by the ligand coupling of triphenylbismuth is promoted by Pd(OAc)$_2$[418].

7 Synthetic Reactions Based on the Chelation of Heteroatoms

7.1 *ortho*-Palladation of Aromatic Compounds

Aromatic compounds which have substituents containing heteroatoms such as N, P, S, and O at a position suitable for mainly five-membered or sometimes six-membered chelating rings undergo cyclopalladation at an *ortho*-position to form a σ-arylpalladium bond by virtue of the stabilization due to the chelation of these heteroatoms. The *ortho*-palladation products are stable and can be isolated[419,420]. After the first report on the preparation of the azobenzene and *N,N*-dimethylbenzylmaine complexes **463** and **466** of PdCl$_2$[421], numerous complexes have been prepared. Some representative *ortho*-palladation products are shown (**463–477**)[421–433]. The amide carbonyl of acetanilide is capable of forming a six-membered chelate ring **477**[431]. In addition to sp^2

aromatic carbons, even an sp^3 carbon of the methyl group of 8-methylquinoline (**475**)[426,432] and *o-N,N*-dimethylaminotoluene (**476**)[433] form σ-bonds between the Pd and sp^3 carbons.

463[421)   464[432)   465[423)   466[421)

467[424)   468[426)   469[427)   470[428)

471[438)   472[429)   473[430)   474[425)

475[426,432)   476[433)   477[431)

The asymmetric cyclopalladation of dimethylaminomethylferrocene takes place in the presence of an optically active carboxylic acid (e.g, *N*-acetylvaline), giving the cyclopalladation product **478** in 78% *ee*, from which optically active ferrocene derivatives were prepared[434].

The σ-arylpalladium bonds in these complexes are reactive and undergo insertion and substitution reactions, and the reactions offer useful methods for the regiospecific functionalization of the aromatic rings, although the reac-

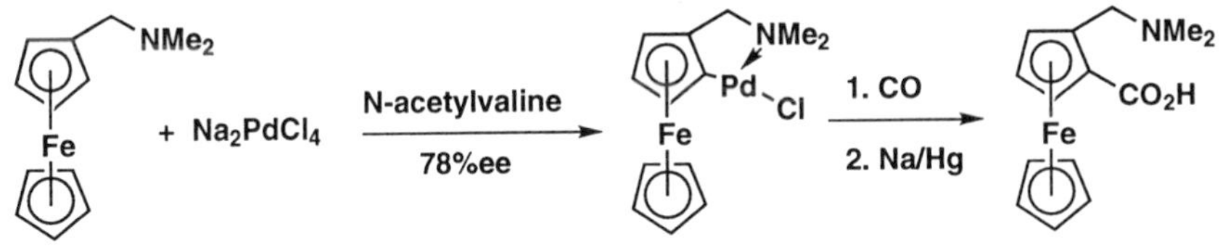

478 [α]$_D$ **37.5°**

tions are difficult to make catalytic in most cases. The first example is the insertion of styrene into the *N,N*-dimethylbenzylamine complex **466** to form a stilbene derivative **479**, which takes place smoothly at room temperature in AcOH[12]. A similar insertion of acrylate and enones takes place[435–437]. Alkenes can be introduced to the *ortho*-palladated ferrocene complex **478** in 20–59% yields in the presence of Et$_3$N[438]. The reaction has been extended to the functionalization of the dopamine analogue (*N,N*-dimethyl-2-arylethyla-mine) **480** via the six-membered *ortho*-palladated complex [439].

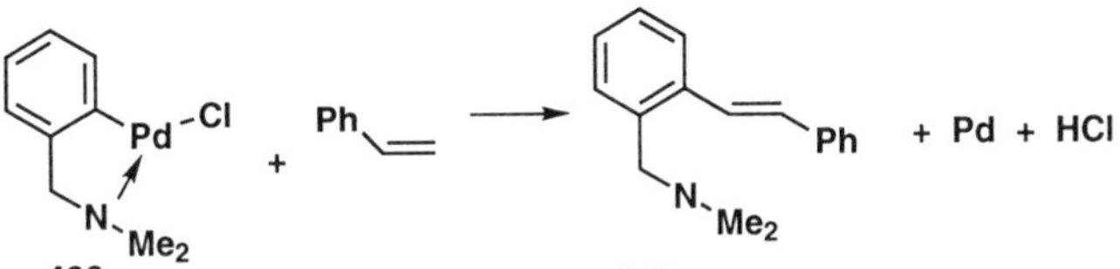

466   **479**

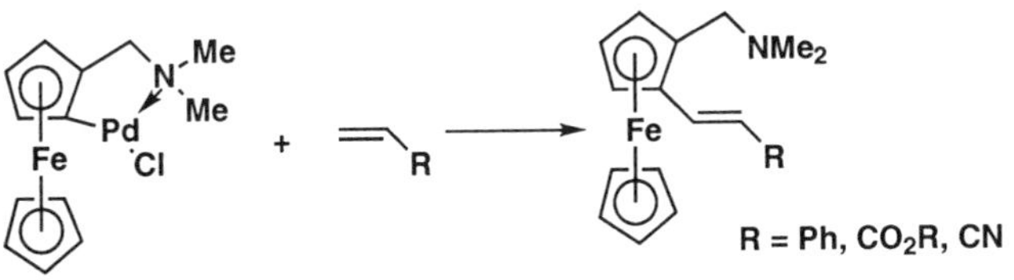

478   R = Ph, CO$_2$R, CN

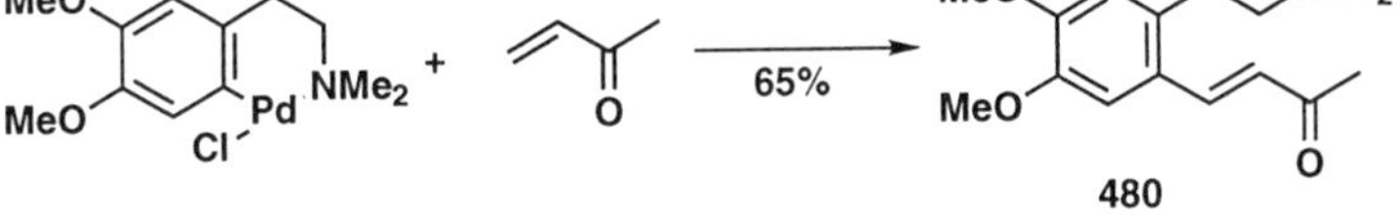

480

Two moles of diphenylacetylene insert into the benzyl methyl sulfide com-plex **481** to afford the eight-membered heterocycle **482**[440]. The cinnolinium salt **483** is prepared by the insertion of alkynes into the azobenzene com-plex[441].

The facile insertion of CO takes place. The 2-aryl-3-indazolone **484** is obtained in high yields from the azobenzene complex **463** in alcohol or water[442]. For unsymmetrically substituted 4-methyl, 4-chloro-, and 4-methoxyazobenzenes, σ-bond formation with the Pd was found to take place mainly with the substituted benzene ring, because the cobalt-catalyzed carbonylation of 484 and subsequent hydrolysis afford aniline and 5-methylanthranilic acid (485). The results show that the σ-bond formation is an electrophilic substitution of PdCl$_2$ on the benzene ring[442,443]. The carbonylation of the Pd acetate complex of azobenzene (486) gives the indazolone 487 and lactone 488 by dicarbonylation in a lower yield. The Schiff base complex 489 affords 3-acetoxy-3-phenylphthalimidine (490)[444]. The chromanone oxime 491 is carbonylated in the aromatic ring via the oxime-stabilized palladated complex 492[445]. The acetanilide complex 493 undergoes carbonylation and alkene insertion. The results show that the amide carbonyl is effective for the chelation to Pd[431]. Similarly, an ester group can be introduced into the phenyl ring in the phenyloxazole 494 via its chelation complex[446]. Isocyanide is isoelectronic with CO and behaves similarly. The 3-imino-2-phenylimidazoline 495 was obtained by the insertion of isocyanide into the azobenzene 463[447].

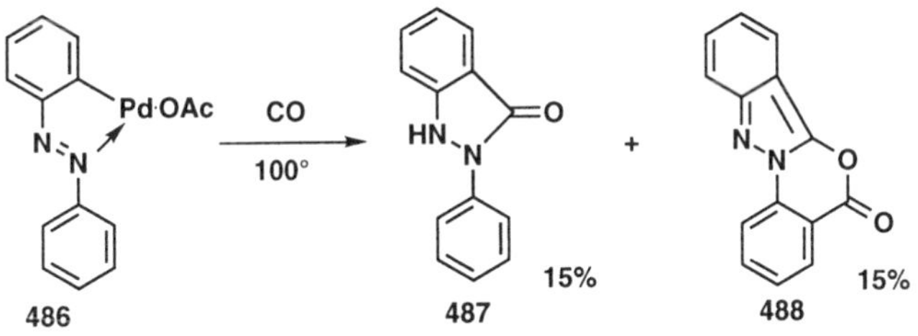
Pd·OAc
N
N
CO
100°
HN
N
O
+
N
N
O
O
15%
15%
486
487
488

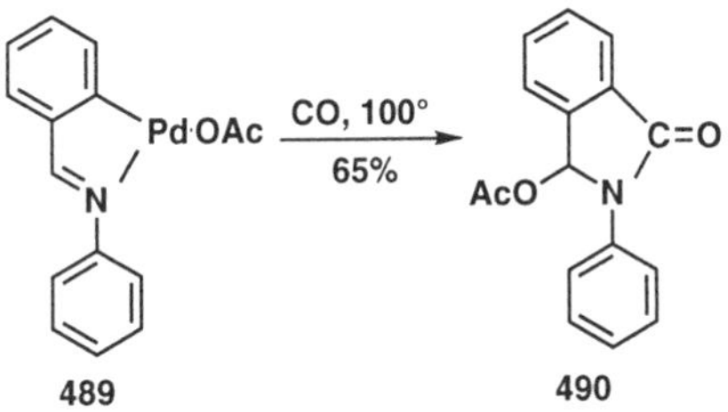
Pd·OAc
N
CO, 100°
65%
C=O
AcO
N
489
490

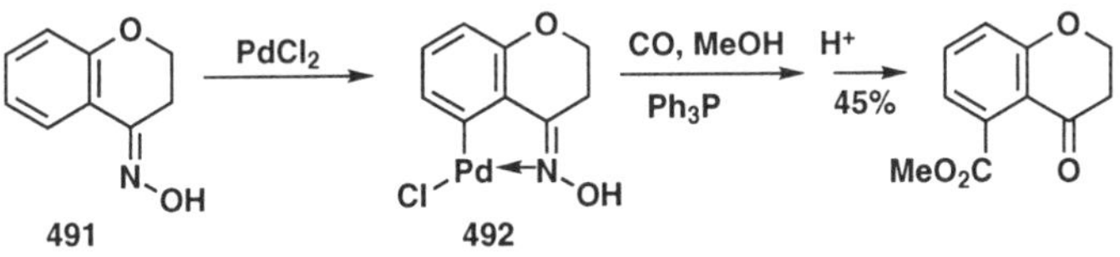
O
N
OH
PdCl2
O
Cl
Pd
N
OH
CO, MeOH
Ph3P
H+
45%
O
MeO2C
O
491
492

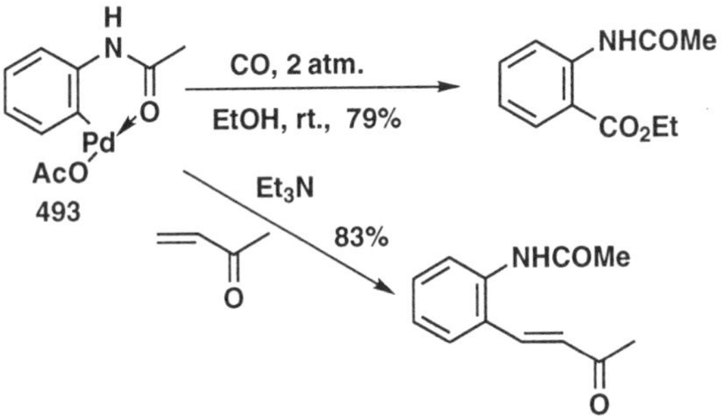
H
N
O
Pd
AcO
493
CO, 2 atm.
EtOH, rt., 79%
NHCOMe
CO2Et
Et3N
83%
O
NHCOMe
O

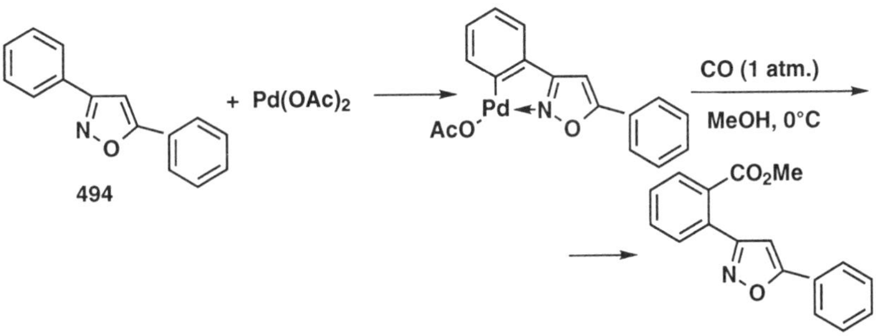
N
O
494
+ Pd(OAc)2
Pd
AcO
N
O
CO (1 atm.)
MeOH, 0°C
CO2Me
N
O

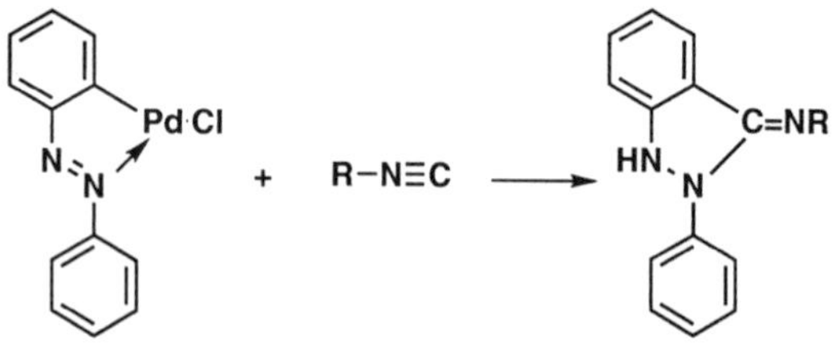

463   495

The alkylation at the *ortho* carbon is possible by the reaction of chelate complexes with Grignard reagents or organolithium reagents in the presence of Ph₃P. Benzaldehyde is *ortho*-methylated via the Schiff base complex **464** and its reaction with MeLi[448,449]. Methylation of 1-phenylazonaphthalene at *peri* and *ortho* positions is possible with MeLi after its cyclopalladation. Treatment of the tetramethylated product **496** with a catalytic amount of Na₂PdCl₄ affords the 2-arylbenzo[g]indazole **497** via cyclopalladation of the methyl group, insertion of an N=N bond, and elimination[450].

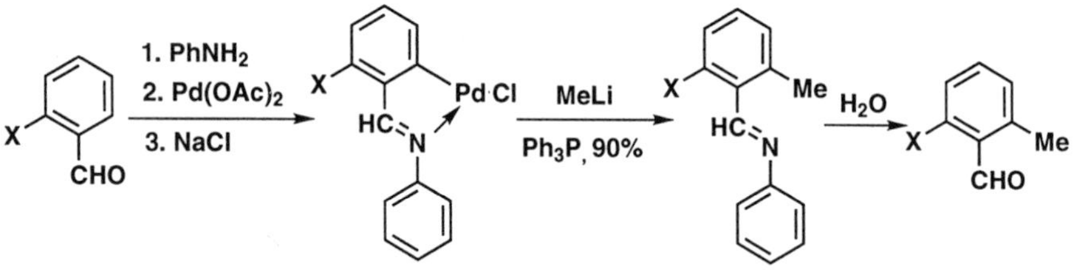

464

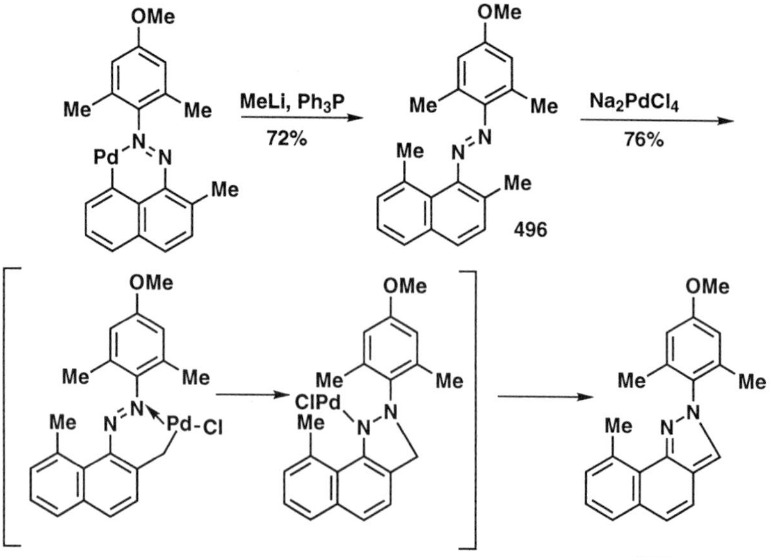

496

497

ortho-Alkylation is possible even with alkyl halides. The treatment of acetanilide (**498**) with three equivalents of Pd(OAc)$_2$ and an excess of MeI affords 2,6-dimethylacetanilide (**499**) by stepwise *ortho*-palladation and methylation twice[451]. However, treatment of the Schiff base complex **500** with trifluoroacetic acid and alkyl iodide produces 2,6-dialkylbenzaldehyde **501** after hydrolysis. The reaction is semicatalytic[452]. Reaction of the benzylamine complex with acetyl chloride produces the acetophenone derivative **502**[453,454].

Chlorination of the azobenzene complex **463** with chlorine produces monochloroazobenzene with regeneration of PdCl$_2$. Then complex formation takes place again with the chlorinated azobenzene. By this sequence, finally tetrachloroazobenzene (**503**) is obtained using a catalytic amount of PdCl$_2$. The reaction, carried out by passing chlorine gas into an aqueous dioxane solution of azobenzene and PdCl$_2$ for 16 h, gives a mixture of polychlorinated azobenzenes[455].

Oxygenation takes place with peracids. The cyclopalladated benzylamine complex **466** is converted into the salicylaldamine complex **504** by the treatment with MCPBA[456] or *t*-BuO$_2$H[457]. Similarly, azobenzene is oxidized with MCPBA at the *ortho* position[458].

The chelated complex of the benzylamine derivative **505** underwent a remarkable oxidative transformation by treatment with thallium trifluoroacetate to give narwedine (**506**) in one step by biomimetic oxidation[459].

7.2 Reactions of Allylic and Homoallylic Amines and Sulfides

The facile cyclopalladation of allylamine proceeds due to a chelating effect of the nitrogen. In MeOH, methoxypalladation takes place to give the five-membered chelating complex **507**[460]. The CO insertion takes place readily in EtOH, giving ethyl 3-methoxy-4-dimethylaminobutyrate (**508**) in 50% yield[461]. The insertion of alkenes also proceeds smoothly, giving the aminoalkenes **509**[462].

The carbopalladation of allylamine with malonate affords the chelating complex **510**, which undergoes insertion of methyl vinyl ketone to form the amino enone **511**[463]. The allylic sulfide **512** has the same chelating effect to give the five-membered complex **513** by carbopalladation[463,464].

An ingenious application of the facile palladation by the chelating effect of allylic amines is the synthesis of a prostaglandin derivative starting from 3-(dimethylamino)cyclopentene (**514**)[465]. The key steps in this synthesis are the facile and stereoselective introductions of a carbanion and an oxy anion into the cyclopentene ring by virtue of the stabilizing chelating effect of the amino group, and the alkene insertion to the Pd—C σ-bond. The first step is the stereo-defined carbopalladation with malonate and the subsequent elimination of β-hydrogen to form 3-substituted 4-aminocyclopentene **515** in 92% yield. The attack of the malonate is *anti* to the amino group. Further treatment of the amino ester **515** with Li_2PdCl_4, 2-chloroethanol, and diisopropylethylamine in DMSO gives rise to the oxypalladation product **516**, which is immediately treated with pentyl vinyl ketone. The insertion of this alkene affords the desired enone **517** in 50% yield. This enone **517** is converted into an important intermediate for prostaglandin synthesis.

The carbopalladation is extended to homoallylic amines and sulfides[466]. Treatment of 4-dimethylamino-1-butene (**518**) with diethyl malonate and Li_2PdCl_4 in THF at room temperature leads to the oily carbopalladated complex **519**, hydrogenation of which affords diethyl 4-(dimethylamino) butylmalonate (**520**) in an overall yield of 91%. Similarly, isopropyl 3-butenyl sulfide (**521**) is carbopalladated with methyl cyclopentanonecarboxylate and Li_2PdCl_4. Reduction of the complex affords the alkylated keto ester **522** in 96% yield. Thus functionalization of alkenes is possible by this method.

An unactivated methyl group can be functionalized by the cyclopalladation of oximes. The equatorial methyl of geminal methyls in steroids or hexapyranosides is selectively acetoxylated by the reaction of the palladation complex **523** of the 3-oxime with lead tetraacetate[467,468].

8 Reactions of Alkynes

Alkynes undergo stoichiometric oxidative reactions with Pd(II). A useful reaction is oxidative carbonylation. Two types of the oxidative carbonylation of alkynes are known. The first is a synthesis of the alkynic carboxylates **524** by oxidative carbonylation of terminal alkynes using PdCl$_2$ and CuCl$_2$ in the presence of a base[469]. Dropwise addition of alkynes is recommended as a preparative-scale procedure of this reation in order to minimize the oxidative dimerization of alkynes as a competitive reaction[470]. Also efficient carbonylation of terminal alkynes using PdCl$_2$, CuCl and LiCl under CO–O$_2$ (1 : 1) was reported[471]. The reaction has been applied to the synthesis of the carbapenem intermediate **525**[472]. The steroidal acetylenic ester **526** formed by this reaction undergoes the hydroarylation of the triple bond (see Chapter 4, Section 1) with aryl iodide and formic acid to give the lactone **527**[473].

$$Ph\text{---}\!\!\equiv\ +\ CO\ +\ MeOH\ \xrightarrow[\text{AcONa, 74\%}]{\text{PdCl}_2,\ \text{CuCl}_2}\ Ph\text{---}\!\!\equiv\text{---}CO_2Me$$
$$524$$

As the second type, the oxidative dicarbonylation of acetylene with PdCl$_2$ in benzene produces chlorides of maleic, fumaric, and muconic acids **528**[474]. Methyl muconate is obtained in MeOH containing thiourea by passing acetylene and oxygen using a catalytic amount of PdCl$_2$[475]. Intramolecular oxidative cyclization–carbonylation of the dipropargylamine **529** produces the dimethylpyrrole-3,4-diacetate **530** after isomerization of the initially produced muconate derivative[476], and the naphthofuranoneacetate **532** is produced by the oxidative carbonylation of 1,2-di(1-hydroxy-2-propynyl)benzene (**531**) using a thiourea complex[477]. The maleate and fumarate derivatives **533** and **534** are obtained as the main products at room temperature by the catalysis of PdCl$_2$, CuCl$_2$, and HCl under an oxygen atmosphere in alcohol[478].

Aconitate was obtained as a minor product in the carbonylation of propargyl alcohol[479]. However, in the two-step synthesis of methyl aconitate (**536**) from propargyl alcohol in 70% overall yield, the first step is the oxidative carbonylation under CO and air using PdI_2 and KI to give dimethyl hydroxymethylbutenedioate (**535**), which is carbonylated further to give trimethyl aconitate (**536**) by use of $[Pd(Tu)_4]I_2$ as a catalyst[480].

Oxidative carbonylation can sometimes be achieved even in the absence of any oxidizing agent. As an example, unexpectedly diphenylcrotonolactone (**537**) was obtained as a major product by the carbonylation of diphenylacetylene in EtOH together with diethyl diphenylmaleate (538) using $PdCl_2$ and HCl. In this oxidative carbonylation, apparently reduction takes place to form the lactone[481].

Ph——Ph + CO + EtOH $\xrightarrow[\text{100°, 100 atm.}]{\text{PdCl}_2}$

Ph Ph

O O

537
66%

+

Ph Ph

EtO_2C CO_2Et

538
26%

As an application of maleate formation, the carbonylation of silylated 3-butyn-1-ol affords the γ-butyrolactone 539[482]. Oxidative carbonylation is possible via mercuration of alkynes and subsequent transmetallation with Pd(II) under a CO atmosphere. For example, chloromercuration of propargyl alcohol and treatment with $PdCl_2$ (1 equiv.) under 1 atm of CO in THF produced the β-chlorobutenolide 540 in 96% yield[483]. Dimethyl phenylmaleate is obtained by the reaction of phenylacetylene, CO, $PdCl_2$, and $HgCl_2$ in MeOH[484,485].

TMS——⟍⟍OH + CO $\xrightarrow[\text{MeC(OEt)}_3,\ \text{MeOH}]{\text{PdCl}_2,\ \text{CuCl}_2}$

O

O CO_2Me

TMS

539

⟍⟍≡⟍ + $HgCl_2$ ⟶

HO

Cl H

HO HgCl

$\xrightarrow[\text{CO, 96\%}]{\text{Li}_2\text{PdCl}_4}$

Cl

O O

540

$\xrightarrow{\text{R}_2\text{CuLi}}$

R

O O

Diphenylacetylene is converted in to *cis*- and *trans*-dimethylstilbenes (541) in 65% yield by treatment with MeMgBr and 3 equivalents of $PdCl_2(PhCN)_2$ in THF at −60 °C, or $PdCl_2$ and LiCl in HMPA[486,487].

Ph——Ph + 2 MeMgBr + $PdCl_2$ ⟶

Ph Ph

Me Me

541

+ Pd + 2 MgBrCl

Propargyl esters **542** undergo an interesting oxidative rearrangement using a catalytic amount of PdBr$_2$ under an oxygen atmosphere to form α-acyloxy-α,β-unsaturated aldehydes **543**. Interestingly, reoxidation of Pd(0) takes place smoothly under oxygen without using other reoxidants. [488, 488a] The reaction offers a very useful synthetic method for corticosteroids **545** from 17-keto steroids **544**[489].

The oxidative coupling of alkynes to 1,3-diynes is possible with Pd(II) salts. Terminal arylalkynes are dimerized smoothly with Pd(0) and CuI in the presence of chloroacetone as a reoxidant[490]. The reaction of aliphatic alkynes gives diynes and the 3-alkyl-4-(1-alkynyl)hexa-1,5-diyn-3-enes **546**. The oxidative homocoupling of 3-hydroxy-4-benzyloxybutyne (**547**) proceeds smoothly with PdCl$_2$ and CuI as catalysts under an oxygen atmosphere[491]. The cross-coupling of two terminal alkynes is carried out using 3 equiv. of one alkyne by the same catalyst[492].

9 Reactions of Allenes

Two monomeric and dimeric 2-substituted π-allylic complexes (**548** and **549**) are obtained by treatment of allene with PdCl₂(PhCN)₂. They are formed by the nucleophilic attack at the central carbon of allene[493, 494].

Oxidative cleavage of the complex **549** with CuCl₂ affords 2,3-bis(chloro-methyl)-1,3-butadiene (**550**) and regenerates PdCl₂. Thus the preparation of this interesting dimerization product **550** can be carried out with a catalytic amount of PdCl₂ and two equivalents of CuCl₂ in MeCN[495]. Similarly, treatment of allene with PdBr₂ affords the dimeric complex **551**. Treatment of this complex with 2 equiv. of bromine yields the dibromide **552**. The tetra-bromide **553** is obtained by the reaction of an excess of bromine[496]. Similarly,

2,3-bis(acetoxymethyl)-1,3-butadiene was obtained with other products by the reaction of allene with Pd(OAc)$_2$[497].

Carbonylation of the complex **548** proceeds in ethanol gives ethyl 3-chloro-3-butenoate (**554**). The lactone **555** and the two esters **556** and **557** are obtained by carbonylation of the dimeric complex **549**. The oxidative carbonylation of allene in ethanol with PdCl$_2$ gives ethyl itacoante (**558**), although the yield is low[498].

The catalytic oxidative carbonylation of allene with PdCl$_2$ and CuCl$_2$ in MeOH affords methyl α-methoxymethacrylate (**559**)[499]. The intramolecular oxidative aminocarbonylation of the 6-aminoallene **560** affords the unsaturated β-amino ester **561**. The reaction has been applied to the enantioselective synthesis of pumiliotoxin (**562**)[500]. A similar intramolecular oxycarbonylation of 6-hydroxyallenes affords 2-(2-tetrahydrofuranyl)acrylates[501].

10 Enone Formation from Ketones, and Oxidation of Alcohols

Carbonyl compounds are dehydrogenated to give α,β-unsaturated carbonyl compounds with $PdCl_2$. Ketones are converted into α,β-unsaturated ketones by the treatment with $PdCl_2$[502,503]. However, the reaction is not efficient. The useful enone formation from ketones is carried out via the silyl enol ethers. The transmetallation of the silyl enol ether **563** with $Pd(OAc)_2$ gives the oxo-π-allylpalladium complex **564** (palladium enolate) (see Sections 2 and 6), which undergoes the elimination of β-hydrogen to give the enone **565**. As an application, the 1,4-addition of Me_2CuLi to cyclohexenone and trapping the resulting enolate with Me_3SiCl form the silyl enol ether **566**. Treatment of this silyl enol ether with a catalytic amount of $Pd(OAc)_2$ and benzoquinone as a reoxidant affords 3-methyl-2-cyclohexenone (**567**) in 91% yield[504]. The enone formation has been applied to a number of natural product syntheses such as aphidicolin[505], helenalin[506], and diosphenols[507]. Application to the synthesis of the clavulone precursor **568** is shown as an example[508]. The dehydrosilylation of silyl enol ethers and even enol acetates to form unsaturated ketones can be carried out in DMF solution using a Pd catalyst supported on silica under an oxygen atmosphere[509]. Oxidative desilylation also proceeds smoothly under anhydrous conditions, involving a Pd—OOH as a catalytically active species[100]. Similarly, the enone **569** was converted into the conjugated dienone **570** via silyl dienol ether[510].

Alcohols are oxidized slowly with $PdCl_2$. Oxidation of secondary alcohols to ketones is carried out with a catalytic amount of $PdCl_2$ under an oxygen atmosphere[73,74]. Also, selective oxidation of the allylic alcohol **571** without attacking saturated alcohols is possible with a stoichiometric amount of $Pd(OAc)_2$ in aqueous DMF (1% H_2O)[511].

11 Oxidative Carbonylation Reactions

Oxidative carbonylation of alcohols with $PdCl_2$ affords the carbonate **572** and oxalate **573**[512–514]. The selectivity of the mono- and dicarbonylation depends on the CO pressure and reaction conditions. In order to make the reaction catalytic, Cu(II) and Fe(III) salts are used. Under these conditions, water is formed and orthoformate is added in order to trap the water. Di-*t*-butyl peroxide is also used for catalytic oxidative carbonylation to give carbonates and oxalates in the presence of 2,6-dimethylpyridine[515].

$$CO + 2ROH + PdCl_2 \longrightarrow O=C\begin{smallmatrix}OR\\OR\end{smallmatrix} + Pd + 2HCl$$
$$\textbf{572}$$

$$2CO + 2ROH + PdCl_2 \longrightarrow \begin{smallmatrix}CO_2R\\|\\CO_2R\end{smallmatrix} + Pd + 2HCl$$
$$\textbf{573}$$

An industrial process for oxalate production from CO, alcohol, and oxygen catalyzed by Pd has been developed[516]. The most ingenious point in this process is use of an alkyl nitrite **574** as a reoxidant of Pd(0). The reaction can be expressed by the equations shown. Although the mechanism of the reaction is not completely clear, formally Pd(0) is oxidized to Pd(II) easily with the alkyl nitrite and then oxidative carbonylation proceeds to give oxalate and Pd(0). NO gas is generated from alkyl nitrite. In turn, NO is reconverted into the alkyl nitrite by reaction of oxygen and alcohol. Most importantly from the standpoint of the commercial process, the water, formed by the oxidation, can be separated easily from alkyl nitrite. Dialkyl carbonates are produced under a lower pressure of CO. The industrial production of dimethyl carbonate by oxidative carbonylation has been developed by the same company using an alkyl nitrite or $CuCl_2$ as the reoxidant. Another promising reaction is the preparation of commercially important diphenyl carbonate (**575**) by oxidative carbonylation of phenol[517].

$$2RONO + 2CO \longrightarrow \begin{smallmatrix}CO_2R\\|\\CO_2R\end{smallmatrix} + 2NO$$
$$\textbf{574}$$

$$2NO + 2ROH + 1/2\,O_2 \longrightarrow 2RONO + H_2O$$

$$\textbf{575}$$

The alkylurea **576** and oxamide **577** are formed by oxidative carbonylation of amines under CO pressure using Pd/C as a catalyst[518]. The urea formation proceeds under atmospheric pressure using $PdCl_2$ and $CuCl_2$[519]. The mono- and double carbonylations of β-aminoethanol (**578** and **579**) afford the cyclic carbamate (oxazolidinones) **580** and oxamide (morpholinediones) **581** [520,521].

Carbamates are produced by the oxidative carbonylation of amines in alcohol, and active research on the commercial production of carbamates as a precursor of isoyanates based on this reaction has been carried out. As an example, ethyl phenylcarbamate (**582**) is produced in a high yield (95%) with

$$CO + RNH_2 \xrightarrow{Pd/C} RNHCONHR + \underset{\underset{CONHR}{|}}{CONHR}$$

$$576 \qquad\qquad 577$$

$$578 \quad + CO \xrightarrow[\text{rt., 80 atm., 86\%}]{PdCl_2, CuCl_2, O_2}$$

(578 structure: HO–CH₂CH₂–NHEt)

$$579 \quad + CO \xrightarrow[\text{AcONa}]{PdCl_2, CuCl_2} \quad 580 \quad + \quad 581$$

a selectivity higher than 97% by the reaction of aniline with CO in EtOH at 150 °C and 50 atm. Pd on carbon is the catalyst and KI is added as a promoter[522]. The reaction proceeds even at room temperature and 1 atm of CO[523].

$$PhNH_2 + CO + EtOH + 1/2\,O_2 \xrightarrow[KI]{Pd/C} PhNHCO_2Et + H_2O$$

$$582$$

As an extension of the oxidative carbonylation with alkyl nitrites, malonate can be prepared by the oxidative carbonylation of ketene (**583**)[524]. Also, the acetonedicarboxylate **585** is prepared by the Pd-catalyzed, alkyl nitrite-mediated oxidative carbonylation of diketene (**584**)[525].

$$CH_2{=}C{=}O + CO + 2\,RONO \xrightarrow{PdCl_2} \underset{CO_2R}{\overset{CO_2R}{<}} + 2\,NO$$

$$583$$

$$584 \quad + CO + 2\,RONO \xrightarrow{PdCl_2} RO_2C{\overset{O}{\diagup\!\!\backslash}}CO_2R + 2\,NO$$

$$585$$

Although turnover of the catalyst is low, even unreactive cyclohexane[526] and its derivatives are oxidatively carbonylated to cyclohexanecarboxylic acid using $K_2S_2O_8$ as a reoxidant in 565% yield based on Pd(II)[527]. Similarly, methane and propane are converted into acetic acid in 1520% yield based on Pd(II) and butyric acid in 5500% yield [528].

12 References

1. J. Smidt, W. Hafner, R. Jira, J.Sedlmeier, R. Sieber, R. Ruttinger, and H. Kojer, *Angew. Chem.*, **71**, 176 (1959); J. Smidt, W. Hafner, R. Jira, R. Sieber, J. Sedlmeier, and J. Sabel, *Angew. Chem., Int. Ed. Engl.*, **1**, 80 (1962); J. Smidt, *Chem. Ind. (London)*, *54 (1962)*.
2. K. I. Matveev, L. I. Kaznetsova, and T. Basalaeva, *Belg. Pat.*, 828 603; *Chem. Abstr.* **85**, 192173 (1976).
3. K. Matsui, S. Uchiumi, A. Iwayama, and T. Umezu (Ube Industries Ltd), *Eur. Pat.Appl.*, EP 55 108; *Chem. Abstr.*, **97**, 162364 (1982).
4. R. C. Larock and T. R. Hightower, *J. Org. Chem.*, **58**, 5298 (1993).
5. R. A. T. M. van Benthem and W. N. Speckamp, *J. Org. Chem.*, **57**, 6083 (1992); R. A. T. M. van Benthem, H. Hiemstra, J. J. Michels, and W. N. Speckamp, *Chem. Commun.*, 357 (1994).
6. C. Chapman, *Analyst (London)*, **29**, 348 (1904).
7. R. Jira and W. Freiesleben, *Organomet. React.*, **3**, 5 (1972).
8 K. Liu, D. Hwai, K. Fujimoto, and T. Kunugi, *Ind. Eng. Chem., Prod. Res. Dev.*, **16**, 223 (1977).
9. A. B. Evnin, J. A. Rabo, and P. H. Kasai, *J. Catal.*, **30**, 109 (1973).
10. W. Schwerdtel, *Chem. Ind. (London)*, 1559 (1968); *Hydrocarbon Process.*, **47**, 187 (1968).
11. P. M. Henry, *Palladium Catalyzed Oxidation of Hydrocarbons*, Reidel, Dordrecht, 1980.
12. J. Tsuji, *Acc. Chem. Res.*, **2**, 144 (1969).
13. Reviews: L. S. Hegedus, in *Comprehensive Organic Synthesis*, Vol. 4, Pergamon Press, Oxford, 1991, pp. 551 and 571; *J. Mol. Catal.*, **19**, 201 (1983); *Angew. Chem., Int. Ed. Engl.*, **27**, 1113 (1988).
14. Reviews: T. Hosokawa and S. Murahashi, *Heterocycles*, **33**, 1079 (1992); *Acc. Chem. Res.*, **23**, 49 (1990).
15. F. C. Phillips, *Am. Chem. J.*, **16**, 255 (1894).
16. S. C. Ogburn and W. C. Brastow, *J. Am. Chem. Soc.*, **55**, 1307 (1933).
17. W. Hafner, R. Jira, J. Sedlmeier, and J. Smidt, *Chem. Ber.*, **95**, 1575 (1962).
18. R. Jira, J. Sedlmeier, and J. Smidt, *Liebigs Ann. Chem.*, **693**, 99 (1966).
19. P. M. Henry, *J. Org. Chem.*, **38**, 2415 (1973); **32**, 2575 (1967); *J. Am. Chem. Soc..*, **86**, 324 (1964); **88**, 1595(1966); **94**, 4437(1972).
20. I. I. Moiseev, O. G. Levanda, and M. N. Vargaftik, *J. Am. Chem. Soc.*, **96**, 1003 (1974).
21. J. E. Bäckvall, B. Akermark, and S. O. Ljunggren, *J. Am. Chem. Soc.*, **101**, 2411 (1979).
22. J. S. Coe and J. B. J. Unsworth, *J. Chem. Soc., Dalton Trans.*, 645 (1975).
23. Reviews: J. Tsuji, *Synthesis*, 369 (1984); in *Comprehensive Organic Synthesis*, Vol. 7, Pergamon Press, Oxford, 1991, p.449.

24. J. Tsuji, I. Shimizu, and K. Yamamoto, *Tetrahedron Lett.*, 2975 (1976).
25. J Tsuji, I. Shimizi, and Y. Kobayashi, *Isr. J. Chem.*, **24**, 153 (1984).
26. W. H. Clement and C. M. Selwitz, *J. Org. Chem.*, **29**, 241 (1964).
27. D. R. Fahey and E. A. Zuech, *J. Org. Chem.*, **39**, 3276 (1974).
28. D. G. Lloyd and B. J. Luberoff, *J. Org. Chem.*, **34**, 3949 (1969).
29. H. Alper, K. Januszkiewicz, and D. J. H. Smith, *Tetrahedron Lett.*, **26**, 2263 (1985).
30. H. A. Zahalka, K. Januszkiewicz, and H. Alper, *J. Mol. Catal.*, **35**, 249 (1986).
31. A. Harada, Y. Hu, and S. Takahashi, *Chem. Lett.*, 2083 (1986).
32. Y. Kusunoki and H. Okazaki, *Hydrocarbon Process.*, 129 (1974).
33. F. J. McQuillin and D. G. Parker, *J. Chem. Soc., Perkin Trans.* 1, 809 (1974).
34. J. Tsuji, H. Nagashima, and H. Nemoto, *Org. Synth.*, **62**, 9 (1984).
35. M. Gouedard, F. Gaudemer, and A. Gaudemer, *Bull. Soc. Chim. Fr.*, 577 (1973).
36. H. H. Horowitz, *J. Appl. Electrochem.*, **14**, 779 (1984).
37. B. S. Tovrog, F. Mares, and S. E. Diamond, *J. Am. Chem. Soc.*, **102**, 6616 (1980).
38. A. Heumann, M. Reglier, and B. Waegell, *Angew. Chem., Int. Ed. Engl.*, **18**, 866, 867 (1979); A. Heumann, F. Chauvet, and B. Waegell, *Tetrahedron Lett.*, **23**, 2767 (1982).
39. M. A. Andrews and K. P. Kelly, *J. Am. Chem. Soc.*, **103**, 2894 (1981); M. A. Andrews, T. C. T. Chang, C. F. Cheng, and K. P. Kelly, *Organometallics*, **3**, 1777 (1984).
40. T. Hosokawa, T. Takahashi, T. Ohta, and S. Murahashi, *J. Organomet. Chem.*, 334, C5 (1987).
41. F. Derdar, J. Martin, C. Martin, J. M. Bregeault, and J. Mercier, *J. Organomet. Chem.*, **338**, C21 (1988).
42. N. H. Kiers, B. L. Feringa, and P. W. N. M. van Leeuwen, *Tetrahedron Lett.*, **33**, 2403 (1992).
43. J. E. Bäckvall and A. Heumann, *J. Am. Chem. Soc.*, **109**, 6396 (1987).
44. I. I. Matveev, *et al.*, *Kinet. Katal.*, 18, 380(1977); *Chem. Abstr.*, **87**, 22002 (1977).
45. B. El Ali, J. M. Bregeault, and J. Martin, *J. Organomet. Chem.*, **327**, C9 (1987).
46. S. F. Davidson, B. E. Mann, and P. M. Maitlis, *J. Chem. Soc., Dalton Trans.*, 1223 (1984).
47. I. I. Moiseev, M. N. Vargaftik, and Y. K. Syrkin, *Dokl. Akad. Nauk SSSR*, **130**, 820 (1960); *Chem. Abstr.*, **54**, 24350 (1960).
48. J. Tsuji and M. Minato, *Tetrahedron Lett.*, **28**, 3683 (1987).
49. J. E. Bäckvall and R. B. Hopkins, *Tetrahedron Lett.*, **29**, 2885 (1988).
50. S. Srinivasan and W. T. Ford, *J. Mol. Catal.*, **64**, 291 (1991).
51. T. Inokuchi, L. Ping, F. Hamaue, M. Izawa, and S. Torii, *Chem. Lett.*, 121 (1994).
52. I. I. Moiseev, M. N. Vargaktik, and Y. K. Syrkin, *Dokl. Akad. Nauk SSSR*, **130**, 820 (1960); *Chem. Abstr.*, **54**, 24350 (1960).
53. M. Roussel and H. Mimoun, *J. Org. Chem.*, **45**, 5387 (1980).
54. J. Tsuji, H. Nagashima, and K. Hori, *Chem. Lett.*, 257 (1980).
55. H. Mimoun, *J. Mol. Catal.*, **7**, 1 (1980); M. Roussel and H. Mimoun, *J. Org. Chem.*, **45**, 5387 (1980); *J. Am. Chem. Soc.*, **102**, 1047 (1980).
56. C. S. Subramaniam, P. J. Thomas, V. R. Mamdapur, and M. S. Chadha, *Synthesis*, 468 (1978).
57. T. Hirao, K. Hayashi, Y. Fujihara, Y. Ohshiro, and T. Agawa, *J. Org. Chem.*, **50**, 279 (1985).
58. C. C. Lenznoff, C. R. McArthur, and M. Whittaker, *Synth. Commun.*, **16**, 225 (1986).
59. P. Knochel and J. F. Normant, *Tetrahedron Lett.*, **27**, 4431 (1986).

60. T. Takahashi, K. Kasuga, M. Takahashi, and J. Tsuji, *J. Am. Chem. Soc.*, **101**, 5072 (1979).
61. P. D. Magnus and M. S. Nobbs, *Synth. Commun.*, **10**, 273 (1980).
62. D. Pauley, F. Anderson, and T. Hudlicky, *Org. Synth.*, **67**, 121 (1988).
63. G. Balme, *Tetrahedron Lett.*, **26**, 2309 (1985).
64. A. J. Pearson and T. Ray, *Tetrahedron Lett.*, **26**, 2981 (1985).
65. N. Yamazaki and C. Kibayashi, *Tetrahedron Lett.*, **29**, 5767 (1988).
66. F. Derdar, J. Martin, C. Martin, J. M. Bregeault, and J. Mercier, *J. Organomet. Chem.*, **338**, C21 (1988).
67. G. Mehta and K. S. Rao, *J. Org. Chem.*, **53**, 425 (1988).
68. G. Rosini, R. Ballini, and E. Marotta, *Tetrahedron*, **45**, 5935 (1989).
69. T. Antonsson, S. Hansson, and C. Moberg, *Acta Chem. Scand., Ser B.*, **39**, 593 (1985).
70. T. A. Hase and K. McCoy, *Synth. Commun.*, **9**, 63 (1979).
71. K. Iseki, M. Yamazaki, M. Shibasaki, and S. Ikegami, *Tetrahedron,* **37**, 4411 (1981).
72. W. G. Lloyd, *J. Org. Chem.*, **32**, 2816 (1977).
73. T. E. Nalesnik and N. L. Holy, *J. Org. Chem.*, **42**, 372 (1977).
74. T. F. Blackburn and J. Schwartz, *Chem. Commun.*, 157 (1977).
75. H. Yatagai, Y. Yamamoto, and K. Maruyama, *J. Am. Chem. Soc.*, **102**, 4548 (1980).
76. A. J. Poss and B. K. Belter, *Synth. Commun.*, **18**, 417(1988).
77. Reviews: J. Tsuji, *Pure Appl. Chem.*, **53**, 2371 (1981); *Top. Curr. Chem.*, **91**, 30 (1980).
78. G. Mehta and K. S. Rao, *J. Am. Chem. Soc.*, **108**, 8015 (1986).
79. T. Imanishi, M. Matsui, M. Yamashita, and C. Iwata, *Tetrahedron Lett.*, **27**, 3163 (1986).
80. J. Tsuji, T. Yamada, and I. Shimizu, *J. Org. Chem.*, **45**, 5209 (1980).
81. A. Hosomi, H. Kobayashi, and H. Sakurai, *Tetrahedron Lett.*, **21**, 955 (1980).
82. H. Yamada, K. Shimizu, M. Nisar, T. Takahashi, and J. Tsuji, *Tetrahedron Lett.*, **31**, 2407 (1990).
83. A. K. Bose, L. Krishnan, D. R. Wagle, and M. S. Manhas, *Tetrahedron Lett.*, **27**, 5955 (1986).
84. G. R. Clark and S. Thiensathit, *Tetrahedron Lett.*, **26**, 2503 (1985).
85. J. E. Byrd, L. Cassar, P. E. Eaton, and J. Halpern, *Chem. Commun.*, 40 (1971).
86. P. Boontanonda and R. Grigg, *Chem. Commun.*, 583 (1977).
87. I. E. Beck, E. V. Gusevskaya, A. V. Golovin, and V. A. Likholobov, *J. Mol. Catal.*, **83**, 301 (1993).
88. H. Takayama, T. Koike, N. Aimi, and S. Sakai, *J. Org. Chem.*, **57**, 2173 (1992).
89. J. Tsuji and T. Mandai, *Tetrahedron Lett.*, 1817 (1978).
90. T. Takahashi, K. Kasuga, and J. Tsuji, *Tetrahedron Lett.*, 4917 (1978).
91. D. G. Miller and D. D. M. Wayner, *J. Org. Chem.*, **55**, 2924 (1990).
92. H. Ogawa, H. Fujinami, K. Taya, and S. Teratani, *Chem. Commun.*, 1274 (1981); *Bull. Chem. Soc. Jpn.*, **57**, 1908 (1984).
93. K. Takehira, H. Orita, I. H. Ho, C. R. Leobardo, G. C. Martinez, M. Shimadzu, T. Hayakawa, and T. Ishikawa, *J. Mol. Catal.*, **42**, 247 (1987); K. Takehira, T. Hayakawa, and H. Orita, *Chem. Lett.*, 1853 (1985); K. Takehira, I. H. Ho, V. C. Martinez, R. S. Chavira, T. Hayakawa, H. Orita, M. Shimadzu, and T. Ishikawa, *J. Mol. Catal.*, **42**, 237 (1987).
94. S. X. Auclair, M. Morris, and M. A. Sturgess, *Tetrahedron Lett.*, **33**, 7739 (1992).
95. J. Tsuji and S. Imamura, *Bull. Chem. Soc. Jpn.*, **40**, 197 (1967).

96. H. Nagashima, K. Sakai, and J. Tsuji, *Chem. Lett.*, 859 (1982).
97. T. J. Grattan and J. S. Whitehurst, *Chem. Commun.*, 43 (1988).
98. A. J. Poss and M. S. Smith, *Synth. Commun.*, **17**, 1735 (1987).
99. J. Tsuji, H. Nagashima, and K. Hori, *Tetrahedron Lett.*, **23**, 2679 (1982).
100. T. Hosokawa, T. Nakahira, M. Takano, and S. Murahashi, *J. Mol. Catal.*, **74**, 486 (1992).
101. T. Hosokawa and S. Murahashi, *Acc. Chem. Res.*, **23**, 49 (1990); T. Hosokawa, T. Uno, S. Inui, and S. Murahashi, *J. Am. Chem. Soc.*, **103**, 2318 (1981).
102. T. Hosokawa, S. Aoki, M. Takano, T. Nakahira, Y. Yoshida, and S. Murahashi, *Chem. Commun.*, 1559 (1991).
103. P. M. Henry, *J. Org. Chem.*, **32**, 2575 (1967).
104. H. Stangl and R. Jira, *Tetrahedron Lett.*, 3589 (1970).
105. J. Y. Lai, F. S. Wang, G. Z. Guo, and L. X. Dai, *J. Org. Chem.*, **58**, 6944 (1993).
106. I. I. Moiseev, M. N. Vargaftik, and Ya. K. Syrkin, *Dokl. Akad. Nauk SSSR*, **133**, 377 (1960).
107. E. W. Stern and M. L. Spector, *Proc. Chem. Soc.*, 370 (1961).
108. I. I. Moiseev and M. N. Vargaftik, *Izv. Akad. Nauk SSSR*, 759 (1965).
109. N. T. Byrom, R. Grigg, and B. Kongkathip, *Chem. Commun.*, 216 (1976); *J. Chem. Soc., Perkin Trans.*, 1643 (1984).
110. K. Mori and Y. B. Seu, *Tetrahedron*, **41**, 3429 (1985).
111. B. Kongkathip, R. Sookkho, and N. Kongkathip, *Chem. Lett.*, 1849 (1985).
112. T. Hosokawa, Y. Makabe, T. Shinohara, and S. Murahashi, *Chem. Lett.*, 1529 (1985).
113. T. Hosokawa, T. Ohta, and S. Murahashi, *J. Org. Chem.*, **52**, 1758 (1987).
114. T. Hosokwa, T. Yamanaka, and S. Murahashi, *Chem. Commun.*, 117 (1993).
115. T. Hosokawa, S. Aoki, and S. Murahashi, *Synthesis*, 558 ((1992).
116. T. Hosokawa, F. Nakajima, S. Iwasa, and S. Murahashi, *Chem. Lett.*, 1387 (1990).
117. J. Nokami, H. Ogawa, S. Miyamoto, T. Mandai, S. Wakabayashi, and J. Tsuji, *Tetrahedron Lett.*, **29**, 5181 (1988).
118. T. M. Meulemans, N. H. Kiers, B. L. Feringa, and P. W. N. M. Leeuwen, *Tetrahedron Lett.*, **35**, 455 (1994).
119. J. Y. Lai, X. X. Shi, and L. X. Dai, *J. Org. Chem.*, **57**, 3485 (1992).
120. K. Fugami, K. Oshima, and K. Utimoto, *Tetrahedron Lett.*, **28**, 809 (1987).
121. R. C. Larock and N. H. Lee, *J. Am. Chem. Soc.*, **113**, 7815 (1991).
122. M. F. Semmelhack and W. R. Epa, *Tetrahedron Lett.*, **34**, 7205 (1993).
123. R. B. Gammill and S. A. Nash, *Tetrahedron Lett.*, **25**, 2953 (1984); *J. Org. Chem.*, **51**, 3116 (1986).
124. T. Hosokawa, M. Hirata, S. Murahashi, and A. Sonoda, *Tetrahedron Lett.*, 1821 (1976).
125. M. F. Semmelhack, C. R. Kim, W. Dobler, and M. Meier, *Tetrahedron Lett.*, **37**, 4925 (1989).
126. S. Saito, T. Hata, N. Takahashi, M. Hirai, and T. Moriwake, *Synlett*, 237 (1992).
127. A. Kasahara, T. Izumi, and M. Ooshima, *Bull. Chem. Soc. Jpn.*, **47**, 2526 (1976).
128. T. Hosokawa, K. Maeda, K. Koga, and I. Moritani, *Tetrahedron Lett.*, 739 (1973); T. Hosokawa, S. Yamashita, S. Murahashi, and A. Sonoda, *Bull. Chem. Soc. Jpn.*, **49**, 3662 (1976); T. Hosokawa, H. Ohka, and I. Moritani, *Bull. Chem. Soc. Jpn.*, **48**, 1533 (1975); T. Hosokawa, S. Miyagi, S. Murahashi, and A. Sonoda, *J. Org. Chem.*, **43**, 2752 (1978).
129. G. Casiraghi, G. Casnati, G. Puglia, G. Sartori, and C. Tenenghi, *Synthesis*, 122 (1977).

130. B. A. Pearlman, J. M. McNamara, I. Hasan, S. Hatakeyama, H. Sekizaki, and Y. Kishi, *J. Am. Chem. Soc.*, **103**, 4248 (1981).
131. O. L. Chapman, M. R. Engel, J. P. Springer, and J. C. Clardy, *J. Am. Chem. Soc.*, **93**, 6696 (1971).
132. R. J. Kimar, G. L. D. Krupadanam, and G. Srimanarayana, *Synthesis*, 535 (1990).
133. K. Maeda, T. Hosokawa, S. Murahashi, and I. Moritani, *Tetrahedron Lett.*, 5075 (1973).
134. T. Hosokawa, N. Shimo, K. Maeda, A. Sonoda, and S. Murahashi, *Tetrahedron Lett.*, 383 (1976).
135. P. M. Henry, *J. Am. Chem. Soc.*, **86**, 3246 (1964).
136. I. I. Moiseev, M. N. Vargaftik, and Ya. K. Syrkin, *Izv. Akad. Nauk SSSR*, 930 (1962); 1144, 1527 (1963).
137. R. van Helden, C. F. Kohll, D. Medema, G. Verberg, and T. Jonkhoff, *Recl. Trav. Chim. Pays-Bas*, **87**, 961 (1968).
138. W. H. Clement and C. M. Selwitz, *Tetrahedron Lett.*, 1081 (1962).
139. ICI, *Br. Pat.*, 964 001 (1964).
140. Bayer, *Ger. Pat.*, 1 185 604 (1965).
141. National Distillers, *US Pat.*, 3 190 912.
142. W. Kitching, Z. Rappoport, S. Winstein, and W. C. Young, *J. Am. Chem. Soc.*, **88**, 2054 (1966).
143. N. Nagato, K. Maki, K. Uematsu, and R. Ishioka (Showa Denko), *Jpn. Pat. Kokai*, 60-32 747 (1985).
144. S. Nakamura and A. Yasui, *Synth. Commun.*, 1, 137 (1971).
145. C. B. Anderson and S. Winstein, *J. Org. Chem.*, **28**, 605 (1963).
146. N. Green, R. N. Haszeldine, and J. Lindley, *J. Organomet. Chem.*, **6**, 107 (1966).
147. S. Wolfe and P. G. C. Campbell, *J. Am. Chem. Soc.*, **93**, 1497, 1499 (1971).
148. P. M. Henry, *J. Am. Chem. Soc.*, **94**, 7305 (1972);
149. P. M. Henry and G. Ward, *J. Am. Chem. Soc.*, **93**, 1494 (1971).
150. S. Hansson, A. Heumann, T. Rein, and B. Akermark, *J. Org. Chem.*, **55**, 975 (1990); *Angew. Chem., Int. Ed. Engl.*, **23**, 453 (1984); S. E. Bystrom, E. M. Lasson, and B. Akermark, *J. Org. Chem.*, **55**, 5674 (1990).
151. E. M. Larsson and B. Akermark, *Tetrahedron Lett.*, **34**, 2523 (1993).
152. A. Heumann, B. Akermark, S. Hansson, and T. Rein, *Org. Synth.*, **68**,109 (1990).
153. J. E. McMurry and P. Kocovsky, *Tetrahedron Lett.*, **25**, 4187 (1984); **26**, 2171 (1985).
154. J. Tsuji, K. Sakai, H. Nagashima, and I. Shimizu, *Tetrahedron Lett.*, **22**, 131 (1981).
155. N. Adachi, K. Kikukawa, M. Takagi, and T. Matsuda, *Bull. Chem. Soc. Jpn.*, **48**, 521 (1975).
156. T. Antonsson, C. Moberg, L. Tottie, and A. Heumann, *J. Org. Chem.*, **54**, 4914 (1989); L. Tottie, P. Baeckstrom, C. Moberg, J. Tegenfeldt, and A. Heumann, *J. Org. Chem.*, **57**, 6579 (1992).
157. P. M. Henry, M. Davis, G. Ferguson, S. Phillips, and R. Restive, *Chem. Commun.*, 112 (1974).
158. Du Pont, *Br. Pat.*, 1 058 995 (1967).
159. Shell, *US Pat.*, 3 349 118 (1967).
160. Kuraray, *Fr. Pat.*, 1 509 372 (1967); M. Tamura and A. Yasui, *Chem. Commun.*, 1209 (1968); *J. Ind. Chem.*, **72**, 575, 578 (1969) (in Japanese).

161. M. Tamura and A. Yasui, *J. Ind. Chem.*, **72**, 581 (1969) (in Japanese); M. Tamura, M. Tsutsumi, and A. Yasui, *J. Chem. Soc. Jpn.*, **72**, 585 (1969) (in Japanese).
162. *Br. Pat.*, 1 124 862; *ChemTech.*, 219 (1976).
163. W. C. Baird, Jr, *J. Org. Chem.*, **31**, 2411 (1966).
164. A. Heumann, S. Kaldy, and A. Tenaglia, *Tetrahedron*, **50**, 539 (1994).
165. M. Sakai, *Tetrahedron Lett.*, 347 (1973).
166. Review: G. Cardillo and M. Orena, *Tetrahedron*, **46**, 3321 (1990).
167. A. Kasahara, T. Izumi, K. Sato, K. Maemura, and T. Hayasaka, *Bull. Chem. Soc. Jpn.*, **50**, 1899 (1977).
168. U. Annby, M. Stenkula, and C. M. Andersson, *Tetrahedron Lett.*, **34**, 8545 (1993).
169. T. Izumi and A. Kasahara, *Bull. Chem. Soc. Jpn.*, **48**, 1673 (1975).
170. D. E. Korte, L. S. Hegedus, and R. K. Wirth, *J. Org. Chem.*, **42**, 1329 (1977).
171. E. Gretz, T. F. Oliver, and A. Sen, *J. Am. Chem. Soc.*, **109**, 8109 (1987).
172. A. Punanzi, A. De Renzi, and G. Paiaro, *J. Am. Chem. Soc.*, 92, 3488 (1970).
173. B. Akermark and J. E. Bäckvall, *Tetrahedron Lett.*, 819 (1975); B. Akermark, J. E. Bäckvall, K. Siirala-Hansen, K. Sjoberg, and K. Zetterberg, *Tetrahedron Lett.*, 1363 (1974); B. Akermark, J. E. Bäckvall, L. S. Hegedus, K. Zetterberg, K. Siirala-Hansen, and K. Sjoberg, *J. Organomet. Chem.*, **72**, 127 (1974).
174. J. E. Bäckvall, *Tetrahedron Lett.*, 2225 (1975).
175. J. E. Bäckvall, *Tetrahedron Lett.*, 163 (1978).
176. J. E. Bäckvall, *Chem. Commun.*, 413 (1977).
177. L. S. Hegedus and K. Siirala-Hansen, *J. Am. Chem. Soc.*, **97**, 1184 (1975).
178. L. S. Hegedus, G. F. Allen, and E. L. Waterman, *J. Am. Chem. Soc.*, **98**, 2674 (1976); L. S. Hegedus, G. F. Allen, J. J. Bozell, and E. L. Waterman, *J. Am. Chem. Soc.*, **100**, 5800 (1978); L. S. Hegedus, G. F. Allen, and D. J. Olsen, *J. Am. Chem. Soc.*, **102**, 3583 (1980).
179. L. S. Hegedus and J. M. McKearin, *J. Am. Chem. Soc.*, **104**, 2444 (1982).
180. P. J. Harrington, L. S. Hegedus, and K. F. McDaniel, *J. Am. Chem. Soc.*, **109**, 4335 (1987).
181. S. Saito, T. Hara, N. Takahashi, M. Hirai, and T. Moriwake, *Synlett*, 237 (1992).
182. C. H. Heathcock, J. A. Stafford, and D. L. Clark, *J. Org. Chem.*, **57**, 2575 (1992).
183. A. Kasahara, T. Izumi, S. Murakami, H. Yanai, and M. Takatori, *Bull. Chem. Soc. Jpn.*, **59**, 927 (1986).
184. H. Hirai and H. Sawai, *Bull. Chem. Soc. Jpn.*, **43**, 2208 (1970).
185. T. Hosokawa, M. Takano, Y. Kuroki, and S. Murahashi, *Tetrahedron Lett.*, **33**, 6643 (1992).
186. A. Kasahara and T. Saito, *Chem. Ind. (London)*, 745 (1975).
187. A. Kasahara, *Chem. Ind. (London)*, 1032 (1976).
188. A. Kasahara and N. Fukuda, *Chem. Ind. (London)*, 485 (1976).
189. A. Kasahara, T. Izumi, and O. Saito, *Chem. Ind. (London)*, 666 (1980); A. Kasahara, T. Izumi, H. Watanbe, and S. Takahashi, *Chem. Ind. (London)*, 121 (1981).
190. M. Kimura, K. Fugami, S. Tanaka, and Y. Tamaru, *J. Org. Chem.*, **57**, 6377 (1992).
191. J. Tsuji and H. Takahashi, *J. Am. Chem. Soc.*, **87**, 3275 (1965); **90**, 2387 (1968).
192. T. Hayashi and L. S. Hegedus, *J. Am. Chem. Soc.*, **99**, 7093 (1977); L. S. Hegedus, T. Hayashi, and W. H. Darlington, *J. Am. Chem. Soc.*, **100**, 7747 (1978); L. S.

Hegedus, R. E. Williams, M. A. McGuire, and T. Hayashi, *J. Am. Chem. Soc.*, **102**, 4973 (1980).
193. G. M. Wieber, L. S. Hegedus, B. Akermark, and E. T. Michalson, *J. Org. Chem.*, **54**, 4649 (1989); J. Montgomery, G. M. Wieber, and L. S. Hegedus, *J. Am. Chem. Soc.*, **112**, 6255 (1990).; J. J. Masters, L. S. Hegedus, and J. Tamariz, *J. Org. Chem.*, **56**, 5666 (1991).
194. J. J. Masters and L. S. Hegedus, *J. Org. Chem.*, **58**, 4547 (1993).
195. L. V. Dunkerton and A. J. Serino, *J. Org. Chem.*, **47**, 2812 (1982).
196. S. Uemura and K. Ichikawa, *Bull. Chem. Soc. Jpn.*, **40**, 1016 (1967).
197. N. Luong-Thi and H. Riviere, *Chem. Commun.*, 918 (1978).
198. S. Murahashi, M. Yamamura, and N. Mita, *J. Org. Chem.*, **42**, 2870 (1977).
199. Y. Ito, H. Aoyama, T. Hirao, A. Mochizuki, and T. Saegusa, *J. Am. Chem. Soc.*, **101**, 494 (1979); Y. Ito, H. Aoyama, and T. Saegusa, *J. Am. Chem. Soc.*, **102**, 4519 (1980); **104**, 5808 (1982).
200. A. S. Kende, B. Roth, P. J. Santilippo, and T. J. Blacklock, *J. Am. Chem. Soc.*, **104**, 1784 (1982); A. S. Kende and D. J. Wustrow, *Tetrahedron Lett.*, **26**, 5411 (1985).
201. A. S. Kende and P. J. Sanfilippo, *Synth. Commun.*, **13**, 715 (1983).
202. M. Shibasaki, T. Mase, and S. Ikegami, *J. Am. Chem. Soc.*, **108**, 2090 (1986).
203. M. Toyota, T. Seishi, and K. Fukumoto, *Tetrahedron Lett.*, **34**, 5947 (1993).
204. M. Toyota, Y. Nishikawa, K. Motoki, N. Yoshida, and K. Fukumoto, *Tetrahedron Lett.*, **34**, 6099 (1993).
205. R. C. Larock and N. H. Lee, *Tetrahedron Lett.*, **32**, 5911 (1991).
206. J. Tsuji, M. Morikawa, and J. Kiji, *Tetrahedron Lett.*, 1061 (1963); *J. Am. Chem. Soc.*, **86**, 8451 (1964).
207. T. Yukawa and S. Tsutsumi, *J. Org. Chem.*, **34**, 738 (1969).
208. D. M. Fenton and P. J. Steinwand, *J. Org. Chem.*, **37**, 2034 (1972).
209. R. F. Heck, *J. Am. Chem. Soc.*, **94**, 2712 (1972).
210. J. K. Stille and R. Divakarumi, *J. Org. Chem.*, **44**, 3474 (1979).
211. K. Inomata, S. Toda, and H. Kinoshita, *Chem. Lett.*, 1567 (1990).
212. K. L. Olivier, D. M. Fenton, and J. Biale, *Hydrocarbon Process.*, No.11, 95 (1972).
213. G. Cometti and G. P. Chiusoli, *J. Organomet. Chem.*, **181**, C14 (1979).
214. K. Wada and Y. Hara, (Mitsubishi Kasei), *Jpn. Pat., Tokukaisho*, 56-71039 (1981); 57-21342 (1982).
215. S. C. A. Nefkens, M. Sperrle, and G. Consiglio, *Angew. Chem., Int. Ed. Engl.*, **32**, 1719 (1993).
216. P. R. Ashton, G. R. Brown, N. S. Isaacs, D. Giuffrida, F. H. Kohnke, J. P. Mathias, A. M. Z. Slawin, D. R. Smith, J. F. Stoddart, and D. J. Williams, *J. Am. Chem. Soc.*, **114**, 6330 (1992).
217. J. K. Stille, D, E. James, and L. F. Hines, *J. Am. Chem. Soc.*, **95**, 5062 (1973); D. E. James, L. F. Hines, and J. K. Stille, *J. Am. Chem. Soc.*, **98**, 1806 (1976); D. E. James and J. K. Stille, *J. Am. Chem. Soc.*, **98**, 1810 (1976).
218. Ube Industries, *US Pat.*, 4 138 580 (1979); 4 234 740 (1980); *Ger. Pat.*, 2 853 178; *Chem. Abstr.*, 91, 123415 (1979).
219. P. Brechot, Y. Chauvin, D. Commereuc, and L. Saussine, *Organometallics*, **9**, 26 (1990).
220. J. K. Stille and D. E. James, *J. Am. Chem. Soc.*, **97**, 674 (1975); *J. Organomet. Chem.*, **108**, 401 (1976).
221. J. K. Stille and R. Divakaruni, *J. Am. Chem. Soc.*, **100**, 1303 (1978); *J. Organomet. Chem.*, **169**, 239 (1979).

222. H. Urata, A. Fujita, and T. Fuchikami, *Tetrahedron Lett.*, **29**, 4435 (1988).
223. M. F. Semmelhack, C. Bodurow, and M. Baum, *Tetrahedron Lett.*, **25**, 3171 (1984); *J. Am. Chem. Soc.*, **106**, 1496 (1984).
224. M. F. Semmelhack and A. Zask, *J. Am. Chem. Soc.*, **105**, 2034 (1983); M. F. Semmelhack, J. J. Bozell, T. Sato, W. Wulff, E. Spiess, and A. Zask, *J. Am. Chem. Soc.*, **104**, 5850 (1982).
225. G. Kraus, J. Li, M. S. Gordon, and J. H. Jensen, *J. Am. Chem. Soc.*, **115**, 5859 (1993).
226. Y. Tamaru, T. Kobayashi, S. Kawamura, H. Ochiai, M. Hojo, and Z. Yoshida, *Tetrahedron Lett.*, **26**, 3207 (1985).
227. J. S. Yadav, V. S. Rao, and B. M. Choudary, *Tetrahedron Lett.*, **31**, 2491 (1990); *J. Mol. Catal.*, **49**, L61 (1989).
228. Y. Tamaru, T. Kobayashi, S. Kawamura, H. Ochiai, and Z. Yoshida, *Tetrahedron Lett.*, **26**, 4479 (1985); Y. Tamaru, M. Hojo, and Z. Yoshida, *J. Org. Chem.*, **53**, 5731 (1988); Y. Tamaru, M. Hojo, H. Higashimura, and Z. Yoshida, *J. Am. Chem. Soc.*, **110**, 3994 (1988).
229. Y. Tamaru, H. Tanigawa, S. Itoh, M. Kimura, T. Sekiyama, and Z. Yoshida, *Tetrahedron Lett.*, **33**, 631 (1992).
230. I. Moritani and Y. Fujiwara, *Tetrahedron Lett.*, 1119 (1967); Y. Fujiwara, I. Moritani, and M. Matsuda, *Tetrahedron*, **24**, 4819 (1968).
231. Review, I. Moritani and Y. Fujiwara, *Synthesis*, 524 (1973).
232. R. Asano, I. Moritani, A. Sonoda, Y. Fujiwara, and S. Teranishi, *J. Chem. Soc. C*, 3691 (1971).
233. R. Asano, I. Moritani, Y. Fujiwara, and S. Teranishi, *Bull. Chem. Soc. Jpn.*, **46**, 663 (1973).
234. I. Kozhevnikov, *React. Kinet. Catal. Lett.*, **5**, 439 (1976).
235. Y. Fujuwara, R. Asano, I. Moritani, and S. Teranishi, *Chem. Lett.*, 1061 (1978).
236. Y. Fujiwara, R. Asano, I. Moritani, and S. Teranishi, *J. Org. Chem.*, **41**, 1681 (1976).
237. A. J. Bingham, L. K. Dyall, R. O. C. Norman, and C. B. Thomas, *J. Chem. Soc. C*, 1879 (1970).
238. K. Yamamura, *J. Chem. Soc., Perkin Trans. 1*, 988 (1975).
239. Y. Fujiwara, M. Yoshidomi, H. Kuromaru, and H. Taniguchi, *J. Organomet. Chem.*, **226**, C36 (1982).
240. B. M. Trost, S. A. Godleski, and J. P. Genet, *J. Am. Chem. Soc.*, **100**, 3930 (1978); B. M. Trost and J. M. D. Fortunak, *Organometallics*, **1**, 7 (1982).
241. T. D. Cushing, J. F. Sanz-Cervera, and R. M. Williams, *J. Am. Chem. Soc.*, **115**, 9323 (1993).
242. R. Benhaddou, S. Czernecki, and G. Ville, *J. Org. Chem.*, **57**, 4612 (1992).
243. S. Czernecki and V. Dechavanne, *Can. J. Chem.*, **61**, 533 (1983).
244. T. Itahara and F. Ouseto, *Synthesis*, 488 (1984).
245. A. Kasahara, T. Izumi, M. Yodono, R. Saito, T. Takeda, and T. Sugawara, *Bull. Chem. Soc. Jpn.*, **46**, 1220 (1973).
246. Y. Fujiwara, O. Maruyama, M. Yoshidomi, and H. Taniguchi, *J. Org. Chem.*, **46**, 851 (1981).
247. T. Itahara, *J. Org. Chem.*, **50**, 5546 (1985).
248. R. S. Shue, *J. Catal.*, **26**, 112 (1972); *Chem. Commun.*, 1510 (1971).
249. Y. Fujiwara, I. Moritani, M. Matsuda, and S. Teranishi, *Tetrahedron Lett.*, 3863 (1968).
250. J. Tsuji and H. Nagashima, *Tetrahedron*, **40**, 2699 (1984).

251. B. M. Choudary, R. M. Sarma, and K. K. Rao, *Tetrahedron*, **48**, 719 (1992).
252. Y. Odaira, T. Oishi, T. Yukawa, and S. Tsutsumi, *J. Am. Chem. Soc.*, **88**, 4105 (1966).
253. N. Kominami, H. Nakajima, T. Kimura, M. Chono, and T. Sakurai, *J. Ind. Chem.*, **74**, 2269, 2272 (1971) (in Japanese).
254. H. A. Tayim, *Chem. Ind. (London)*, 1468 (1970).
255. R. Hüttel, J. Kratzer, and M. Bechter, *Angew. Chem.*, **71**, 456 (1959); *Chem. Ber.*, **94**, 766 (1961).
256. H. C. Volger, *Recl. Trav. Chim. Pays-Bas*, **86**, 677 (1967); **88**, 225 (1969).
257. T. Matsuda and Y. Nakamura, *J. Ind. Chem.*, **72**, 1756 (1969) (in Japanese).
258. S. Uemura, T. Okada, and K. Ichikawa, *J. Chem. Soc. Jpn.*, **89**, 692 (1968) (in Japanese).
259. C. F. Kohll and R. van Helden, *Recl. Trav. Chim. Pays-Bas*, **86**, 193 (1967).
260. K. Hirota, Y. Isobe, Y. Kitade, and Y. Maki, *Synthesis*, 495 (1987).
261. P. Mushak and M. A. Battiste, *Chem. Commun.*, 1146 (1969).
262. W. Menzenmaier and H. Straub, *Synthesis*, 49 (1976).
263. K. Kaneda, T. Uchiyama, H. Kobayashi, Y. Fujiwara,T. Imanaka, and S. Taranishi, *Tetrahedron Lett.*, 2005 (1977).
264. W. Munzenmaier and H. Straub, *Chem. Ztg.*, **98**, 419 (1974).
265. J. Tsuji, H. Takahashi, and M. Morikawa, *Tetrahedron Lett.*, 4387 (1965).
266. R. Huttel, J. Kratzer, and M. Bechter, *Chem. Ber.*, **94**, 766 (1961); R. Huttel and H. Christ, *Chem. Ber.*, **96**, 3101 (1963); **97**, 1439 (1964).
267. H. C. Volger, *Ind. Eng. Chem., Prod. Res. Dev.*, **9**, 311 (1970); *Recl. Trav. Chim. Pays-Bas*, **88**, 225 (1969).
268. D. Morelli, R. Ugo, F. Conti, and M. Donati, *Chem. Commun.*, 801 (1967); *J. Organomet. Chem.*, **30**, 421 (1971).
269. Review, S. A. Godleski, *Comprehensive Organic Synthesis*, Vol. 4, Pergamon Press, Oxford, 1991, p.585.
270. Review: B. M. Trost, *Acc. Chem. Res.*, **13**, 385 (1980).
271. B. M. Trost and T. J. Fullerton, *J. Am. Chem. Soc.*, **95**, 292 (1973).
272. B. M. Trost, W. P. Conway, P. E. Strege, and T. J. Dietsche, *J. Am. Chem. Soc.*, **96**, 7165 (1974).
273. P. S. Manchand, H. S. Wong, and J. F. Blount, *J. Org. Chem.*, **43**, 4769 (1978).
274. B. M. Trost, T. J. Dietsche, and T. J. Fullerton, *J. Org. Chem.*, **39**, 737 (1974).
275. B. M. Trost and L. Weber, *J. Org. Chem.*, **40**, 3617 (1975); B. M. Trost, L. Weber, P. E. Strege, T. J. Fullerton, and T. J. Dietsche, *J. Am. Chem. Soc.*, **100**, 3426 (1978).
276. G. W. Parshall and G. Wilkinson, Chem. Ind. (London), 261 (1962); *Inorg. Chem.*, **1**, 896 (1962).
277. W. R. Jackson and I. U. Straus, *Tetrahedron Lett.*, 2591 (1975); *Aust. J. Chem.*, **30**, 553 (1977); **31**, 1073 (1978).
278. R. W. Howsam and F. J. McQuillin, *Tetrahedron Lett.*, 3667 (1968); *Chem. Commun.*, 15 (1978).
279. D. J. Collins, W. R. Jackson, and R. N. Timms, *Tetrahedron Lett.*, 495 (1976); *Aust. J. Chem.*, **30**, 2167 (1977).
280. L. S. Hegedus, W. H. Darlington, and C. E. Russel, *J. Org. Chem.*, **45**, 5193 (1980).
281. H. M. R. Hoffmann, A. R. Otte, and A. Wilde, *Angew. Chem., Int. Ed. Engl.*, **31**, 234 (1992).
282. A. Wilde, A. R. Otte, and H. M. R. Hoffmann, *Chem. Commun.*, 615 (1993).

283. J. S. Temple and J. Schwartz, *J. Am. Chem. Soc.*, **102**, 7381 (1980); **104**, 1310 (1982); M. Riediker and J. Schwartz, *Tetrahedron Lett.*, **22**, 4655 (1981).
284. J. Tsuji, J. Kiji, and M. Morikawa, *Tetrahedron Lett.*, 1811 (1963).
285. J. Tsuji, S. Imamura, and J. Kiji, *J. Am. Chem. Soc.*, **86**, 4491 (1964).
286. M. Donati and F. Conti, *Tetrahedron Lett.*, 4953 (1966).
287. I. T. Harrison, E. Kimura, E. Bohme, and J. H. Fried, *Tetrahedron Lett.*, 1589 (1969).
288. K. Dunne and F. J. McQuillin, *J. Chem. Soc. C*, 2196, 2200 (1970).
289. R. K. Haynes, W. R. Jackson, and A. Stragalinou, *Aust. J. Chem.*, **33**, 1537 (1980).
290. T. Suzuki and J. Tsuji, *Bull. Chem. Soc. Jpn.*, **46**, 655 (1973).
291. D. N. Jones and S. D. Knox, *Chem. Commun.*, 166 (1975).
292. K. Jitsukawa, K. Kaneda, and S. Teranishi, *J. Org. Chem.*, **48**, 389 (1983).
293. J. Muzart, P. Pale, and J. P. Pete, *Tetrahedron Lett.*, **24**, 4567 (1983).
294. A. Riahi, J. Cossy, J. Muzart, and J. P. Pete, *Tetrahedron Lett.*, **26**, 839 (1985); J. Muzart and A. Riahi, *Organometallics*, **11**, 3478 (1992).
295. D. H. R. Barton and H. Patin, *Chem. Commun.*, 799 (1977).
296. B. L. Shaw, *Chem. Ind. (London)*, 1190 (1962).
297. S. D. Robinson and B. L. Shaw, *J. Chem. Soc.*, 4806 (1963); 5002 (1964).
298. M. Donati and F. Conti, *Tetrahedron Lett.*, 1219 (1966).
299. R. M. Rowe and D. A. White, *J. Chem. Soc. A*, 1451 (1967).
300. Reviews: J. E. Bäckvall, *Acc. Chem. Res.*, **16**, 335 (1983); *Pure Appl. Chem.*, **55**, 1669 (1983); *New J. Chem.*, **14**, 447 (1990); *Adv. Met. Org. Chem.*, **1**, 135 (1989).
301. J. Tsuji and S. Hosaka, *J. Am. Chem. Soc.*, **87**, 4075 (1965), J. Tsuji, J. Kiji, and S. Hosaka, *Tetrahedron Lett.*, 605 (1964).
302. Mitsuibishi Kasei, *ChemTech*, 759 (1988).
303. R. G. Brown and J. M. Davidson, *J. Chem. Soc. A*, 1321 (1971).
304. J. E. Bäckvall and R. E. Nordberg, *J. Am. Chem. Soc.*, **103**, 4959 (1981).
305. J. E. Bäckvall, S. E. Bystrom, and R. E. Nordberg, *J. Org. Chem.*, **49**, 4619 (1984).
306. J. E. Bäckvall, J. Vagberg, and R. E. Nordberg, *Tetrahedron Lett.*, **25**, 2717 (1984).
307. J. E. Nystrom, T. Rein, and J. E. Bäckvall, *Org. Synth.*, **67**, 105 (1988).
308. J. E. Bäckvall and J. O. Vagberg, *Org. Synth.*, **69**, 38(1990).
309. J. E. Nystrom, T. Rein, and J. E. Bäckvall, *Org. Synth.*, **67**, 105 (1988); J. E. Bäckvall, J. E. Nystrom, and R. E. Nordberg, *Tetrahedron Lett.*, **23**, 1617 (1982); *J. Am. Chem. Soc.*, **107**, 3676 (1985); J. E. Nystrom and J. E. Bäckvall, *J. Org. Chem.*, **48**, 3947 (1983).
310. J. E. Bäckvall, R. Gatti, and H. E. Schink, *Synthesis*, 341 (1993); J. E. Bäckvall, K. L. Granberg, P. G. Andersson, R. Gatti, and A. Gogoll, *J. Org. Chem.*, **58**, 5445 (1993).
311. J. E. Bäckvall, Z. D. Renko, and S. E. Bystrom, *Tetrahedron Lett.*, **28**, 4199 (1987); H, E. Schink, H. Pettersson, and J. E. Backvall, *J. Org. Chem.*, **56**, 2769 (1991).
312. Review: J. E. Bäckvall, *Pure Appl. Chem.*, **64**, 429 (1992).
313. J. E. Bäckvall, P. G. Andersson, G. B. Stone, and A. Gogoll, *J. Org. Chem.*, **56**, 2988 (1991); J. E. Bäckvall and P. G. Andersson, *J. Am. Chem. Soc.*, **112**, 3683 (1990).
314. P. G. Andersson and J. E. Bäckvall, *J. Am. Chem. Soc.*, **114**, 8696 (1992).
315. J. E. Bäckvall and J. O. Vagberg, *J. Org. Chem.*, **53**, 5695 (1988).

316. P. G. Andersson and J. E. Bäckvall, *J. Org. Chem.*, **56**, 5349 (1991).
317. J. E. Bäckvall and P. G. Andersson, *J. Am. Chem. Soc.*, **114**, 6374 (1992).
318. J. E. Bäckvall and P. G. Andersson, *J. Org. Chem.*, 56, 2274 (1991).
319. P. G. Andersson, Y. I. M. Nilsson, and J. E. Bäckvall, *Tetrahedron*, **50**, 559 (1994).
320. J. E. Bäckvall, A. K. Awasthi, and Z. D. Renko, *J. Am. Chem. Soc.*, **109**, 4750 (1987); J. E. Bäckvall, R. B. Hopkins, H. Greenberg, M. M. Mader, and A. K. Awasthi, *J. Am. Chem. Soc.*, **112**, 5160 (1990); H. Greenberg, A. Gogoll, and J. E. Bäckvall, *J. Org. Chem.*, **56**, 5808 (1991).
321. D. Y. Waddar, P. F. Todd, D. J. Royall, and A. Pattison, *Br. Pat.*, 1 497 414, *Chem. Abstr.*, **89**, 42481a (1978).
322. R. C. Larock, L. W. Harrison, and M. H. Su, *J. Org. Chem.*, **49**, 3664 (1984).
323. R. C. Larock, S. Varaprath, H. H. Lau, and G. A. Fellow, *J. Am. Chem. Soc.*, **104**, 1900 (1982); **106**, 5274 (1984); R. C. Larock, C. L. Liu, H. H. Lau, and S. Varaprath, *Tetrahedron Lett.*, **25**, 4459 (1984).
324. R. van Helden and G. Verberg, *Recl. Trav. Chim. Pays-Bas*, **84**, 1263 (1965).
325 J. M. Davidson and C. Triggs, *Chem. Ind. (London)*, 457 (1966); *J. Chem. Soc. A*, 1324 (1968).
326. J. M. Davidson and C. Triggs, *Chem. Ind. (London)*, 1361 (1967); *J. Chem. Soc. A*, 1331 (1968).
327. H. Itatani and H. Yoshimoto, *Chem. Ind. (London)*, 674 (1971); *J. Org. Chem.*, **38**, 76 (1973); *Bull. Chem. Soc. Jpn*, **46**, 2490 (1973); *J. Catal.*, **31**, 8 (1973); M. Kashima, H. Yoshimoto, and H. Itatani, *J. Catal.*, **29**, 92 (1973).
328. R. R. S. Clark, R. O. C. Norman, C. B. Thomas, and J. S. Wilson, *J. Chem. Soc., Perkin Trans. 1*, 1289 (1974).
329. J. Tsuji and H. Nagashima, *Tetrahedron*, **40**, 2699 (1984).
330. T. Itahara, *Chem. Ind. (London)*, 599 (1982).
331. A. D. Ryabov, S. A. Deiko, A. K. Yatsimirsky, and I. V. Berezin, *Tetrahedron Lett.*, **22**, 3793 (1981).
332. A. Shiotani, H. Itatani, and T. Inagaki, *J. Mol. Catal.*, **34**, 57 (1986); A. Shiotani, M. Yoshikiyo, and H. Itatani, *J. Mol. Catal.*, **18**, 23 (1983).
333. B. Akermark, L. Eberson, E. Jonsson, and E. Petterson, *J. Org. Chem.*, **40**, 1365 (1975).
334. A. Shiotani and H. Itatani, *J. Chem. Soc., Perkin Trans. 1*, 1236 (1976).
335. G. Bringmann and H. Reuscher, *Tetrahedron Lett.*, **30**, 5249 (1989).
336. T. Itahara and T. Sakakibara, *Synthesis*, 607 (1978).
337. W. Harris, C. H. Hill, E. Keech, and P. Malsher, *Tetrahedron Lett.*, **34**, 8361 (1993).
338. I. V. Kozhevnikov, *React. Kinet. Catal. Lett.*, **6**, 401 (1977); *Chem. Abstr.*, **87**, 117357 (1977); **5**, 415 (1976); *Chem. Abstr.*, **86**, 120465 (1977).
339. T. Itahara, *Chem. Commun.*, 49 (1980); 254 (1981); *Heterocycle*, **14**, 100 (1980).
340. D. L. Boger and M. Patel, *J. Org. Chem.*, **53**, 1405 (1988).
341. T. Itahara, *J. Org. Chem.*, **50**, 5272 (1985).
342. Review: D. J. Rawlinson and G. Sosnovsky, *Synthesis*, 567 (1973).
343. L. Eberson and L. Gomez-Gonzalez, *Acta Chem. Scand., Ser. B*, **27**, 1162 (1973).
344. D. R. Bryant, J. E. McKeon, and B. C. Ream, *Tetrahedron Lett.*, 3371 (1968).
345. C. H. Bushweller, *Tetrahedron Lett.*, 6123 (1968).
346. J. M. Holovka and E. Hurley (Marathon Oil), *Ger. Pat.*, 2 047 845.
347. L. Eberson and L. Jonsson, *Acta Chem. Scand., Ser. B*, **28**, 771 (1974).

348. L. Eberson and L. Jonsson, *Acta Chem. Scand., Ser. B*, **30**, 361 (1976); *Liebigs Ann. Chem.*, 233 (1977); *Chem. Commun.*, 885 (1974).

349. G. U. Mennen, A. I. Rudenkov, K. I. Matveev, and I. V. Kozhvnikox, *React. Kinet. Catal. Lett.*, **5**, 401 (1976); *Chem. Abstr.*, **86**, 105719 (1977).

350. A. I. Rudenkov, H. Mennenga, L. N. Rachkovskaya, K. I. Matveev, and I. V. Kozhevnikov, *Kinet. Katal.*, **18**, 915 (1977); *Chem. Abstr.*, **87**, 200503z (1977).

351. T. Jintoku, K. Nishimura, K. Takaki, and Y. Fujiwara, *Chem. Lett.*, 1687 (1990); 1865 (1987).

352. L. Eberson and L. Gomez-Gonzalez, *Chem. Commun.*, 263 (1971); *Acta Chem. Scand., Ser. B*, **27**, 1249, 1255 (1973).

353. D. R. Bryant, J. E. McKeon, and B. C. Ream, *J. Org. Chem.*, **33**, 4123 (1968); **34**, 1106 (1969).

354. E. Benazzi, C. J. Cameron, and H. Mimoun, *J. Mol. Catal.*, **69**, 299 (1991).

355. G. G. Arzoumanidis and F. C. Rauch, *J. Org. Chem.*, **38**, 4443 (1973).

356. K. Ichikawa, S. Uemura, and T. Okada, *J. Chem. Soc. Jpn.*, **90**, 212 (1969) (in Japanese).

357. P. M. Henry, *Tetrahedron Lett.*, 2285 (1968); *J. Org. Chem.*, **36**, 1886 (1971).

358. R. O. C. Norman, W. J. E. Parr, and C. B. Thomas, *J. Chem. Soc., Perkin Trans. 1*, 369 (1974).

359. T. Tisue and W. J. Downs, *Chem. Commun.*, 410 (1969).

360. T. Jintoku, H. Taniguchi, and Y. Fujiwara, *Chem. Lett.*, 1159 (1987); Y. Fujiwara, I. Kawata, H. Sugimoto, and H. Taniguchi, *J. Organomet. Chem.*, **256**, C35 (1983).

361. Y. Fujiwara, I. Kawata, T. Kawauchi, and H. Taniguchi, *Chem. Commun.*, 220 (1980); 132 (1982).

362. R. Ugo, A. Chiesa, P. Nardi, and R. Psaro, *J. Mol. Catal.*, **59**, 23 (1990).

362a. T. Itahara, *Chem. Lett.*, 1151(1982).

363. T. Sakakibara and Y. Odaira, *J. Org. Chem.*, **41**, 2049 (1976).

364. Review: R. F. Heck, *Org. React.*, **27**, 345 (1982).

365. R. C. Larock, *Organomercury Compounds in Organic Synthesis*, Springer, Berlin, 1985.

366. I. I. Moiseev and M. N. Vargaftik, *Dokl. Akad. Nauk SSSR*, **166**, 370 (1966); *Chem. Abstr.*, **64**, 11248 (1966).

367. R. F. Heck, *J. Am. Chem. Soc.*, **90**, 5518, 5526, 5531, 5535, 5538, 5542, 5546 (1968); **91**, 6707 (1969); **93**, 6896 (1971); *J. Organomet. Chem.*, **37**, 389 (1972).

368. P. M. Henry and G. A. Ward, *J. Am. Chem. Soc.*, **94**, 673 (1972).

369. A. Kasahara, T. Izumi, G. Saito, M. Yodono, R. Saito, and Y. Goto, *Bull. Chem. Soc. Jpn.*, **45**, 894 (1972); A. Kasahara and T. Izumi, *Bull. Chem. Soc. Jpn.*, **45**, 951, 1256 (1972); **46**, 665 (1973); A. Kasahara, T. Izumi, M. Yodono, R. Saito, T. Takeda, and T. Sugawara, *Bull. Chem. Soc. Jpn.*, **46**, 1220 (1973); A. Kasahara, T. Izumi, T. Taskeda, and H. Imamura, *Bull. Chem. Soc. Jpn.*, **47**, 183 (1974); A. Kasahara, T. Izumi, K. Endo, T. Takeda, and M. Ookita, *Bull. Chem. Soc. Jpn.*, **47**, 1967 (1974).

370. Review, R. C. Larock, *J. Organomet. Chem. Library*, **1**, 257 (1976).

371. R. Saito, T. Izumi, and A. Kasahara, *Bull. Chem. Soc. Jpn.*, **46**, 1776 (1973).

372. H. Horino and N. Inoue, *Chem. Commun.*, 500 (1976).

373. D. D. Narkhede, P. R. Iyer, and C. S. R. Iyer, *Tetrahedron*, **46**, 2031 (1990).

374. R. C. Larock, K. Narayanan, R. K. Carlson, and J. A. Ward, *J. Org. Chem.*, **52**, 1364 (1987).

375. R. C. Larock, D. R. Leach, and S. M. Bjorge, *J. Org. Chem.*, **51**, 5221 (1986).

376. Review: D. Bergstrom, X. Lin, G. Wang, D. Rotstein, P. Beal, K. Norris, and J. Ruth, *Synlett*, 179 (1992).
377. D. E. Bergstrom and J. L. Ruth, *J. Am. Chem. Soc.*, **98**, 1587 (1976); *J. Org. Chem.*, **43**, 2870 (1978); D. E. Bergstrom and M. K. Ogawa, *J. Am. Chem. Soc.*, **100**, 8106 (1978).
378. I. Arai and G. D. Daves, Jr, *J. Am. Chem. Soc.*, **100**, 287 (1978); *J. Org. Chem.*, **43**, 4110 (1978); T. D. Lee and G. D. Daves, Jr, *J. Org. Chem.*, **48**, 399 (1983).;
379. R. C. Larock and M. Mitchell, *J. Am. Chem. Soc.*, **100**, 180 (1978).
380. R. C. Larock, D. J. Leuck, and L. W. Harrison, *Tetrahedron Lett.*, **28**, 4977 (1987).
381. R. C. Larock and S. Ding, *J. Org. Chem.*, **58**, 2081 (1993); R. C. Larock and S. K. Stolz-Dunn, *Tetrahedron Lett.*, **29**, 5069 (1988).
382. R. F. Heck, *J. Am. Chem. Soc.*, **90**, 5542 (1968); F. G. Stakem and R. F. Heck, *J. Org. Chem.*, **45**, 3584 (1980).
383. R. C. Larock and K. Takagi, *Tetrahedron Lett.*, **24**, 3457 (1983).
384. M. O. Unger and R. A. Fouty, *J. Org. Chem.*, **34**, 18 (1969).
385. R. A. Kretchmer and R. Glowinski, *J. Org. Chem.*, **41**, 2661 (1976).
386. R. C. Larock and B. Riefling, *J. Org. Chem.*, **41**, 2241 (1976); **43**, 1468 (1978).
387. R. F. Heck, *J. Am. Chem. Soc.*, **90**, 5546 (1968).
388. T. Izumi, T. Iino, and A. Kasahara, *Bull. Chem. Soc. Jpn.*, **46**, 2251 (1973).
389. P. M. Henry, *Tetrahedron Lett.*, 2285 (1968).
390. J. K. Stille and P. K. Wong, *J. Org. Chem.*, **40**, 335 (1975).
391. R. Jaouhari and P. H. Dixneuf, *Tetrahedron Lett.*, **27**, 6315 (1986).
392. R. C. Larock, K. Oertle, and K. M. Beatty, *J. Am. Chem. Soc.*, **102**, 1966 (1980).
393. S. Uemura, Y. Ikeda, and K. Ichikawa, *Chem. Commun.*, 390 (1971).
394. T. Spencer and F. G. Thorpe, *J. Organomet. Chem.*, **99**, C8 (1975).
395. R. A. Kjonass, *J. Org. Chem.*, **51**, 3708 (1986).
396. S. Uemura, K. Zushi, M. Okano, and K. Ichikawa, *Chem. Commun.*, 234 (1972).
397. R. C. Larock and C. A. Fellows, *J. Am. Chem. Soc.*, **104**, 1900 (1982)
398. R. C. Larock and C. A. Fellows, *J. Org. Chem.*, **45**, 363 (1980).
399. H. A. Dieck and R. F. Heck, *J. Org. Chem.*, **40**, 1083 (1975).
400 H. Yatagai, Y. Yamamoto, and K. Maruyama, *Chem. Commun.*, 852 (1977).
401. N. Miyaura and A. Suzuki, *Chem. Lett.*, 879 (1981); N. Yamashina, S. Hyuga, S. Hara, and A. Suzuki, *Tetrahedron Lett.*, **30**, 6555 (1989).
402. T. Ohe, K. Ohe, S. Uemura, and N. Sugita, *J. Organomet. Chem.*, **344**, C5 (1988).
403. V. V. R. Rao, C. V. Kumar, and D. Devaprabhakara, *J. Organomet. Chem.*, **179**, C7 (1979).
404. R. A. Outten and G. D. Daves, *J. Org. Chem.*, **54**, 29 (1989).
405. L. S. Liebeskind and S. W. Riesinger, *Tetrahedron Lett.*, **32**, 5681 (1991).
406. S. Kanemoto, S. Matsubara, K. Oshima, K. Utimoto, and H. Nozaki, *Chem. Lett.*, 5 (1987).
407. G. A. Tolstikov, M. S. Miftakhov, N. A. Danilova, Y. L. Vel'der, and L. V. Spirikhin, *Synthesis*, 633 (1989).
408. G. J. Boons, D. A. Entwistle, S. V. Ley, and M. Woods, *Tetrahedron Lett.*, **34**, 5649 (1993).
409. H. L. Elbe and G. Kobrich, *Chem. Ber.*, **107**, 1654 (1974).
410. R. C. Larock, *J. Org. Chem.*, **41**, 2241 (1976); R. C. Larock and B. Riefling, *J. Org. Chem.*, **43**, 1468 (1978).

411. W. P. Weber, R. A. Felix, A. K. Willard, and K. E. Koenig, *Tetrahedron Lett.*, 4701 (1971).
412. I. S. Akhrem, N. M. Chistovalova, E. I. Mysov, and M. E. Volpin, *J. Organomet. Chem.*, **72**, 163 (1974).
413. J. Yoshida, K. Tamao, M. Takahashi, and M. Kumada, *Tetrahedron Lett.*, 2161 (1978); K. Tamao, T. Kakui, and M. Kumada, *Tetrahedron Lett.*, 619 (1979); *Organometallics*, **1**, 542 (1982).
414. Review, N. Petragnani and J. V. Comasseto, *Synthesis*, 897 (1992).
415. K. Ohe, H. Takahashi, S. Uemura, and N. Sugita, *J. Org. Chem.*, **52**, 4859 (1987); *J. Organomet. Chem.*, **326**, 35 (1987); **350**, 227 (1988).
416. S. Uemura, M. Wakasugi, and M. Okano, *J. Organomet. Chem.*, **194**, 277 (1980).
417. D. H. R. Barton, N. Ozbalik, and M. Ramesh, *Tetrahedron Lett.*, **29**, 3533 (1988).
418. D. H. R. Barton, N. Ozbalik, and M. Ramesh, *Tetrahedron*, **44**, 5661 (1988).
419. Reviews: M. I. Bruce, *Angew. Chem., Int. Ed. Engl.*, **16**, 73 (1977); I. Omae, *Chem. Rev.*, **79**, 287 (1987); G. R. Newkome, W. E. Puckatt, V. K. Gupta, and G. E. Kiefer, *Chem. Rev.*, **86**, 451 (1986); A. D. Rybov, *Synthesis*, 233 (1985).
420. I. Omae, *Organometallic Intramolecular Coordination Compounds*, Elsevier, Amsterdam, 1986.
421. A. C. Cope and R. D. W. Siekman, *J. Am. Chem. Soc.*, **87**, 3272 (1965); A. C. Cope and E. C. Friedrich, *J. Am. Chem. Soc.*, **90**, 909 (1968).
422. S. P. Molnar and M. Orchin, *J. Organomet. Chem.*, **16**, 196 (1969).
423. H. Onoue, K. Nakagawa, and I. Moritani, *J. Organomet. Chem.*, **43**, 431 (1972); **35**, 217 (1972).
424. A, Kasahara, *Bull. Chem. Soc. Jpn.*, **41**, 1272 (1968).
425. D. M. Fenton, *J. Org. Chem.*, **38**, 3192 (1973).
426. G. E. Hartwell, R. V. Lawrence, and M. J. Smas, *Chem. Commun.*, 912 (1970).
427. J. M. Thompson and R. F. Heck, *J. Org. Chem.*, **40**, 2667 (1975).
428. A. Kasahara, T. Izumi, and M. Maemura, *Bull. Chem. Soc. Jpn.*, **50**, 1878 (1977); T. Izumi, K. Endo, O. Saito, I. Shimizu, M. Maemura, and A. Kasahara, *Bull. Chem. Soc. Jpn.*, **51**, 663 (1978).
429. T. Komatsu and M. Nonoyama, *Inorg. Nucl. Chem.*, **39**, 1161 (1977).
430. N. D. Cameron and M. Kilner, *Chem. Commun.*, 687 (1975).
431. H. Horino and N. Inoue, *J. Org. Chem.*, **46**, 4416 (1981).
432. A. J. Deeming and I. P. Rothwell, *Chem. Commun.*, 344 (1978); *J. Organomet. Chem.*, **205**, 117 (1981).
433. M. Pfeffer, E. Wehman, and G. van Koten, *J. Organomet. Chem.*, **282**, 127 (1985).
434. V. I. Sokolov, L. L. Troitskaya, and O. A. Reutov, *J. Organomet. Chem.*, **182**, 537 (1979).
435. M. Julia, M. Duteil, and J. Y. Lallemand, *J. Organomet. Chem.*, **102**, 239 (1975).
436. R. A. Holton, *Tetrahedron Lett.*, 355 (1977).
437. B. J. Brisdon, P. Nair, and S. F. Dyke, *Tetrahedron*, **37**, 173 (1981).
438. R. A. Holton and R. G. Davis, *J. Am. Chem. Soc.*, **99**, 4175 (1977), R. A. Holton and R. V. Nelson, *J. Organomet. Chem.*, **201**, C35 (1980).
439. C. D. Liang, *Tetrahedron Lett.*, **27**, 1971 (1986).
440. J. Dupont and M. Pfeffer, *J. Organomet. Chem.*, **321**, C13 (1987).
441. G. Wu, A. L. Rheingold, and R. F. Heck, *Organometallics*, **6**, 2386 (1987).
442. H. Takahashi and J. Tsuji, *J. Organomet. Chem.*, **10**, 511 (1967).
443. M. I. Bruce, B. L. Goodall, and F. G. A. Stone, *Chem. Commun.*, 558 (1973).

444. J. M. Thompson and R. F. Heck, *J. Org. Chem.*, **40**, 2667 (1975).
445. T. Izumi, T. Katou, A. Kasahara, and K. Hanaya, *Bull. Chem. Soc. Jpn.*, **51**, 3407 (1978).
446. T. Sakakibara, T. Kume, T. Ohyabu, and T. Hase, *Chem. Pharm. Bull.*, **37**, 1694 (1989).
447. Y. Yamamoto and H. Yamazaki, *Synthesis*, 750 (1976).
448. S. Murahashi, Y. Tamba, M. Yamamura, and I. Moritani, *Tetrahedron Lett.*, 3749 (1974); M. Yamamura, I. Moritani, and S. Murahashi, *Chem. Lett.*, 1923 (1974); S. Murahashi, Y. Tamba, M. Mamamura, and N. Yoshimura, *J. Org. Chem.*, **43**, 4099 (1978).
449. G. D. Hartman, W. Halczenko, and B. T. Phillips. *J. Org. Chem.*, **51**, 142 (1986).
450. M. Hugentobler, A. J. Klaus, P. Ruppen, and P. Rys, *Helv. Chim. Acta*, **67**, 113 (1984).
451. S. J. Tremont and H. R. Rahman, *J. Am. Chem. Soc.*, **106**, 5759 (1984).
452. J. S. McCallum, J. R. Gasdaska, L. S. Liebeskind, and S. J. Tremont, *Tetrahedron Lett.*, **30**, 4085 (1989).
453. R. A. Holton and K. J. Natalie, Jr, *Tetrahedron Lett.*, **22**, 267 (1981).
454. P. W. Clark, H. J. Dyke, S. F. Dyke, and G. Perry, *J. Organomet. Chem.*, **253**, 399 (1983).
455. D. R. Fahey, *Chem. Commun.*, 417 (1970); *J. Organomet. Chem.*, **27**, 283 (1971).
456 B. A. Grigor and A. J. Nielson, *J. Organomet. Chem.*, **129**, C17 (1977).
457. P. L. Alsters, H. T. Teunissen, J. Boersma, A. L. Spek, and G. van Koten, *Organometallics*, **12**, 4691 (1993).
458. A. Mahapatra, D. Bandyopadhyay, P. Bandyopadhyay, and A. Chakravorty, *Chem. Commun.*, 999 (1984).
459. R. A. Holton, M. P. Sibi, and W. S. Murphy, *J. Am. Chem. Soc.*, **110**, 314 (1988).
460. A. C. Cope, J. M. Kliegman, and E. C. Friedrich, *J. Am. Chem. Soc.*, **89**, 287 (1967).
461. D. Medema, R. van Helden, and C. F. Kohll, *Inorg. Chim. Acta*, **3**, 255 (1969).
462. R. A. Holton and R. A. Kjonaas, *J. Organomet. Chem.*, **133**, C5 (1977).
463. R. A. Holton and R. A. Kjonaas, *J. Am. Chem. Soc.*, **99**, 4177 (1977).
464. Y. Takahashi, A. Tokuda, S. Sakai, and Y. Ishii, *J. Organomet. Chem.*, **35**, 415 (1972).
465. R. A. Holton, *J. Am. Chem. Soc.*, **99**, 8083 (1977).
466. R. A. Holton and R. A. Kjonaas, *J. Organomet. Chem.*, **142**, C15 (1977).
467. J. E. Baldwin, C. Najera, and M. Yus, *Chem. Commun.*, 126 (1985).
468. V. Rocherolle, J. C. Lopez, A. Olesker, and G. Lukacs, *Chem. Commun.*, 513 (1988).
469. J. Tsuji, M. Takahashi, and T. Takahashi, *Tetrahedron Lett.*, **21**, 849 (1980).
470. S. F. Vasilevsky, B. A. Trofimov, A. G. Mal'kina, and L. Brandsma, *Synth. Commun.*, **24**, 85 (1994).
471. T. T. Zung, L. G. Bruk, and O. N. Temkin, *Mendeleev Commun.*, 2 (1994).
472. J. S. Prasad and L. S. Liebeskind, *Tetrahedron Lett.*, **28**, 1859 (1987).
473. A. Arcadi, E. Bernocchi, A. Burini, S. Cacchi, F. Marinelli, and B. Pietroni, *Tetrahedron*, **47**, 481 (1988).
474. J. Tsuji, N. Iwamoto, and M. Morikawa, *J. Am. Chem. Soc.*, **86**, 2095 (1964).
475. G. P. Chiusoli, C. Venturello, and S. Merzoni, *Chem. Ind. (London)*, 977 (1968).
476 G. P. Chiusoli, M. Costa, and S. Reverberi, *Synthesis*, 262 (1989).
477. M. Costa, L. D. Santos, G. P. Chiusoli, B. Gabriele, and G. Salerno, *J. Mol. Catal.*, **78**, 151 (1993).

478. H. Alper, B. Despeyroux, and J. B. Woell, *Tetrahedron Lett.*, **24**, 5691 (1983); D. Zargarian and H. Alper, *Organometallics*, **10**, 2914 (1991).
479. T. Nogi and J. Tsuji, *Tetrahedron*, **25**, 4099 (1969).
480. B. Gabriele, M. Costa, G. Salerno, and G. P. Chiusoli, *Chem. Commun.*, 1007 (1992).
481. J. Tsuji and T. Nogi, *J. Am. Chem. Soc.*, **88**, 1289 (1966).
482. Y. Tamaru, M. Hojo, and Z. Yoshida, *J. Org. Chem.*, **56**, 1099 (1991).
483. R. C. Larock and B. Riefling, *Tetrahedron Lett.*, 4661 (1976); R. C. Larock, B. Riefling, and C. A. Fellows, *J. Org. Chem.*, **43**, 131 (1978).
484. R. F. Heck, *J. Am. Chem. Soc.*, **94**, 2712 (1972).
485. A. Kasahara, T. Izumi, and A. Suzuki, *Bull. Chem. Soc. Jpn.*, **50**, 1639 (1977).
486. N. Garty and M. Michman, *J. Organomet. Chem.*, **36**, 391 (1972).
487. C. Broquet and H. Riviere, *J. Organomet. Chem.*, **226**, 1 (1982).
488. H. Kataoka, K. Watanabe, and K. Goto, *Tetrahedron Lett.*, **31**, 4181 (1990).
488a.R. Mahrwald and H. Schick, *Angew. Chem. Int. Ed. Engl.*, **30**, 593 (1991).
489. H. Kataoka, K. Watanabe, K. Miyazaki, S. Tahara, K. Ogu, R. Matsuoka, and K. Goto, *Chem. Lett.*, 1705 (1990).
490. B. Rossi, A. Carpita, and C. Bigelli, *Tetrahedron Lett.*, **26**, 523 (1985).
491. S. Takano, T. Sugihara, and K. Ogasawara, *Synlett*, 453 (1990).
492. S. Takano, T. Sugihara, and K. Ogasawara, *Tetrahedron Lett.*, **32**, 2797 (1991).
493. R. G. Schultz, *Tetrahedron Lett.*, 301(1964), *Tetrahedron*, **20**, 2809 (1964).
494. M. S. Lupin and B. L. Shaw, *Tetrahedron Lett.*, 883 (1964); M. S. Lupin, J. Powell, and B. L. Shaw, *J. Chem. Soc. A*, 1687 (1966)
495. L. S. Hegedus, N. Kambe, Y. Ishii, and A. Mori, *J. Org. Chem.*, **50**, 2240 (1985).
496. A. M. Ali, S. Tanimoto, and T. Okamoto, *J. Org. Chem.*, **53**, 3639 (1988).
497. G. D. Shier, *J. Organomet. Chem.*, **10**, P15 (1967).
498. J. Tsuji and T. Susuki, *Tetrahedron Lett.*, 3027 (1965).
499. H. Alper, F. W. Hartstock, and B. Despeyroux, *Chem. Commun.*, 905 (1984); *J. Mol. Catal.*, **34**, 381 (1986).
500. D. Lathbury, P. Vernon, and T. Gallagher, *Tetrahedron Lett.*, **27**, 6009 (1986); D. N. A. Fox, D. Lathbury, M. F. Mahon, K. C. Molloy, and T. Gallagher, *J. Am. Chem. Soc.*, **113**, 2652 (1991).
501. R. D. Walkup and G. Park, *Tetrahedron Lett.*, **28**, 1023 (1987).
502. E. Mincione, G. Ortaggi, and A. Sirna, *Synthesis*, 773 (1977).
503. R. J. Theissen, *J. Org. Chem.*, **36**, 752 (1971).
504. Y. Ito, T. Hirao, and T. Saegusa, *J. Org. Chem.*, **43**, 1011 (1978).
505. B. M. Trost, Y. Nishimura, K. Yamamoto, and S. S. McElvain, *J. Am. Chem. Soc.*, **101**, 1328 (1979).
506. M. R. Roberts and R. H. Schlessinger, *J. Am. Chem. Soc.*, **101**, 7626 (1979).
507. M. Kim, L. A. Applegate, O. S. Park, S. Vasudeva, and D. S. Watt, *Synth. Commun.*, **20**, 989 (1990).
508. M. Shibasaki and Y. Ogawa, *Tetrahedron Lett.*, **26**, 3841 (1985).
509. T. Baba, K. Nakano, S. Nishiyama, S. Tsuruya, and M. Masai, *Chem. Commun.*, 1697 (1989); T. Baba, K. Izumi, S. Nishiyama, S. Tsuruya, and M. Masai, *J. Mol. Catal.*, **62**, L5 (1990); *J. Chem. Soc., Perkin Trans. 2*, 1113 (1990).
510. M. Ihara, S. Suzuki, N. Taniguchi, and K. Fukumoto, *Synlett*, 435 (1993).
511. V. Bellosta, R. Benhaddou, and S. Czernecki, *Synlett*, 861 (1993).
512. D. M. Fenton and P. J. Steinwand, *J. Org. Chem.*, **39**, 701 (1974).
513. M. Graziani, P. Uguagliati, and G. Carturan, *J. Organomet. Chem.*, **27**, 275 (1971).
514. S. P. Current, *J. Org. Chem.*, **48**, 1779 (1983).

515. G. E. Morrris, D. Oakley, D. A. Pippard, and D. J. H. Smith, *Chem. Commun.*, 410 (1987).
516. Ube Industries, Ltd., *Belg. Pat.*, 870 268 (1979).
517. J. E. Hallgren and R. O. Mattews, *J. Organomet. Chem.*, **175**, 135 (1979); J. E. Hallgren, G. M. Lucas, and R. O. Mathews, *J. Organomet. Chem.*, **204**, 135 (1981); **212**, 135 (1981).
518. J. Tsuji and N. Iwamoto, *Chem. Commun.*, 380 (1966).
519. P. Giannoccaro, *J. Organomet. Chem.*, **336**, 271 (1987).
520. W. Tam, *J. Org. Chem.*, **51**, 2977 (1986).
521. S. Murahashi, Y. Mitsue, and K. Ike, *Chem. Commun.*, 125 (1987).
522. S. Fukuoka, M. Chono, and M. Kohno, *J. Org. Chem.*, **49**, 1458 (1984); *Chemtech*, 670 (1984).
523. H. Alper, G. Vasapollo, F. W. Hartstock, M. Mlekuz, D. J. H. Smith, and G. E. Morris, *Organometallics*, **6**, 2391 (1987); H. Alper and F. W. Hartstock, *Chem. Commun.*, 1141 (1985).
524. K. Nishimura, S. Furusaki, Y. Shiomi, K. Fujii, K. Nishihara, and M. Yamashita (Ube Industries) *Eur. Pat.*, EP 77 542, *Chem. Abstr.*, **99**, 87654 (1983).
525. S. Uchiumi, K. Ataka, and K. Sataka, *Eur. Pat. Appl.*, EP 1 088 332; *Chem. Abstr.*, **101**, 90431 (1984).
526. Y. Fujiwara, K. Takaki, J. Watanabe, Y. Uchida, and H. Tamiguchi, *Chem. Lett.*, 1687 (1989).
527. K. Satoh, J. Watanabe, K. Takaki, and Y. Fujiwara, *Chem. Lett.*, 1433 (1991); K. Nakata, J. Watanabe, K. Takaki, and J. Fujiwara, *Chem. Lett.*, 1437 (1991).
528. T. Nishiguchi, K. Nakata, K. Takaki, and Y. Fujiwara, *Chem. Lett.*, 1141 (1992); 1005 (1993).

Chapter 4

Catalytic Reactions with Pd(0) and Pd(II)

Several Pd(0) complexes are effective catalysts of a variety of reactions, and these catalytic reactions are particularly useful because they are catalytic without adding other oxidants and proceed with catalytic amounts of expensive Pd compounds. These reactions are treated in this chapter. Among many substrates used for the catalytic reactions, organic halides and allylic esters are two of the most widely used, and they undergo facile oxidative additions to Pd(0) to form complexes which have σ-Pd—C bonds. These intermediate complexes undergo several different transformations. Regeneration of Pd(0) species in the final step makes the reaction catalytic. These reactions of organic halides except allylic halides are treated in Section 1 and the reactions of various allylic compounds are surveyed in Section 2. Catalytic reactions of dienes, alkynes, and alkenes are treated in other sections. These reactions offer unique methods for carbon–carbon bond formation, which are impossible by other means.

1 Reactions of Organic Halides and Pseudo-Halides Catalyzed by Pd(0)

In Grignard reactions, Mg(0) metal reacts with organic halides of sp^3 carbons (alkyl halides) more easily than halides of sp^2 carbons (aryl and alkenyl halides). On the other hand, Pd(0) complexes react more easily with halides of sp^2 carbons. In other words, alkenyl and aryl halides undergo facile oxidative additions to Pd(0) to form complexes **1** which have a Pd—C σ-bond as an initial step. Then mainly two transformations of these intermediate complexes are possible: insertion and transmetallation. Unsaturated compounds such as alkenes, conjugated dienes, alkynes, and CO insert into the Pd—C bond. The final step of the reactions is reductive elimination or elimination of β-hydrogen. At the same time, the Pd(0) catalytic species is regenerated to start a new catalytic cycle. The transmetallation takes place with organometallic compounds of Li, Mg, Zn, B, Al, Sn, Si, Hg, etc., and the reaction terminates by reductive elimination.

Facile oxidative addition is possible with iodides and bromides. The reactions of iodides can be carried out even in the absence of a phosphine ligand,

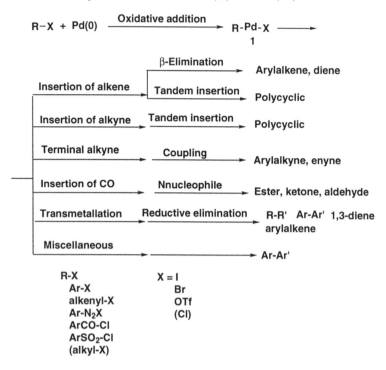

and $Pd_2(dba)_3$, $Pd(OAc)_2$, or even Pd on carbon can be used as catalysts. The reactions of bromides generally require phosphine ligands. A phosphonium salt is formed to some extent by the reaction of Ph_3P with iodides and bromides. In this case, the use of hindered tri(o-tolyl)phosphine is recommended[1]. Chlorides are inert under normal reaction conditions. They react by use of more electron-donating bidentate ligands such as dppp under fairly severe conditions. Chlorobenzene undergoes palladium-catalyzed reactions by the coordination of $Cr(CO)_3$, which activates the Cl—C bond by its strongly electron-attracting property[2,3]. In order to neutralize HX formed by the reactions, bases such as tertiary amines, sodium or potassium acetate and carbonate are added.

In addition to halides, the pseudo-halides R—X = ArCO—Cl, $ArSO_2$—Cl, Ar—$N_2^+X^-$, R—OP(O)(OR)$_2$, R—OSO_2CF_3 (OTf), R—OSO_2R_f, (R_f=perfluoroalkyl) R—OSO_2F, R—OSO_2CH_3, and Ar—ArI$^+$ are good leaving groups and undergo oxidative addition to Pd(0) to form aryl or alkenyl palladium complexes as intermediates. It should be noted that these leaving groups have different reactivities toward Pd(0). Some of them react only under special conditions with special reactants.

The most useful pseudo-halides are aryl triflates (trifluoromethylsulfonates) of phenols and enol triflates derived from carbonyl compounds[4,5,6].

Aromatic acyl halides and sulfonyl halides undergo oxidative addition, followed by facile elimination of CO and SO_2 to form arylpalladium complexes. Benzenediazonium salts are the most reactive source of arylpalladium complexes.

Oxidative addition of alkyl halides to Pd(0) is slow. Furthermore, alkyl–Pd complexes, formed by the oxidative addition of alkyl halides, undergo facile elimination of β-hydrogen and the reaction stops at this stage without undergoing insertion or transmetallation. Although not many examples are available, alkynyl iodides react with Pd(0) to form alkynylpalladium complexes.

1.1 Aryl, Alkenyl, Benzyl, and Alkynyl Halides, and Their Pseudo-Halides

1.1.1 Reactions with Alkenes, and 1,2-, 1,3-, and 1,4-Dienes

1.1.1.1 Intermolecular reactions with alkenes. Organopalladium halides, formed by oxidative addition and lacking β-sp^3-bonded hydrogens, react with unhindered alkenes to form new alkenes in which an original vinyl hydrogen is replaced by the organic group of the intermediate Pd complexes. This transformation discussed in this section involves three elemental reactions: (1) oxidative addition of an organic halide to form **1**, (2) insertion of an alkene to form **2** (or carbopalladation of alkene), and (3) *syn* elimination of β-hydrogen to give an arylalkene or conjugated diene. When two β-hydrogens are present, generally *syn* elimination takes place to lead to (*E*)-alkenes. Since strong acids are formed, the reaction must be carried out in the presence of bases. Alkali metal carbonates, acetates, and secondary or tertiary amines are usually used. After being discovered by two groups[7,8], this reaction, which is now called the Heck reaction, has been used extensively in organic synthesis. [9,10] This very useful transformation is possible only with Pd; other transition metals are almost ineffective in giving similar products in acceptable yields.

Success of the reactions depends considerably on the substrates and reaction conditions. Rate enhancement in the coupling reaction was observed under high pressure (10 kbar)[11]. The oxidative addition of aryl halides to Pd(0) is a highly disfavored step when powerful electron donors such as OH and NH_2 reside on aromatic rings. Iodides react smoothly even in the absence of a ligand, and bromides in the presence of a phosphine ligand. The reaction of bromoiodobenzene (3) with two different alkenes can be carried out stepwise in the absence and then in the presence of a phosphine to afford the dialkenyl-benzene 4[12].

Stereochemical features in the oxidative addition and the elimination of β-hydrogen of cyclic and acyclic alkenes are different. The insertion (palladation) is *syn* addition. The *syn* addition (carbopalladation) of R—Pd—X to an acyclic alkene is followed by the *syn* elimination of β-hydrogen to give the *trans*-alkene 6, because free rotation of 5 is possible with the acyclic alkene. On the other hand, no rotation of the intermediate 7 is possible with a cyclic alkene and the *syn* elimination of β-hydrogen gives the allylic compound 8 rather than a substituted alkene.

Chlorides are inert. However, the reaction of *p*-chlorobenzophenone (9) with a styrene derivative proceeds satisfactorily at 150 °C by using dippb [1,4-bis(-diisopropylphosphino)butane] as a ligand to give the stilbene derivative 10. However, dippp [1,3-bis(diisopropylphosphino)propane] is an ineffective ligand[13]. On the other hand, the coupling of chlorobenzene with styrene proceeds in the presence of Zn under base-free conditions to afford the *cis*-stilbene 11 as a main product with evolution of H_2. As the ligand, dippp is active and dippb is ineffective. Zn is a reducing agent of (dippp)PdCl$_2$, formed by the coupling reaction, to generate (dippp)Pd(0)[14]. Aryl chlorides are activated by the coordination of the strongly electron-withdrawing Cr(CO)$_3$ **12**, and react with an alkene. Deprotection with iodine affords the coupled product **13**[2,3].

Chlorobenzene reacts with alkenes with bimetallic catalyses of Ni and Pd. Chlorobenzene is converted *in situ* into iodobenzene (**14**) by the Ni-catalyzed reaction of NaI at 140 °C. NiBr$_2$, rather than the Ni(0) complex, is found to be a good catalyst. Then the Pd-catalyzed reaction of the iodobenzene with acrylate takes place[15].

For insertion, alkenes bearing EWGs (electron-withdrawing groups) are most reactive. Extensive studies have been carried out to find the optimum conditions of the insertion. The reaction of triflates and alkenes bearing an EWG proceeds smoothly using bidentate nitrogen ligands, such as bipyridyl and phenanthroline[16]. The reaction proceeds smoothly by the addition of KHCO$_3$ and Bu$_4$NCl in DMF at room temperature[17,18]. Ag salts accelerate the reaction, possibly by removing halide ions strongly attached to Pd from a coordination sphere and replacing them with weak ligands such as acetate ion, making the insertion easier[19,20]. Ag_2CO_3 is used to avoid double bond isomerization. Thallium salts also accelerate the reaction[21,22] and suppress the double bond isomerization[23]. Also, addition of some alkali metal halides such as LiCl and KBr has a favorable effect on the reaction. DMF, MeCN, and other polar solvents are preferred solvents. Evidence has been accumulated to show that water is a good solvent in the presence of $NaHCO_3$ or K_2CO_3 as a base and in the absence of a phosphine ligand for some aryl iodides such as *o*-iodobenzoic acid (**15**). Reaction of acrylic acid with *o*- or *m*-iodobenzoic acids catalyzed by $Pd(OAc)_2$ is carried out smoothly in water as a sole solvent to give cinnamic acid derivatives in high yields. The reaction of acrylonitrile proceeds as a slurry in water to give the substituted cinnamnitrile **16** in 94% yield[24]. The addition of K_2CO_3 and quaternary ammonium salts (Bu_4NCl, Bu_4NBr, Bu_4NHSO_4) is important for increasing the rate of the reaction in aqueous solution (MeCN–H_2O, 10 : 1), and the reaction of water-insoluble substrates can be carried out even in pure water by the addition of these salts[25]. The reaction of iodides of nitrogen heterocycles with 2,3-dihydrofuran proceeds only in aqueous DMF or EtOH (1:1); no reaction takes place in pure DMF or EtOH[26]. The reaction can also be carried out in an aqueous solution by use of a water-soluble sulfonated phosphine (TMSPP)[27].

16, trans:cis = 4:1

The reaction of iodobenzene with acrylate is a good synthetic method for the cinnamate **17**[7]. In the competitive reaction of acrylate with a mixture of *o*- and *p*-iodoanisoles (**18** and **19**), the *o*-methoxycinnamate **20** was obtained selectively owing to the molecular recognition by interlamellar montmorillonite ethylsilyldiphenylphosphine (L) as a heterogenized homogeneous catalyst used as a ligand[28].

In the reaction of *o*-dibromobenzene, disubstitution is faster than monosubstitution, and the disubstituted product is obtained as a major product. Similarly, 1,2,4,5-tetrabromobenzene reacts with styrene to give the tetrasubstituted product[29]. The bridged, annulated [2,2]paracyclophanediene **22** was

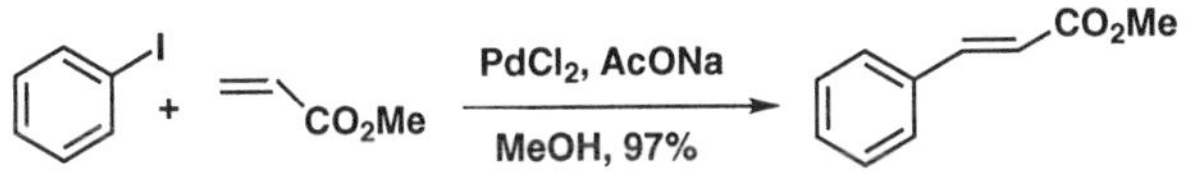

17

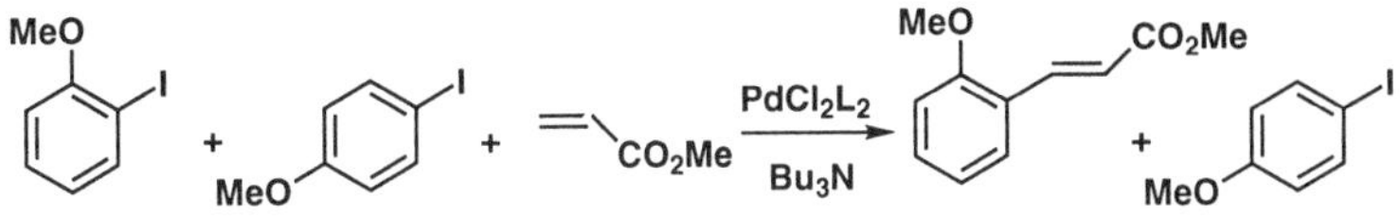

18   19   20   19

100 : 95

prepared based on the facile reaction of the tetrabromide **21** with styr-ene[30,31]. Furthermore, complete substitution of all bromines in hexabromo-benzene with styrene took place in good yields to give **23** and **24**[10].

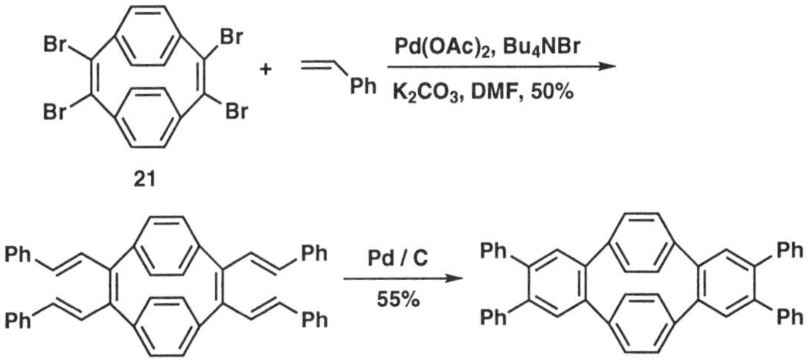

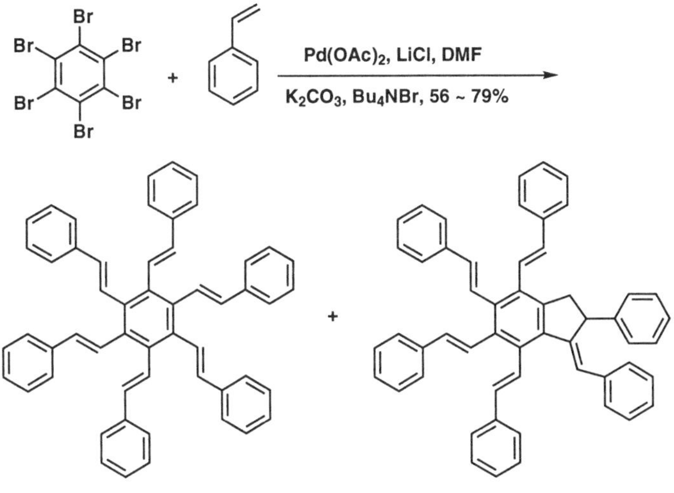

21

22

23   24

The coupling of 1,8-diiodonaphthalene (**25**) with acenaphthylene (**26**) affords acenaphth[1,2-a]acenaphthylene (**27**). It should be noted that the reaction involves unusual *trans* elimination of H—Pd—I[32]. This tetrasubstituted double bond in **27** reacts further with iodobenzene to give the [4, 3, 3]propellane **28** in 72%. This unusual reaction may be accelerated by strain activation, although it took 14 days[33].

2,4-Diene carboxylates can be prepared by the reaction of alkenyl halides with acrylates[34]. For example, pellitorine (**30**) is prepared by the reaction of 1-heptenyl iodide (**29**) with an acrylate[35]. Enol triflates are reactive pseudohalides derived from carbonyl compounds, and are utilized extensively for novel transformations. The 3,5-dien-3-ol triflate **31** derived from a 4,5-unsaturated 3-keto steroid is converted into the triene **32** by the reaction of methyl acrylate[36].

A mixture of *E* and *Z* forms is obtained by the reaction of the (*Z*)-alkenyl bromide **33**. *Z* to *E* isomerization takes place. However, the reaction is remarkably accelerated by using K_2CO_3 instead of $KHCO_3$[17] and Bu_4NCl in DMF, and the reaction of the (*Z*)-iodide **34** proceeds rapidly at room temperature without isomerization[37].

Cyclic alkenes give different regioisomers depending on the reaction conditions owing to double bond isomerization caused by *syn* elimination of Pd—H species and its readdition. The following three reaction conditions were tested for the reaction of cycloheptene (**35**)[18,38]:

Procedure A: 2.5% $Pd(OAc)_2$, n-Bu_4NCl, 3 AcOK
Procedure B: 3% $Pd(OAc)_2$, 9% Ph_3P, 2 Ag_2CO_3
Procedure C: 2.5% $Pd(OAc)_2$, 2.5% Ph_3P, n-Bu_4NCl, 3 AcOK.

Addition of Ag_2CO_3 cleanly suppresses the double bond isomerization[20]. Thus, with procedure B, no isomerization of double bond is observed. Addition of TlOAc also suppresses the double bond isomerization[23].

On the other hand, in the stepwise reaction of 2,3-dihydrofuran (**36**) with two different aryl iodides by procedure C, complete isomerization of the double bond around the ring takes place. The reaction was applied to the synthesis of an inhibitor of a blood platelet-activating factor **37**[39]. The asymmetric arylation of 2,3-dihydrofuran (**36**) with phenyl triflate using BINAP as a chiral ligand gave 2-phenyl-2,3-dihydrofuran (**39**) as a major product in 96% *ee* in the presence of 1,8-bis(dimethylamino)naphthalene (**38**) as a base[40,41]. The coupling of the 7-heteroatom norbornadiene derivative **40** with an aryl iodide proceeds with C—O bond cleavage rather than elimination of β-hydrogen to afford **41**, generating Pd(II), which is reduced to Pd(0) with Zn[42].

3-Butenoic acid and 4-pentenoic acid (**42**) react with alkenyl halides or tri-flates to afford γ-alkenyl-γ-lactones and the δ-alkenyl-δ-valerolactone **44** via the π-allylpalladium intermediate **43** formed by the elimination of Pd—H and its readdition in opposite regiochemistry using a phosphine-free Pd cata-lyst[43].

By the reaction of bromotoluene with ethylene under pressure, *p*-methylstyrene and stilbene (**45**) are obtained[44,45]. A polymer **47** is obtained by the reaction of *p*-bromostyrene (**46**) with ethylene. The reaction has been applied to polymer synthesis[46]. One example is the reaction of 1,4-divinylbenzene (**48**) with 9,10-dibromoanthracene to give the oligo(arylenevinylene)s **49**[47].

For the clean synthesis of styrene derivatives, vinyltrimethylsilane (**50**) is used. Vinyltrimethylsilane (**50**) reacts with aryl and alkenyl iodides[48] and arenediazonium tetrafluorobotrates[49] to give a 1,4-diene and the styrene **51** by the elimination of the silyl group under the usual conditions. This is a better synthetic method for styrene derivatives than the reaction of ethylene itself. On the other hand, alkenylsilanes are obtained by the reaction of vinyltrimethylsilane (**50**) in the presence of AgNO$_3$[50]. The alkenylsilane **52** is also obtained by the reaction of vinylsilane bearing electronegative substituents on Si without using AgNO$_3$[51]. (*Z*)-Trimethyl(2-phenylvinyl)silane (**55**) is obtained stereoselectively by the phenylation of (*E*)-1,2-bis(trimethylsilyl)ethylene (**53**)[52]. The regioselectivity and stereochemistry of this reaction are explained by the *syn* desilylpalladation of the intermediate complex **54**.

In the reaction of α,β-unsaturated ketones and esters, sometimes simple Michael-type addition (insertion and hydrogenolysis, or hydroarylation, and hydroalkenylation) of alkenes is observed[53,54]. For example, a simple addition product **56** to methyl vinyl ketone was obtained by the reaction of the heteroaromatic iodide **55**[55]. The corresponding bromide affords the usual insertion–elimination product. Saturated ketones are obtained cleanly by hydroarylation of α,β-unsaturated ketones with aryl halides in the presence of sodium formate, which hydrogenolyses the R—Pd—I intermediate to R—Pd—H[56]. Intramolecular hydroarylation is a useful reaction. The diiodide **57** reacts smoothly with sodium formate to give a model compound for the aflatoxin **58**. (see Section 1.1.6)[57]. Use of triethylammonium formate and Bu₄NCl gives better results.

The reactivity of alkenes with electron-donating groups is lower than those with EWGs. The reaction of alkenes with electron-donating groups such as vinyl ethers, vinyl esters, enamides, and enamines gives two regioisomers depending on the substrates and reaction conditions[58,59]. The α-substituted products **59**, **60**, and **61** seem to be common products, which can be converted into ketones by hydrolysis[60–64]. Triflates are the best leaving group, and nearly complete α-selectivity is observed by use of bidentate ligands such as dppe, dppp, and dppb, among which, dppp is the best. With this ligand, complete α-substitution is observed even with allyl alcohol. Ph_3P gives a lower α-selectivity. The reaction is slower with aryl iodides and bromides, and can be accelerated by the addition of $AgNO_3$, AgOTf, and TlOAc. The high α-regioselectivity with aryl triflates is explained in the following way. The oxidative addition of aryl triflate affords a cationic complex and an electronic effect of substituted alkene is more important than a steric effect in the subsequent insertion step, and hence the α-substitution becomes a main path. On the other hand, aryl iodides are more reactive than triflates toward electron-withdrawing alkenes such as acrylate or acrylonitrile. In this case, both steric and electronic effects favor β-substitution[64]. 3,4-Dihydro-2H-pyran (**62**) reacts smoothly with aryl iodides, and the aryl group is introduced at α regioselectively to give **63** after double bond migration using $Pd(OAc)_2$ and $Ph_3P[65]$.

Under certain conditions, the β-substitution products are obtained as major products[66]. Methyl vinyl ether reacts with bromonitrobenzene to give the β-methoxystyrene **64** in good yield in toluene at 120 °C by using Pd on carbon as a catalyst without a phosphine ligand. This reaction is used for the preparation of metoprolol, which is an important β-blocker[67].

Vinyl acetate reacts with the alkenyl triflate **65** at the β-carbon to give the 1-acetoxy-1,3-diene **66**[68]. However, the reaction of vinyl acetate with 5-iodo-pyrimidine affords 5-vinylpyrimidine with elimination of the acetoxy group[69]. Also stilbene (**67**) was obtained by the reaction of an excess of vinyl acetate with iodobenzene when interlamellar montmorillonite ethylsilyl-diphenylphosphine (L) palladium chloride was used as an active catalyst[70]. Commonly used $PdCl_2(Ph_3P)_2$ does not give stilbene.

Methyl α-acetamidoacrylate (**68**) reacts with aryl or alkenyl halides and triflates to give β-aryl or alkenyl dehydroamino acid analogs, which are hydro-genated to give the α-amino acids **69**[71–73]. Arylation and alkenylation of methyl α-methoxyacrylate (**70**) offers a good synthetic method for β-aryl- or β-alkenyl-α-keto esters **71** after coupling and hydrolysis[74]. Reaction of *N*-vinyl amides with aryl halides proceeds smoothly to give a mixture of two regio-isomers[75]. The reaction of the *N*-vinyloxazolone **72** with iodobenzene and subsequent hydrogenation of the product offer a synthetic method for β-phe-nethylamine (**73**)[75a].

An α-arylalkanoate is prepared by the reaction of aryl halide or triflate with the ketene silyl acetal **74** as an alkene component. However, the reaction is explained by transmetallation of Ph—Pd—Br with **74** to generate the Pd enolate **75**, which gives the α-arylalkanoate by reductive elimination[76].

Benzyl chloride reacts with alkenes bearing an EWG[8]. The reaction with acrylate proceeds smoothly to give γ-phenylcrotonate (**76**) in the presence of Bu$_3$N without a ligand. No reaction takes place with Pd(Ph$_3$P)$_4$[77].

Trifluoroacetimidoyl iodide **77** can be used as a ketone synthon and reacts smoothly with acrylate to give **78**, which is hydrolyzed to give an unsaturated ketone **79**[78].

Norbornene (**80**) and related cyclic alkenes are reactive alkenes and undergo sequential reactions specific to these systems with aryl halides to give polycyclic compounds. After the first insertion, further reactions to form varieties of products take place depending on the reaction conditions, because there is no possibility of the *syn* elimination of β-hydrogen in **81**. As examples, reaction of an alkenyl halide in the presence of formic acid gives **82** by intermolecular insertion and hydrogenolysis, and the cyclopropane **83** by subsequent intramolecular insertion and hydrogenolysis[79,80]. The four-membered ring compound **84** and the polycycle **85** and other products were obtained by the reaction of aryl halides in anisole using Pd(Ph₃P)₄ and *t*-BuOK[81,82].

Under different conditions [Pd(OAc)$_2$, K$_2$CO$_3$, Bu$_4$NBr, NMP], the 1 : 3 coupling product **86** with 4-aryl-9,10-dihydrophenanthrene units was obtained. The product **86** was transformed into a variety of polycyclic aromatic compounds such as **87** and **88**[83]. The polycyclic heteroarene-annulated cyclopentadiene **90** is prepared by the coupling of 3-iodopyridine and dicyclopentadiene (**89**), followed by retro-Diels–Alder reaction on thermolysis[84].

Three-component coupling with vinylstannane, norbornene (**80**), and bromobenzene affords the product **91** via oxidative addition, insertion, transmetallation, and reductive elimination[85]. Asymmetric multipoint control in the formation of **94** and **95** in a ratio of 10 : 1 was achieved by diastereo-differentiative assembly of norbornene (**80**), the (*S*)-(*Z*)-3-siloxyvinyl iodide **92** and the alkyne **93**, showing that the control of four chiralities in **94** is possible by use of the single chirality of the iodide **92**. The double bond in **92** should be *Z*; no selectivity was observed with *E* form[86].

When allylic alcohols are used as an alkene component in the reaction with aryl halides, elimination of β-hydrogen takes place from the oxygen-bearing carbon, and aldehydes or ketones are obtained, rather than γ-arylated allylic alcohols[87,88]. The reaction of allyl alcohol with bromobenzene affords dihydrocinnamaldehyde. The reaction of methallyl alcohol (**96**) with aryl halides is a good synthetic method for dihydro-2-methylcinnamaldehyde (**97**).

The reaction of allylic alcohols with alkenyl halides is complicated by the elimination of Pd—H and its readdition in the reverse direction to form a sufficiently stable π-allylpalladium complex. The reaction of an alkenyl halide with a primary allylic alcohol to give the 4-enal **98** cleanly can be achieved in MeCN in the presence of Ag_2CO_3, Bu_4NHSO_4, and a catalytic amount of $Pd(OAc)_2$[89]. On the other hand, the conjugated dienol **99**, rather than the 4-enal, is obtained when the alkenyl halide is treated with allylic alcohol in the presence of AgOAc, or Ag_2CO_3 and $Pd(OAc)_2$ in DMF[90]. The conjugated dienol **100** was obtained with high selectivity by the reaction of primary and secondary allylic alcohols with enol triflates by using of $Pd(OAc)_2$, tri(o-tolyl)phosphine, and Et_3N in DMF[91].

$$C_6H_{13}\diagup\!\!\!\diagdown I \;+\; \diagup\!\!\!\diagdown OH \xrightarrow[\text{Bu}_4\text{NHSO}_4,\ \text{MeCN, 54\%}]{\text{Pd(OAc)}_2,\ \text{Ag}_2\text{CO}_3} \; C_6H_{13}\diagup\!\!\!\diagdown\!\!\!\diagdown CHO$$

98

$$C_6H_{13}\diagup\!\!\!\diagdown I \;+\; \diagup\!\!\!\diagdown OH \xrightarrow[\text{DMF, 61\%}]{\text{Pd(OAc)}_2,\ \text{AgOAc}} \; C_6H_{13}\diagup\!\!\!\diagdown\!\!\!\diagdown OH$$

99

$$\text{(steroid, C}_8\text{H}_{17}\text{, TfO–)} \;+\; \diagup\!\!\!\diagdown OH \xrightarrow[\text{Et}_3\text{N, K}_2\text{CO}_3,\ \text{DMF, 74\%}]{\text{Pd(OAc)}_2,\ \text{(o-tol)}_3\text{P}} \; \text{(steroid, C}_8\text{H}_{17}\text{, HO–)}$$

100

The prochiral *meso* form of 2-cyclopenten-1,4-diol (**101**) reacts with the (*Z*)-alkenyl iodide **102** to give the 3-substituted cyclopentanone **103** with nearly complete diastereoselectivity (98 : 2)[92]. The reaction is used for the synthesis of prostaglandin. The alkenyl iodide **102** must be in the *Z* form in order to obtain the high diastereoselectivity. The selectivity is low when the corresponding (*E*)-alkenyl iodide is used[93].

101   OTBDMS 102

$$\xrightarrow[\text{MeCN}]{\text{Pd(OAc)}_2} \; \left[\; \text{I-Pd}\diagup\!\!\!\diagdown\text{O-TBDMS, C}_5\text{H}_{11}\;\right] \xrightarrow[\text{83\%}]{\text{Na}_2\text{CO}_3} \; 103$$

The reaction of a halide with 2-butene-1,4-diol (**104**) affords the aldehyde **105**, which is converted into the 4-substituted 2-hydroxytetrahydrofuran **106**, and oxidized to the 3-aryl-γ-butyrolactone **107**[94]. Asymmetric arylation of the cyclic acetal **108** with phenyl triflate[95] using Pd–BINAP afforded **109**, which was converted into the 3-phenyllactone **110** in 72% *ee*[96]. Addition of a molecular sieve (MS3A) shows a favorable effect on this arylation. The reaction of the 3-siloxycyclopentene **111** with an alkenyl iodide affords the silyl enol ether **112** regioselectively[97]. Similarly, an enol ether is obtained as a main product from 3-methoxycyclohexene[98].

In addition to allylic alcohols, other unsaturated alcohols react with halides to give carbonyl compounds. For example, although the reaction is slow (3 days), 10-undecenyl alcohol (**113**) reacts with iodobenzene to give 11-phenyl-undecanal (**117**). The migration of the Pd by the elimination of H—Pd—I to generate **114** and its readdition to the double bond with different regiochem-istry to give **115** are repeated until the Pd reaches to the terminal position **116**, forming the aldehyde **117** irreversibly[99]. The surprisingly efficient and repeated elimination of H—Pd—I and its readdition clearly show that 'H—Pd—I' is a living species.

As expected, the formation of a carbonyl group is not possible with *tert*-allylic alcohols. Although the aromatic ring bears electron-donating groups, the 2,2-disubstituted chromene **119** was formed smoothly with the *tert*-allylic alcohol **118**[100].

Aryl or alkenyl halides attack the central carbon of the allene system in the 2,3-butadien-1-ol **120** to form the π-allyl intermediate **121**, which undergoes elimination reaction to afford the α,β-unsaturated ketone **122** or aldehyde. The reaction proceeds smoothly in DMSO using dppe as a ligand[101].

Allylic amines are coupled to halides giving either allylic amines or enamines depending on the reaction condition. Reaction of steroidal dienyl triflate with Boc-diprotected allylamine affords allylamine. Use of AcOK as a base is crucial for the clean coupling[102]. The *tert*-allylic amine **123** reacts with an aryl halide to give the enamine **125** in DMF and allylic amine **124** in nonpolar solvents[103].

The alkenyloxirane **126** in excess reacts with aryl and alkenyl halides or triflates in the presence of sodium formate to afford the allylic alcohol **127**[104]. Similarly, the reaction of the alkenyloxetane **128** gives the homo-allylic alcohol **130**[105]. These reactions can be explained by insertion of the double bond in the Ar—Pd bond, followed by ring opening (or β-elimination) to form the allylic or homoallylic alkoxypalladium **129**, which is converted into the allylic **127** or homoallylic alcohol **130** by the reaction of formate. The 3-alkenamide **132** was obtained by the reaction of the 4-alkenyl-2-azetizinone **131** with aryl iodide and sodium formate [106].

The unconjugated alkenyl oxirane **133** reacts with aryl halides to afford the arylated allylic alcohol **134**. The reaction is explained by the migration of the Pd via the elimination and readdition of H—Pd—I[107].

PhI + + HCO₂Li

$$\xrightarrow[\text{DMF, i-Pr}_2\text{NEt, 1.5 day, 62\%}]{\text{Pd(OAc)}_2,\ \text{Bu}_4\text{NCl, LiCl}}$$

133

Ph $\diagup\diagup\diagup\diagup$ OH

E / Z = 76 / 24   134

An efficient carboannulation proceeds by the reaction of vinylcyclopropane (**135**) or vinylcyclobutane with aryl halides. The multi-step reaction is explained by insertion of alkene, ring opening, diene formation, formation of the π-allylpalladium **136** by the readdition of H—Pd—I, and its intramolecular reaction with the nucleophile to give the cyclized product **137**[108].

The arylpalladium halide complex **139** is an intermediate of the decarbonylation of aroyl halides (see Section 1.2)[109] and it undergoes facile alkene insertion. Therefore, similarly to aryl halides, acyl halides can be used for the alkene insertion[110]. Addition of a tertiary amines having a pK value in the range 7.5–11, such as tributylamine and benzyldimethylamine, as a base is crucial. The reaction is carried out without a phosphine ligand, because Ph₃P was found to inhibit the reaction. Since aroyl chlorides are more reactive than aryl iodides, the following chemoselective reaction is possible. The aroyl chloride moiety in *p*-bromobenzoyl chloride (**138**) is converted into the arylpalladium **139**, which reacts first with methyl acrylate using Pd(OA)₂ as a catalyst to give methyl *p*-bromocinnamate (**140**). Then acrylonitrile can be introduced at the *para*-position to give **141** when tri(*o*-tolyl)phosphine is added. Similarly, chlorides of α, β-unsaturated acids undergo decarbonylation–insertion. The reaction of acrylate with fumaroyl dichloride (**142**) affords the octatriendioate **143**[111].

Oxidative addition of the sulfonyl chlorides **144** is followed by facile generation of SO_2 to form arylpalladium complexes which undergo alkene insertion[112,113].

The diazonium salts **145** are another source of arylpalladium complexes[114]. They are the most reactive source of arylpalladium species and
the reaction can be carried out at room temperature. In addition, they can
be used for alkene insertion in the absence of a phosphine ligand using
$Pd_2(dba)_3$ as a catalyst. This reaction consists of the indirect substitution reaction of an aromatic nitro group with an alkene. The use of diazonium salts is
more convenient and synthetically useful than the use of aryl halides, because
many aryl halides are prepared from diazonium salts. Diazotization of the
aniline derivative **146** in aqueous solution and subsequent insertion of acrylate
catalyzed by $Pd(OAc)_2$ by the addition of MeOH are carried out as a one-pot
reaction, affording the cinnamate **147** in good yield[115]. The *N*-nitroso-*N*-
arylacetamide **148** is prepared from acetanilides and used as another precursor
of arylpalladium intermediate. It is more reactive than aryl iodides and
bromides and reacts with alkenes at 40 °C without addition of a phosphine
ligand[116].

PhN$_2$BF$_4$ + Ph—CH=CH$_2$ →[Pd$_2$(dba)$_3$ / aq. MeCN, 94%] Ph—CH=CH—Ph

145

146

NH$_2$ / OMe + HNO$_2$ →[HBF$_4$ / H$_2$O] →[CH$_2$=CH—CO$_2$Et , Pd(OAc)$_2$ / MeOH, 68%] MeO—C$_6$H$_4$—CH=CH—CO$_2$Et

147

NHAc / Br →[N$_2$O$_3$] N(NO)Ac / Br **148** + Ph—CH=CH$_2$ →[Pd(dba)$_2$ / MeCN, 68%] Br—C$_6$H$_4$—CH=CH—CH=CH—Ph

A σ-aryl–Pd bond is formed by the transfer of an aryl group even from arylphosphines to Pd and alkene insertion takes place[117–119]. This reaction is slow and it is not a serious problem when triarylphosphine is used as a ligand. The cinnamate **149** is obtained by the reaction of Ph$_3$P with acrylate in the presence of Pd(OAc)$_2$ in AcOH.

Ph$_3$P + CH$_2$=CH—CO$_2$Me →[Pd(OAc)$_2$] Ph—CH=CH—CO$_2$Me

149

The alkynyl iodide **150** undergoes the oxidative addition to form an alkynylpalladium iodide, and subsequent insertion of an alkene gives the conjugated enyne **151** under phase-transfer conditions[120].

C$_6$H$_{13}$—C≡C—I + CH$_2$=CH—CO$_2$Me →[Pd(OAc)$_2$, K$_2$CO$_3$ / Bu$_4$NCl, DMF, 55%] C$_6$H$_{13}$—C≡C—CH=CH—CO$_2$Me

150   **151**

The alkenyl moiety, rather than the aryl moiety, in the aryl(alkenyl)iodonium salt **152** reacts smoothly with alkenes under mild conditions[121].

Ph—CH=CH—I$^+$—Ph, OTs$^-$ + CH$_2$=CH—CO$_2$Me →[Pd(OAc)$_2$, DMF / NaHCO$_3$, rt., 75%] Ph—CH=CH—CH=CH—CO$_2$Me

152

1.1.1.2 Intramolecular reactions with alkenes. While the intermolecular reaction is limited to unhindered alkenes, the intramolecular version permits the participation of even hindered substituted alkenes, and various cyclic compounds are prepared by the intramolecular Heck reaction. Particularly the tandem insertion reaction of double bonds of halo polyenes lacking β-sp^3-bonded hydrogens produces polycyclic compounds. Extensive studies have been carried out on this useful cyclization reaction. In early studies, indoles, the indoleacetate **153**, and the indolohydroquinones **154a** and **154b** were produced[122–124].

The stereospecific synthesis of an A ring synthon of 1α-hydroxyvitamin D has been carried out. The (*E*)-alkene is cyclized to give the (*E*)-*exo*-diene **155**, and the (*Z*)-alkene affords the (*Z*)-*exo*-diene **156** stereospecifically[125,126]. These results can be understood by the *cis* addition and *syn* elimination mechanism.

Various bicyclic and polycyclic compounds are produced by intramolecular reactions[127]. In the syntheses of the decalin systems **157** [38] and **158** [128], *cis* ring junctions are selectively generated. In the formation of **158**, allylic silyl ether remains intact. A bridged bicyclo[3.3.1]nonane ring **159** was constructed in a good yield[129]. Cyclization to form the seven-membered ring **160**[130] and eight-membered ring **161**[131] with an *exo*-methylene group proceeds with high yields. In the latter case, oxidation of alcohol to ketone takes place during the cyclization.

The intramolecular version for synthesizing cyclic and polycyclic compounds offers a powerful synthetic method for naturally occurring macrocyclic and polycyclic compounds, and novel total syntheses of many naturally occurring complex molecules have been achieved by synthetic designs based on this methodology. Cyclization by the coupling of an enone and alkenyl iodide has been applied to the synthesis of a model compound of 16-membered car-bomycin B **162** in 55% yield. A stoichiometric amount of the 'catalyst' was used because the reaction was carried out under high dilution conditions[132].

Synthesis of camptothecin (**163**) is another example[133]. The iboga alkaloid analog **164** has been synthesized smoothly by the intramolecular coupling of iodoindole and unsaturated ester to form an eight-membered ring. *N*-Methyl protection of the indole is important for a smooth reaction[134]. An efficient construction of the multifunctionalized skeleton **165** of congeners of FR900482 has been achieved[135].

The cyclization to form very congested quaternary carbon centers involving the intramolecular insertion of di-, tri-, and tetrasubstituted alkenes is particularly useful for natural products synthesis[136–138]. In the total synthesis of gelsemine, the cyclization of **166** has been carried out, in which very severe steric hindrance is expected. Interestingly, one stereoisomer **167** was obtained in 95% yield with high stereoselectivity when $Pd_2(dba)_3$ without added phosphine ligand was used. On the other hand, the other isomer **168** was obtained almost exclusively when Ag_3PO_4 was added. This high stereoselectivty is due to the coordination of the angular vinyl group during the insertion step, which is enhanced by the dissociation of bromide anion from Pd by the added Ag salt. The importance of the vinyl group is demonstrated by the fact that an equimolar mixture of the cyclized products corresponding to **167** and **168** is obtained when the vinyl group is displaced with an ethyl group. Addition of phosphine ligands such as dppe or dppf also gives poor stereoselection[139].

Total syntheses of tazettine and 6a-epi-pretazettine have been carried out by application of the Pd-catalyzed cyclization of **169**, in which a single pentacyclic product **170** was obtained, establishing a preference for an eclipsed orientation of the Pd—C σ-bond and alkene π-bond in the key intramolecular insertion step[140].

In the synthesis of morphine, bis-cyclization of the octahydroisoquinoline precursor **171** by the intramolecular Heck reaction proceeds using palladium trifluoroacetate and 1,2,2,6,6-pentamethylpiperidine (PMP). The insertion of the diene system forms the π-allylpalladium intermediate **172**, which attacks the phenol intramolecularly to form the benzofuran ring (see Section 1.1.1.3). Based on this method, elegant total syntheses of (−)- and (+)-dihydrocodeinone and (−)- and (+)-morphine (**173**) have been achieved[141].

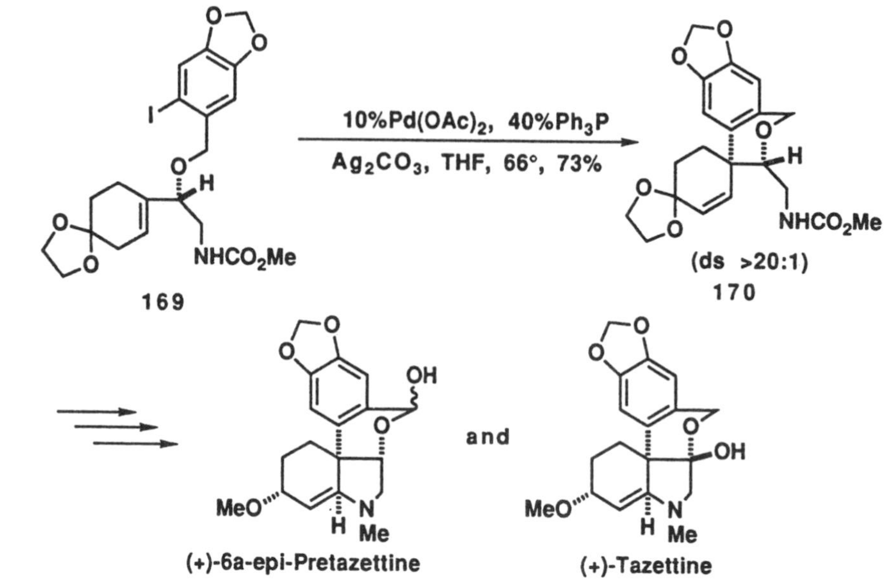

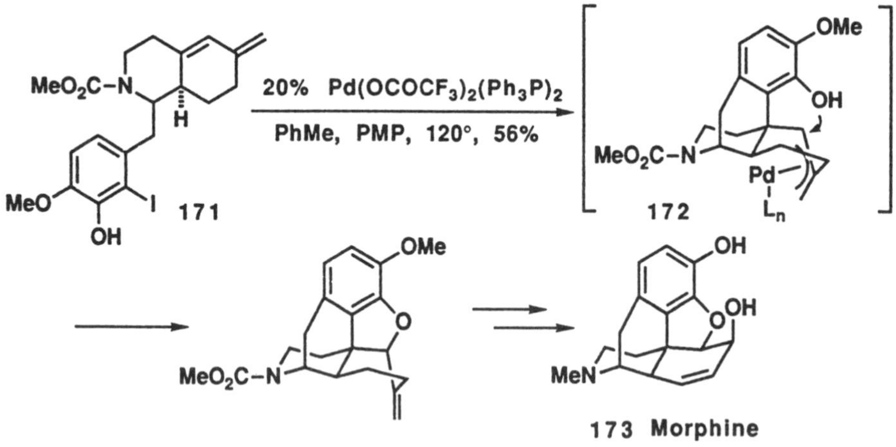

10%Pd(OAc)$_2$, 40%Ph$_3$P

Ag$_2$CO$_3$, THF, 66°, 73%

169

(ds >20:1)
170

(+)-6a-epi-Pretazettine   and   (+)-Tazettine

20% Pd(OCOCF$_3$)$_2$(Ph$_3$P)$_2$

PhMe, PMP, 120°, 56%

171

172

173 Morphine

Asymmetric cyclization using chiral ligands has been studied. After early attempts[142–144], satisfactory optical yields have been obtained. The hexahydropyrrolo[2,3-*b*]indole 176 has been constructed by the intramolecular Heck reaction and hydroarylation[145]. The asymmetric cyclization of the enamide 174 using (*S*)-BINAP affords predominantly (98 : 2) the (*E*)-enoxysilane stereoisomer of the oxindole product, hydrolysis of which provides the (*S*)-oxindole aldehyde 175 in 84% yield and 95% *ee*, and total synthesis of (−)-physostigmine (176) has been achieved[146].

Synthesis of a *cis*-decalin system by the asymmetric cyclization[38] has been carried out with high enantioselectivity[142,143,147,148]. Using BINAP as a chiral ligand, 91% *ee* was achieved in the asymmetric cyclization of **177** to give **178**. In order to achieve an efficient asymmetric cyclization, selection of the reaction conditions is crucial, and sometimes added Ag salts play an important role[148]. A catalytic asymmetric cyclization of **179** to prepare the key intermediate enone **180** for vernolepin synthesis has been carried out[149]. Highly efficient asymmetric cyclization of **181** to give the tetralin system **182** has been applied to the synthesis of (−)-eptazocine (**183**)[150]. Hydrindans are synthesized in 86% *ee*[151].

Asymmetric Heck reaction of the conjugated diene **184** and subsequent acetate anion capture of the π-allylpalladium intermediate afforded **185** in 80% *ee*, which was converted into the key intermediate **186** for the capnellenols. The choice of solvents is crucial in these asymmetric cyclizations. In this case, DMSO gives the best results[152]. An indolizidine derivative **188** was prepared in 86% *ee* by the asymmetric cyclization of the enamide **187** using (*R,S*)-BPPFOH in the presence of Ag-exchanged zeolite[147,153].

In these cyclizations, the reaction can be terminated in other ways than elimination of β-hydrogen. Typically the reaction ends by an anion capture process[154]. The following anion transfer agents are known: H^-, OAc^-, CN^-, SO_2Ph, $CH(CO_2R)_2$, NHR_2, CO/ROH, and RM [M = Sn(IV), B(III), Zn(II)]. Trapping with an amine after alkene insertion to give **189** and **190** is an example. *N*-Acetyl protection is important in this reaction[155].

In the presence of a carbon nucleophile in an alkene, the alkylpalladium, formed by the insertion, attacks the nucleophile intramolecularly without undergoing elimination of β-hydrogen to give the cyclized product. In this way, a new cyclopentanation of alkenes is possible by trapping with carbon nucleophiles. This reaction is understood also as intramolecular carbopalladation of alkene with σ-arylpalladium species as shown by **191**, followed by reductive elimination to give the cyclopentane derivative **192**[156]. This reaction involves the alkylative cyclization of malonate. As a completely intramolecular version of this strategy, the tricyclic triquinane skeleton **195** was constructed in one step from **193** by the intramolecular carbopalladation of the substituted cyclopentene with σ-alkenylpalladium as shown by **194**. A total synthesis of capnellene has been achieved based on this bis-cyclization reaction[157]. The cyclization of the cyclopentene even without a C-3-methyl group in **193** proceeds satisfactorily by trapping with the cyanoacetate instead of the malonate without undergoing elimination of β-hydrogen, thus offering a general solution to the synthesis of triquinanes[158]. As another possibility, cyclization–carbonylation proceeds smoothly at 1 atm in the presence of thallium(I) acetate[159]. Furthermore, the 2-iodoaniline derivative **196** undergoes insertion, carbonylation and trapping with carbon nucleophile to give the spirocycle **197**. Addition of rate-accelerating thallium(I) acetate enables this cascade process to proceed under mild conditions (1 atm, 80 °C)[160].

The alkylpalladium intermediate **198** cyclizes on to an aromatic ring, rather than forming a three-membered ring by alkene insertion[161]. Spirocyclic compounds are easily prepared[162]. Various spiroindolines such as **200** were prepared. In this synthesis, the second ring formation involves attack of an alkylpalladium species **199** on an aromatic ring, including electron-rich or -poor heteroaromatic rings[163].

A interesting and useful reaction is the intramolecular polycyclization reaction of polyalkenes by tandem or domino insertions of alkenes to give polycyclic compounds[138]. In the tandem cyclization, an intermediate in many cases is a neopentylpalladium formed by the insertion of 1,1-disubstituted alkenes, which has no possibility of β-elimination. The key step in the total synthesis of scopadulcic acid is the Pd-catalyzed construction of the tricyclic system **202** containing the bicyclo[3.2.1]octane substructure. The single tricyclic product **202** was obtained in 82% yield from **201**[20,164]. The benzyl chloride **203** undergoes oxidative addition and alkene insertion. Formation of the spiro compound **204** by the intramolecular double insertion of alkenes is an example[165].

In some cases, cyclopropanation is observed. Halodienes of the following types undergo the cyclopropanation. In case A, straightforward cyclopropanation of the first insertion product **205** generates **206** as a final product. In case B, the neopentylpalladium **207** is generated and undergoes alkene insertion to give the cyclopropane **208**, or trapped with nucleophile as another possibility. The intermediate **208** undergoes cyclopropylcarbinyl–homoallyl rearrangement from **208** to **210** and β-elimination to give the ring-expanded product **211**[166]. Intermolecular insertion of an alkene into the neopentylpalladium **208** to give **209** may also take place. As examples of case A, the strained three-membered ring compounds **213** and **215** are formed in the bicyclization reactions of **212** and **214**[161].

The product **211** may be formed directly by the *endo-trig* cyclization of the halodiene. The fact that the product **211** is formed by the cyclopropylcarbinyl–homoallyl rearrangement of **208**, not by direct *endo-trig* cyclization, is supported by the following observation. The reaction of the 3-indolyl triflate **216** with 1,4-pentadien-3-ol (**217**) afforded the aldehyde **221** as an unexpected product formed by skeletal change[167]. The reaction is explained by the formation of the cyclopropane **218** by alkene insertion, followed by the above-mentioned rearrangement to generate **219**, which is converted into the π-allyl intermediate **220** via the dienol. The origin of *E* stereochemistry of the double bond in the product **221** is understood by the more stable *syn* form of the π-allylpalladium intermediate **220**.

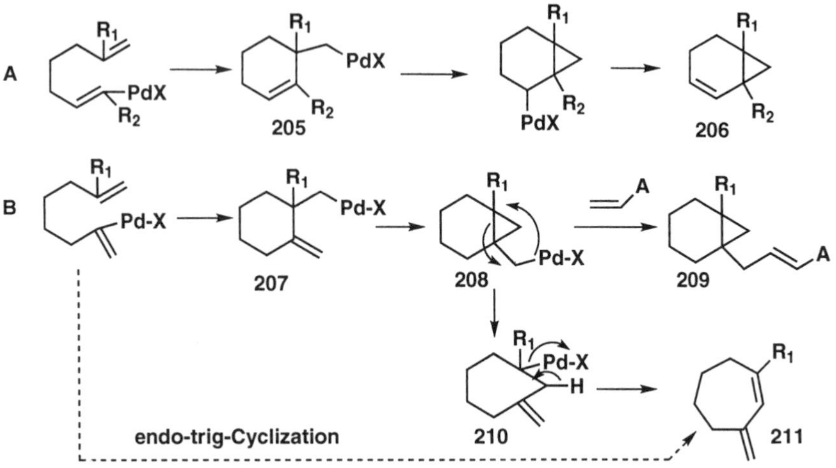
A
205
206
B
207
208
209
210
211
endo-trig-Cyclization

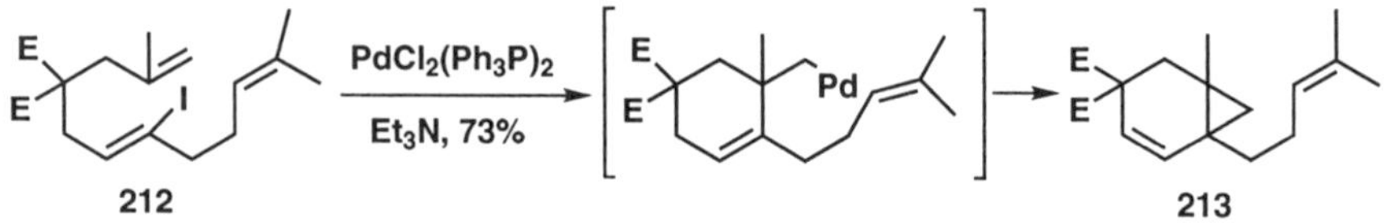
PdCl2(Ph3P)2
Et3N, 73%
212
213

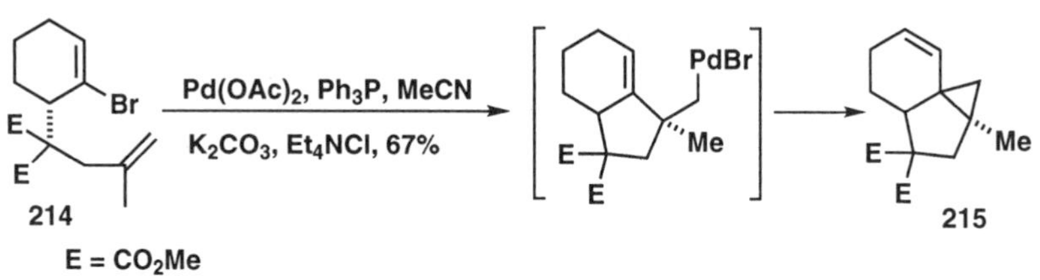
Pd(OAc)2, Ph3P, MeCN
K2CO3, Et4NCl, 67%
214
215
E = CO2Me

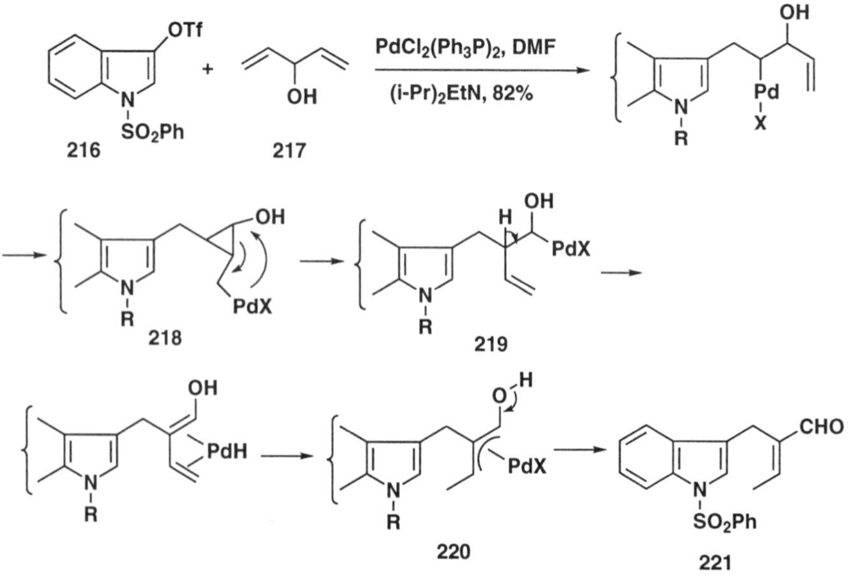
PdCl2(Ph3P)2, DMF
(i-Pr)2EtN, 82%
216
217
218
219
220
221

In an efficient diastereo-differentiative assembly of three components of norbornene, *cis*-alkenyl iodide, and KCN, the isomerization of the *cis* to the *trans* double bond takes place to give the coupled product **224**. The isomerization is explained by the formation of the cyclopropane **222**, its rearrangement to give a *trans* double bond in **223**, and trapping with CN anion to give **224**[168].

Furthermore, the cyclization of the iododiene **225** affords the six-membered product **228**. In this case too, complete inversion of the alkene stereochemistry is observed. The (Z)-allylic alcohol **229** is not the product. Therefore, the cyclization cannot be explained by a simple *endo* mode cyclization to form **229**. This cyclization is explained by a sequence of (i) *exo*-mode carbopalladation to form the intermediate **226**, (ii) cyclopropanation to form **227**, and (iii) cyclopropylcarbinyl to homoallyl rearrangement to afford the (E)-allylic alcohol **228**[166]. (For further examples of cyclopropanation and *endo* versus *exo* cyclization, see Section 1.1.2.2.)

A concise synthesis of the strychnos alkaloid (dehydrotubifoline) skeleton has been carried out. The skeleton **231** was constructed in a good yield by the intramolecular Heck reaction of **230** as a key reaction[169]. However, the cyclization of the indoline carbamate **232**, instead of the unprotected **230**, afforded an unexpected product, **235**. In this case, the σ-alkylpalladium intermediate **233** does not undergo elimination of β-hydrogen, owing to stabilization by the chelating effect of the carbamate group. Instead, in this case also, the cyclopropane **234** is formed, followed by 120° rotation and rearrangement to give the unexpected product **235**[170].

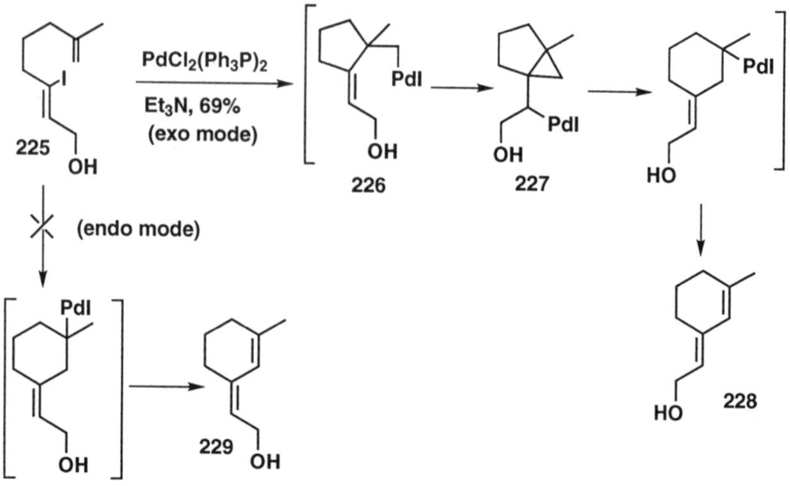
PdCl2(Ph3P)2
Et3N, 69%
(exo mode)
(endo mode)
PdI
I
OH
HO
225
226
227
228
229

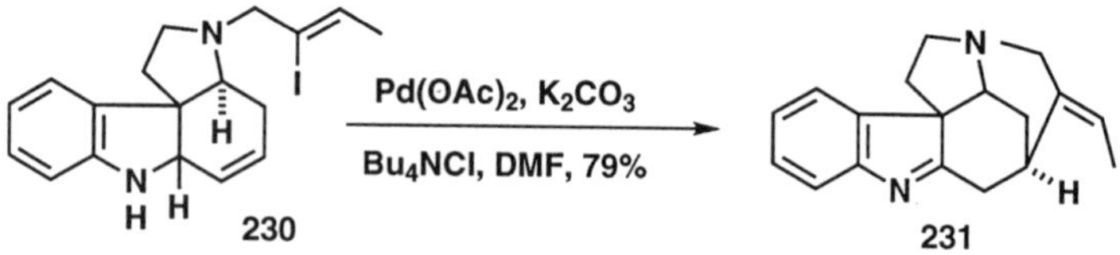
Pd(OAc)2, K2CO3
Bu4NCl, DMF, 79%
N
H
I
230
231

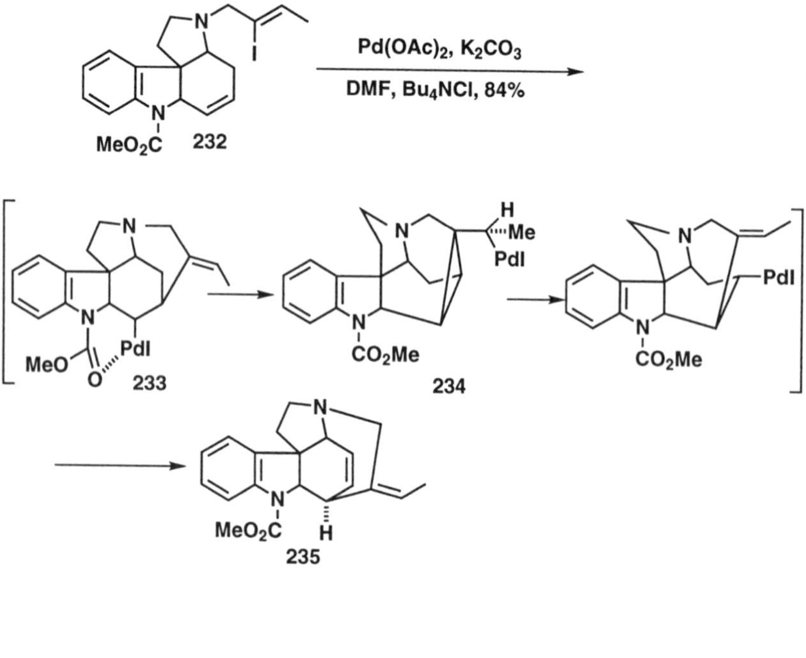
Pd(OAc)2, K2CO3
DMF, Bu4NCl, 84%
MeO2C
CO2Me
PdI
Me
232
233
234
235

In the synthesis of benzoprostacyclin analogue, the first step is the reaction of the ene oxide **237** with the phenol **236** (see Section 2. 2.1.2). Then inter- and intramolecular tandem insertions of two alkenes take place to give **240**. The important aspect is that *syn* elimination of β-hydrogen from the Pd–alkyl intermediate **238**, formed by the intramolecular insertion of the alkene, is impossible, because there is no hydrogen of the same stereochemistry (α-oriented) as the Pd at the β-positions of Pd, and hence the intermolecular insertion of 1-octen-3-one (**239**) proceeds smoothly[171]. As another example, a neopentylpalladium is trapped with organo tin, boron, and zinc reagents. Phenylation of the neopentylpalladium **241** with sodium tetraphenylborate afforded **242**[172]. Hydrogenolysis takes place by use of formic acid.

1.1.1.3 Reactions with 1,2-, 1,3-, and 1,4-dienes. The reaction of conjugated dienes with aryl and alkenyl halides can be explained by the following mechanism. Insertion of a conjugated 1,3-diene into an aryl or alkenylpalladium bond gives the π-allylpalladium complex **243** as an intermediate, which reacts further in two ways. As expected, nucleophiles such as carbon nucleophiles, amines, and alcohols attack the π-allylpalladium intermediate to form the 1,4-addition product **244**. In the absence of the nucleophiles, the elimination of β-hydrogen takes place to give the substituted 1,3-diene **245**. In some cases, the substituted 1,3-diene **245** reacts again with the aryl halide to form the π-allylpalladium **246**. Subsequent elimination affords the 1,4-diarylated 1,3-diene **247**.

In the reaction of aryl and alkenyl halides with 1,3-pentadiene (**248**), amine and alcohol capture the π-allylpalladium intermediate to form **249**. In the reactions of *o*-iodoaniline (**250**) and *o*-iodobenzyl alcohol (**253**) with 1,3-dienes, the amine and benzyl alcohol capture the π-allylpalladium intermediates **251** and **254** to give **252** and **255**[173–175]. The reaction of *o*-iodoaniline (**250**) with 1,4-pentadiene (**256**) affords the cyclized product **260** via arylpalladium formation, addition to the diene **256** to form **257**, palladium migration (elimination of Pd—H and readdition to give **258**) to form the π-allylpalladium **259**, and intramolecular displacement of π-allylpalladium with the amine to form **260**[176]. *o*-Iodophenol reacts similarly.

The π-allylpalladium complexes formed as intermediates in the reaction of 1,3-dienes are trapped by soft carbon nucleophiles such as malonate, cyanoacetate, and malononitrile[177–179]. The reaction of (*o*-iodophenylmethyl) malonate (**261**) with 1,4-cyclohexadiene is terminated by the capture of malonate via Pd migration to form **262**. The intramolecular reaction of **263** generates π-allylpalladium, which is trapped by malononitrile to give **264**. *o*-Iodophenylmalonate (**265**) adds to 1,4-cyclohexadiene to form a π-allylpalladium intermediate via elimination of H—Pd—X and its readdition, which is trapped intramolecularly with malonate to form **266**[176].

In the absence of nucleophiles, the elimination of β-hydrogen from π-allyl-palladium intermediates to give the (*E, E*)-1-aryl-1,3-dienes **267** and **268** takes place[173,180]. 1-Aryl-1,3,5-hexatriene **269** reacts with an aryl bromide to give the 1,6-diaryl-1,3,5-hexatriene **270**. The best results are obtained with aryl halides having EWGs in the *para*-position[181].

When allene derivatives are treated with aryl halides in the presence of Pd(0), the aryl group is introduced to the central carbon by insertion of one of the allenic bonds to form the π-allylpalladium intermediate **271**, which is attacked further by amine to give the allylic amine **272**. A good ligand for the reaction is dppe[182]. Intramolecular reaction of the γ-aminoallene **273** affords the pyrrolidine derivative **274**[183].

The reaction of the *o*-iodophenol **275** with an alkylallene affords the benzo-furan derivative **276**[184]. Similarly, the reactions of the 6-hydroxyallenes **277** and **279** with iodobenzene afford the tetrahydrofurans **278** and **280**. Under a CO atmosphere, CO insertion takes place before the insertion of the allenyl bond, and a benzoyl group, rather than a phenyl group, attacks the allene carbon to give **280**. Reaction of iodobenzene with 4,5-hexadienoic acid (**281**) affords the furanone derivative **282**[185].

Allenes also react with aryl and alkenyl halides, or triflates, and the π-allyl-palladium intermediates are trapped with carbon nucleophiles. The formation of **283** with malonate is an example[186]. The steroid skeleton **287** has been constructed by two-step reactions of allene with the enol triflate **284**, followed by trapping with 2-methyl-1,3-cyclopentanedione (**285**) to give **286**[187]. The inter- and intramolecular reactions of dimethyl 2,3-butenylmalonate (**288**) with iodobenzene afford the 3-cyclopentenedicarboxylate **289** as a main pro-duct[188].

The reaction of 1,2-heptadiene with iodobenzene in the absence of nucleo-philes affords 2-phenyl-1,3-heptadiene (**290**) by β-elimination of the π-allylpal-ladium intermediate[182].

1.1.2 Reactions with Alkynes

1.1.2.1 Terminal alkynes to form alkenyl- and arylalkynes. Direct introduction of sp^2 carbon into alkynes is very difficult, but it can be done easily by using Pd catalysts. Terminal alkynes undergo two Pd-catalyzed reactions with organic halides. One reaction is the substitution of alkynic hydrogen to form disubstituted alkynes in the presence of CuI as a co-catalyst. Another is insertion to form alkenyl compounds in the absence of CuI (see Section 1.1.2.2). Facile palladium-catalyzed coupling of aryl and alkenyl halides with terminal alkynes is used extensively for the construction of aryl alkyne or enyne systems[189]. Two methods of coupling are known, the direct coupling of terminal alkynes and the coupling of metal acetylides. The direct coupling reaction is catalyzed by Pd phosphine catalyst in the presence of amines[190,191]. The addition of CuI as a co-catalyst gives better results[192,193]. CuI activates alkynes by forming a copper acetylide **291**[194], which undergoes transmetallation with arylpalladium halide to form the alkynylarylpalladium species **292**, and reductive elimination is the final step. Pd on carbon and Ph_3P is also an active catalyst[195]. However, the coupling proceeds smoothly without CuI when the water-soluble TMSPP is used as a ligand in aqueous solution[27]. Coupling without CuI also gives the arylalkynes or enynes **293** in high yields when the reaction is carried out in piperidine or pyrrolidine. Use of these amines is crucial, and poor results are obtained without CuI when Et$_3$N, Pr$_2$NH, Et$_2$NH, and morpholine are used[196]. Interestingly, the Pd-catalyzed reaction of terminal alkynes with alkenyl chlorides, which are inert in many other Pd-catalyzed reactions, proceeds smoothly without special activation of the chlorides. In addition, it should be mentioned that the coupling of halides with some terminal alkynes (**294**) proceeds smoothly using CuI (5 mol%) and Ph$_3$P (10 mol%) as catalysts and K$_2$CO$_3$ as a base in DMF or DMSO without using Pd catalyst[197].

Numerous applications have been reported. A derivative of the (alkyn-1-yl)nucleosides **295**, which have anticancer and antiviral activities, has been synthesized by this reaction. They are also used as chain-terminating nucleosides for DNA sequencing[198,199]. In this reaction, use of DMF as the solvent is most important for successful operation[200]. Only the alkenyl bromide moiety in 2-bromo-3-acetoxycycloheptene (**296**) reacts with alkynes without attacking the allylic acetate moiety[201].

Monosubstitution of acetylene itself is not easy. Therefore, trimethylsilyl-acetylene (**297**)[202–206] is used as a protected acetylene. The coupling reaction of trimethylsilylacetylene (**297**) proceeds most efficiently in piperidine as a solvent[207]. After the coupling, the silyl group is removed by treatment with fluoride anion. Hexabromobenzene undergoes complete hexasubstitution with trimethylsilylacetylene to form hexaethynylbenzene (**298**) after desilylation in total yield of 28% for the six reactions[208,209]. The product was converted into tris(benzocyclobutadieno)benzene (**299**). Similarly, hexabutadiynylbenzene was prepared[210].

Acetylene is also protected as propargyl alcohol (**300**)[211], which is deprotected by hydrolysis with a base, or oxidation with MnO_2 and alkaline hydrolysis. Sometimes, propargyl alcohols are isomerized to enals. Propargyl alcohol (**300**) reacts with 3-chloropyridazine (**301**) and Et_2NH to give 3-diethylaminoindolizine (**303**) in one step via the enal **302**[212]. Similarly, propargyl alcohol reacts with 2-halopyridines and secondary amines. 2-Methyl-3-butyn-2-ol (**304**) is another masked acetylene, and is unmasked by treatment with KOH or NaOH in butanol[205,206,213–215] or *in situ* with a phase-transfer catalyst[216].

In addition to coupling via Cu acetylides generated *in situ* mentioned above, the coupling of terminal alkynes has been carried out smoothly using actylides of Zn and other metals as an alternative method of arylation and alkenylation of alkynes[217,218]. Sn[219,219a], Zn[220,221], and Mg[222] acetylides are used frequently as activated alkynes rather than alkynes themselves, and their reactions with halides proceed without using CuI. Coupling with the tin acetylide **305** proceeds smoothly and competes with the CuI-catalyzed direct coupling method. However, sometimes the coupling takes place only with tin acetylides. The highly strained neocarzinostatin framework can be constructed by the intramolecular reaction of alkenyl bromide with alkynes[223]. In this case, the tributyltin acetylide **306** must be used. No cyclization takes place with the terminal alkyne itself. However, this is not always true, and in some cases free alkynes give better results than the corresponding tin acetylides[224].

The neocarzinostatin chromophore is prepared by two-step reactions of the 1,1-diiodoalkene **307** with three different alkynes based on interesting differences in chemoselectivity. The first step is the oxidative addition of Pd(0) to the less reactive (Z)-iodide in the absence of CuI, followed by the favored intramolecular insertion (or carbopalladation) of the terminal triple bond to generate alkenylpalladium. Then the coupling of the tin acetylide with the alkenylpalladium iodide takes place to give **308**. Finally, the intermolecular alkyne coupling of another alkyne **309** takes place using CuI catalyst to give **310**. On the other hand, it was reported that no intermolecular coupling of a simple 1,1-diiodoalkene with terminal alkynes takes place[225,226].

Zinc acetylides, prepared *in situ* by the treatment of lithium acetylides with $ZnCl_2$, are widely used. The zinc acetylide **311**, prepared *in situ*, reacts with (Z)-3-iodo-2-buten-1-ol (**312**) with nearly complete retention of stereochemistry to afford an important intermediate **313** for carotenoid synthesis[227].

Both *cis*- and *trans*- 1,2-dibromoethylenes react with metal acetylides, but with different reactivity. For the competitive reaction of *cis*- and *trans*-1,2-dibromoethylenes with the zinc acetylide **314** shows that the *trans* isomer is more reactive than the *cis* isomer[228,229]. It was also found by a competitive reaction with the zinc acetylide **317** that the *trans* vinyl monobromide **315** is more reactive than the *cis* isomer **316**[230].

The terminal diyne **320** is prepared by coupling of the zinc acetylide **318** with *trans*-1-iodo-2-chloroethylene (**319**), followed by elimination of HCl with sodium amide[231]. Similarly, terminal di- and triynes are prepared by using *cis*-1,2-dichloroethylene[232]. The 1-alkenyl or 1-aryl-2-(perefluoroalkyl) acetylene **321** is prepared by the reaction of a zinc acetylide with halides[233].

Mg acetylides can be used for the coupling. As an example, the thiophenylenyne **322** was prepared by repeating the coupling of Mg acetylides with halides[222].

The coupling reaction has widespread use in the construction of enediyne systems present in naturally occurring anticancer antibiotics[234]. The Pd–CuI catalyzed coupling reaction of the alkenyl bromides **323** and **326** with the terminal alkynes **324** and **325**, followed by hydrogenation of the triple bonds to *cis* double bonds, is extremely useful for the synthesis of the polyene structures of the lipoxins **327** and related eicosanoids[235]. The extensive applications of the coupling reaction are summarized in reviews[236].

The novel intramolecular reaction of the alkenyl bromide with the terminal alkyne in **328**, followed by intramolecular Diels–Alder reaction, afforded the highly strained dynemicin A structure **329** in one step[237].

Interestingly, 1,2-dichloroethylene can be used for coupling without activation of the chlorides. The reaction of *cis*-1,2-dichloroethylene (**331**) has wide synthetic applications, particularly for the synthesis of an enediyne structure[238]. Typically, this reaction is useful for the construction of the highly strained enediyne structure **333** present in naturally occurring anticancer antibiotics such as espermicin and calichemicin. As shown, both the chlorines in (Z)-1,2-dichloroethylene (**331**) are displaced stepwise with two different terminal alkynes, **330** and **332**[239–244]. Both chlorines in *trans*-1,2-dichloroethylene (**334**) react smoothly[245]. The polyene structure of lipoxin B was constructed by the stepwise coupling of (E)-1,2-dichloroethylene (**334**) with the alkenylstannane **336** and terminal alkyne **335** to generate the trieneyne **337**, followed by hydrogenation of the alkyne to the *cis* -alkene[246–248]. Alkenyl monochlorides are less reactive than 1,2-dichloroethylene, but the *trans*-chloroalkene **338** reacts rapidly with a terminal alkyne to afford the enyne **339** in the presence of piperidine using PdCl$_2$(PhCN)$_2$, which is more active than PdCl$_2$(Ph$_3$P)$_2$. Other amines such as Et$_3$N give poor yields[249].

Both chlorines of 1,1-dichloroethylene (**340**) react stepwise with different terminal alkynes to form the unsymmetrical enediyne **341**[250]. The coupling of the dichloroimine **342** with tin acetylide followed by hydrolysis affords the dialkynyl ketone **343**[251]. The phenylthioimidoyl chloride **344** undergoes stepwise reactions with two different tin acetylides to give the dialkynylimine **345**[252].

Chlorobenzenes activated by coordination of Cr(CO)$_3$ react with terminal alkynes[253]. The 1-bromo-1,2-alkadiene **346** reacts with a terminal alkyne to afford the alka-1,2-dien-4-yne **347**[254]. Enol triflates are used for the coupling with terminal alkynes. Formation of **348** in the syntheses of ginkgolide[255] and of vitamin D are examples[256]. Aryl and alkenyl fluorides are inert. Only bromide or iodide is attacked when the fluoroiodoalkene **349** or fluoroiodoarene is subjected to the Pd-catalyzed coupling with alkynes[257–259].

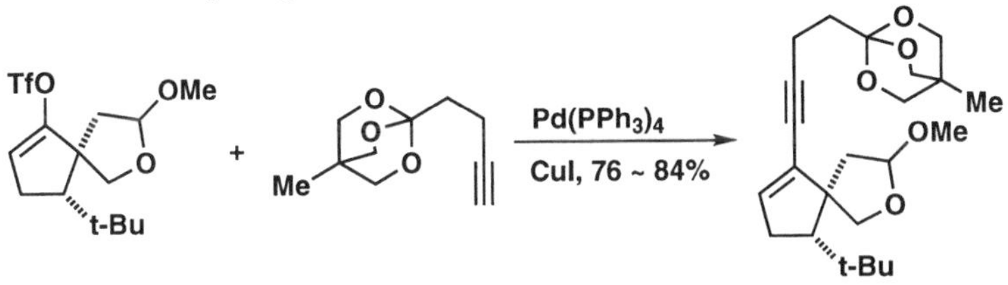

348

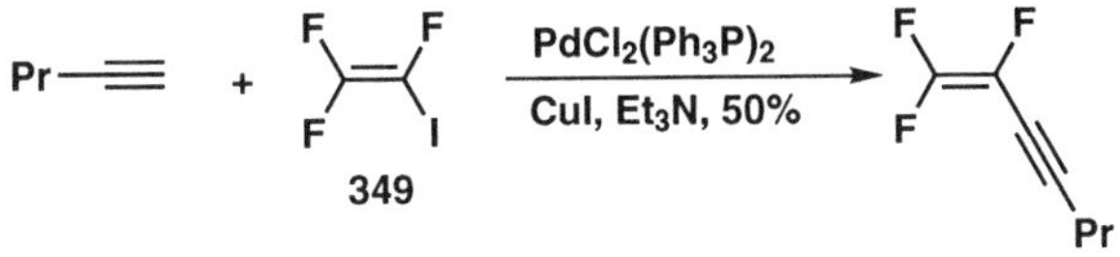

349

Alkynes with EWGs are poor substrates for the coupling with halides. Therefore, instead of the inactive propynoate, triethyl orthopropynoate (**350**) is used for the coupling with aryl halides to prepare the arylpropynoate **351**. The coupling product **353** of 3,3-diethoxy-1-propyne (**352**) with an aryl halide is the precursor of an alkynal[260]. The coupling of ethoxy(tributylstannyl)acetylene (**354**) with aryl halides is a good synthetic method for the arylacetate **355**[261].

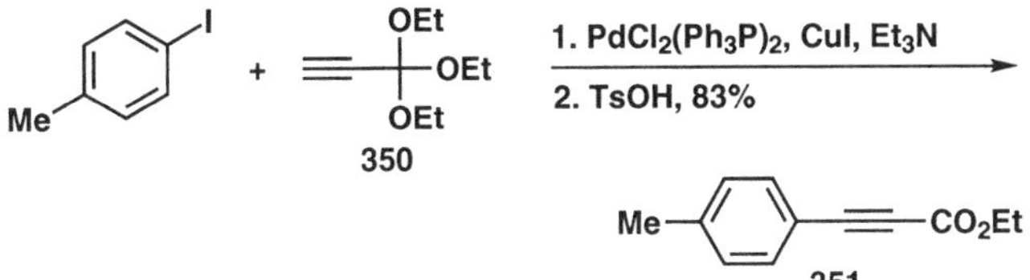

350

351

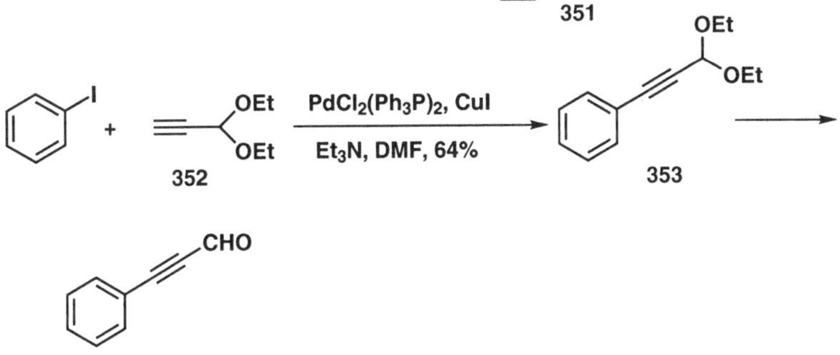

352   **353**

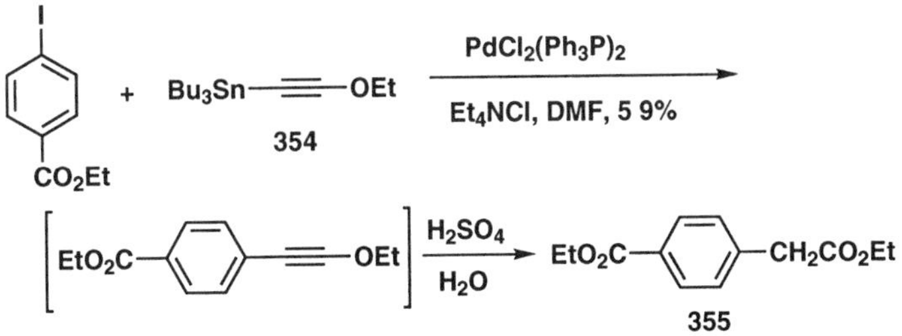

354

355

The formation of disubstituted alkynes by coupling of terminal alkynes, followed by intramolecular attack of an alcohol or amine, is used for the preparation of benzofurans and indoles. The benzo[*b*]furan **356** can be prepared easily by the reaction of *o*-iodophenol with a terminal alkyne[262]. The 2-substituted indole **358** is prepared by the coupling of 2-ethynylaniline (**357**) with aryl and alkenyl halides or triflates, followed by Pd(II)-catalyzed cyclization[263].

The alkynyl iodide **359** undergoes cross-coupling with a terminal alkyne to give the 1,3-diyne **360**[264]. No homocoupling product is formed. This reaction offers a good synthetic method for unsymmetrical 1,3-diynes.

1.1.2.2 Internal alkynes (including reactions with terminal alkynes by insertion mechanism). Internal alkynes insert to some Pd—C bonds to generate the alkenyl–Pd bonds (*syn* addition of organopalladium species to alkynes, or carbopalladation of alkynes). The tandem insertion of internal alkynes has been proposed for the first time in the cyclotrimerization of internal alkynes. [265] Whereas alkene insertion is followed by facile dehydropalladation (elimination of β-hydrogen, whenever there is a β-hydrogen), and generation of Pd(0) catalytic species, alkyne insertion produces the thermally stable alkenylpalladium species **361**, and further transformations are required for the regeneration of Pd(0) species for catalytic recycling. In other words, the generated alkenylpalladium species **361** cannot undergo β-elimination (formation of alkynes or allenes) even in the presence of a β-hydrogen. In other words, the alkyne insertion is a 'living' process. Thus the alkenylpalladium species **361** are capable of undergoing further insertion or anion capture before termination.

Terminal alkynes undergo the above-mentioned substitution reaction with aryl and alkenyl groups to form arylalkynes and enynes in the presence of CuI as described in Section 1.1.2.1. In addition, the insertion of terminal alkynes also takes place in the absence of CuI, and the σ-alkenylpalladium complex **362** is formed as an intermediate, which cannot terminate by itself and must undergo further reactions such as alkene insertion or anion capture. These reactions of terminal alkynes are also treated in this section.

The alkenylpalladium intermediate **364**, formed by the intramolecular insertion of **363**, is terminated by hydrogenolysis with formic acid to give the terminal alkene **365**[266]. The intramolecular insertion of **366** is terminated by the reaction of the alkynylstannane **367** to afford the conjugated dienyne system **368**[267].

The alkenyl iodide or triflate **369** reacts in the absence of CuI with two moles of terminal alkyne **370** to form the substituted fulvene **371**. The reaction can be explained by the intermolecular insertion of the alkyne twice, followed by the intramolecular insertion of the alkene, and β-elimination to form the fulvene **371**[268].

In the alkylative cyclization of the 1,6-enyne **372** with vinyl bromide, formation of both the five-membered ring **373** by *exo* mode carbopalladation and isomerization of the double bonds and the six-membered ring **374** by *endo* mode carbopalladation are observed[269]. Their ratio depends on the catalytic species. Also, the cyclization of the 1,6-enyne **375** with β-bromostyrene (**376**) affords the *endo* product **377**. The *exo* mode cyclization is commonly observed in many cases, and there are two possible mechanistic explanations for that observed in these examples. One is direct *endo* mode carbopalladation. The other is the *exo* mode carbopalladation to give **378** followed by cyclopropanation to form **379**, and the subsequent cyclopropylcarbinyl-homoallyl rearrangement affords the six-membered ring **380**. Careful determination of the *E* or *Z* structure of the double bond in the cyclized product **380** is crucial for the mechanistic discussion.

The reaction of the 1,6-enyne **375** with the alkenyl bromide **381** gives the vitamin D skeleton **383**. The six-membered ring **383** is claimed to be the product of direct 6-*endo-trig* cyclization of **382**, based on the reported stereochemistry of the triene **383**. However, the *exo* mode cyclization, followed by rearrangement to give the six-membered ring with a different stereochemistry, may be another possibility. In the construction of the conjugated triene system in vitamin D, the alkenylpalladium formed from **385** undergoes intermolecular insertion into the terminal triple bond of the 1,7-enyne **384** first to form the alkenylpalladium **386**, which then undergoes further intramolecular insertion of the double bond, yielding the triene system **387** in 76% yield[270].

Many examples of insertions of internal alkynes are known. Internal alkynes react with aryl halides in the presence of formate to afford the trisubstituted alkenes[271,272]. In the reaction of the terminal alkyne **388** with two molecules of iodobenzene, the first step is the formation of the phenylacetylene **389**. Then the internal alkyne bond, thus produced, inserts into the phenyl–Pd bond to give **390**. Finally, hydrogenolysis with formic acid yields the trisubstituted alkene **391**[273,274]. This sequence of reactions is a good preparative method for trisubstituted alkenes from terminal alkynes.

Intramolecular reaction can be used for polycyclization reaction[275]. In the so-called Pd-catalyzed cascade carbopalladation of the polyalkenyne **392**, the first step is the oxidative addition to alkenyl iodide. Then the intramolecular alkyne insertion takes place twice, followed by the alkene insertion twice. The last step is the elimination of β-hydrogen. In this way, the steroid skeleton **393** is constructed from the linear diynetriene **392**[276].

The dienyne **394** undergoes facile polycyclization. Since the neopentylpalladium **395** is formed which has no hydrogen β to the Pd after the insertion of the disubstituted terminal alkene, the cyclopropanation takes place to form the π-allylpalladium intermediate **396**, which is terminated by elimination to form the diene **397**[275]. The dienyne **398** undergoes remarkable tandem 6-*exo-dig*, 5-*exo-trig*, and 3-*exo-trig* cyclizations to give the tetracycle **399** exclusively[277].

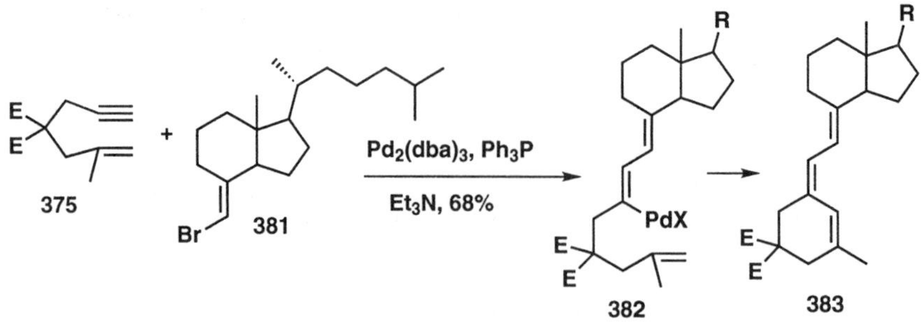
E
E
375
+
381
Br
Pd2(dba)3, Ph3P
Et3N, 68%
R
PdX
E
E
382
R
E
E
383

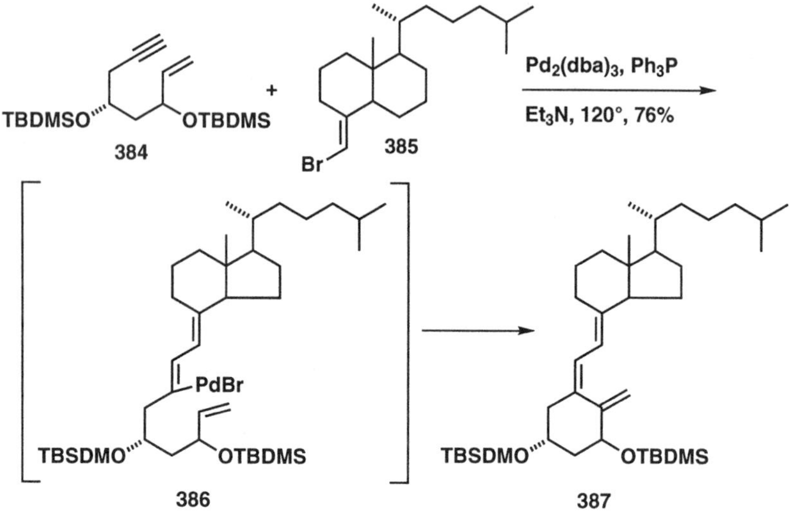
TBDMSO
OTBDMS
384
+
Br
385
Pd2(dba)3, Ph3P
Et3N, 120°, 76%
PdBr
TBSDMO
OTBDMS
386
TBSDMO
OTBDMS
387

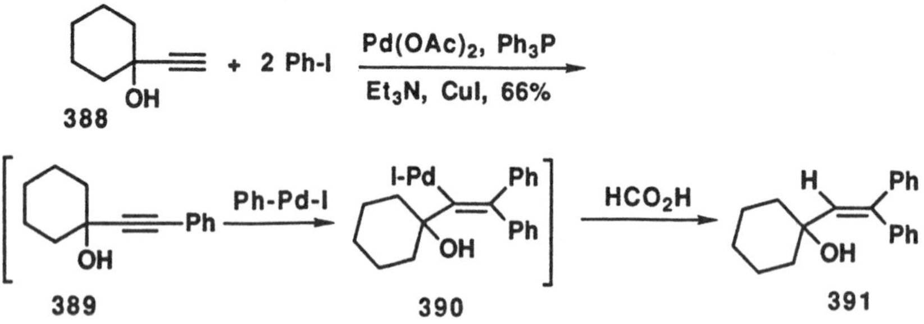
OH
388
+ 2 Ph-I
Pd(OAc)2, Ph3P
Et3N, CuI, 66%
Ph
OH
389
Ph-Pd-I
I-Pd
Ph
Ph
OH
390
HCO2H
H
Ph
Ph
OH
391

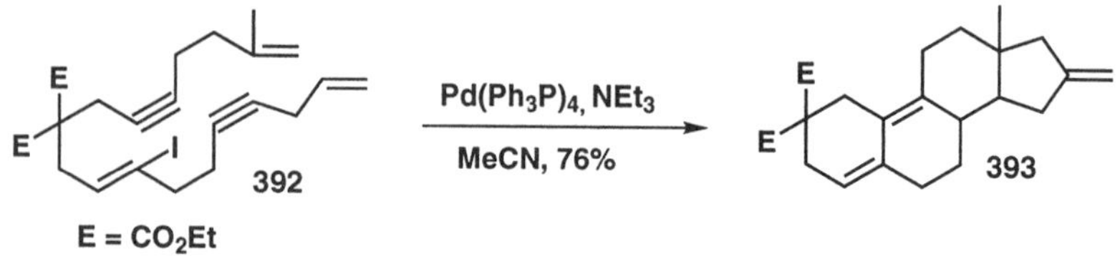
E
E
I
392
E = CO2Et
Pd(Ph3P)4, NEt3
MeCN, 76%
E
E
393

The σ-alkenylpalladium **400**, formed as an intermediate, is converted into the benzene derivative **401** by the intermolecular insertion of acrylate[278]. A formal [2 + 2 + 2] cycloaddition takes place by the reaction of 2-iodonitrobenzene with the 1,6-enyne **402**. The neopentylpalladium intermediate **403** undergoes 6-*endo-trig* cyclization on to the aromatic ring to give **404**[279].

The naphthalene ring compounds **405** and **406** are formed by the intermolecular reaction of *m*-iodoanisole with diphenylacetylene[280].

The benzene derivative **409** is synthesized by the Pd-catalyzed reaction of the haloenyne **407** with alkynes. The intramolecular insertion of the internal alkyne, followed by the intermolecular coupling of the terminal alkyne using Pd(OAc)$_2$, Ph$_3$P, and CuI, affords the dienyne system **408**, which cyclizes to the aromatic ring **409**[281]. A similar cyclization of **410** with the terminal alkyne **411** to form benzene derivatives **412** and **413** without using CuI is explained by the successive intermolecular and intramolecuar insertions of the two triple bonds and the double bond[282]. The angularly bisannulated benzene derivative **415** is formed in one step by a totally intramolecular version of polycyclization of bromoenediyne **414**[283,284].

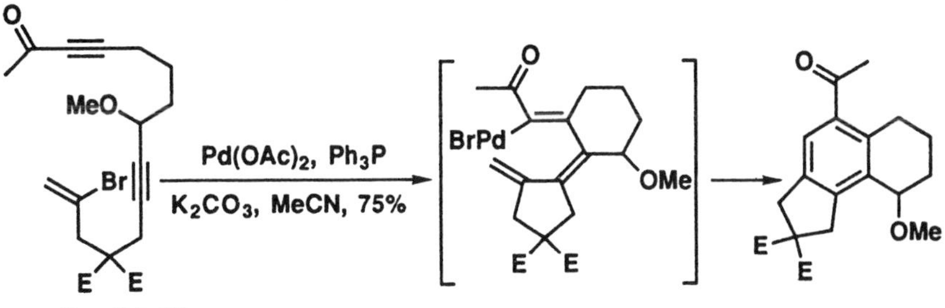

E = CO₂Et

414

415

The insertion of internal alkynes is terminated by anion capture. In the reaction of **416**, the final step is the capture by organotin or zinc reagents to give **417**[285,286]. This reaction is overall *cis* addition to the triple bond. Comparative studies with various metal reagents in the tandem cyclic carbo-palladation and cross-coupling reaction of *o*-iodo-3-pentynylbenzene (**418**) have been carried out[287]. The carbometallation to give **419** is competitive with the direct cross-coupling to afford **420**. The direct cross-coupling takes place with the Zn reagent to give **420**. However, Zr, Sn, Al, and B reagents undergo initial cyclic carbopalladation, followed by cross-coupling to give **419** as desired, but their reactivity is different depending on the organic parts. Zr seems to be most effective for introducing alkenyl groups, but not for aryl and alkynyl groups. Sn is good for introduction of alkenyl and alkynyl groups, but not for aryl groups. Al is most effective for cyclization–arylation.

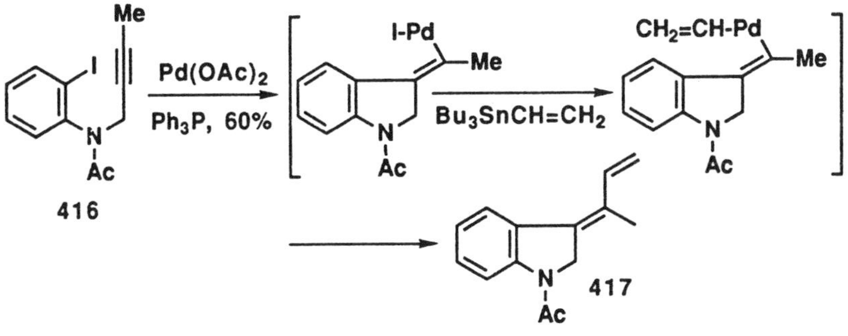

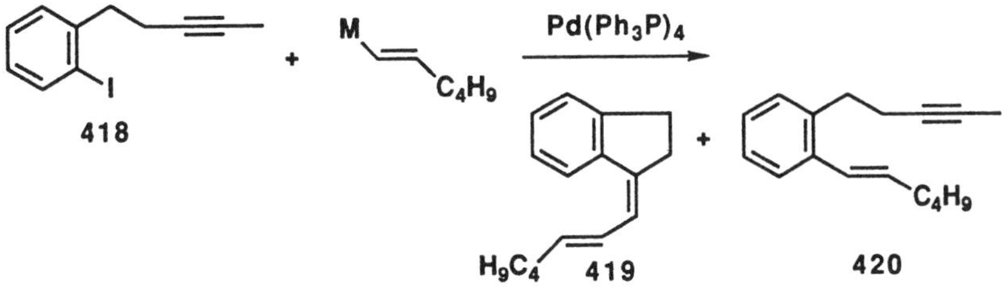

416

417

418

419   **420**

		419	**420**
ZnCl	THF, r.t.,	19	68
Cp₂ZrCl	THF, refl.	84	3
SnMe₃	THF-MMPA, refl.	69	2

Interesting formation of the fulvene **422** takes place by the reaction of the alkenyl bromide **421** with a disubstituted alkyne[288]. The indenone **425** is prepared by the reaction of *o*-iodobenzaldehyde (**423**) with internal alkyne. The intermediate **424** is formed by oxidative addition of the C—H bond of the aldehyde and its reductive elimination affords the enone **425**[289,290].

The *N*-methylbenzo[*de*]quinoline **426** was prepared by trapping the insertion product of an internal alkyne with a tertiary dimethylamine. One methyl group is eliminated. The dimethylaminonaphthalene–Pd complex **427** is an active catalyst and other Pd compounds are inactive[290a].

Pyrrole derivatives are prepared by the coupling and annulation of *o*-iodoanilines with internal alkynes[291]. The 4-amino-5-iodopyrimidine **428** reacts with the TMS-substituted propargyl alcohol **429** to form the heterocondensed pyrrole **430**, and the TMS is removed[292]. Similarly, the tryptophane **434** is obtained by the reaction of *o*-iodoaniline (**431**) with the internal alkyne **432** and deprotection of the coupled product **433**[293]. As an alternative method, the 2,3-disubstituted indole **436** is obtained directly by the coupling of the *o*-alkynyltrifluoroacetanilide **435** with aryl and alkenyl halides or triflates[294].

The reactions discussed above afford the *cis* addition products to triple bonds. On the other hand, the *trans* addition of aryl or alkenyl groups and soft nucleophiles to terminal and internal alkynes is known. In this reaction, aryl or alkenylpalladium intermediates activate the triple bond to induce addition of an oxygen or a soft carbon nucleophile from the rear side to form the *trans* adducts. This reaction can be understood as oxypalladation or carbopalladation of alkynes with σ-aryl or σ-alkenyl palladium species. The reaction of the 4-alkynoic acid 437 with aryl, alkenyl[295], and alkynyl halides 439[296] produces the γ-alkylidene butyrolactones 438 and 440. The reaction of 4-pentyn-1-ol or o-(2-propynyl)phenol (441) with aryl halides affords 2-alkylidenedihydrobenzofuran 442 and the benzofuran 443[297]. Similarly, 2,3,5-trisubstituted furans are obtained by the reaction of 2-propargyl-1,3-dicarbonyl compounds[298]. The reaction of a nitrogen analogue 444 affords pyrrolidine derivative 445[299]. The stereodefined benzylidenecyclopentane 447 is obtained by the *trans* addition of an aryl group and carbanion to the terminal alkyne 446[300].

1.1.3 Carbonylation

1.1.3.1 Formation of carboxylic acids and their derivatives. Aryl and alkenyl halides undergo Pd-catalyzed carbonylation under mild conditions, offering useful synthetic methods for carbonyl compounds. The facile CO insertion into aryl- or alkenylpalladium complexes, followed by the nucleophilic attack of alcohol or water affords esters or carboxylic acids. Aromatic and α,β-unsaturated carboxylic acids or esters are prepared by the carbonylation of aryl and alkenyl halides in water or alcohols[301–305].

In the total synthesis of zearalenone (**451**), the ester **450** was prepared by the carbonylation of the crowded aryl iodide **448**. The alkyl iodide moiety in the alcohol molecule **449** is not attacked[306]. Methyl trifluoromethacrylate (**453**) was prepared by the carbonylation of 3,3,3-trifluoro-2-bromopropylene (**452**). The carbonylation in the presence of alkylurea affords **454**, which is converted into the trifluoromethyluracil **455**[307].

Usually, iodides and bromides are used for the carbonylation, and chlorides are inert. However, oxidative addition of aryl chlorides can be facilitated by use of bidentate phosphine, which forms a six-membered chelate structure and increases the electron density of Pd. For example, benzoate is prepared by the carbonylation of chlorobenzene using bis(diisopropylphosphino)propane (dippp) (**456**) as a ligand at 150 °C[308]. The use of tricyclohexylphosphine for the carbonylation of neat aryl chlorides in aqueous KOH under biphasic conditions is also recommended[309,310].

The benzoic acid derivative **457** is formed by the carbonylation of iodobenzene in aqueous DMF (1 : 1) without using a phosphine ligand at room temperature and 1 atm[311]. As optimum conditions for the technical synthesis of the anthranilic acid derivative **458**, it has been found that *N*-acetyl protection, which has a chelating effect, is important[312]. Phase-transfer catalysis is combined with the Pd-catalyzed carbonylation of halides[313]. Carbonylation of 1,1-dibromoalkenes in the presence of a phase-transfer catalyst gives the geminal dicarboxylic acid **459**. Use of a polar solvent is important[314]. Interestingly, addition of trimethylsilyl chloride (2 equiv.) increased yield of the lactone **460** remarkably[315]. Formate esters as a CO source and NaOR are used for the carbonylation of aryl iodides under a nitrogen atmosphere without using CO[316]. Chlorobenzene coordinated by $Cr(CO)_3$ is carbonylated with ethyl formate[317].

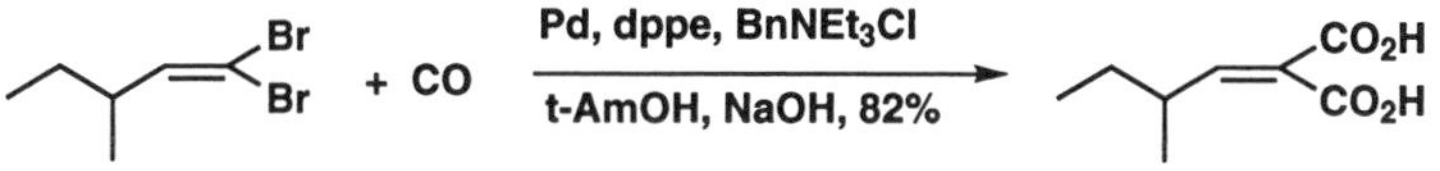

459

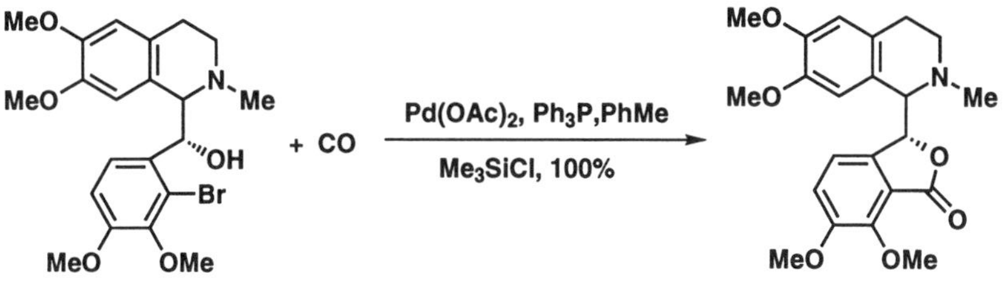

460

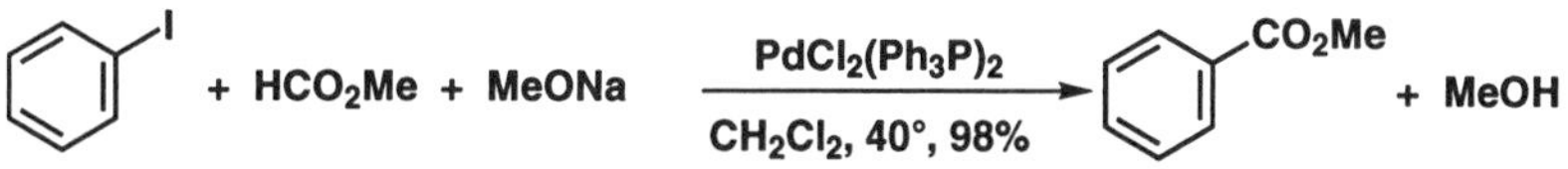

Carbonylation of benzyl chloride affords phenylacetate[318]. Preparation of phenylacetic acid (**461**) in a high yield by the carbonylation of benzyl chloride in two-phase solvents (water and heptane) using water-soluble sulfonated phosphine (DPMSPP) as a ligand has been carried out[313,319–321]. *t*-Butyl phenylacetate is prepared at 1 atm using triethylbenzylammonium chloride as a phase-transfer catalyst[322]. Carbonylation of benzyl halides can also be carried out under neutral conditions using molecular sieves or tetramethylurea[323]. In the total synthesis of the macrolide curvularin (**464**), the Pd-catalyzed carbonylation of the benzyl chloride **462** to give the ester **463** has been applied[324]. Although benzyl acetate cannot be carbonylated, chiral 1- and 2(1–acetoxymethyl)naphthalenes (**465**) undergo the formate-mediated carbonylation to give free carboxylic acids **466** and **467** with extensive racemization in the presence of sodium formate using dppp. The corresponding 1-naphthylethyl formate is also carbonylated[325].

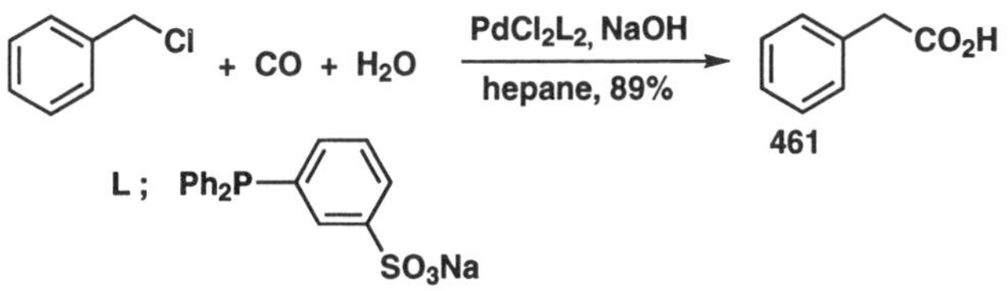

461

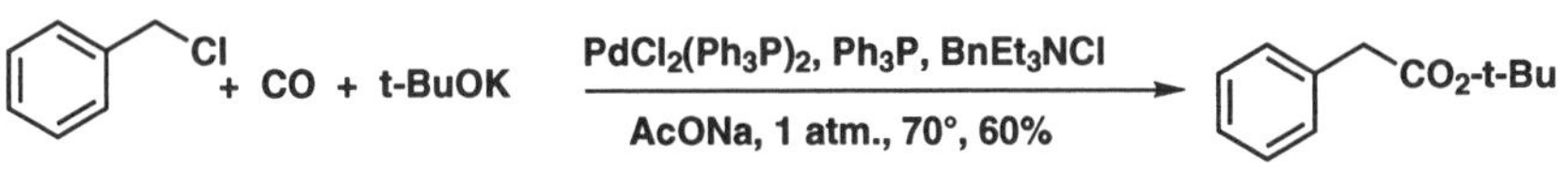

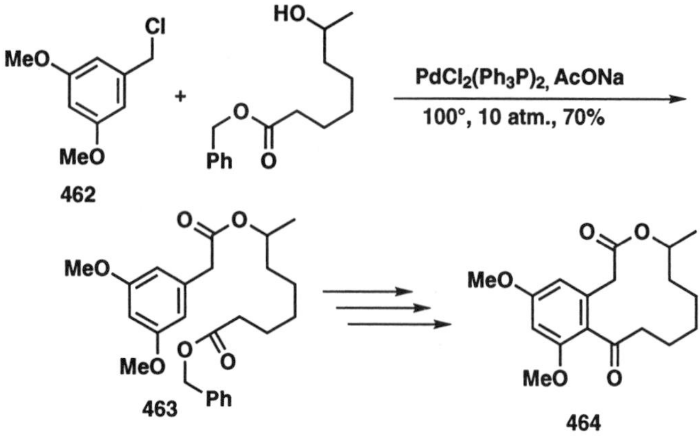

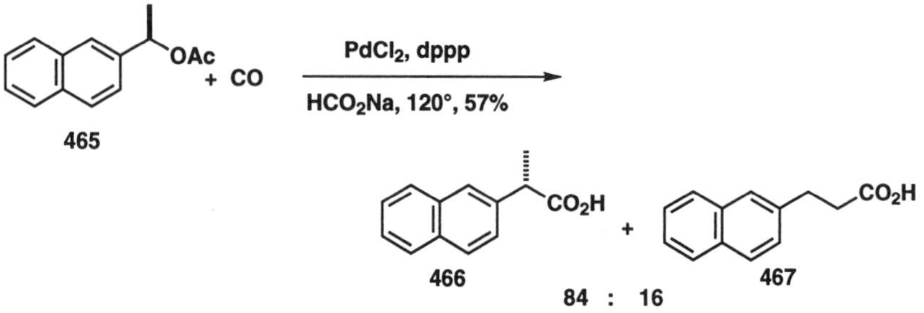

462

463

464

465

466   467

84 : 16

The phenylacetic acid derivative **469** is produced by the carbonylation of the aromatic aldehyde **468** having electron-donating groups[326]. The reaction proceeds at 110 °C under 50–100 atm of CO with the catalytic system Pd–Ph₃P–HCl. The reaction is explained by the successive dicarbonylation of the benzylic chlorides **470** and **471** formed *in situ* by the addition of HCl to aldehyde to form the malonate **472**, followed by decarboxylation. As supporting evidence, mandelic acid is converted into phenylacetic acid under the same reaction conditions[327].

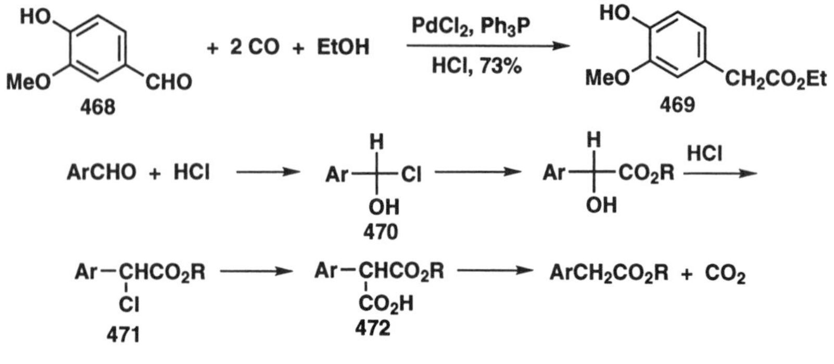

468   469

$$\text{ArCHO} + \text{HCl} \longrightarrow \text{Ar}\overset{\text{H}}{\underset{\text{OH}}{\text{C}}}\text{Cl} \longrightarrow \text{Ar}\overset{\text{H}}{\underset{\text{OH}}{\text{C}}}\text{CO}_2\text{R} \xrightarrow{\text{HCl}}$$

470

$$\text{Ar}-\underset{\text{Cl}}{\text{CHCO}_2\text{R}} \longrightarrow \text{Ar}-\underset{\text{CO}_2\text{H}}{\text{CHCO}_2\text{R}} \longrightarrow \text{ArCH}_2\text{CO}_2\text{R} + \text{CO}_2$$

471   472

Triflates of phenols are carbonylated to form aromatic esters by using Ph_3P[328]. The reaction is 500 times faster if dppp is used[329]. This reaction is a good preparative method for benzoates from phenols and naphthoates (**473**) from naphthols. Carbonylation of the bis-triflate of axially chiral 1,1'-binaphthyl-2,2'-diol (**474**) using dppp was claimed to give the monocarboxylate **475**[330]. However, the optically pure dicarboxylate **476** is obtained under similar conditions[331]. The use of 4.4 equiv. of a hindered amine (ethyldiisopropylamine) is crucial for the dicarbonylation. The use of more or less than 4.4 equiv. of the amine gives the monoester **475**.

Carbonylation of enol triflates derived from ketones and aldehydes affords α,β-unsaturated esters[332]. Steroidal esters are produced via their aryl and enol triflates[328]. The enol triflate in **477** is more reactive than the aryl triflate and the carbonylation proceeds stepwise. First, carbonylation of the enol triflate affords the amide **478** and then the ester **479** is obtained in DMSO using dppp[333].

Carboxylic acids are produced in water. Selection of solvents is crucial and the carbonylation of the enol triflate **480** can be carried out in aqueous DMF, and that of the aryl triflate **481** in aqueous DMSO using dppf as a ligand[328,334]. The carbonylation of the enol triflate **482** to form the α, β-unsaturated acid **483** using dppf as a ligand in aqueous DMF has been applied in the total synthesis of multifunctionalized glycinoeclepin[335].

Carboxylic acids are obtained from aryl iodides by the reaction of chloroform under basic conditions without using CO[336].

Other pseudo-halides are used for carbonylation. Phenyl fluorosulfonate (**484**) can be carbonylated to give benzoate[337]. Aryl(aryl)iodonium salts[338], aryl(alkenyl)iodonium salts (**485**)[339], and aryl(alkynyl)iodonium salts (**486**)[340] are reactive compounds and undergo carbonylation under mild conditions (room temperature, 1 atm) to give aryl, alkenyl, and alkynyl esters. Iodoxybenzene (**487**) is carbonylated under mild conditions in water[341]. Benzoic acid derivatives are prepared by the carbonylation of arene-diazonium salts under mild conditions[342]. For example, the acid anhydride **489** is prepared by the carbonylation of the benzenediazonium salt **488** in acetic acid. By this method, nitrobenzene can be converted into benzoic acid indirectly[343]. In the presence of tetraalkylstannane, ketones are obtained[344].

The carbonylation of aryl halides under mild conditions in the presence of CsF affords the acid fluoride **490** in good yields. Unlike acyl chlorides, acyl fluorides are inert toward Pd(0) catalyst[345]. Benzenesulfonyl chloride (**491**) undergoes desulfonylation–carbonylation to give the benzoate **492** in the presence of titanium tetralkoxide at 160 °C[346].

Heteroaromatic esters such as **493** and amides are produced by the carbonylation of heterocyclic bromides[347,348]. Even dichloropyrazine (**494**) and chloropyridine are carbonylated under somewhat severe conditions (120 °C, 40 atm)[349]. The carbonylation of trifluoroacetimidoyl iodide (**495**) proceeds under mild conditions, and can be used for the synthesis of the trifluoromethylglycine derivatives **496** and **497**[350].

Carbonylation of halides in the presence of primary and secondary amines at 1 atm affords amides[351]. The intramolecular carbonylation of an aryl bromide which has amino group affords a lactam and has been used for the synthesis of the isoquinoline alkaloid **498**[352]. The naturally occurring seven-membered lactam **499** (tomaymycin, neothramycin) is prepared by this method[353]. The α-methylene-β-lactam **500** is formed by the intramolecular carbonylation of 2-bromo-3-alkylamino-1-propene[354].

The carbonylation of *o*-diiodobenzene with a primary amine affords the phthalimide **501**[355,356]. Carbonylation of iodobenzene in the presence of *o*-diaminobenzene (**502**) and DBU or 2,6-lutidine affords 2-phenylbenzimidazole (**503**)[357]. The carbonylation of aryl iodides in the presence of pentafluoroaniline affords 2-arylbenzoxazoles directly. 2-Arylbenzoxazole is prepared indirectly by the carbonylation of *o*-aminophenol[358]. The optically active aryl or alkenyl oxazoline **505** is prepared by the carbonylation of the aryl or enol triflates in the presence of the opticaly active amino alcohol **504**, followed by treatment with thionyl chloride[359].

Alkenyl chlorides are generally not very reactive, but vinyl chloride is exceptionally reactive and its carbonylation with NH_3 at 100 °C gave the Michael adduct of acrylamide **506** in high yields[360].

Under certain conditions, double carbonylation takes place to give derivatives of α-keto carboxylic acids. It is competitive with monocarbonylation. The carbonylation of iodobenzene in water using $Ca(OH)_2$ as a base and alkylphosphine as a ligand gives the α-keto acid (phenylglyoxylic acid) 507[361]. In aqueous isopropyl alcohol in the presence of $Ca(OH)_2$, double carbonylation and reduction proceed to afford mandelic acid (508) using Me_3P as a ligand[362]. The α-keto amide 509 is formed with high chemoselectivity by use of secondary amines and alkylphosphines or bidentate phosphines such as dppb[363,364]. The product of the double carbonylation of *o*-iodoacetanilide is converted into isatin (510) and the quinoline 511[365]. Chlorobenzene (512), coordinated by $Cr(CO)_3$, undergoes double carbonylation using $PdCl_2(MePh_2P)_2$ to give α-keto amide and benzamide. $Cr(CO)_3$ is decoordinated during the reaction[366].

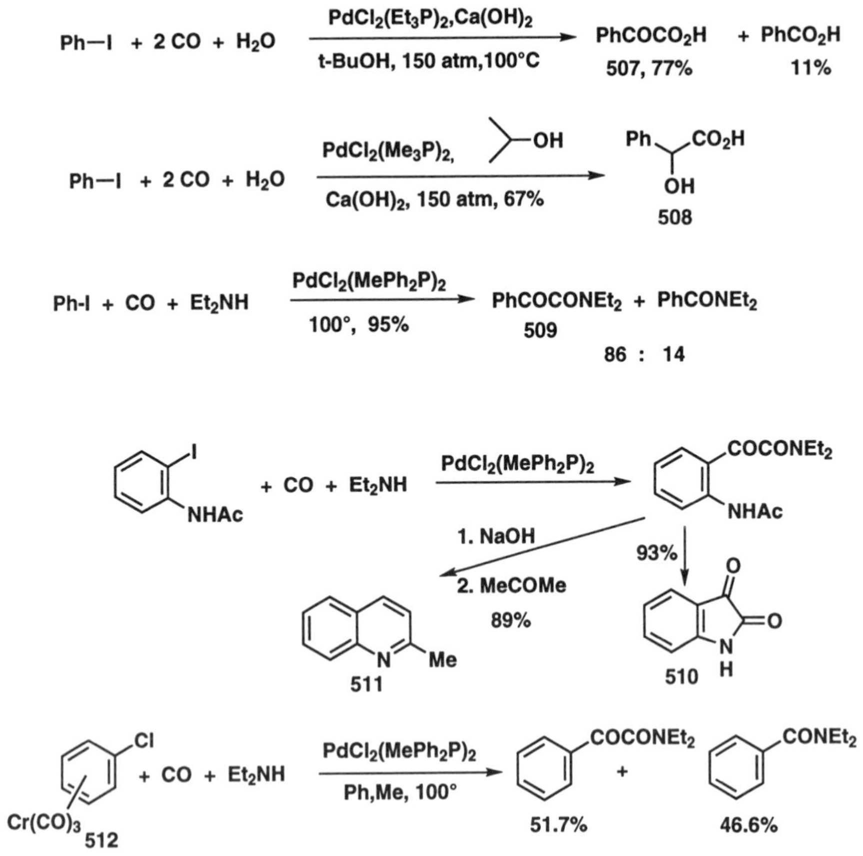

$$Ph-I + 2CO + H_2O \xrightarrow[\text{t-BuOH, 150 atm,100°C}]{PdCl_2(Et_3P)_2,Ca(OH)_2} PhCOCO_2H + PhCO_2H$$

507, 77%   11%

$$Ph-I + 2CO + H_2O \xrightarrow[\text{Ca(OH)}_2\text{, 150 atm, 67%}]{PdCl_2(Me_3P)_2,}$$

508

$$Ph-I + CO + Et_2NH \xrightarrow[\text{100°, 95%}]{PdCl_2(MePh_2P)_2} PhCOCONEt_2 + PhCONEt_2$$

509

86 : 14

The α-keto ester **513** is formed from a bulky secondary alcohol using tricyclohexylphosphine or triarylphosphine, but the selectivity is low[367–369]. Alkenyl bromides are less reactive than aryl halides for double carbonylation[367]. α-Keto amides are obtained from aryl and alkenyl bromides, but α-keto esters are not obtained by their carbonylation in alcohol[370]. A mechanism for the double carbonylation was proposed[371,372].

$$\text{PhI} + \text{CO} + \text{sec-BuOH} \xrightarrow[\text{Et}_3\text{N, 91\%conv.}]{\text{PdCl}_2(\text{PCy}_3)_2} \underset{\textbf{513}}{\text{PhCOCO}_2\text{Bu}} + \text{PhCO}_2\text{Bu}$$
$$64 : 36$$

1.1.3.2 Formation of aldehydes. Aldehydes can be prepared by the carbonylation of halides in the presence of various hydride sources. The carbonylation of aryl and alkenyl iodides and bromides with CO and H_2 (1:1) in aprotic solvents in the presence of tertiary amines affords aldehydes[373,374]. Aryl chlorides, as tricarbonylchromium derivatives, are converted into aldehydes at 130 °C[366]. Sodium formate can be used as a hydride source to afford aldehydes. Chlorobenzene (**514**) was carbonylated at 150 °C to give benzaldehyde with CO and sodium formate by using dippp as a ligand[375,376].

Aldehydes can also be prepared by the carbonylation of aryl and alkenyl halides and triflate, and benzyl and allyl chlorides using tin hydride as a hydride source and $Pd(Ph_3P)_4$ as a catalyst[377]. Hydrosilanes are used as another hydride source[378]. The arenediazonium tetrafluoroborate **515** is converted into a benzaldehyde derivative rapidly in a good yield by using Et_3SiH or PHMS as the hydride source[379].

1.1.3.3 Formation of ketones.   Ketones can be prepared by the carbonylation of halides and pseudo-halides in the presence of various organometallic compounds of Zn, B, Al, Sn, Si, and Hg, and other carbon nucleophiles, which attack acylpalladium intermediates (transmetallation and reductive elimination).

The carbonylation of aryl iodides in the presence of alkyl iodides and Zn–Cu couple affords aryl alkyl ketones via the formation of alkylzinc species from alkyl iodides followed by transmetallation and reductive elimination[380]. The Pd-catalyzed carbonylation of the diaryliodonium salts **516** under mild conditions in the presence of Zn affords ketones **517** via phenylzinc. The α-diketone **518** is formed as a byproduct[381].

Organoboranes can be used for ketone synthesis under basic conditions. The aryl ethyl ketone **519** is obtained by the carbonylation of aryl iodide with triethylborane in the presence of Zn(acac)$_2$ as a base[382]. The cyclic ketone **521** is prepared from the iododiene **520** by hydroboration of the terminal double bond, followed by carbonylation[383]. The phenylboronic acid **522** can be used for aryl ketone formation. Carbonylation of benzyl bromide with **522** afforded the ketone **523**, which was converted into the flavone **524**[384]. Isonitrile can be used for ketone formation instead of CO. The coupling reaction of aryl iodide, *t*-butyl isonitrile and 9-alkyl-9-BBN affords aryl alkyl ketones after hydrolysis of the product **525**. Use of 9-alkyl-9-BBN is crucial[385]. Interestingly, the dialkyl ketone **526** can be prepared by the carbonylative coupling of alkyl iodides and alkylboranes (typically 9-alkyl-9-BBN) without elimination of β-hydrogen. Irradiation with visible light is important to achieve satisfactory yields[386]. Although this is not the carbonylation of aryl or alkenyl halides, this ketone formation from alkyl halides is treated here for convenience and a better understanding (see Section 1.3).

Organoaluminum compounds such as triphenylaluminum (**527**) are used for ketone synthesis[387]. On the other hand, the reaction of *i*-Bu₃Al affords the corresponding alcohol **528** by reductive carbonylation[388].

Organotin compounds such as aryl-, alkenyl-, and alkynylstannanes are useful for the ketone synthesis by transmetallation of acylpalladium **529** and reductive elimination of **530** as shown[389–393]. Acetophenone (**531**) is obtained by the carbonylation of iodobenzene with Me₄Sn. Diaryl ketones can be prepared at room temperature under 1 atm of CO by the reaction of aryl halides with phenyltrimethyltin using π-allylpalladium chloride as a catalyst and HMPA as a solvent[394]. The aryl triflates **532** are shown to be excellent substrates, and ketones are obtained in good yields under 1 atm of CO at 70 °C by use of dppf as a ligand and LiCl as a promoter. The acylpalladium **533**, formed by the insertions of CO, alkene, and CO, is converted into the alkenyl-1,4-diketone **534** by being trapped with a vinyltin compound[395]. Reaction of the arenediazonium salt **535** with Me_4Sn is used for the synthesis of aryl methyl ketones[396].

The reaction of alkenyl iodides or triflates, alkenylstannanes, and CO affords divinyl ketones[397,398]. Thus the capnellene skeleton **538** has been synthesized by the carbonylation of the cyclopentenyl triflate **536** with the alkenyltin **537**[392]. The macrocyclic divinyl ketone **540** has been prepared in a moderate yield by the carbonylative cyclization of **539**[399].

539 + CO →[PdCl₂(MeCN)₂ / DMF, LiCl, 53%]→ **540**

The aryl- and heteroarylfluorosilanes **541** can be used for the preparation of the unsymmetrical ketones **542**[400]. Carbonylation of aryl triflate with the siloxycyclopropane **543** affords the γ-keto ester **545**. In this reaction, transmetallation of the siloxycyclopropane **543** with acylpalladium and ring opening generate Pd homoenolate as an intermediate **544** without undergoing elimination of β-hydrogen[401].

541 + CO + (I-C₆H₄-CHO) →[π-C₃H₅PdCl, KF, DMI / 100°, 1 atm., 72%]→ **542**

543 + (aryl-OTf) + CO →[Pd(Ph₃P)₄ / HMPA, 87%]→ **544** → **545**

Alkyl- and arylmercury(II) halides are used for the ketone formation[402].

When active methylene compounds such as β-keto esters or malonates are used instead of alcohols, acylated β-keto esters and malonates **546** are produced. For this reaction, dppf is a good ligand[403]. The intramolecular version of the reaction proceeds by trapping the acylpalladium intermediate with enolate to give five- and six-membered rings smoothly. Formation of **547** by intramolecular trapping with malonate is an example[404].

(Ph-I) + CO + CH(CO₂Et)₂ →[Pd(0), dppf / Et₃N, 120°, 75%]→ **546**

When an enolizable carbonyl group is present, competition between C and O trappings takes place. Generally, O trapping is favored over C trapping in the cyclization of the same ring size[405]. The carbonylative cyclization of *o*-bromophenyl butyl ketone (**548**) gives the enol lactone **549** by *O*-enolate trapping. The carbonylation in the presence of Ti isocyanate complex, prepared *in situ* from TiCl$_4$, N$_2$, and Mg, affords the lactam **550**[406]. The intermolecular *O*-enolate trapping and cyclization take place by the carbonylation of an alkenyl iodide with 1,3-cyclohexanedione to give the enol lactone **551**[407].

In the presence of a double bond at a suitable position, the CO insertion is followed by alkene insertion. In the intramolecular reaction of **552**, different products, **553** and **554**, are obtained by the use of different catalytic species[408,409]. Pd(dba)$_2$ in the absence of Ph$_3$P affords **554**. PdCl$_2$(Ph$_3$P)$_3$ affords the spiro β-keto ester **553**. The carbonylation of *o*-methallylbenzyl chloride (**555**) produced the benzoannulated enol lactone **556** by CO, alkene, and CO insertions. In addition, the cyclobutanone derivative **558** was obtained as a byproduct via the cycloaddition of the ketene intermediate **557**[410]. Another type of intramolecular enone formation is used for the formation of the heterocyclic compounds **559**[411]. The carbonylation of the 1-iodo-1,4-diene **560** produces the cyclopentenone **561** by CO, alkene, and CO insertions[409,412].

The reaction of *o*-iodophenol, norbornadiene and CO proceeds via alkene and CO insertions to afford the lactone **562**, which is converted into coumarin (**563**) by the retro-Diels–Alder reaction. In this coumarin synthesis, norbonadiene behaves as a masked acetylene[413].

Carbonylation of halides in the presence of terminal and internal alkynes produces a variety of products. The substituted indenone **564** is formed by the reaction of *o*-diiodobenzene, alkyne, and CO in the presence of Zn[414].

The carbonylation of aryl iodides in the presence of terminal alkynes affords the acyl alkynes **565**. Bidentate ligands such as dppf give good results. When Ph₃P is used, phenylacetylene is converted into diphenylacetylene as a main product[415]. Triflates react similarly to give the alkynyl ketones **566**[416]. In the presence of Zn–Cu couple as a reducing agent, aryl alkenyl ketones **567** are obtained by co-catalysis of Pd and Cp$_2$TiCl$_2$[417].

The alkynyl ketones formed by the carbonylation react further intramolecularly with nucleophiles. Methyl *o*-iodophenylacetate reacts with terminal alkynes and CO to produce the alkynyl ketone **568,** which undergoes Pd(II)-catalyzed intramolecular carbopalladation to afford the 3-substituted 2-benzylidene-2,3-dihydro-1*H*-inden-1-one (**569**) in a one-pot reaction. Pd(II) species is generated *in situ* in some way[418]. The carbonylation of *o*-iodophenol (**570**) with phenylacetylene in anisole using AcOK as a base and Pd(Ph$_3$P)$_4$ gives (*Z*)-aurone (**571**) in 82% yield[413]. On the other hand, carbonylation under pressure (20 atm) in Et$_2$NH using dppf affords the flavone **572**[419]. The chemoselectivity for the formation of either **571** or **572** is dependent on the substituents in both reactants and the reaction conditions[413,420]. Similarly, the 2-substituted 1,4-dihydro-4-oxoquinoline **573** is prepared from iodoaniline[421,422]. In the presence of Et$_2$NH, a 2-aryl-4-dialkylaminoquinoline is formed[423]. 6-Phenyl-4-pyridone-3-carboxylate (**575**) was obtained by the carbonylative coupling of the bromoenamine **574** with phenylacetylene[424].

The 2-substituted 3-acylindoles **579** are prepared by carbonylative cycliza-tion of the 2-alkynyltrifluoroacetanilides **576** with aryl halides or alkenyl tri-flates. The reaction can be understood by the aminopalladation of the alkyne with the acylpalladium intermediate as shown by **577** to generate **578**, followed by reductive elimination to give **579**[425].

The carbonylation of iodobenzene with the benzylacetylene **580** affords the (*E*)-3-arylidenebutenolide **582** by carbonylation of the benzoyl alkyne formed as a primary product[426]. The vinylpalladium **581** is formed by the addition of

Pd hydride. Subsequent enolate formation, double bond isomerization, and carbonylation give the butenolide **582**.

The reaction of iodobenzene and the substituted propargyl alcohol **583** under CO (10 atm) and CO₂ (10 atm) affords the 3(2*H*)-furanone **586**[427]. The carbonate **585** is formed by the addition of CO₂ to the acylalkynyl alcohol **584**, and its decarboxylative rearrangement affords the 3(2*H*)-furanone **586**.

In the presence of KCN, cyanocarbonylation of iodobenzene takes place to form benzoyl cyanide (**587**)[428].

1.1.4 Reactions with Organometallic Compounds of the Main Group Metals via Transmetallation

Aryl- or alkenylpalladium halide complexes formed by the oxidative addition undergo transmetallation with alkyl, aryl, alkenyl, allyl, and benzyl compounds of main group elements, and then carbon–carbon bond formation takes place by reductive elimination as a final step. This process offers powerful methods for cross-coupling and is very useful in organic synthesis[429,430]. Depending on the metals and the reaction conditions, the homocoupling products are obtained as byproducts. The reactions of alkynyl metal compounds with halides are treated in Section 1.1.2. Li and Mg reagents are widely used in organic synthesis, but neither tolerates a wide variety of functional groups on either coupling partner, and often homocoupling of the organic halides is observed. Organometallic reagents containing metals of intermediate electropositive character generally lead to higher yields of the coupled products and fewer side-reactions. In addition, these organometallic reagents tolerate a wide range of functional groups in either or both of the coupling partners. From these standpoints, organozinc, -tin, and -boron compounds are particularly useful. Easy preparation of stereo- and regiodefined alkyl- and alkenylmetals via hydrometallation or carbometallation of alkenes and alkynes enhances the usefulness of the coupling.

$$Ar-X + Pd(0) \longrightarrow Ar-Pd-X \xrightarrow{MR} Ar-Pd-R \longrightarrow Ar-R + Pd(0) + (Ar-Ar + R-R)$$

Another important reaction via transmetallation is carbon–metal bond formation by reaction with bimetallic reagents. This is a useful synthetic method for various main group organometallic reagents.

$$Ar-X + Pd(0) \longrightarrow Ar-Pd-X \xrightarrow{RM-M'R'} Ar-Pd-MR + XM'R'$$
$$Pd(0) + Ar-MR$$

1.1.4.1 Organolithium and -magnesium compounds. Compared with extensive studies carried out on the Ni-catalyzed transmetallation reaction of Grignard reagents[431,432], few examples of the Pd-catalyzed reactions of Mg are known. Li and Mg reagents are used mainly for the preparation of other metal compounds such as Sn, Zn, and B compounds.

The Li compound **588** formed by the *ortho*-lithiation of *N,N*-dimethylaniline reacts with vinyl bromide to give the styrene derivative **589**[433]. The 2-phenylindole **591** is formed by the coupling of 1-methyl-2-indolylmagnesium formed *in situ* from the indolyllithium **590** and MgBr$_2$, with iodobenzene using dppb[434]. 2-Furyl- and 2-thienyllithium in the presence of MgBr$_2$ react with alkenyl halides[435]. The arylallenes **592** and 1,2,4-alkatrienes are prepared by the coupling reaction of the allenyllithium with aryl or alkenyl halides[436].

The coupling of Grignard reagents with aryl and alkenyl iodides and bromides is catalyzed by Pd[437–439]. The Pd-catalyzed coupling of alkenyl halides with alkenylmagnesium affords the conjugated diene **593**[440]. A striking difference in reactivity between the (*E*)- and (*Z*)-alkenyl bromides **594** and **595** is observed. The *E*-isomer **594** is more reactive than the *Z*-isomer **595**[441]. The coupling of aryl Grignard reagents with aryl iodides offers a good method for biphenyl formation[442]. In the coupling of *p*-iodophenol with aryl Grignard reagent, 2–3 equiv. of the Grignard reagent are used without the protection of the phenol[443].

The coupling of *n*-alkyl Grignard reagents proceeds smoothly with Pd(Ph₃P)₄[444]. Branched Grignard reagents such as isobutylmagnesium bromide (**596**) can be coupled satisfactorily by using of dppf as a ligand[445]. Poor results are obtained with Ph₃P. The coupling of bromostyrene (**597**) with *s*-butylmagnesium without isomerization took place with dppf as a ligand, and isomerization was observed when Ph₃P was used[446]. The selective monoalkylation or arylation of *p*-dichlorobenzene with Grignard reagents to give **598** takes place using dppf or dppb as a ligand[447]. The coupling of 1,1-dichloro-2-phenylethylene with phenylmagnesium bromide proceeds stereoselectively to give (*Z*)-1-chloro-1,2-diphenylethylene (**599**). Then **599** reacts with another Grignard reagent to give a triarylethylene[448].

The coupling of alkyl Grignard reagents with alkyl iodides to afford alkanes by use of dppf as a ligand has been reported[449], but re-examination of the reaction has shown that only reduction takes place, and no coupling was observed[450].

1.1.4.2 Organozinc compounds. Cross-coupling of organozinc reagents with alkenyl and aryl halides proceeds generally with high yields and tolerates a wide range of functionality. Organozinc reagents are prepared most conveniently *in situ* by the reactions of organolithium, magnesium, or aluminum compounds with $ZnCl_2$ and used for the coupling with alkenyl and aryl halides[451]. The reaction of reactive halides with Zn–Cu couple is another method. Organozincs are useful reagents for selective carbon–carbon bond formation[430,452]. In order to clarify the scope of the Pd-catalyzed cross-coupling with respect to metals in the organometallic reagents, the reactions of various (E)-octenylmetals **600** with (E)-1-hexenyl iodide using $Pd(Ph_3P)_4$ as a catalyst have been carried out [453]. Zn, Cd, Al, and Zr gave the desired cross-coupling product **601** in higher than 70% yields within 1–6 h at room temperature, whereas Li, Mg, Hg, B, Si, Sn, Ti, and Ce did not. Furthermore, Zn reagents are far more efficient than those containing Al and Zr. Unreactive α-(phenylthio)alkenylstannane is converted into the zinc reagent and then used for coupling[454]. However, it should be noted that each metal reagent has its own optimum conditions, and different results may be obtained by selecting different conditions. This is particularly true for B and Sn compounds. Alkyl groups can be introduced without β-elimination by use of alkyl zinc reagents. Organozinc reagents are inert to ketones and esters, and used without their protection. These features are another advantage of organozinc reagents. A variety of zinc reagents, such as *n*-alkyl, benzyl, 2-phenylethyl, homoallyl, and homopropargylzinc reagents, have been shown to couple with alkenyl halides satisfactorily using $PdCl_2(Ph_3P)_2$ + DIBAL, or $Pd(Ph_3P)_4$ as a catalyst[455–457].

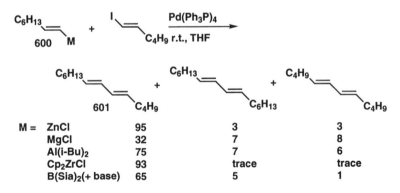

M =		601		
ZnCl		95	3	3
MgCl		32	7	8
Al(i-Bu)₂		75	7	6
Cp₂ZrCl		93	trace	trace
B(Sia)₂(+ base)		65	5	1

The couplings of aryl- and alkenylzinc reagents with various halides have widespread uses for the cross-coupling of aromatic rings[430,458–460]. The reaction tolerates the presence of some functional groups such as amino and cyano groups[461]. The reaction of zinc derivatives of aromatic and heteroaromatic compounds with aryl and heterocyclic halides has wide synthetic applications in the cross-coupling of these rings[462]. The coupling of the imidazolylzinc chloride **602** and 2-bromopyridine (**603**) is one example[463]. For aryl–aryl coupling, Zn seems to be among the most satisfactory metals[460].

Stereospecific alkenyl–alkenyl coupling to form conjugated dienes is possible by the Zn method. The cross-coupling of the alkenylaluminum **604**, prepared by the hydroalumination of terminal alkynes, with aryl and alkenyl halides to give aryl alkenes and dienes proceeds with a Pd catalyst in the presence of $ZnCl_2$. The reaction proceeds via double transmetallation from Al to Zn and then to Pd. The reaction of alkenyl halides with alkenylaluminums is stereospecific and useful for diene synthesis with definite E and Z structures[451]. The method has been applied to stereoselective syntheses of pheromones[464]. The sex pheromone(8Z,10Z)-8,10-dodecadienyl acetate (**605**) has been prepared in a 97% isomerically pure form by stereospecific alkenyl–alkenyl coupling[465]. Disubstituted alkenylaluminums are prepared by the hydroalumination of internal alkynes or carboalumination of terminal alkynes. The coupling of these alkenylaluminums with halides proceeds smoothly with Pd catalysts in the presence of $ZnCl_2$[466]. The coupling of the alkenylcuprates **606** with aryl and alkenyl halides is also carried out in the presence of $ZnBr_2$[467].

Regiocontrolled α- or β-alkenylation and arylation of cyclic enones are possible without protection of the ketone by applying the coupling reaction of the α- or β-halo enones **607** and **608** with aryl and alkenylzinc reagents[468,469].

Triflates are used for the reaction[470]. The 5-phenyltropone **609** is prepared by coupling of the triflate with phenylzinc chloride[471]. Instead of the expensive triflate, phenyl fluoroalkanesulfonate as a triflate equivalent is used for coupling[472]. Phenyl fluorosulfonate (**610**) is another reagent used for coupling[473].

In the reaction of the 1,1-dichloro-1-alkene **611** with phenylzinc chloride, only monoarylation takes place regioselectively to give the (Z)-1-chloro-1-phenylalkene **612**[468,474].

The reaction of aryl and alkenyl halides and triflates with the Reformatsky reagent **613** in polar solvents affords the α-aryl carboxylic esters **614**[475,476]. Facile elimination of β-hydrogen takes place with α-alkylated Reformatsky reagents. Iodides of *N*-heterocyclic compounds such as pyridine, quinoline, and pyrimidine react smoothly with Reformatsky reagents. The position of iodine with respect to the ring nitrogen determines the ratio of the cross- and homocoupling products. The cross-coupling of **613** takes place smoothly with 4-iodo-2,6-dimethyl- and 2-iodo-4,6-dimethylpyrimidine (**615**), but no reaction takes place with 5-iodo-2,4-dimethylpyrimidine[477].

Interestingly, alkylzinc reagents which have β-hydrogens undergo coupling smoothly without the elimination of β-hydrogen to give alkylarenes or alky-lalkenes. 1,5-Dienes and 1,5-enynes are prepared by the coupling of homo-allylic and homopropargylic organozincs with alkenyl halides[456]. This means that reductive elimination is faster than the elimination of β-hydrogen. The zinc homoenolate **616** reacts with aryl and other halides without β-elim-ination to give the 3-arylpropionate **617**[478]. Tri(*o*-tolyl)phosphine and dppf are good ligands. The zinc homoenolate **619** is generated by the reaction of 1-alkoxy-1-siloxycyclopropane **618** with $ZnCl_2$, and react with halides[479]. Furthermore, the 3-, 4-, 5-, 6-, and 7-iodozincalkyl ketones **620** react smoothly with various halides and triflates[470]. Alkyl–alkenyl coupling provides a start-ing compound for the sex pheromone, (2*E*,13*Z*)-2,13-octadecadienyl acetate **621**[480]. The Pd-catalyzed coupling of the (dialkoxyboryl)methylzinc reagent **622** with alkenyl iodide or bromide affords synthetically useful stereo-defined allylic boronate, which reacts without isolation intramolecularly with aldehyde to give a spiro compound[481].

Alkylzinc reagents are prepared *in situ* from organolithium or magnesium and $ZnCl_2$ in the presence of a catalyst and halides. Preparation from alkyl halides and Zn gives lower yields[457,482–484]. Alkylzinc formation from alkyl iodides and $ZnEt_2$ proceeds more efficiently by the catalysis of Pd, and intramolecular carbozincation of the alkene **623**, followed by coupling with alkenyl iodide, is catalyzed by $PdCl_2(dppf)$[485]. Perfluoroalkylzinc iodide generated *in situ* from the iodide **624** and Zn powder under ultrasound irradiation couples with alkenyl bromides[486].

93% *ee* was obtained by the cross-coupling of the Grignard reagent **625** with vinyl bromide in the presence of $ZnCl_2$ using the ferrocenylphosphine BPPFA as a chiral ligand[487]. The bromide in (2-bromoethenyl)diisopropoxyborane (**626**) reacts with alkenylzinc reagent first under neutral conditions, then the produced alkenylborane **627**, without isolation, is attacked by iodobenzene by the addition of LiOH to afford the 1,4-disubstituted diene **628** in a high yield[488].

The optically active 1-bromoallene **629** reacts with diphenylzinc to give the phenylallene **630** in a high *ee* by the inversion of the configuration[489]. Allenyl halides react with organozinc reagents to give substituted allenes[490]. The reaction of the optically active 1-bromoallene **631** proceeds with a high degree of inversion of the configuration around the allenyl moiety, but retention was observed with the allenyl iodide[489, 491,492]. Various allenyl metals (Mg, Cu, Li, Zn) react with halides to give allenyl derivatives[493]. Marasin (nona-6,8-diyne-3,4-dienol) **(634)** has been synthesized by the coupling of the alkynyl bromide **632** with the allenylzinc reagent **633**[494].

Diketene **(635)** is converted into 3-phenyl-3-butenoic acid **(636)** by the reaction of phenylzinc, magnesium, and aluminum reagents via C—O bond cleavage[495].

1.1.4.3 Organoboron compounds.   The coupling of organoboron compounds with aryl, alkenyl, and alkynyl halides is one of the most useful coupling reactions. These cross-coupling reactions are called the Suzuki reactions or Suzuki–Miyaura coupling. The reaction proceeds via transmetallation in the presence of bases[496–499]. No reaction takes place under neutral conditions. This is a characteristic feature of boron chemistry, which is different from that of other organometallic reagents. The role of the base is explained by activation of either Pd or boranes. Most likely, the formation of Ar—Pd—OR (OR = base) from Ar—Pd—X facilitates the transmetallation with organoboranes[500]. Various aryl, alkenyl, and even alkylborane reagents of different reactivity can be used for the coupling with aryl, alkenyl, alkynyl, and some alkyl halides, offering very useful synthetic methods.

$$\text{Ar-Pd-X} \xrightarrow{\text{NaOR'}} \text{Ar-Pd-OR'} \longrightarrow \text{Ar-Pd-R} \; + \; \text{R'OB} \Big\langle$$
$$\underset{\text{R-B}-}{}$$

The coupling reaction of the aryl and heteroarylboronic acids **637** and **638** with aryl and heterocyclic halides and triflates is a very useful method for the cross-coupling of various aromatic and heteroaromatic rings. Numerous examples are known[501–505d]. The coupling of arylboronic acids can be carried out smoothly in water or aqueous DMF as a solvent using $Pd(OAc)_2$ without Ph_3P. The reaction is slower with the addition of a ligand[506]. Similar coupling can be carried out using Pd on charcoal as a catalyst in EtOH in the presence of Na_2CO_3 without a ligand to give coupling products in good yields[507]. The reaction has been successfully applied to large-scale productions of medicinal compounds such as the 4-arylphenylalanine **640** by the coupling of tyrosine triflate (**639**) in toluene in the presence of K_2CO_3 under heterogeneous conditions[508]. Similarly, the substituted β-lactam **641** is produced on a large scale by the coupling of an enol triflate[509]. Boron compounds after the reactions can be handled and disposed more easily than Zn and Sn compounds used for similar cross-coupling. For this reason, the boron method is superior to Zn and Sn methods in coupling on a commercial scale. Chlorobenzene is activated by coordination of $Cr(CO)_3$. The activated chlorine in **642**, although it is deactivated by an electron-donating methoxy group, is more reactive than aryl bromide and reacts with *p*-bromophenylboronic acid. The bromide remains intact[510].

Polyphenylene polymers can be prepared by this coupling. For example, the preparation of poly(*p*-quaterphenylene-2,2′-dicarboxylic acid) (**643**) was carried out using aqueous $NaHCO_3$ and a water-soluble phosphine ligand (DPMSPP)[511]. Branched polyphenylene was also prepared[512].

Poor yields are obtained in the coupling of *ortho*-substituted arylboronic acids[506]. $Ba(OH)_2$ as a base gives good results for the coupling of these sterically hindered compounds[513], but unsatisfactory results are observed

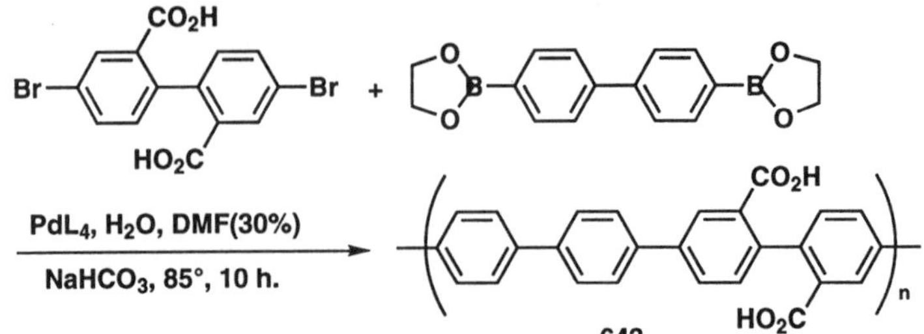

643

L = P(Ph)₂(m-C₆H₄SO₃Na)

with unhindered arylboronic acids. Good yields are obtained in the coupling of arylboronic acids with sterically hindered methyl 5-bromo-4,6-dimethylnicotinate (**644**) only with the use of tri(*p*-tolyl)phosphine. Arylboronate dianion is the reactive intermediate, and the use of boronic acid free from diphenyllboronic acid and triphenylboron is important[506]. The coupling of arylboronic acids with the monocyclic heteroaryl chlorides **645** proceeds smoothly using Pd(dppb)Cl₂ as a catalyst, whereas Pd(Ph₃P)₄ is a good catalyst for the coupling of chloroquinoline derivatives[514,515]. The reaction is useful for the introduction of an aryl group into tropone to yield **646**[516]. Protonolysis of arylboronic acids is observed in some cases, particularly by *ortho*-substitution with methoxy or Cl, but it can be avoided by converting the boronic acids into the corresponding esters of 1,3-propanediol **647** and using K₃PO₄ or Cs₂CO₃ in anhydrous DMF[513].

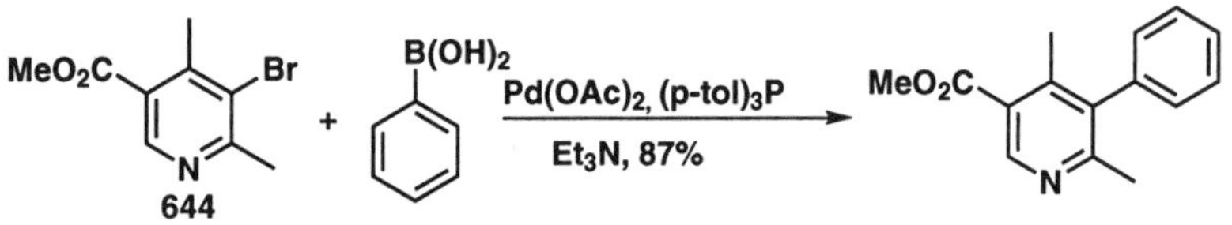

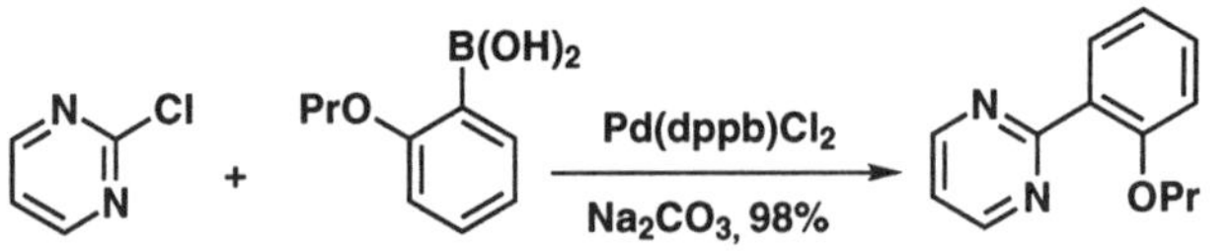

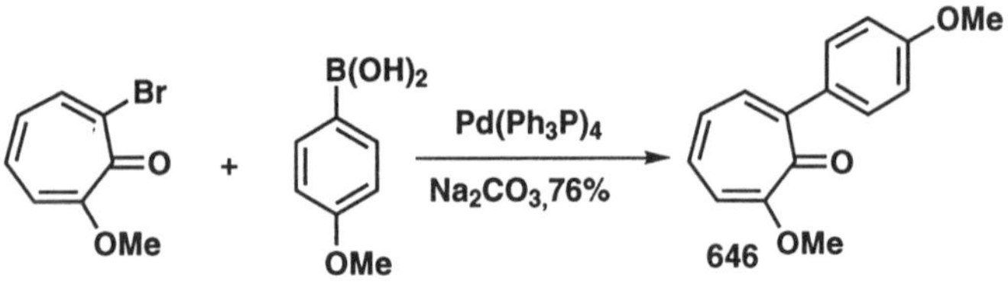

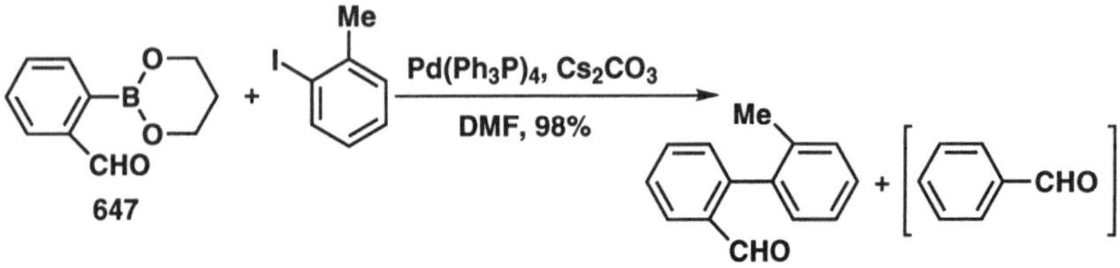

The coupling of alkenylboranes with alkenyl halides is particularly useful for the stereoselective synthesis of conjugated dienes of the four possible double bond isomers[499]. The *E* and *Z* forms of vinylboron compounds can be prepared by hydroboration of alkynes and haloalkynes, and their reaction with (*E*) or (*Z*)-vinyl iodides or bromides proceeds without isomerization, and the conjugated dienes of four possible isomeric forms can be prepared in high purity.

However, in the reaction of 1-alkenylboranes with aryl- or 1-alkenyl iodides, 2-aryl-1-alkenes **648** are obtained as the main products. When Pd metal produced from Pd(OAc)$_2$ as a catalyst and Et$_3$N as a weak base are used, abnormal products are formed. On the other hand, normal products **649** are obtained by using NaOH[517].

base: Et$_3$N   96 : 4 (94%)
NaOH   44 : 56 (86%)

This method of diene formation with definite *E* and *Z* structures has wide synthetic applications [518], particularly for the syntheses of natural products with conjugated polyene structures. Bombykol and its isomers (**650** and **651**) have been prepared by this method[519]. The synthesis of chlorothricolide is another example[520]. The (*Z, E, E*)-triene parts of the leukotriene **652** and DiHETE were constructed by coupling of (*Z, E*)-dienylborane with an (*E*)-alkenyl iodide[236,521,522].

In the total synthesis of the naturally occurring big molecule of palytoxin, which has numerous labile functional groups, this coupling is the most useful for the creation of *E, Z*-conjugated diene part **653**. In this case, thallium hydroxide as a base accelerates the reaction 1000 times more than KOH[523]. Even Tl_2CO_3 can be used instead of a strong base in other cases[524].

Triflates are sensitive to bases, but they can be applied to cross-coupling using K_3PO_4 in dioxane[525]. The primary alkylboranes **654** derived by the hydroboration of terminal alkenes with 9-BBN are coupled with aryl and alkenyl halides and triflates under properly selected conditions. The reaction proceeds smoothly without elimination of β-hydrogen using $PdCl_2(dppf)$ or $Pd(Ph_3P)_4$ and K_3PO_4 in dioxane or DMF[526]. The alkenyl sulfide **655** was prepared by this method[527]. Instead of 9-BBN, the stable oxygenated derivative of trimethylsilylmethyl-9-BBN (**656**) can be used for the preparation of the allylic silane **657** by coupling with alkenyl bromide. Benzyl- and propargyl-trimethylsilanes are prepared similarly[528].

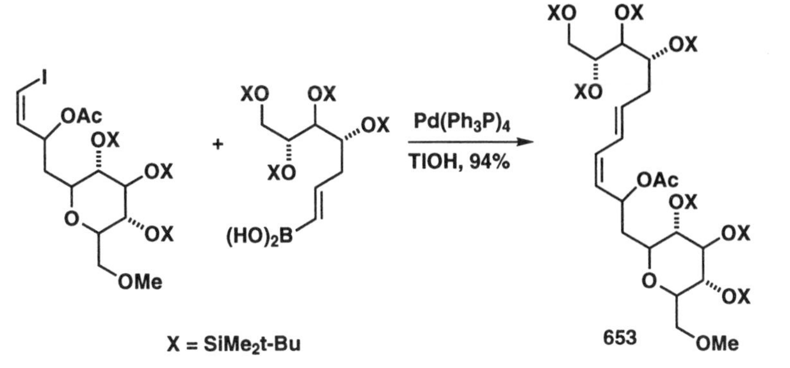
XO
OX
OX
OAc
Pd(Ph3P)4
TlOH, 94%
(HO)2B
OMe
X = SiMe2t-Bu
653

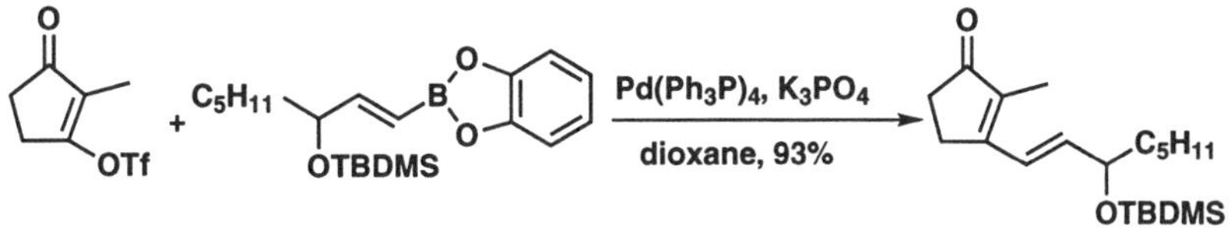
OTf
C5H11
OTBDMS
Pd(Ph3P)4, K3PO4
dioxane, 93%

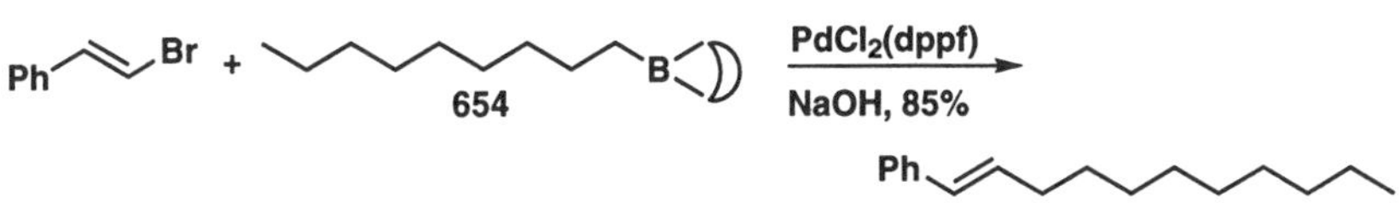
Ph
Br
654
PdCl2(dppf)
NaOH, 85%

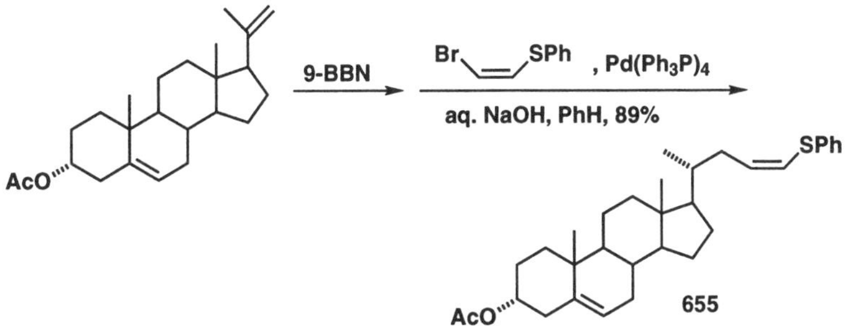
AcO
9-BBN
Br
SPh
, Pd(Ph3P)4
aq. NaOH, PhH, 89%
655

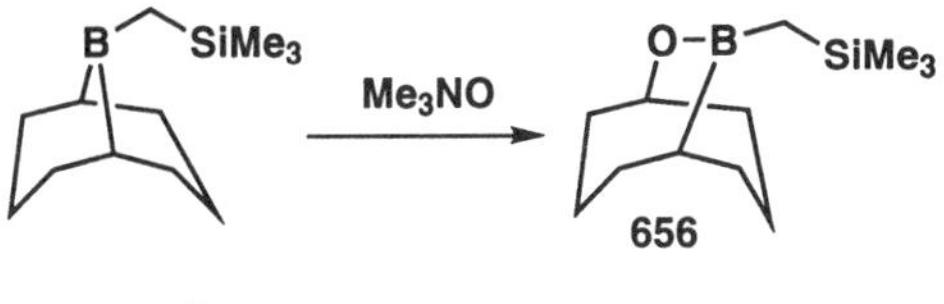
B
SiMe3
Me3NO
O–B
656

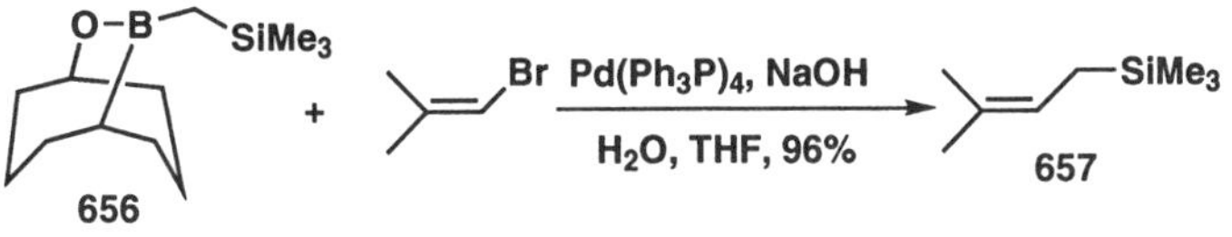
O–B
SiMe3
656
Br
Pd(Ph3P)4, NaOH
H2O, THF, 96%
657

The intramolecular cross-coupling of alkenyl triflate with alkylborane, prepared by *in situ* hydroboration of the double bond in **658** with 9-BBN, is applied to the annulation reaction to give **659** using K_3PO_4 as a base[525]. Sodium tetraphenylborate (**660**) is used for the cross-coupling with aryl and alkenyl halides and triflates. Under basic conditions, two phenyl groups of tetraphenylborate are utilized for the coupling[529,530]. The stereo-defined conjugated enyne **662** is prepared by the coupling of alkynyl bromide **661** with alkenylborane[499].

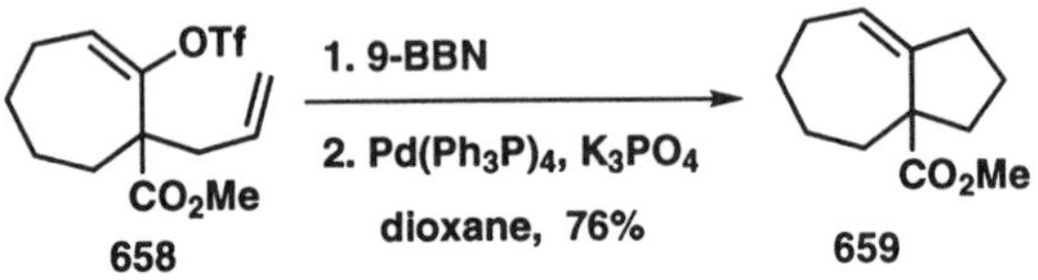

Organoborons react only in the presence of bases. In the following reactions, butylzinc reacts with the alkenyl bromide, obtained by bromoboration of 1-octyne, under neutral conditions to give **663** without attacking the organoboron part, which then reacts with aryl halide by the addition of a base to give a styrene derivative[531]. Similarly, α, β-unsaturated ketones are prepared by two step reactions of the α-methoxyvinylzinc reagent **664** as a methyl ketone precursor under neutral conditions and organoborane moiety under basic conditions[532]. When both alkylborane and alkyltin functions are present in the same molecule, the borane moiety reacts selectively under basic conditions [533]. Under basic conditions, organoborane is more reactive and reacts chemoselectively with halides to afford **665** without attacking the organotin moiety in the same molecule. Similarly **666** was prepared chemoselectively.

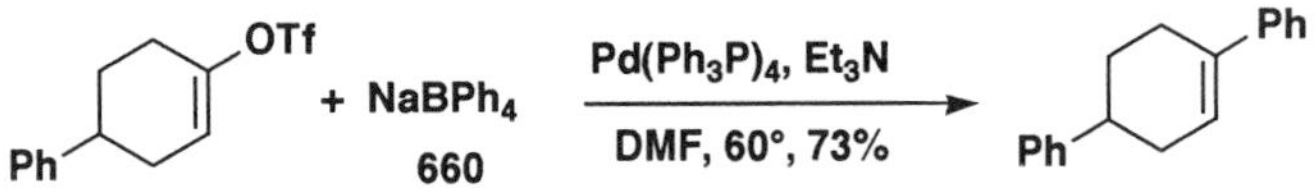

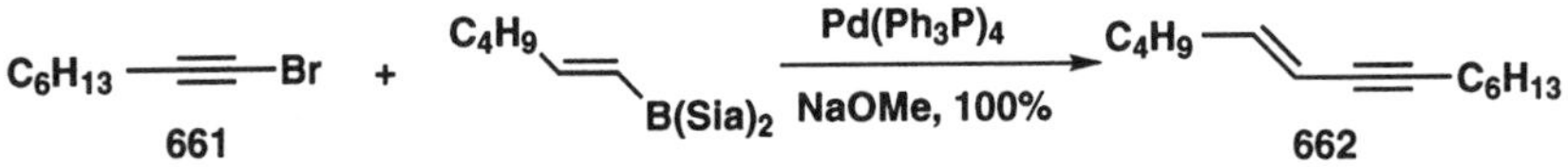

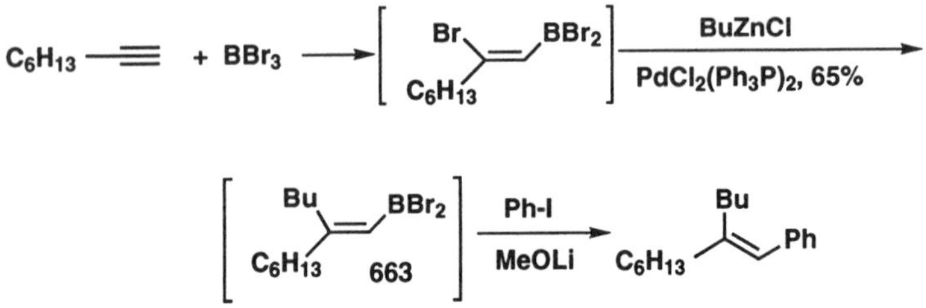

The thioboration of terminal alkynes with 9-(alkylthio)-9-borabicyclo[3.3.1]-nonanes (9-RS-9-BBN) proceeds regio- and stereoselectively by catalysis of Pd(Ph₃P)₄ to produce the 9-[(Z)-2-(alkylthio)-1-alkenyl]-9-BBN derivative **667** in high yields. The protonation of the product **667** with MeOH affords the Markownikov adduct **668** of thiol to 1-alkyne. One-pot synthesis of alkenyl sulfide derivatives **669** via the Pd-catalyzed thioboration–cross-coupling sequence is also possible. Another preparative method for alkenyl sulfides is the Pd-catalyzed cross-coupling of 9-alkyl-9-BBN with 1-bromo-1-phe-nylthioethene or 2-bromo-1-phenylthio-1-alkene[534].

Benzyl halides can be used for the coupling with organoboranes. The Pd-catalyzed reaction of the organoboroxine **670** with *o*-bis(bromomethyl)benzene (**671**) produced the cross-coupled product **673** and corresponding bifuran **676**. These unusual reactions are explained by the following mechanism. The oxidative addition and transmetallation generate the complex **672**. The reductive elimination and intramolecular coupling of the benzylpalladium with the furan give **673**. Also, **672** undergoes oxidative addition and transmetallation of another benzyl bromide to give the bis(organopalladium) complex **674**. Although no example is known, the disproportionation of **674** may give the difurylpalladium complex **675**, reductive elimination of which affords difuran **676**[535].

The most interesting and difficult cross-coupling is alkyl–alkyl coupling, because oxidative addition of alkyl halides having β-hydrogen is slow. In addition, easy elimination of β-hydrogen is expected after the oxidative addition.

Although this is not a reaction of aryl or alkenyl halides, the alkyl–alkyl coupling of alkyl halides with alkylboranes is treated here for convenience and better understanding (also treated in Section 1.3) Successful alkyl–alkyl coupling has been achieved only by using alkylboranes. Hexyl iodide reacts with 9-octyl-9-BBN smoothly in the presence of K_3PO_4 in dioxane to give tetradecane in 64% yield[536,537]. It is remarkable that hexyl iodide undergoes smooth oxidative addition and transmetallation without β-elimination.

$$9\text{-}C_8H_{17}\text{-}9\text{-BBN} + \quad \diagdown\diagup\diagdown\diagup\diagdown\diagup\diagdown I \xrightarrow[\text{dioxane, 64\%}]{Pd(Ph_3P)_4,\ K_3PO_4}$$

$$\diagup\diagdown\diagup\diagdown\diagup\diagdown\diagup\diagdown\diagup\diagdown\diagup\diagdown\diagup\diagdown$$

1.1.4.4 Organoaluminum and -zirconium compounds. Coupling of alkenyl-alanes with alkenyl halides to give *E, E-* and *E, Z*-conjugated dienes is an early example of the Pd-catalyzed cross-coupling[451]. The enol phosphate **678** derived from **677** is displaced with a methyl group of Me_3Al using Pd catalyst in dichloroethane. 4-*t*-Butylcyclohexanone (**677**) can be converted into 2-methyl-5-*t*-butylcyclohexanone (**679**) using this reaction[538]. When Et_3Al is used, the displacement of triflate with an ethyl group takes place in polar solvents to give **680**[539]. However, the hydrogenolysis takes place in non-polar solvents[540,541]. In this case, the ethylpalladium undergoes rapid elimination of β-hydrogen to give the Pd hydride, and the reductive elimination affords the hydrogenolyzed product (Section 1.1.6) The allyltri-methylsilane **682** can be introduced by the cross-coupling of enol triflates with dimethyl(trimethylsilylmethyl)aluminum (**681**)[542]. Aryl halides and triflates are coupled with alkyl groups by the treatment with Me_3Al, Et_3Al, and Bu_3Al[543,544]. Alkynyl iodides are also used for the coupling. The alkenylalane **683**, obtained by highly stereo- and regioselective Zr-catalyzed carboalumination of an alkyne, undergoes Pd-catalyzed cross-coupling with 1-iodoalkynes in the presence of $ZnCl_2$ to give the stereodefined enyne **684**. Formation of alkenylzinc by transmetallation is expected in the reaction[466].

The alkenylzirconium **685**, prepared by hydrozirconation of a terminal alkyne with hydrozirconocene chloride, reacts with alkenyl halide to afford the conjugated diene **686**[545]. The Zr reagent can be used even in the presence of the carbonyl group in **687**, which is sensitive to Al and Mg reagents.

1.1.4.5 Organotin compounds. Organotin compounds (organostannanes) are used extensively for cross-coupling, which is sometimes called Stille coupling or Migita-Kosugi-Stille coupling. Aryl-, alkenyl- and alkylstannanes are used for coupling with aryl and alkenyl halides, pseudo-halides, and arenediazonium salts[390,391]. The reaction of allyltin with aryl iodides is the first example of the Pd-catalyzed cross-coupling of organotin reagents[546]. Generally only one of four groups on the tin enters into the coupling reaction.

$$\text{R-X} + \text{R'SnR''}_3 \xrightarrow{\text{Pd(0)}} \text{R-R'} + \text{X-SnR''}_3$$

Different groups are transferred with different selectivities from tin. A simple alkyl group has the lowest transfer rate. Thus unsymmetrical organotin reagents containing three simple alkyl groups (usually methyl or butyl) are chosen, and the fourth group, which undergoes transfer, is usually alkynyl, alkenyl, aryl, benzyl, and allyl groups. The cross-coupling of these groups with aryl, alkenyl, alkynyl, and benzyl halides affords a wide variety of cross-coupled products, which are difficult to prepare by the uncatalyzed reactions. Usually Ph_3P is used as a ligand. However, a large rate acceleration is observed in some couplings of stannanes when tri-2-furylphosphine and triphenylarsine are used[547]. It is claimed that the ligandless Pd complex prepared *in situ* from $LiPdCl_3$, RX, and $R'SnMe_3$ is an active catalyst for cross-coupling at room temperature[548].

Aryl halides react with a wide variety of aryl-, alkenyl- and alkylstannanes[548–550]. Coupling of an aryl triflate with an arylstannane is a good preparative method for diaryls such as **688**. The coupling of alkenylstannanes with alkenyl halides proceeds stereospecifically to give conjugated dienes **689**. The allylstannane **690** is used for allylation[397,546,551–553]. Aryl and enol triflates react with organostannanes smoothly in the presence of LiCl[554].

The cross-coupling of aromatic and heteroaromatic rings has been carried out extensively[555]. Tin compounds of heterocycles such as oxazolines[556,557], thiophene[558,559], furans[558], pyridines[558], and selenophenes [560] can be coupled with aryl halides. The syntheses of the phenyloxazoline **691**[552], dithiophenopyridine **692**[561] and 3-(2-pyridyl)quinoline **693**[562] are typical examples.

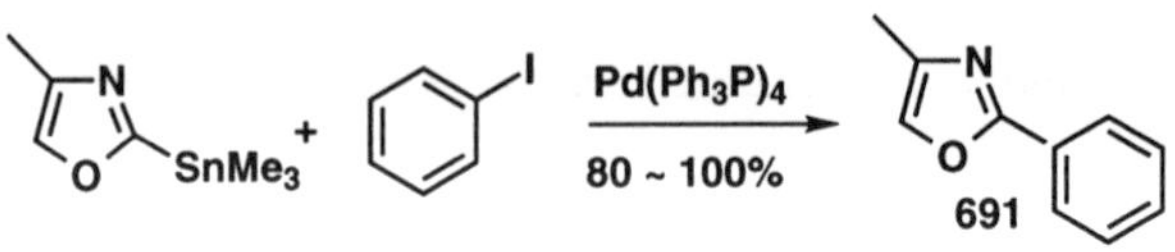

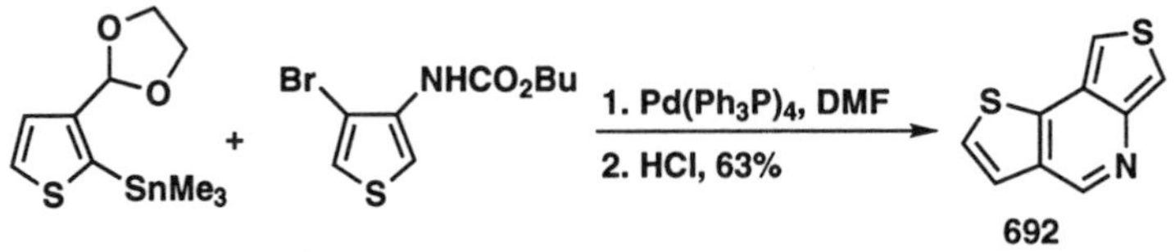

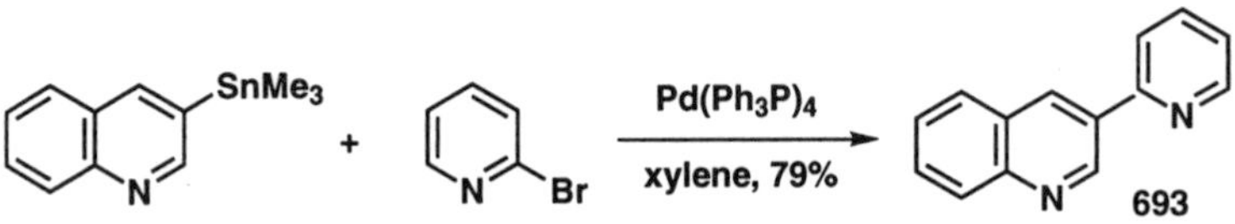

691

692

693

Arenediazonium salts are also used for the coupling[563]. (*Z*)-Stilbene was obtained unexpectedly by the reaction of the α-stannylstyrene **694** by addition–elimination. This is a good preparative method for *cis*-stilbene[564]. The rather inactive aryl chloride **695** can be used for coupling with organostannanes by the coordination of Cr(CO)₃ on aromatic rings[3,565].

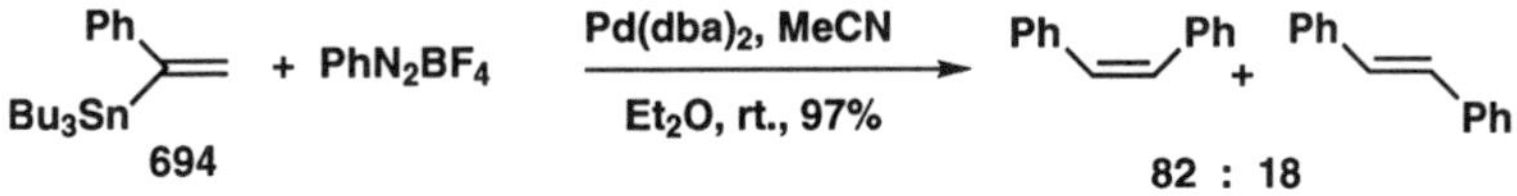

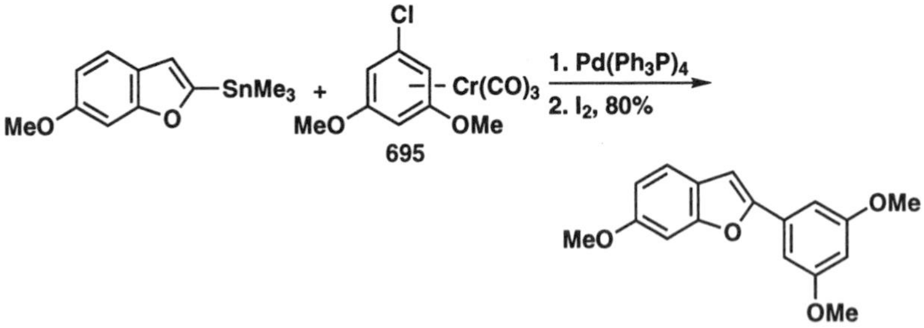

694   82 : 18

695

The effect of some additives and ligands on the cross-coupling of the stannylpyridine **696** has been studied. However, it seems likely that these effects are observed case by case. Faster reaction and higher yields were observed on addition of CuO and Ag₂O[566–568]. Optimum conditions for the coupling of alkenyl triflates with arylstannanes have been studied. Ligandless Pd complexes such as Pd(dba)₂ are the most active catalysts in the reaction of enol triflate. Ph₃P inhibits the reaction. NMP as a polar solvent gives the best results[569]. The use of tri(2-furyl)phosphine and Ph_3As in the coupling of stannanes with halides and triflates **697** increases the rate of the transmetallation of the stannanes to Pd, which is thought to be the rate-determining step of the catalytic cycle. The relative rates are Ph_3P : $(fur)_3P$: Ph_3As : Ph_3As + $ZnCl_2$ = 1 : 3.5 : 95 : 151[547,570,].

CuO	75%	70 min.
Ag_2O	73%	25 min.
none	47%	4 hr.

The use of Ph_3As with $Pd_2(dba)_3$ in the absence of LiCl in DMF gave the best results in the coupling of 3-tributylstannylindole with enol triflate to give **698**[571]. On the other hand, the coupling of highly hindered less reactive electron-rich aryl triflate **699** with methyl-, allyl-, and alkenylstannanes proceeds smoothly by using $PdCl_2(Ph_3P)_2$, Ph_3P, and LiCl in boiling DMF[572]. A procedure for the reaction of p-nitrophenyl triflate with p-methoxyphenylstannane to give the coupled product **700** in 48% yield is available[573]. The coupling of the sterically hindered bromo-α-naphthyl triflate **701** or 2,6-disubstituted phenyl triflates with aryltrimethylstannane proceeds in a moderate yield in the presence of CuBr (0.05 equiv.) as a co-catalyst and LiCl (3 equiv.) without attacking the bromide in the same molecule. Under the usual conditions, the transfer of methyl group takes place to give **702** as a main reaction path[574,575].

The coupling of the enol triflate **703** with the vinylstannane **704**[397] has been applied to the synthesis of glycinoeclepin[576]. The introduction of a (Z)-propenyl group in the β-lactam derivative **705** proceeds by use of tri-2-furylphosphine[577]. However, later a smooth reaction to give the propenyl-lactam in 82% yield was achieved simply by treating with Pd(OAc)₂ in NMP or CH₂Cl₂ for 3–5 min without addition of LiCl and the phosphine ligand[578].

The intramolecular coupling of organostannanes is applied to macrolide synthesis. In the zearalenone synthesis, no cyclization was observed between arylstannane and alkenyl iodide. However, intramolecular coupling takes place between the alkenylstannane and aryl iodide in **706**. A similar cyclization is possible by the reaction of the alkenylstannane **707** with enol triflate[579]. The coupling was applied to the preparation of the bicyclic 1,3-diene system **708**[580].

The intramolecular coupling of an alkenylstannane and alkenyl bromide[397] has been applied to the construction of the polyene macrolactam unit **709** found in leinamycin. The cyclization takes place by using Pd(Ph₃As)₄. The coupling fails when Pd(Ph₃P)₄ is used[581]. In the total synthesis of rapamycin (**712**), its macroring was constructed by intermolecular coupling of two alkenyl iodides in **711** with the vinylenedistannane (**710**) in the presence of PdCl₂(MeCN)₂ and Hunig's base in 28% yield together with unreacted starting material (30%) and an iodostannane intermediate (30%)[582].

For coupling, the cheaper aryl fluorosulfonate **713** is used as an alternative to the expensive aryl triflates to give the same results[473]. The arenesulfonates **714** are active for the reaction with vinylstannanes when dppp and LiCl are used in DMSO[583]. The bromide **715** attacks the arylstannane moiety selectively without reacting with the organoboron moiety in **716** in the absence of a base[584].

The arylation of the 1-tributylstannyl glycal **717** offers a synthetic route to chaetiacandin[585,586]. The Pd-catalyzed reactions of the 3-stannylcyclobute-nedione **718** with iodobenzene, and benzoyl chloride[587], and alkenylation with alkenyl(phenyl)iodonium triflates proceed smoothly by the co-catalysis by CuI[588,589].

β-Alkenylation of cyclopentenone with the alkenylstannane **719** has been used for the introduction of an ω-chain into a prostaglandin derivative[590]. Even the vinyl mesylate (methanesulfonate) **720** can be used for coupling with alkenylstannanes[591].

The reaction of the 1- and (2-ethoxyvinyl)tributylstannanes (**721**) and (**723**) as masked carbonyls with aryl halides proceeds smoothly and the products **722** and **724** are used for further reactions[592,593].

The alkenyl(phenyl) iodonium salt **725** undergoes the facile cross-coupling with vinylstannane to form the conjugated diene **726**[594].

The intramolecular insertion of an internal alkyne into an aryl or alkenyl halide **727** generates aryl- or alkenylpalladium as an intermediate, which is trapped with an organozinc or organostannane to give **728**. Overall *cis* addition to the alkyne takes place[595,596]. The reaction of the alkenylstannane **730** with the 2-bromomethylfuran **729** is used for the introduction of a prenyl group[597].

Benzyl bromide can be converted into ethylbenzene (**731**) by the reaction of Me_4Sn. The use of HMPA as a solvent is important. Overall inversion of configuration takes place at the chiral center of deuterated benzyl bromide[598]. The cyanomethylation[599] and methoxymethylation[600] of aromatic rings are carried out by the reaction of cyanomethyltributyltin (**732**) and methoxymethyltributyltin.

Tin enolates of ketones can be generated by the reaction of the enol acetate **733** with tributyltin methoxide[601] and they react with alkenyl halides via transmetallation to give **734**. This reaction offers a useful method for the introduction of an aryl or alkenyl group at the α-carbon of ketones[602]. Tin enolates are also generated by the reaction of silyl enol ethers with tributyltin fluoride and used for coupling with halides[603].

The ketimine **736** is prepared by the reaction of the imidoyl chloride **735** with organotin reagents[604,605].

Aryl and alkenyl phenyl sulfides are prepared by the reaction of aryl and alkenyl halides and triflates with tributylstannyl phenyl sulfide. 2-Chloropyrimidine (**737**) is used for the coupling[606,607]. The diaryl or divinyl sulfide **739** is prepared by the reaction of distannyl sulfide (**738**)[548]. *N,N*-Diethylaminotributyltin (**740**) reacts with aryl halides to give arylamines[608].

Pd-catalyzed coupling of $(Me_3Sn)_2$ with halides provides a unique synthetic method for organotin reagents. The oxidative addition of halides, transmetallation, and reductive elimination afford the organostannane **741**. Aryl, alkenyl, benzyl, and allyl halides react with $(Me_3Sn)_2$ to afford aryl, alkenyl, benzyl, and allylstannanes. $(Bu_3Sn)_2$ is unreactive[609–612]. The intramolecular coupling between aryl iodides takes place in the presence of $(Me_3Sn)_2$. In this reaction, at first, the aryltin **742** is formed by transmetallation of one of the iodides, and reacts further with another aryl iodide by transmetallation and reductive elimination to give the coupling product **743**[613]. As another example, the spiro ring in **745** is formed by alkene insertion to form the neopentyl-palladium **744**, stannylation, and coupling with aryl iodide using $(Me_3Sn)_2$[614]. The silylstannane **746** is used for the *in situ* generation of vinyl-stannane from vinyl triflate, and used for intramolecular coupling to form **747**[615].

Unexpected *cine* substitution to afford **749**, rather than *ipso* substitution, was observed in the reaction of the vinylstannane **748** derived from camphor with phenyl bromide[616].

1.1.4.6 Organosilicon compounds. Organosilicon compounds are less active than organic compounds of other main group metals mentioned above, and activation is necessary to bring them into Pd-catalyzed reactions. Exceptionally, iodotrimethylsilane behaves as an active halide and undergoes the Heck-type reaction with alkenes to afford the alkenylsilane **750**[617].

The transmetallation of the siloxycyclopropane **751** with the aryl- or alkenylpalladium **752** generates the Pd homoenolate **753**, and subsequent reductive elimination gives the β-aryl or alkenyl ketone **754**[618]. It should be noted that the Pd homoenolate **753** generated in this reaction undergoes reductive elimination without β-elimination.

The transmetallation of Si attached to sp^2 carbons (aryl and alkenyl) is possible in the presence of TASF[619]. Trimethylsilylethylene (**755**) activated by TASF is an ethylene equivalent and reacts with aryl and alkenyl halides to afford the styrene derivative **756** and a diene[620]. The reactivity for the transmetallation is enhanced by using Si compounds attached to alkoxy group or fluoride anion. The reaction proceeds more easily by the action of F$^-$ supplied from TASF and TBAF, forming five-coordinated silicate compounds. The cross-coupling reactions of aryl- and alkenylsilyl compounds **755** and **757** with aryl or alkenyl halides and triflates offer good synthetic methods for biaryls, alkenyl arenes, and conjugated dienes[621,622]. The cross-coupling of halides or triflates with the allyltrifluorosilane **758** in the presence of TBAF is γ-selective, and offers a new approach to regiochemical control in allylic systems[623]. The optically active benzylic trifluorosilane **759** undergoes cross-coupling with aryl triflate with complete retention of chirality at 50 °C. Racemization takes place at higher temperatures[624].

A trialkylsilyl group can be introduced into aryl or alkenyl groups using hexaalkyldisilanes. The Si—Si bond is cleaved with a Pd catalyst, and trans-metallation and reductive elimination afford the silylated products. In this way, 1,2-bis-silylethylene **761** is prepared from 1,2-dichloroethylene (**760**)[625,626]. The facile reaction of $(Me_3Si)_2$ to give **762** proceeds at room temperature in the presence of fluoride anion[627]. Alkenyl- and arylsilanes are prepared by the reaction of $(Me_3Si)_3Al$ (**763**)[628].

Substituted aroyl- and heteroaroyltrimethylsilanes (acylsilanes) are prepared by the coupling of an aroyl chloride with $(Me_3Si)_2$ without decarbonylation, and this chemistry is treated in Section 1.2[629]. Under certain conditions, aroyl chlorides react with disilanes after decarbonylation. Thus the reaction of aroyl chlorides with disilane via decarbonylation is a good preparative method for aromatic silicon compounds. As an interesting application, trimellitic anhydride chloride (**764**) reacts with dichlorotetramethyldisilane to afford 4-chlorodimethylsilylphthalic anhydride (**765**), which is converted into **766** and used for polymerization[630]. When the reaction is carried out in a non-polar solvent, biphthalic anhydride (**767**) is formed[631]. Benzylchlorodimethylsilane (**768**) is obtained by the coupling of benzyl chloride with dichlorotetramethyl-disilane[632,633].

1.1.4.7 Organocopper compounds. Organocopper reagents are used for Pd-catalyzed coupling with halides[548]. The coupling of alkenyl triflate with a 3–4-fold excess of the 1,3-dienyl organocopper reagent **769** gives the coupled product in 80% corrected yield for the recovered triflate. This coupling is more effective than that of organotin reagents[634]. The vinylborane **770**, unreactive under neutral conditions, is converted *in situ* into Cu reagent **771** and used for coupling with aryl iodides[635]. The (α-ethoxycarbovinyl)(dicyclohexylamido)cuprates **773**, prepared from the 2-butynoate **772**, couples with alkenyl iodide to afford the unsaturated ester **774**[636]. Ullmann-type cross-coupling of 3-iodopyridine (**775**) with the Cu reagent **777**, formed *in situ* from *o*-bromonitrobenzene, proceeds smoothly in DMSO by the catalysis of Pd(Ph$_3$P)$_4$ and copper bronze. It is interesting that the Pd(0) reacts chemoselectively with iodopyridine to give **776**, and Cu with bromonitrobenzene to generate **777**[637].

1.1.4.8 Organophosphorus compounds. Phosphorus–carbon bond formation takes place by the reaction of various phosphorus compounds containing a P—H bond with halides or triflates. Alkylaryl- or alkenylalkylphosphinates are prepared from alkylphosphinate[638]. The optically active isopropyl alkenyl-methylphosphinate **778** is prepared from isopropyl methylphosphinate with retention[639]. The monoaryl and symmetrical and asymmetric diarylphosphinates **780**, **781**, and **782** are prepared by the reaction of the unstable methyl phosphinate **779** with different amounts of aryl iodides. Trimethyl orthoformate is added to stabilize the methyl phosphinate[640].

Dialkyl arylphosphonates and alkenylphosphonates are prepared by the coupling of halides or triflates with the dialkyl phosphonate **783**[641–643].

Phosphine oxides are prepared similarly[644]. Selective monophosphinylation of 2,2'-bis[(trifluoromethanesulfonyl)oxy]-1,1'-binaphthyl (**784**) with diphenylphosphine oxide using dppb or dppp as a ligand takes place to give optically active 2-(diarylphosphino)-1,1'-binaphthyl (**785**). No bis-substitution is observed[645,646].

The mixed triarylphosphine **787** can be prepared by the reaction of (trimethylsilyl)diphenylphosphine (**786**) with aryl halides[647]. Ph₃P is converted into the alkenylphosphonium salt **788** by the reaction of alkenyl triflates[648].

1.1.5 Displacement Reactions with Carbon, Oxygen, and Sulfur Nucleophiles

Arylation or alkenylation of soft carbon nucleophiles such as malonate is carried out by using a copper catalyst, but it is not a smooth reaction. The reaction of malononitrile, cyanoacetate, and phenylsulfonylacetonitrile with aryl iodide is possible by using a Pd catalyst to give the coupling products.

The presence of a cyano group seems to be important[649]. The reaction has been successfully applied to halides of pyridine, quinoline, isoquinoline, and oxazoles[650]. An interesting application is the synthesis of tetracyanoquino-dimethane (**789**) by the reaction of *p*-diiodobenzene with malononitrile[651].

No intermolecular reaction of malonate or β-keto esters with halides has been reported, but the intramolecular reaction of β-diketones such as **790** and malonates proceeds smoothly[652,653]. Even the simple ketone **791** can be arylated or alkenylated intramolecularly. In this reaction, slow addition of a base is important to prevent alkyne formation from the vinyl iodide by elimination[654].

α-Naphthylmethyl acetate (**792**) undergoes the displacement reaction with malonate. The reaction is explained by the formation of the π-allylpalladium complex **793**[655].

Aryl, heteroaryl, and alkenyl cyanides are prepared by the reaction of halides[656-658] or triflates[659,660] with KCN or LiCN in DMF, HMPA, and THF. Addition of crown ethers[661] and alumina[662] promotes efficient aryl and alkenyl cyanation. Iodobenzene is converted into benzonitrile (**794**) by the reaction of trimethylsilyl cyanide in Et$_3$N as a solvent. No reaction takes place with aryl bromides and chlorides[663]. The reaction was employed in an estradiol synthesis. The 3-hydroxy group in **796** was derived from the iodide **795** by converting it into a cyano group[664].

The rather unreactive chlorine of vinyl chloride can be displaced with nucleophiles by the catalytic action of PdCl$_2$. The conversion of vinyl chloride to vinyl acetate (**797**) has been studied extensively from an industrial standpoint[665-671]. DMF is a good solvent. 1,2-Diacetoxyethylene (**798**) is obtained from dichloroethylene[672]. The exchange reaction suffers steric hindrance. The alkenyl chloride **799** is displaced with an acetoxy group whereas **800** and **801** cannot be displaced[673,674]. Similarly, exchange reactions of vinyl chloride with alcohols and amines have been carried out[668].

Aryl sulfides are prepared by the reaction of aryl halides with thiols and thiophenol in DMSO[675,676] or by the use of phase-transfer catalysis[677]. The alkenyl sulfide **803** is obtained by the reaction of lithium phenyl sulfide (**802**) with an alkenyl bromide[678].

A Pd-catalyzed reaction of amines with halides is expected, but actually little is known about the reaction. The CDE ring system of lavendamycin (**805**) has been constructed by the intramolecular reaction of aryl bromide with aniline derivative in **804**, but 1.2 equiv. of Pd(Ph$_3$P)$_4$ is required[679].

1.1.6 Hydrogenolysis with Various Hydrides

Hydrogenolysis of aryl and alkenyl halides and triflates proceeds by the treatment with various hydride sources. The reaction can be explained by the transmetallation with hydride to form palladium hydride, which undergoes reductive elimination. Several borohydrides are used for this purpose[680]. Deuteration of aromatic rings is possible by the reaction of aryl chlorides with $NaBD_4$[681].

Formate is an excellent hydride source for the hydrogenolysis of aryl halides[682]. Ammonium or triethylammonium formate[683] and sodium formate are mostly used[684,685]. Dechlorination of the chloroarene **806** is carried out with ammonium formate using Pd charcoal as a catalyst[686]. By the treatment of 2,4,6-trichloroaniline with formate, the chlorine atom at the *para*-position is preferentially removed[687]. The dehalogenation of 2,4-dihaloestrogene is achieved with formic acid, KI, and ascorbic acid[688].

The enone **807** is converted into the dienol triflate **808** and then the conjugated diene **809** by the hydrogenolysis with tributylammonium formate[689,690]. Naphthol can be converted into naphthalene by the hydrogenolysis of its triflate **810**[691–693] or sulfonates using dppp or dppf as a ligand[694]. Aryl tetrazoyl ether **811** is cleaved with formic acid using Pd on carbon as a catalyst[695].

Another method for the hydrogenolysis of aryl bromides and iodides is to use MeONa[696]. The removal of chlorine and bromine from benzene rings is possible with MeOH under basic conditions by use of dippp as a ligand[697]. The reduction is explained by the formation of the phenylpalladium methoxide **812**, which undergoes elimination of β-hydrogen to form benzene, and MeOH is oxidized to formaldehyde. Based on this mechanistic consideration, reaction of alcohols with aryl halides has another application. For example, cyclohexanol (**813**) is oxidized smoothly to cyclohexanone with bromobenzene under basic conditions[698].

Grignard reagents or alkylaluminum compounds bearing a β-hydrogen can be used for hydrogenolysis. The hydrogenolyses of the alkenyl sulfone **814** with isopropylmagnesium bromide (**815**)[699] and the alkenyl phosphate **817** with Et_3Al[541] to give alkenes **816** and **820** are examples. The reaction is explained by transmetallation to give **818**, followed by β-elimination to afford the Pd—H species **819**, which undergoes reductive elimination to afford the alkene **820**. Hydrosilanes are used for the hydrogenolysis of halides[700].

1.1.7 Homocoupling (Reductive Coupling) of Organic Halides

Homocoupling of aryl halides promoted by Cu metal is called the Ullman reaction. The Ullmann coupling of aryl and alkenyl halides using Pd to give symmetrical biphenyl and conjugated dienes proceeds smoothly with Pd(0) species, which is oxidized to PdX$_2$, and some reducing agents are used to regenerate Pd(0) from Pd(II), making the reaction catalytic[701,702]. Hence this reductive coupling of the halides is mechanistically different from all the reactions of aryl halides discussed so far. However, for convenience, reductive coupling catalyzed by Pd(0) and reducing agents is treated here. Coupling of iodobenzene is possible with a catalytic amount of Pd(OAc)$_2$ without adding a reducing agent to give biphenyl in 75% yield. Similarly, although the reaction took 7 days, perylene (**822**) was obtained in 85% yield by the intramolecular coupling of the aryl diiodide **821** with Pd(OAc)$_2$[32]. PdI$_2$ generated by the coupling may be reduced to Pd(0) with an amine derived from DMF or Bu$_4$NBr during a long period of time. Smooth intramolecular coupling of the alkenyl bromide **823** to give the 1,3-diene **824** proceeds using a catalytic amount of Pd(OAc)$_2$ and a stoichiometric amount of Ph$_3$P in the presence of bases. Ph$_3$P reduces PdBr$_2$ formed by the coupling to Pd(0)[703]. The coupling of vinyl chloride to afford butadiene proceeds using Pd(Ph$_3$P)$_4$ as a catalyst in the presence of an amine as a reducing agent for Pd(II)[360]. Many patents have been applied for the efficient homocoupling of aryl halides, particularly chloro- or bromophthalic acid to biphenyltetracarboxylic acid (**825**) using formic acid or alkali metal alkoxides as reducing agents[704]. The coupling of the aryl halides or triflates **826** is carried out in the presence of Zn as the reducing agent[705] or by electrochemical reduction[706,707].

821   Pd(OAc)$_2$, Bu$_4$NBr, DMF   K$_2$CO$_3$, 7 days, 85%   822   + (PdI$_2$)

823   + Ph$_3$P   Pd(OAc)$_2$ K$_2$CO$_3$, 92%   824   + Ph$_3$PO   E = CO$_2$Me

Pd/C, reducing agent   825

Efficient homocoupling of the aryl iodonium salt **827** using Zn is catalyzed by Pd(acac)$_2$[708]. Homocoupling of the arylsulfonyl chloride **828** as a pseudo-halide takes place in the presence of 2 equiv. of Ti tetraisopropoxide[709].

1.1.8 Reactions of Aromatic Compounds

The Pd-catalyzed coupling of aryl halides or triflates with aromatic rings to give biaryl compounds proceeds only intramolecularly[710]. The reaction is carried out with PdCl$_2$ or Pd(OAc)$_2$ as a catalyst in the presence of bases. The oxidative addition to the halide is followed by the palladation of aromatic ring with Pd(II) species to generate diarylpalladium. Finally, reductive elimination gives the coupled product. The synthesis of the naphthylisoquinoline alkaloid **829** is an example[711]. The reaction has been applied to the synthesis of gilvocarcin M (**830**)[712,713]. The regioselective coupling of the aryl triflate **831** has been applied to the syntheses of fluoranthene and benzofluoranthene[714]. The intramolecular coupling of aryl halides with heteroaromatic rings, such as the imidazole **832**[715] and indole, takes place smoothly[716]. The corresponding intermolecular reaction is not possible.

o-Iodoanisole (**833**) undergoes an interesting coupling to give the 6*H*-dibenzo[*b,d*]pyran **838** in a high yield (90%) in one step. The key steps of the reaction are formation of the five-membered *ortho*-palladation products **834** and **836** involving intramolecular C—H activation, and their reaction with iodoanisole to give **835** and **837**[717]. Facile orthopalladation is a driving force.

1.2 Acyl and Sulfonyl Chlorides and Related Compounds

Acyl halides are reactive compounds and react with nucleophiles without a catalyst, but they are activated further by forming the acylpalladium intermediates, which undergo insertion and further transformations. The decarbonylative reaction of acyl chlorides as pseudo-halides to form the arylpalladium is treated in Section 1.1.1.1. The reaction without decarbonylation is treated in this section.

Butyl vinyl ether reacts with aroyl chlorides using $Pd(OAc)_2$ without a ligand to give the unsaturated ketone **839**, which is a precursor of a 1-aryl-1,3-dicarbonyl compound. The reaction is regioselective β-attack. Addition of Ph_3P inhibits the reaction[718].

The alkynyl ketones **840** can be prepared by the reaction of acyl chlorides with terminal alkynes. CuI in the presence of Et_3N is the cocatalyst[719]. (1-Alkynyl)tributylstannanes are also used for the alkynyl ketone synthesis[720]. The α, β-alkynic dithio and thiono esters **842** can be prepared by the reaction of the corresponding acid chloride **841** with terminal alkynes[721,722].

The alkylphenylacetyl chloride **843** and benzoyl chloride undergo decarbonylative cross-condensation to give the enone **845** in the presence of Et_3N[723]. The reaction is explained by the insertion of the ketene **844** into the Pd–aryl bond and β-elimination. To support this mechanism, α, β-unsaturated ketones are obtained by the reaction of ketenes with aroyl chlorides[724].

Acyl halides react with organometallic reagents without catalysts, but sometimes the Pd-catalyzed reactions give higher yields and selectivity than the uncatalyzed reactions. Acyl halides react with Pd(0) to form the acylpalladium complexes **846**, which undergo facile transmetallation.

Organozincs are good reagents for ketone synthesis[451]. The ketone synthesis is carried out by the reaction of alkyl, alkenyl, and arylzinc reagents with acyl chlorides[725,726]. The β-keto ester **848** is prepared by the reaction of an aroyl halide with the Reformatsky reagent **847**[727]. In the examples shown, the alkylzinc reagents **849** and **851** can be coupled with acyl chlorides to give the ketones **850** and **853**. No elimination of β-hydrogen of the alkylpalladium **852**, formed by transmetallation with alkylzinc, takes place, because the reductive elimination is faster than elimination of β-hydrogen[728].

The Pd-catalyzed coupling of an acyl chloride with benzyl chloride to form the benzyl ketone **854** proceeds in the presence of an excess of Zn. In this reaction, benzyl chloride reacts with Zn to form benzylzinc, which undergoes transmetallation with acylpalladium complex[729]. The reaction has been applied to the synthesis of riccardin B (**855**)[730].

The dienone **858** is synthesized by coupling of the alkenyl copper reagent **856** with crotyl chloride (**857**) in the presence or absence of ZnCl$_2$[731]. Tetrabutyllead (**859**) reacts with benzoyl chloride to afford butyl phenyl ketone[732].

Various organotin reagents react with acyl and aroyl halides under mild conditions without decarbonylation to give carbonyl compounds[390,391]. Alkyl- or alkenyltin reagents react with acyl and aroyl chlorides to give ketones[548,733,734]. One example is the preparation of the α,β-unsaturated γ-keto esters **860** and **861**, carried out under a CO atmosphere[735]. The reaction has been applied intramolecularly to the synthesis of the macrocyclic keto lactone **863** from **862**[736]. The intramolecular coupling of an acyl chloride with an alkenyltin reagent was applied to the synthesis of the key intermediate of pyrenophorin[733,737].

The α-diketone **865** can be prepared by the coupling of the acylstannane **864** with acyl chlorides[738,739]. The α-keto ester **868** is prepared by the coupling of (α-methoxyvinyl)tributylstannane (**866**) with acyl chloride, followed by ozonization of the coupled product **867**[740,741].

The ester **870** is prepared by the cross-coupling of the chloroformate **869** with an organotin reagent. Some chloroformates are easily decomposed by a Pd catalyst, and hence the reaction should be carried out by slow addition of the chloroformates. Similarly, the amide **872** is prepared by the reaction of the carbamoyl chloride **871**[742]. The coupling of alkylcopper with ethyl chloroformate catalyzed by Pd affords esters[743].

$$\text{869} + \text{ClCO}_2\text{C}_8\text{H}_{17} \xrightarrow[\text{HMPA, 100°, 70\%}]{\text{PdCl}_2(\text{Ph}_3\text{P})_2, \text{PhMe}} \text{870}$$

$$\text{871} + \text{ClCONMePh} \xrightarrow[\text{71\%}]{\text{PdCl}_2(\text{Ph}_3\text{P})_2} \text{872}$$

The Pd-catalyzed hydrogenolysis of acyl chlorides with hydrogen to give aldehydes is called the Rosenmund reduction. Rosenmund reduction catalyzed by supported Pd is explained by the formation of an acylpalladium complex and its hydrogenolysis[744]. Aldehydes can be obtained using other hydrides. For example, the Pd-catalyzed reaction of acyl halides with tin hydride gives aldehydes[745]. This is the tin form of Rosenmund reduction. Aldehydes are formed by the reaction of the thio esters **873** with hydrosilanes[746,747].

$$\underset{\text{R-C-Cl}}{\overset{\text{O}}{\|}} + \text{Bu}_3\text{SnH} \xrightarrow[\text{90\%}]{\text{Pd(Ph}_3\text{P)}_4} \text{RCHO}$$

The acylstannanes **874** and **875** are prepared by the reaction of acyl chlorides with (Me₃Sn)₂. The symmetrical 1,2-diketones **877** can be prepared by the reaction of an excess of benzoyl chloride with (Et₃Sn)₂. Half of the benzoyl chloride is converted into the benzoyltin reagent **876**, which is then coupled with the remaining benzoyl chloride under a CO atmosphere to afford the α-diketone **877**[748]. Triethyl phosphite is used as a ligand.

$$\text{MeCOCl} + \text{Me}_3\text{SnSnMe}_3 \xrightarrow[\text{70\%}]{\text{Pd(Ph}_3\text{P)}_4} \underset{\text{874}}{\overset{\text{O}}{\underset{\|}{\text{Me-C-SnMe}_3}}}$$

$$\text{PhCOCl} + \text{Me}_3\text{SnSnMe}_3 \xrightarrow[\text{(EtO)}_3\text{P, 110°, 93\%}]{\pi\text{-C}_3\text{H}_5\text{PdCl}} \underset{\text{875}}{\text{PhCOSnMe}_3}$$

$$\text{PhCOCl} + \text{Et}_3\text{SnSnEt}_3 \xrightarrow[\text{(EtO)}_3\text{P, CO, 70\%}]{\pi\text{-C}_3\text{H}_5\text{PdCl}} \underset{\text{876}}{\text{PhCOSnEt}_3} + \text{Et}_3\text{SnCl}$$

$$\text{Pd(0)} \downarrow \text{PhCOCl}$$

$$\underset{\text{877}}{\text{PhCOCOPh}} + \text{Et}_3\text{SnCl}$$

The reaction of benzoyl chloride with $(Me_3Si)_2$ affords benzoyltrimethylsilane (878)[626,749,750]. Hexamethyldigermane behaves similarly. The siloxycyclopropane 879 forms the Pd homoenolate of a ketone and reacts with an acyl halide to form 880. The 1,4-diketone 881 is obtained by reductive elimination of 880 without undergoing elimination of β-hydrogen[751].

$$\text{PhCOCl} + \text{Me}_3\text{SiSiMe}_3 \xrightarrow[\text{(EtO)}_3\text{P, 110°, 93\%}]{\pi\text{-C}_3\text{H}_5\text{PdCl}} \underset{878}{\text{PhCOSiMe}_3}$$

Sulfonyl chloride reacts with an alkenylstannane to give the alkenylsulfone 882[752].

Acyl halides are intermediates of the carbonylations of alkenes and organic halides. Decarbonylation of acyl halides as a reversible process of the carbonylation is possible with Pd catalyst. The decarbonylation of aliphatic acid chlorides proceeds with Pd(0) catalyst, such as Pd on carbon or $PdCl_2$, at around 200 °C[109,753]. The product is a mixture of isomeric internal alkenes. For example, when decanoyl chloride is heated with $PdCl_2$ at 200 °C in a distillation flask, rapid evolution of CO and HCl stops after 1 h, during which time a mixture of nonene isomers was distilled off in a high yield. The decarbonylation of phenylpropionyl chloride (883) affords styrene (53%). In addition, 1,5-diphenyl-1-penten-3-one (884) is obtained as a byproduct (10%), formed by the insertion of styrene into the acyl chloride. Formation of the latter supports the formation of acylpalladium species as an intermediate of the decarbonylation. Decarbonylation of the benzoyl chloride 885 can be carried out in good yields at 360 °C with Pd on carbon as a catalyst, yielding the aryl chloride 886[754].

The decarbonylation–dehydration of the fatty acid **887** catalyzed by $PdCl_2(Ph_3P)_2$ (0.01 mol%) was carried out by heating its mixture with acetic anhydride at 250 °C to afford the terminal alkene **888** with high selectivity and high catalyst turnover number (12 370). The reaction may proceed by the oxidative addition of Pd to the mixed anhydride[755].

The reduction of acyl halides with hydrogen to form aldehydes using Pd catalyst is well known as the Rosenmund reduction[756]. Some acyl chlorides give decarbonylation products rather than aldehydes under Rosenmund conditions. The diene **890** was obtained by decarbonylation in an attempted Rosenmund reduction of acetyloleanolic acid chloride (**889**)[757]. Rosenmund reduction of sterically hindered acyl chlorides such as diphenyl- and triphenylacetyl chloride (**891**) gives the decarbonylated products **892**[758].

$$Ph_3CCOCl + H_2 \xrightarrow{Pd} Ph_3CH + CO + HCl$$
891 **892**

From these facts, a mechanism of the Rosenmund reduction has been proposed, in which the formation of the acylpalladium species **893** is the first step of the aldehyde formation and also the decarbonylation, although the Rosenmund reduction proceeds under heterogeneous conditions[744].

The acylpalladium complex formed from acyl halides undergoes intramolecular alkene insertion. 2,5-Hexadienoyl chloride (**894**) is converted into phenol in its attempted Rosenmund reduction[759]. The reaction is explained by the oxidative addition, intramolecular alkene insertion to generate **895**, and β-elimination. Chloroformate will be a useful compound for the preparation of α, β-unsaturated esters if its oxidative addition and alkene insertion are possible. An intramolecular version is known, namely homoallylic chloroformates are converted into α-methylene-γ-butyrolactones in moderate yields[760]. As another example, the homoallylic chloroformamide **896** is converted into the α-methylene-γ-butyrolactams **897** and **898**[761]. An intermolecular version of alkene insertion into acyl chlorides is known only with bridgehead acid chlorides. Adamantanecarbonyl chloride (**899**) reacts with acrylonitrile to give the unsaturated ketone **900**[762].

1.3 Alkyl Halides, Polyhalides, α-Halo Ketones and Esters

The palladium-catalyzed reaction of alkyl halides is usually difficult to carry out. In the presence of a base, they are converted into alkenes by β-elimination. However, alkyl–alkyl coupling without β-elimination is possible by using alkyl-boranes. Hexyl iodide reacts smoothly with 9-octyl-9-BBN in the presence of K_3PO_4 to give tetradecane in 64% yield[536,537]. This reaction is treated in Section 1.1.4.3. No coupling takes place when organometallic reagents of Zn, Mg, Al, Sn, Zr, and Hg are used [450]. Only 9-alkyl-9-BBN is an effective reagent.

The α-bromo-γ-lactone **901** undergoes smooth coupling with the acetonyltin reagent **902** to afford the α-acetonyl-γ-butyrolactone **903**[763]. The α-chloro ether **904**, which has no possibility of β-elimination after oxidative addition, reacts with vinylstannane to give the allyl ether **905**. The α-bromo ether **906** is also used for the intramolecular alkyne insertion and transmetallation with allylstannane to give **907**[764].

On the other hand, the halohydrin (chloro and bromo) **908** is converted into a ketone via oxidative addition and β-elimination in boiling benzene with catalysis by Pd(OAc)$_2$ and tri(*o*-tolyl)phosphine in the presence of K_2CO_3[765,766].

The Pd-catalyzed elimination of the mesylate **909** at an anomeric center, although it is a saturated pseudo-halide, under mild conditions is explained by the facile oxidative addition to the mesylate C—O bond, followed by elimination of β-hydrogen to give the enol ether **910**[767].

The carbonylation of some alkyl halides such as iodocyclohexane (**911**) can be carried out under neutral conditions in the presence of *N,N,N,N*-tetramethylurea (TMU), which is a neutral compound, but catches generated hydrogen halide. Molecular sieves (MS-4A) are used for the same purpose[768]. Very reactive ethyl β-iodobutyrate (**912**) is carbonylated to give ethyl methylsuccinate (**913**) in the presence of TMU. The expected elimination of HI to form crotonate, followed by carbonylation, does not occur.

Particularly alkyl halides which have a perfluoroalkyl group at the β-position undergo smooth carbonylation. Probably the coordination of fluorine to form a five-membered chelate ring accelerates the reaction. Double carbonylation to give the α-keto amide **915** is possible in Et$_2$NH with the fluorine-bearing alkyl iodide **914**[769,770]. The ester **917** is obtained by the carbonylation of the β-perfluoroalkyl iodide **916** in ethanol.

Alkyl ketones can be prepared by the carbonylation of alkyl iodides in the presence of organoboranes. The carbonylation of iodocyclohexane with 9-octyl-9-BBN at 1 atm gives cyclohexyl octyl ketone in 65% yield[386]. This reaction is treated in Section 1.1.3.3. Methyl α-methylacetoacetate (919) is obtained by the reaction of the 2-bromopropionate 918, which has a β-hydrogen, with CO and Me₄Sn. Ph₃As as a ligand gives better results than Ph₃P[771].

Polyhaloalkyl compounds such as CCl_4 and the trichloroacetate 920 undergo addition to alkenes with catalysis by $Pd(OAc)_2$ and tri(o-tolyl)phosphine in the presence of K_2CO_3[772]. The reaction is explained by a free radical mechanism, involving one-electron transfer from Pd(0) to CCl_4 to generate a trichloromethyl radical, which adds to the alkene, rather than the mechanism of oxidative addition to CCl_4, followed by the alkene insertion. Addition of trichloroacetate to 1-octene affords the 2,2,4-trichlorododecanoate 921. The addition of CCl_4 or $BrCCl_3$ to the allylic alcohol 922 affords the γ-trichloro ketone 923[773].

Under CO pressure in alcohol, the reaction of alkenes and CCl_4 proceeds to give branched esters. No carbonylation of CCl_4 itself to give triichloroacetate under similar conditions is observed. The ester formation is explained by a free radical mechanism. The carbonylation of 1-octene and CCl_4 in ethanol affords ethyl 2-(2,2,2-trichloroethyl)decanoate (924) as a main product and the simple addition product 925[774].

Dimethyl iodo(4-pentenyl)malonate (926) undergoes a Pd-catalyzed intramolecular radical-type reaction to form the alkyl iodides 927 and 928, rather than a Heck-type reaction product[775]. The same products are also obtained by a radical reaction promoted by tin hydride[776]. Although yield was low, a similar cyclization of the α-chloro ester 929 to form the seven-membered ring 930 was observed[777].

$$C_8H_{17}\diagup\!\!\!=\quad + \ CO \ + \ CCl_4 \ + \ EtOH \quad \xrightarrow[\text{K}_2\text{CO}_3,\ 50°,\ 40\ \text{atm.}]{\text{Pd(OAc)}_2,\ \text{Ph}_3\text{P}}$$

The reaction of perfluoroalkyl iodides with alkenes affords the perfluoro-alkylated alkyl iodides **931**. α,α-Difluoro-functionalized phosphonates are prepared by the addition of the iododifluoromethylphosphonate (**932**) at room temperature[778]. A one-electron transfer-initiated radical mechanism has been proposed for the addition reaction. Addition to alkynes affords 1-perfluoro-alkyl-2-iodoalkenes (**933**)[779–781]. The fluorine-containing oxirane **934** is obtained by the reaction of allyl alcohol[782]. Under a CO atmosphere, the carbocarbonylation of the alkenol **935** and the alkynol **937** takes place with perfluoroalkyl iodides to give the fluorine-containing lactones **936** and **938**[783].

$$C_8F_{17}I + \text{allyl alcohol} \xrightarrow[\text{K}_2\text{CO}_3, 70\%]{\text{PdCl}_2(\text{Ph}_3\text{P})_2} \left[C_8F_{17}\underset{\text{OH}}{\overset{\text{PdX}}{\diagup}} \right] \longrightarrow C_8F_{17}\diagdown\!\!\triangle\text{O} \quad \mathbf{934}$$

$$C_8F_{17}I + \underset{\mathbf{935}\ \text{OH}}{\diagdown\!\!\diagup\!\!\diagdown} + CO \xrightarrow{48\%} C_8F_{17}\text{-}\mathbf{936}$$

$$C_8F_{17}I + \underset{\mathbf{937}}{\equiv\!\!\diagup\!\!\diagdown\text{OH}} + CO \xrightarrow[\text{MeCN, 80°, 36\%}]{\text{PdCl}_2(\text{Ph}_3\text{P})_2,\ \text{K}_2\text{CO}_3} C_8F_{17}\text{-}\mathbf{938}$$

The silyl enol ether **940** is prepared from the α-bromo ketone **939** by the transmetallation with trimethylsilyltributyltin[784].

$$\underset{\mathbf{939}}{\text{Br}\diagup\!\!\underset{\text{O}}{\overset{\text{Ph}}{\diagdown}}} + \text{Me}_3\text{SiSnBu}_3 \xrightarrow[81\%]{\text{PdCl}_2,\ (\text{MeO})_3\text{P}} \underset{\mathbf{940}}{\overset{\text{Ph}}{\underset{\text{OSiMe}_3}{\diagup\!\!\diagup}}}$$

The α,β-unsaturated ester and amide **942** is prepared by the Pd-catalyzed Wittig-type reaction of the bromoacetate or bromoacetamide **941** with aldehydes and Bu$_3$As[785].

(CHO / NO$_2$-phenyl) + Bu$_3$As + BrCH$_2$CON(piperidine) **941** $\xrightarrow[53\%]{\text{Pd}(\text{Ph}_3\text{P})_4}$ O$_2$N-phenyl-CH=CH-CON(piperidine) **942**

1.4 References

1. C. B. Ziegler and R. F. Heck, *J. Org. Chem.*, **43**, 2941 (1978)
2. Review: J. F. Carpentier, F. Petit, A. Mortreux, V. Dufaud, J. M. Basset, and J. Thivolle-Cazat, *J. Mol. Catal.*, **81**, 1 (1993).
3. W. J. Scott, *Chem. Commun.*, 1755 (1987).
4. Reviews: (a) W. J. Scott and J. E. McMurry, *Acc. Chem. Res.*, **21**, 47 (1988); (b) K. Ritter, *Synthesis*, 735 (1993).
5. S. Cacchi, E. Morrera, and G. Ortar, *Tetrahedron Lett.*, **25**, 2271 (1984).
6. W. J. Scott, M. R. Pena, K. Sward, S. J. Stoessel, and J. K. Stille, *J. Org. Chem.*, **50**, 2302 (1985).
7. T. Mizoroki, K. Mori, and A. Ozaki, *Bull. Chem. Soc. Jpn.*, **44**, 581 (1971); **46**, 1505 (1973).
8. R. F. Heck, and J. P. Nolley, Jr, *J. Org. Chem.*, **37**, 2320 (1972); H. A. Dieck and R. F. Heck, *J. Am. Chem. Soc.*, **96**, 1133 (1974); *J. Org. Chem.*, **40**, 1083

(1975), B. A. Patel, C. H. Ziegler, N. A. Cortese, J. E. Plevyak, T. C. Zebovitz, M. Terpko, and R. F. Heck, *J. Org. Chem.*, **42**, 3903 (1977).

9. Reviews: R. F. Heck, *Org. React.*, **27**, 345 (1982); *Adv. Catal.*, **26**, 323 (1977); *Acc. Chem. Res.*, **12**, 146 (1979); in *Comprehensive Organic Synthesis*, Vol. 4, Pergamon Press, Oxford, 1991, p. 833.

10. Review: A. de Meijere and F. E. Meyer, **33**, 2379 (1994); *Angew. Chem., Int. Ed. Engl.*, **33**, 2379 (1994).

11. K. Voigt, U. Schick, F. E. Meyer, and A. de Meijere, *Synlett*, 189 (1994).

12. J. E. Plevyak, J. E. Dickerson, and R. F. Heck, *J. Org. Chem.*, **44**, 4078 (1979).

13. Y. Ben-David, M. Portnoy, M. Gozin, and D. Milstein, *Organometallics*, **11**, 1995 (1992).

14. M. Portnoy, Y. Ben-David, and D. Milstein, *Organometallics*, **12**, 4734 (1993)

15. J. J. Bozell and C. E. Vogt, *J. Am. Chem. Soc.*, **110**, 2655 (1988).

16. W. Cabri, I. Candiani, A. Bedeschi, and R. Santi, *Synlett*, 871 (1992).

17. T. Jeffery, *Chem. Commun.*, 1287 (1984).

18. R. C. Larock, *Pure Appl. Chem.*, **62**, 653 (1990).

19. K. Karabelas and A. Hallberg, *Tetrahedron Lett.*, **26**, 3131 (1985); *J. Org. Chem.*, **51**, 5286 (1986); **53**, 4909 (1988). K. Karabelas, C. Westerlund, and A. Hallberg, *J. Org. Chem.*, **50**, 3896 (1985).

20. M. M. Abelman, T. Oh, and L. E. Overman, *J. Org. Chem.*, **52**, 4130 (1987); M. M. Abelman and L. E. Overman, *J. Am. Chem. Soc.*, **110**, 2328 (1988).

21. R. Grigg, V. Loganathan, S. Sukirthalingam, and V. Sridharan, *Tetrahedron Lett.*, **31**, 6573 (1990).

22. W. Cabri, I. Candiani, A. Bedeschi, and R. Santi, *Tetrahedron Lett.*, **32**, 1753 (1991).

23. R. Grigg, V. Loganathan, V. Santhakumar, V. Sridharan, and A. Teasdale, *Tetrahedron Lett.*, **32**, 687 (1991).

24. N. A. Bumagin, P. G. More, and I. P. Beletskaya, *J. Organomet. Chem.*, **371**, 397 (1989).

25. T. Jeffery, *Tetrahedron Lett.*, **35**, 3051 (1994).

26. H. C. Zhang and G. D. Daves, *Organometallics*, **12**, 1499 (1993).

27. J. P. Genet, E. Blart, and M. Savignac, *Synlett*, 715 (1992).

28. R. M. Choudary, M. R. Sarma, and K. K. Rao, *Tetrahedron Lett.*, **31**, 5781 (1990).

29. A. Lansky, O. Reiser, and A. de Meijere, *Synlett*, 405 (1990).

30. O Reiser, B. Konig, K. Meerbolz, J. Heinze, T. Wallauer, F. Gerson, B. Frim, M. Rabinovitz, and A. de Meijere, *J. Am. Chem. Soc.*, **115**, 3511 (1993).

31. O. Reiser, S. Reichow, and A. de Meijere, *Angew. Chem. Int. Ed. Engl.*, **26**, 1277 (1987)

32. G. Dyker, *Tetrahedron Lett.*, **32**, 7241 (1991); *J. Org. Chem.*, **58**, 234 (1993).

33. G. Dyker, J. Korning, P. G. Jones, and P. Bubenitschek, *Angew. Chem. Int. Ed. Engl.*, **32**, 1733 (1993).

34. J. I. Kim, B. A. Patel, and R. F. Heck, *J. Org. Chem.*, **46**, 1067 (1981).

35. T. Jeffery, *Synth. Commun.*, **18**, 77(1988).

36. A. Arcadi, E. Benocchi, S. Cacchi, L. Caglioti, and F. Martinelli, *Tetrahedron Lett.*, **31**, 2463 (1990).

37. T. Jeffery, *Tetrahedron Lett.*, **26**, 2667 (1985).

38. R. C. Larock, H. Song, B. E. Baker, and W. H. Gong, *Tetrahedron Lett.*, **29**, 2919 (1988); R. C. Larock, W. H. Gong, and B. E. Baker, *Tetrahedron Lett.*, **30**, 2603 (1989); *J. Org. Chem.*, **54**, 2047 (1989).

39. R. C. Larock and W. H. Gong, *J. Org. Chem.*, **55**, 407 (1990).

40. F. Ozawa, A. Kubo, and T. Hayashi, *J. Am. Chem. Soc.*, **113**, 1417 (1991).

41. F. Ozawa, A. Kubo, and T. Hayashi, *Tetrahedron Lett.*, **33**, 1485 (1992); **34**, 2505 (1993).
42. J. P. Duan and C. H. Cheng, *Tetrahedron Lett.*, **34**, 4019 (1993).
43. R. C. Larock and D. J. Leuck, *Tetrahedron Lett.*, **29**, 6399 (1988).
44. J. E. Plevyak and R. F. Heck, *J. Org. Chem.*, **43**, 2454 (1978).
45. W. Heitz, *Makromol. Chem.*, **189**, 119 (1988); *Makromol. Chem. Rapid Commun.*, **9**, 373, 581 (1988).
46. Review: U. Scherf and K. Mullen, *Synthesis*, 23 (1992).
47. H. P. Weitzel and K. Mullen, *Makromol. Chem.*, **191**, 2837 (1990).
48. A. Hallberg and C. Westerlund, *Chem. Lett.*, 1993 (1982).
49. K. Kikukawa, K. Ikenaga, F. Wada, and T. Matsuda, *Chem. Lett.*, 1337 (1983); *J. Organomet. Chem.*, **270**, 277 (1984); *J. Chem. Soc., Perkin Trans. 1*, 1959 (1986)
50. K. Karabelas and A. Hallberg, *J. Org. Chem.*, **51**, 5286 (1986); **53**, 4909 (1988); *Tetrahedron Lett.*, **26**, 3131 (1985).
51. H. Yamashita, B. L. Roan, and M. Tanaka, *Chem. Lett.*, 2175 (1990).
52. K. Karabelas and A. Hallberg, *J. Org. Chem.*, **54**, 1773 (1989).
53. J. P. Konopelski, K. S. Chu, and G. R. Negrete, *J. Org. Chem.*, **56**, 1355 (1991).
54. G. E. Stokker, *Tetrahedron Lett.*, **28**, 3179 (1987).
55. A. Wada, H. Yasuda, and S. Kanatomo, *Synthesis*, 771 (1988).
56. S. Cacchi and A. Arcadi, *J. Org. Chem.*, **48**, 4236 (1983).
57. H. M. R. Martin, R. Hoffmann, B. Schmidt, and S. Wolff, *Tetrahedron*, **47**, 9357 (1991).
58. Review: G. D. Daves and A. Hallberg, *Chem. Rev.*, **89**, 1433 (1989).
59. L. Larhed, C. M. Andersson, and A. Hallberg, *Tetrahedron*, **50**, 285 (1994).
60. C. M. Andersson and A. Hallberg, *J. Org. Chem.*, **54**, 1502 (1989).
61. L. A. Paquette, C. S. Ra, and S. D. Edmonson, *J. Org. Chem.*, **55**, 2443 (1990).
62. R. A. Haack and K. R. Beck, *Tetrahedron Lett.*, **30**, 1605 (1989).
63. W. Cabri, I. Candiani, A. Bedeschi, and R. Santi, *Tetrahedron Lett.*, **32**, 1753 (1991).
64. W. Cabri, I. Candiani, A. Bedeschi, and S. Penco, *J. Org. Chem.*, **55**, 3654 (1990), **57**, 1481 (1992); W. Cabri, I. Candiani, A. Bedeschi, and R. Santi, *J. Org. Chem.*, **57**, 3558 (1992).
65. I. Arai and G. D. Davis, Jr, *J. Org. Chem.*, **44**, 21 (1979).
66. C. M. Andersson, A. Hallberg, and G. D. Daves, *J. Org. Chem.*, **52**, 3529 (1987); C. M. Andersson and A. Hallberg, *J. Org. Chem.*, **53**, 2112 (1988); A. Hallberg and L. Westfelt, *J. Chem. Soc., Perkin Trans. 1*, 933 (1984); C. M. Andersson, J. Larsson, and A. Hallberg, *J. Org. Chem.*, **55**, 5757 (1990).
67. A. Hallberg, L. Westfelt, and C. M. Andersson, *Synth. Commun.*, **15**, 1131 (1985); A. Hallberg, L. Westfelt, and B. Holm, *J. Org. Chem.*, **46**, 5414 (1981); C. M. Anderson and A. Hallberg, *J. Org. Chem.*, **54**, 1502 (1989).
68. P. G. Ciattini, E. Morera, and G. Ortar, *Tetrahedron Lett.*, **32**, 1579 (1991).
69. I. Arai and G. D. Daves, *J. Heterocycl. Chem.*, **15**, 351 (1978).
70. B. M. Choudary and M. R. Sarma, *Tetrahedron Lett.*, **31**, 1495 (1990); B. M. Choudary, M. R. Sarma, and K. K. Rao, *Tetrahedron*, **48**, 719 (1992).
71. J. J. Bozell, C. E. Vogt, and J. Gozum, *J. Org. Chem.*, **56**, 2584 (1991).
72. A. S. Carlstrom and T. Frejd, *Synthesis*, 414 (1989); *J. Org. Chem.*, **56**, 1289 (1991).
73. A. Arcadi, S. Cacchi, F. Marinelli, E. Morera, and G. Ortar, *Tetrahedron*, **46**, 7151 (1990).
74. S. Cacchi, P. G. Ciattini, E. Morera, and G. Ortar, *Tetrahedron Lett.*, **28**, 3039 (1987).

75. C. B. Ziegler and R. F. Heck, *J. Org. Chem.*, **43**, 2949 (1978).
75a. C. A. Busacca, R. E. Johnson, and J. Swestock, *J. Org. Chem.*, **58**, 3299 (1993).
76. C. Carfagna, A. Musco, G. Sallese, R. Santi, and T. Fiorani, *J. Org. Chem.*, **56**, 261 (1991).
77. P. Yi, Z. Zhuangyu, and H. Hongwen, *Synth. Commun.*, **22**, 2019 (1992).
78. K. Uneyama and H. Watanabe, *Tetrahedron Lett.*, **32**, 1459 (1991).
79. M. Catellani and G. P. Chiusoli, *J. Organomet. Chem.*, **233**, C21 (1982); **239**, C35 (1982); **250**, 509(1983)
80. A. Arcadi, F. Marinelli, E. Bernocchi, S. Cacchi, and G. Ortar, *J. Organomet. Chem.*, **368**, 249 (1989).
81. M. Catellani and G. P. Chiusoli, *J. Organomet. Chem.*, **296**, C11 (1985); **247**, C59 (1983); C. Bocelli, M. Catellani, and G. P. Chiusoli, *J. Organomet. Chem.*, **279**, 225 (1985).
82. G. P. Chiusoli, *J. Mol. Catal.*, **41**, 75 (1987).
83. K. Albrecht, O. Reiser, M. Weber, B. Knieriem, and A. de Meijere, *Tetrahedron*, **50**, 383 (1994).
84. O. Reiser, M. Weber, A. de Meijere, *Angew. Chem., Int. Ed. Engl.*, **28**, 1037 (1989); K. Albrecht, O. Reiser, M. Weber, and A. de Meijere, *Synlett*, 521 (1992).
85. M. Kosugi, H. Tamura, H. Sano, and T. Migita, *Tetrahedron*, **45**, 961 (1989); *Chem. Lett.*, 193 (1987).
86. S. Torii, H. Okumoto, T. Kotani, and S. Nakayasu, *Tetrahedron Lett.*, **33**, 3503 (1992).
87. J. B. Melpolder and R. F. Heck, *J. Org. Chem.*, **41**, 265 (1976); S. A. Buntin and R. F. Heck, *Org. Synth.*, **61**, 82 (1983); *Org. Synth., Coll. Vol.* 7, 361 (1990).
88. A. J. Chalk and S. A. Magennis, *J. Org. Chem.*, **41**, 1206 (1976).
89. T. Jeffery, *Tetrahedron Lett.*, **31**, 6641 (1990).
90. T. Jeffery, *Chem. Commun.*, 324 (1991), *Tetrahedron Lett.*, **32**, 2121 (1991); **34**, 1133 (1993).
91. E. Bernocchi, S. Cacchi, P. G. Ciattini, E. Morera, and G. Ortar, *Tetrahedron Lett.*, **33**, 3073 (1992).
92. S. Torii, H. Okumoto, F. Akahoshi, and T. Kotani, *J. Am. Chem. Soc.*, **111**, 8932 (1989).
93. R. C. Larock, F. Kondo, K. Narayanan, L. K. Sydnes, and M. F. H. Hsu, *Tetrahedron Lett.*, **30**, 5737 (1989).
94. T. Mandai, S. Hasegawa, T. Fujimoto, M. Kawada, and J. Tsuji, *Synlett*, 85 (1990).
95. S. Takano, K. Samizu, and K. Ogasawara, *Synlett*, 393(1993).
96. Y. Koga, M. Sodeoka, and M. Shibasaki, *Tetrahedron Lett.*, **35**, 1227 (1994).
97. S. Torii, H. Okumoto, T. Kotani, and F. Akahoshi, *Chem. Lett.*, 1971 (1989).
98. R. C. Larock, E. K. Yum, and H. Yang, *Tetrahedron*, **50**, 305 (1994).
99. R. C. Larock, W. Y. Leung, and S. Stolz-Dunn, *Tetrahedron Lett.*, **30**, 6629 (1989).
100. X. Garcias, P. Ballester, and J. M. Saa, *Tetrahedron Lett.*, **32**, 7739 (1991).
101. I. Shimizu, T. Sugiura, and J. Tsuji, *J. Org. Chem.*, **50**, 537 (1985).
102. A. Arcadi, E. Bernocchi, S. Cacchi, L. Caglioti, and F. Marinelli, *Tetrahedron Lett.*, **31**, 2463 (1990).
103. L. Filippini, M. Gusmeroli, and R. Riva, *Tetrahedron Lett.*, **34**, 1643 (1993).
104. R. C. Larock and S. Ding, *J. Org. Chem.*, **58**, 804 (1993).
105. R. C. Larock and S. Ding, *Synlett*, 145 (1993).
106. R. C. Larock and S. Ding, *Tetrahedron Lett.*, **34**, 979 (1993).
107. R. C. Larock and W. Y. Leung, *J. Org. Chem.*, **55**, 6244 (1990).
108. R. C. Larock and E. K. Yum, *Synlett*, 529 (1990).

109. J. Tsuji and K. Ohno, *J. Am. Chem. Soc.*, **90**, 94 (1968).
110. H. U. Blaser and A. Spencer, *J. Organomet. Chem.*, **233**, 267 (1982); A. Spencer, *J. Organomet. Chem.*, **240**, 209 (1982); **265**, 323 (1984).
111. A. Kasahara, T. Izumi, and K. Kyuda, *Chem. Ind. (London)*, 467 (1988); *Synthesis*, 704 (1988).
112. M. Miura, H. Hashimoto, K. Itoh, and M. Nomura, *Tetrahedron Lett.*, **30**, 975 (1989).
113. A. Kasahara, T. Izumi, K. Miyamoto, and T. Sakai, *Chem. Ind. (London)*, 51, 728 (1988); 192 (1989).
114. K. Kikukawa and T. Matsuda, *Chem. Lett.*, 159 (1977); K. Kikukawa, K. Nagira, F. Wada, and T. Matsuda, *Tetrahedron*, **37**, 31 (1981); K. Kikukawa, K. Maemura, Y. Kiseki, F. Wada, T. Matsuda, and C. S. Giam, *J. Org. Chem.*, **46**, 4885 (1981).
115. S. Sengupta and S. Bhattacharyya, *J. Chem. Soc., Perkin Trans. 1*, 1943 (1993).
116. K. Kikukawa, M. Naritomi, G. X. He, F. Wada, and T. Matsuda, *J. Org. Chem.*, **50**, 299 (1985).
117. T. Yamane, K. Kikukawa, M. Takagi, and T. Matsuda, *Chem. Commun.*, 695 (1972), *Tetrahedron*, **29**, 955 (1973); *J. Mol. Catal.*, **4**, 449 (1978).
118. D. R. Fahey and J. E. Maha, *J. Am. Chem. Soc.*, **98**, 4459 (1976).
119. R. Asano, I. Moritani, Y. Fujiwara, and S. Teranishi, *Bull. Chem. Soc. Jpn.*, **37**, 2320 (1973).
120. T. Jeffery, *Synthesis*, 70 (1987).
121. R. M. Moriarty, W. R. Epa, and A. K. Awasthi, *J. Am. Chem. Soc.*, **113**, 6315 (1991).
122. M. Mori, K. Chiba, and Y. Ban, *Tetrahedron Lett.*, 1037 (1977), **23**, 5315 (1982).
123. M. O. Terpka and R. F. Heck, *J. Am. Chem. Soc.*, **101**, 5281 (1979).
124. R. Odle, B. Blevins, M. Ratcliff, and L. S. Hegedus, *J. Org. Chem.*, **45**, 2709 (1980). L. S. Hegedus, T. A. Mulhern, and A. Mori, *J. Org. Chem.*, **50**, 4282 (1985).
125. K. Nagasawa, Y. Zako, H. Ishihara, and I. Shimizu, *Tetrahedron Lett.*, **32**, 4937 (1991); *J. Org. Chem.*, **58**, 2523 (1993).
126. A. M. Garcia, L. Castedo, and A. Mourino, *Tetrahedron Lett.*, **33**, 4365 (1992).
127. R. Grigg, V. Sridharan, P. Stevenson, S. Sukirthalingam, and T. Worakun, *Tetrahedron*, **46**, 4003 (1990).
128. S. Laschat, F. Narjes, and L. E. Overman, *Tetrahedron*, **50**, 347 (1994).
129. E. Negishi, Y. Zhang, and B. O'Connor, *Tetrahedron Lett.*, **29**, 2915 (1988).
130. L. F. Tietze and R. Schimpf, *Synthesis*, 876 (1993).
131. J. J. Masters, D. K. Jung, W. G. Bommann, and S. J. Danishefsky, *Tetrahedron Lett.*, **34**, 7253 (1993).
132. F. E. Ziegler, U. R. Chakraborty, and R. B. Weisenfeld, *Tetrahedron*, **37**, 4035 (1981).
133. D. L. Crimin, M. F. Baefsky, and H. Hong, *J. Am. Chem. Soc.*, **114**, 10971 (1992).
134. R. J. Sundverg and R. J. Cherney, *J. Org. Chem.*, **55**, 6028 (1990).
135. K. F. McClure and S. J. Danishefsky, *J. Am. Chem. Soc.*, **115**, 6094 (1993); K. F. McClure, S. J. Danishefsky, and G. K. Schulte, *J. Org. Chem.*, **59**, 356 (1994).
136. R. Grigg, V. Sridharan, P. Stevenson, and T. Worakun, *Chem. Commun.*, 1697 (1986).
137. M. M. Abelman, T. Oh, and L. E. Overman, *J. Org. Chem.*, **52**, 4133 (1987).
138. Reviews: L. E. Overman, M. M. Abelman, D. J. Kucera, V. D. Tran, and D. J. Ricca, *Pure Appl. Chem.*, **64**, 1813 (1992); L. E. Overman, *Pure Appl. Chem.*, **66**, 1423, (1994).

139. W. G. Earley, T. Oh, and L. E. Overman, *Tetrahedron Lett.*, **29**, 3785 (1988); A. Madin and L. E. Overman, *Tetrahedron Lett.*, **33**, 4859 (1992).
140. M. M. Abelman, L. E. Overman, and V. D. Tran, *J. Am. Chem. Soc.*, **112**, 6959 (1990).
141. C. Y. Hong, N. Kado, and L. E. Overman, *J. Am. Chem. Soc.*, **115**, 11028 (1993); C. Y. Hong and L. E. Overman, *Tetrahedron Lett.*, **35**, 3453 (1994).
142. Y. Sato, M. Sodeoka, and M. Shibasaki, *J. Org. Chem.*, **54**, 4738 (1989).
143. Y. Sato, M. Sodeoka, and M. Shibasaki, *Chem. Lett.*, 1953 (1990).
144. N. E. Carpenter, Î. J. Kucera, and L. E. Overman, *J. Org. Chem.*, **54**, 5846 (1989).
145. H. M. R. Hoffmann, B. Schmidt, and S. Wolff, *Tetrahedron*, **45**, 6113 (1989).
146. A. Ashimori, T. Matsuura, L. E. Overman, and D. J. Poon, *J. Org. Chem.*, **58**, 6949 (1993).
147. Y. Sato, S. Watanabe, and M. Shibasaki, *Tetrahedron Lett.*, **33**, 2589 (1992)..
148. Y. Sato, S. Nukui, M. Sodeoka, and M. Shibasaki, *Tetrahedron*, **50**, 371 (1994).
149. K. Kondo, M. Sodeoka, M. Mori, and M. Shibasaki, *Tetrahedron Lett.*, **34**, 4219 (1993), *Synthesis*, 920 (1993).
150. T. Takemoto, M. Sodeoka, H. Sasai, and M. Shibasaki, *J. Am. Chem. Soc.*, **115**, 8477 (1993).
151. Y. Sato, T. Honda, and M. Shibasaki, *Tetrahedron Lett.*, **33**, 2593 (1992).
152. K. Kagechika and M. Shibasaki, *J. Org. Chem.*, **56**, 4093 (1991); K. Kagechika, T. Oshima, and M. Shibasaki, *Tetrahedron*, **49**, 1773 (1993).
153. S. Nukui, M. Sodeoka, and M. Shibasaki, *Tetrahedron Lett.*, **34**, 4965 (1993).
154. B. Burns, R. Grigg, V. Santhakumar, V. Sridharan, P. Stevenson, and T. Worakun, *Tetrahedron*, **48**, 7297 (1992).
155. L. Shi, C. K. Narula, K. T. Mak, L. Kao, Y. Xu, and R. F. Heck, *J. Org. Chem.*, **48**, 3894 (1983); C. K. Narula, K. T. Mak, and R. F. Heck, *J. Org. Chem.*, **48**, 2792 (1983).
156. G. Fourmet, G. Balme, and J. Gore, *Tetrahedron Lett.*, **30**, 69 (1989); D. Bouyssi, G. Balme, G. Fournet, N. Monteiro, and J. Gore, *Tetrahedron Lett.*, **32**, 1641 (1991); G. Balme, D. Bouyssi, R. Faure, J. Gore, and B. Vanhemelryck, *Tetrahedron*, **48**, 3891 (1990); G. Fournet, G. Balme, and J. Gore, *Tetrahedron*, **46**, 7763 (1990).
157. G. Balme and D. Bouyssi, *Tetrahedron*, **50**, 403 (1994).
158. P. Vittoz, D. Bouyssi, C. Traversa, J. Gore, and G. Balme, *Tetrahedron Lett.*, **35**, 1871 (1994).
159. R. Grigg, P. Kennewell, and A. J. Teasdale, *Tetrahedron Lett.*, **33**, 7789 (1992).
160. R. Grigg and V. Sridharan, *Tetrahedron Lett.*, **34**, 7471 (1993).
161. R. Grigg, V. Sridharan, and S. Sukirthanlingam, *Tetrahedron Lett.*, **32**, 3855 (1991)
162. R. Grigg, V. Sridhara, P. Stevenson, and T. Worakun, *Chem. Commun.*, 1697 (1986); R. Grigg, V. Sridharan, P. Stevenson, and S. Sukirthalingam, *Tetrahedron*, **45**, 3557 (1989).
163. R. Grigg, P. Fretwell, C. Meerholts, and V. Sridharan, *Tetrahedron*, **50**, 359 (1994).
164. L. E. Overman, D. J. Ricca, and V. D. Tran, *J. Am. Chem. Soc.*, **115**, 2042 (1993); D. J. Kucera, S. J. O'Connor, and L. E. Overman, *J. Org. Chem.*, **58**, 5304 (1993).
165. G. Z. Wu, F. Lamaty, and E. Negishi, *J. Org. Chem.*, **54**, 2507 (1989).
166. Z. Owczarczyk, F. Lamaty, E. J. Vawter, and E. Negishi, *J. Am. Chem. Soc.*, **114**, 10091 (1992).
167. G. W. Gribble and S. C. Conway, *Synth. Commun.*, **22**, 2129 (1992).

168. S. Torii, H. Okumoto, H. Ozaki, S. Nakayasu, T. Tadokoro, and T. Kotani, *Tetrahedron Lett.*, **33**, 3499 (1992).
169. V. H. Rawal, C. Michoud, and R. F. Monestel, *J. Am. Chem. Soc.*, **115**, 3030 (1993).
170. V. H. Rawal and C. Michoud, *J. Org. Chem.*, **58**, 5582 (1993).
171. R. C. Larock and N. H. Lee, *J. Org. Chem.*, **56**, 6253 (1991).
172. R. Grigg, M. J. Dorrity, J. F. Malone, V. Sridharan, and S. Sukirthalingam, *Tetrahedron Lett.*, **31**, 1343 (1990); R. Grigg, V. Logarathan, S. Sukirthalingam, and V. Sridharan, *Tetrahedron Lett.*, **31**, 6573 (1990).
173. B. A. Patel, J. E. Dickerson, and R. F. Heck, *J. Org. Chem.*, **43**, 5018 (1978); B. A. Patel, L. C. Kao, N. A. Cortese, J. V. Minkiewicz, and R. F. Heck, *J. Org. Chem.*, **44**, 918 (1979); F. G. Stakem and R. F. Heck, *J. Org. Chem.*, **45**, 3584 (1980); W. Fischetti, K. T. Mak, A. L. Rheingold, and R. F. Heck, *J. Org. Chem.*, **48**, 948 (1983).
174. J. M. D'Connor, B. J. Stallman, W. G. Clark, A. Y. L. Shu, R. E. Spada, T. M. Stevenson, and H. A. Dieck, *J. Org. Chem.*, **48**, 807 (1983).
175. R. C. Larock, N. G. Berrios-Pena, and K. Narayanan, *J. Org. Chem.*, **55**, 3447 (1990).
176. R. C. Larock, N. G. Berrios-Pena, C. A. Fried, E. K. Yum, C. Tu, and W. Leong, *J. Org. Chem.*, **58**, 4509 (1993).
177. M. Uno, T. Takahashi, and S. Takahashi, *Chem. Commun.*, 785 (1987).
178. R. C. Larock and C. A. Fried, *J. Am. Chem. Soc.*, **112**, 5882 (1990).
179. R. Grigg and V. Sridharan, *Tetrahedron Lett.*, **30**, 1139 (1989).
180. T. Jeffery, *Tetrahedron Lett.*, **33**, 1989 (1992).
181. T. Mitsudo, W. Fischetti, and R. F. Heck, *J. Org. Chem.*, **49**, 1640 (1984)
182. I. Shimizu and J. Tsuji, *Chem. Lett.*, 233 (1984).
183. I. W. Davies, D. I. C. Scopes, and T. Gallagher, *Synlett*, 85 (1993).
184. R. C. Larock, N. G. Berrios-Pena, and C. A. Fried, *J. Org. Chem.*, **56**, 2615 (1991).
185. R. D. Walkup, L. Guan, M. D. Mosher, S. W. Kim, and Y. S. Kim, *Synlett*, 88 (1993).
186. M. Ahmar, B. Cazes, and J. Gore, *Tetrahedron Lett.*, **25**, 4505 (1984); M. Ahmar, J. J. Barieux, B. Cazes, and J. Gore, *Tetrahedron*, **43**, 513 (1987); B. Friess, B. Cazes, and J. Gore, *Tetrahedron Lett.*, **29**, 4089 (1988); B. Cazes, *Pure Appl. Chem.*, **62**, 1867 (1990).
187. V. Gauthier, B. Cazes, and J. Gore, *Tetrahedron Lett.*, **32**, 915 (1991).
188. M. Ahmar, B. Cazes, and J. Gore, *Tetrahedron Lett.*, **26**, 3795 (1985); *Tetrahedron*, **43**, 3453 (1987)
189. Review, K. Sonogashira, *Comprehensive Organic Synthesis*, Vol. 3, p. 521, Pergamon Press, 1990.
190. L. Cassar, *J. Organomet. Chem.*, **93**, 253 (1975).
191. H. A. Dieck and R. F. Heck, *J. Organomet. Chem.*, **93**, 259 (1975).
192. K. Sonogashira, Y. Tohda, and N. Hagihara, *Tetrahedron Lett.*, 4467 (1975); S. Takahashi, Y. Kuroyama, K. Sonogashira, and N. Hagihara, *Synthesis*, 627 (1980).
193. V. Ratovelomana and G. Linstrumelle, *Synth. Commun.*, **11**, 917   (1981).
194. K. Sonogashira, T. Yatake, Y. Tohda, S. Takahashi, and N. Hagihara, *Chem. Commun.*, 291 (1977).
195. M. A. Dela Rosa, E. Velarde, and A. Guzman, *Syn. Commun.*, **20**, 2059 (1990).
196. M. Alami, F. Ferri, and G. Linstrumelle, *Tetrahedron Lett.*, **34**, 6403(1993).
197. K. Okura, M. Furuune, M. Enna, M. Miura, and M. Nomura, *J. Org. Chem.*, **58**, 4716 (1993); *Tetrahedron Lett.*, **33**, 5363 (1992).

198. F. W. Hobbs, *J. Org. Chem.*, **54**, 3420 (1989).
199. M. J. Robins and P. J. Barr, *Tetrahedron Lett.*, **22**, 421 (1981); *J. Org. Chem.*, **48**, 1854 (1983).
200. M. J. Robins, R. S. Vinayak, and S. G. Wood, *Tetrahedron Lett.*, **31**, 3731 (1990).
201. G. C. Nwokogu, *J. Org. Chem.*, **50**, 3900 (1985); *Tetrahedron Lett.*, **25**, 3263 (1984).
202. W. B. Austin, N. Bilow, W. J. Kelleghan, and K. S. Y. Lau, *J. Org. Chem.*, **46**, 2280 (1981).
203. D. Solooki, V. O. Kennedy, C. A. Tessier, and W. J. Young, *Synlett*, 427 (1990).
204. N. V. Harris, C. Smith, and K. Bowden, *Synlett*, 577 (1990).
205. E. T. Sabourin and A. Onopchenko, *J. Org. Chem.*, **48**, 5135 (1983).
206. C. Huynh and G. Linstrumelle, *Tetrahedron*, **44**, 6337 (1988).
207. L. Brandsma, H. G. M. van den Heuvel, and H. D. Verkruijsse, *Synth. Commun.*, **20**, 1889(1990).
208. R. Diercks and K. P. C. Vollhardt, *J. Am. Chem. Soc.*, **108**, 3150 (1986); H. Schwager, S. Spyroudis, and K. P. C. Vollhardt, *J. Organomet. Chem.*, **382**, 191 (1990).
209. W. Tao, S. Nesbitt, and R. F. Heck, *J. Org. Chem.*, **55**, 63 (1990).
210. R. Boese, J. R. Green, J. Mittendorf, D. L. Mohler, and K. P. C. Vollhardt, *Angew. Chem., Int. Ed. Engl.*, **31**, 1643 (1992).
211. N. A. Bumagin, A. B. Ponomaryov, and I. P. Beletskaya, *Synthesis*, 728 (1984).
212. A. Ohsawa, Y. Abe, and H. Igeta, *Chem. Lett.*, 241 (1979); *Bull. Chem. Soc. Jpn.*, **53**, 3273 (1980).
213. T. Kusumoto, T. Ueda, T. Hiyama, S. Takehara, T. Shoji, M. Osawa, T. Kuriyama, K. Nakamura, and T. Fujisawa, *Chem. Lett.*, 523 (1990).
214. D. E. Ames, D. Bull, and C. Takundwa, *Synthesis*, 364 (1981).
215. A. P. Melissaris and M. H. Litt, *J. Org. Chem.*, **57**, 6998 (1992).
216. S. A. Nye and K. T. Potts, *Synthesis*, 375 (1988).
217. A. O. King, N. Okukado, and E. Negishi, *Chem. Commun.*, 683 (1977); A. O. King, E. Negishi, F. J. Villani, and A. Silveira, *J. Org. Chem.*, **43**, 358 (1978).
218. Review: E. Negishi, *Acc. Chem. Res.*, **15**, 340 (1982).
219. J. K. Stille and J. H. Simpson, *J. Am. Chem. Soc.*, **109**, 2138 (1987); D. E. Rudisill, L. A. Castonguay, and J. K. Stille, *Tetrahedron Lett.*, **29**, 1509 (1988).
219a. A. N. Kashin, I. G. Bumagina, N. A. Bumagin, and I. P. Beletskaya, *Zh. Org. Khim.*, **17**, 21 (1981).
220. F. Tellier, R. Sauvetre, and J. F. Normant, *Tetrahedron Lett.*, **27**, 3147 (1986).
221. N. Yoneda, S. Matsuoka, N. Miyaura, and A. Suzuki, *Bull. Chem. Soc. Jpn.*, **63**, 2124 (1990).
222. R. Rossi, A. Capita, and A. Lezzi, *Tetrahedron*, **40**, 2773 (1984).
223. M. Hirama, K. Fujiwara, K. Shigematsu, and Y. Fukazawa, *J. Am. Chem. Soc.*, **111**, 4120 (1989)
224. D. A. Siesel and S. W. Staley, *Tetrahedron Lett.*, **34**, 3679 (1993).
225. J. M. Nuss, R. A. Rennel, and B. M. Levine, *J. Am. Chem. Soc.*, **115**, 6991 (1993).
226. S. Torii, H. Okumoto, T. Tadokoro, A. Nishimura, and M. A. Rashid, *Tetrahedron Lett.*, **34**, 2139 (1993).
227. E. Negishi, M. Ay, Y. V. Gulevich, and Y. Noda, *Tetrahedron Lett.*, **34**, 1437 (1993).
228. A. G. Meyers, P. M. Harrington, and E. Y. Kuo, *J. Am. Chem. Soc.*, **113**, 694 (1991).
229. A. Carpita and R. Rossi, *Tetrahedron Lett.*, **27**, 4351 (1986); B. P. Andreini, M. Bonetti, A. Carpita, and R. Rossi, *Tetrahedron*, **43**, 4591 (1987).

230. B. P. Andreini, A. Carpita, and R. Rossi, *Tetrahedron Lett.*, **27**, 5533 (1986); B. P. Andreini, A. Carpita, R. Rossi, and B. Scamuzzi, *Tetrahedron*, **45**, 5621 (1989); R. Rossi, A. Carpita, and V. Lippolis, *Synth. Commun.*, **21**, 333 (1991).
231. E. Negishi, N. Okukado, S. F. Lovich, and F. T. Luo, *J. Org. Chem.*, **49**, 2629 (1984).
232. A. S. Kende and C. A. Smith, *J. Org. Chem.*, **53**, 2655 (1988).
233. N. Yoneda, S. Matsuoka, N. Miyaura, T. Fukuhara, and A. Suzuki, *Bull. Chem. Soc. Jpn.*, **63**, 2124 (1990).
234. Review; K. C. Nicolaou and W. M. Dai, *Angew. Chem., Int. Ed. Engl.*, **30**, 1387 (1991).
235. K. C. Nicolaou, T. Ladduwahetty, and E. M. Elisseou, *Chem. Commun.*, 1580 (1985).
236. Reviews: K. C. Nicolaou, J. Y. Ramphal, N. A. Petasis, and C. N. Serhan, *Angew. Chem., Int. Ed. Engl.*, **30**, 1100 (1991); K. C. Nicolaou, J. Y. Ramphal, J. M. Palazon, and R. A. Spanevello, *Angew. Chem., Int. Ed. Engl.*, **28**, 587 (1989).
237. J. A. Porco, F. J. Schoenen, T. J. Stout, J. Clardy, and S. L. Schreiber, *J. Am. Chem. Soc.*, **112**, 7410 (1990).
238. V. Ratovelomanana, D. Guillerm, and G. Limstrumelle, *Tetrahedron Lett.*, **25**, 6001 (1984); **26**, 3811 (1985).
239. P. Magnus, R. T. Lewis, and J. C. Huffman, *J. Am. Chem. Soc.*, **110**, 6921 (1988); P. Magnus, H. Annoura, and J. Hailing, *J. Org. Chem.*, **55**, 1709(1990); P. Magnus, and P. A. Carter, *J. Am. Chem. Soc.*, **110**, 1626 (1988).
240. K. Tomioka, H. Fujita, and K. Koga, *Tetrahedron Lett.*, **30**, 851 (1989).
241. S. L. Schreiber and L. L. Kiessling, *J. Am. Chem. Soc.*, **110**, 631 (1988); **112**, 7416 (1990); J. A. Porco, Jr, F. J. Schoenen, T. J. Stout, J. Clardy, and S. L. Schreiber, *J. Am. Chem. Soc.*, **112**, 7410 (1990).
242. K. C. Nicolaou, C. K. Hwang, A. L. Smith, and S. V. Wendeborn, *J. Am. Chem. Soc.*, **112**, 7416 (1990); **113**, 3106 (1991).
243. A. S. Kende and C. A. Smith, *Tetrahedron Lett.*, **29**, 4217 (1988).
244. M. Alami, B. Crousse, and G. Limstrumelle, *Tetrahedron Lett.*, **35**, 3543 (1994).
245. V. Ratovelomenana and G. Linstrumelle, *Tetrahedron Lett.*, **22**, 315 (1981); D. Cuillerm and G. Linstrumelle, *Tetrahedron Lett.*, **27**, 5857 (1986).
246. Y. Kobayashi, N. Kato, T. Shimazaki, and F. Sato, *Tetrahedron Lett.*, **29**, 6297 (1988).
247. M. A. Tropis and J. R. Pougny, *Tetrahedron Lett.*, **30**, 4951 (1989).
248. K. C. Nicolaou and S. E. Weber, *J. Am. Chem. Soc.*, **106**, 5734 (1984).
249. M. Alami and G. Linstrumelle, *Tetrahedron Lett.*, **32**, 6109 (1991).
250. V. Ratovelomanana, A. Hammond, and G. Linstrumelle, *Tetrahedron Lett.*, **28**, 1649 (1987).
251. Y. Ito, M. Inoue, and M. Murakami, *Tetrahedron Lett.*, **29**, 5379 (1988); Y. Ito, M. Inoue, H. Yokota, and M. Murakami, *J. Org. Chem.*, **55**, 2567 (1990).
252. Y. Ito, M. Inoue, and M. Murakami, *Chem. Lett.*, 1261 (1989).
253. D. Villemin and S. Endo, *J. Organomet. Chem.*, **293**, C10 (1985); M. E. Wright, *J. Organomet. Chem.*, **376**, 353 (1989).
254. T. Jeffery and G. Linstrumelle, *Synthesis*, 32 (1983).
255. E. J. Corey, M. Kang, M. C. Desai, A. K. Ghosh, and I. N. Houpis, *J. Am. Chem. Soc.*, **110**, 649 (1988).
256. M. Torneiro, Y. Fall, L. Castedo, and A. Mourino, *Tetrahedron Lett.*, **33**, 105 (1992); L. Castedo, J. L. Mascarenas, A. Mourino, and L. A. Sarandeses, *Tetrahedron Lett.*, **29**, 1203 (1988).
257. Z. Y. Yang, and D. J. Burton, *Tetrahedron Lett.*, **31**, 1369 (1990).

258. S. Eddarir, C. Francesch, H. Mestdagh, and C. Rolando, *Tetrahedron Lett.*, **31**, 4449 (1990); **32**, 69 (1991).
259. Y. Zhang and J. Wen, *Synthesis*, 727 (1990).
260. T. Sakamoto, F. Shiga, A. Yasuhara, D. Uchiyama, Y. Kondo, and H. Yamanaka, *Synthesis*, 746 (1992).
261. T. Sakamoto, A. Yasuhara, Y. Kondo, and H. Yamanaka, *Synlett*, 502 (1992).
262. A. Arcadi, F. Marinelli, and S. Cacchi, *Synthesis*, 749 (1986).
263. A. Arcadi, S. Cacchi, and F. Marinelli, *Tetrahedron Lett.*, **30**, 2581 (1989).
264. J. Wityak and J. B. Chan, *Synth. Commun.*, **21**, 977 (1991); T. R. Hoye and P. R. Hanson, *Tetrahedron Lett.*, **34**, 5043 (1993).
265. H. Dietl, H. Reinheimer, J. Moffat, and P. M. Maitlis, *J. Am. Chem. Soc.*, **92**, 2276, 2285 (1970).
266. B. Burns, R. Grigg, V. Sridharan, and T. Worakun, *Tetrahedron Lett.*, **29**, 4325 (1988).
267. J. M. Nuss, B. H. Levine, R. A. Rennels, and M. M. Heravi, *Tetrahedron Lett.*, **32**, 5243 (1991).
268. G. C. M. Lee, B. Tobias, J. M. Holmes, D. A. Harcourt, and M. E. Garst, *J. Am. Chem. Soc.*, **112**, 9330 (1990).
269. B. M. Trost and J. Dumas, *Tetrahedron Lett.*, **34**, 19 (1993).
270. B. M. Trost and J. Dumas, *J. Am. Chem. Soc.*, **114**, 1924 (1992); B. M. Trost, J. Dumas, and M.Villa, *J. Am. Chem. Soc.*, **114**, 9836 (1992).
271. S. Cacchi, M. Felici, and B. Pietroni, *Tetrahedron Lett.*, **25**, 3137 (1984).
272. H. Finch, N. A. Pegg, and B. Evans, *Tetrahedron Lett.*, **34**, 8353 (1993).
273. A. Arcadi, S. Cacchi, and F. Marinelli, *Tetrahedron*, **41**, 5121 (1985).
274. A. Arcadi, E. Bernocchi, A. Burini, S. Cacchi, F. Marinelli, and B. Pietroni, *Tetrahedron Lett.*, **30**, 3465 (1989).
275. Y. Zhang and E. Negishi, *J. Am. Chem. Soc.*, **111**, 3454 (1989).
276. Y. Zhang, G. Wu, G. Agnel, and E. Negishi, *J. Am. Chem. Soc.*, **112**, 8590 (1990)
277. F. E. Meyer, P. J. Parsons, and A. de Meijere, *J. Org. Chem.*, **56**, 6487 (1991); F. E. Meyer, J. Brandenbur, P. J. Parsons, and A. de Meijere, *Chem. Commun.*, 390 (1992).
278. P. J. Parsons, M. Stefinovic, P. Willis, and F. Meyer, *Synlett*, 864 (1992).
279. S. Brown, S. Clarkson, R. Grigg, and V. Sridharan, *Tetrahedron Lett.*, **34**, 157 (1993).
280. G. Wu, A. L. Rheingold, S. J. Geib, and R. F. Heck, *Organometallics*, **6**, 1941 (1987).
281. S. Torii, H. Okumoto, and A. Nishimura, *Tetrahedron Lett.*, **32**, 4167 (1991).
282. E. Negishi, M. Ay, and T. Sugihara, *Tetrahedron*, **49**, 5471 (1993).
283. F. E. Meyer and A. de Meijere, *Synlett*, 777 (1991).
284. E. Negishi, L. S. Harring, Z. Owczarczyk, M. M. Mohamud, and A. Ay, *Tetrahedron Lett.*, **33**, 3253 (1992).
285. B. Burns, R. Grigg, P. Ratananukul, and V. Sridharan, *Tetrahedron Lett.*, **29**, 5565 (1988); B. Burns, R. Grigg, V. Sridharan, P. Stevenson, S. Sukanthini, and T. Worakun, *Tetrahedron Lett.*, **30**, 1135 (1989).
286. F. T. Luo and R. T. Wang, *Tetrahedron Lett.*, **32**, 7703 (1991).
287. E. Negishi, Y. Noda, F. Lamaty, and E. J. Vawter, *Tetrahedron Lett.*, **31**, 4393 (1990).
288. L. J. Silverberg, G. Wu, A. L. Rheingold, and R. F. Heck, *J. Organomet. Chem.*, **409**, 411 (1991).
289. W. Tao, L. J. Silverberg, A. L. Rheingold, and R. F. Heck, *Organometallics*, **8**, 2550 (1989).

290. R. C. Larock, M. J. Doty, and S. Cacchi, *J. Org. Chem.*, **58**, 4579 (1993).

290a. N. Beydoun and M. Pfeffer, *Synthesis*, 729 (1990).

291. R. C. Larock and E. K. Yum, *J. Am. Chem. Soc.*, **113**, 6689 (1991).

292. D. Wensbo, A. Eriksson, T. Jeschke, U. Annby, and S. Gronowitz, *Tetrahedron Lett.*, **34**, 2823 (1993).

293. T. Jeschke, D. Wensbo, U. Annby, S. Gronowitz, and L. S. Cohen, *Tetrahedron Lett.*, **34**, 6471 (1993).

294. A. Arcadi, S. Cacchi, and F. Marinelli, *Tetrahedron Lett.*, **33**, 3915 (1992).

295. A. Arcadi, A. Burini, S. Cacchi, M. Delmastro, F. Marinelli, and B. R. Pietroni, *J. Org. Chem.*, **57**, 976 (1992).

296. D. Bouyssi, J. Gore, and G. Balme, *Tetrahedron Lett.*, **33**, 2811 (1992).

297. F. T. Luo, I. Schreuder, and R. T. Wang, *J. Org. Chem.*, **57**, 2213 (1992).

298. A. Arcadi, S. Cacchi, R. C. Larock, and F. Morinelli, *Tetrahedron Lett.*, **34**, 2813 (1993).

299. F. T. Luo and R. T. Wang, *Tetrahedron Lett.*, **33**, 6835 (1992).

300. G. Fournet, G. Balme, B. Vanhemelryck, and J. Gore, *Tetrahedron Lett.*, **31**, 5147 (1990); G. Fournet, G. Balme, and J. Gore, *Tetrahedron*, **47**, 6293 (1991); D. Bouyssi, G. Balme, and J. Gore, *Tetrahedron Lett.*, **32**, 6541 (1991).

301. A. Schoenberg, I. Bartoletti, and R. F. Heck, *J. Org. Chem.*, **39**, 3318 (1974).

302. H. Yoshida, N. Sugita, K. Kudo, and Y.Takesaki, *J. Chem. Soc. Jpn.*, 1386 (1974) (in Japanese).

303. M. Hidai, T. Hikita, Y. Wada, Y. Fujikura, and Y. Uchida, *Bull. Chem. Soc. Jpn.*, **48**, 2075 (1975).

304. T. Ito, K. Mori, T. Mizoroki, and A. Ozaki, *Bull. Chem. Soc. Jpn.*, **48**, 2091 (1975).

305. J. K. Stille and P. K. Wong, *J. Org. Chem.*, **40**, 532 (1975).

306. T. Takahashi, T. Nagashima, and J. Tsuji, *Chem. Lett.*, 369 (1980).

307. T. Fuchikami and I. Ojima, *Tetrahedron Lett.*, **23**, 4099 (1982).

308. Y. Ben-David, M. Portnoy, and D. Milstein, *J. Am. Chem. Soc.*, **111**, 8742 (1989).

309. M. Huser, M. Youinou, and J. A. Osborn, *Angew. Chem., Int. Ed. Engl.*, **28**, 1386 (1989).

310. V. V. Grushin and H. Alper, *Chem. Commun.*, 611 (1992).

311. N. A. Bumagin, K. V. Nikitin, and I. P. Beletskaya, *J. Organomet. Chem.*, **358**, 563 (1988).

312. J. W. Tilley and R. A. LeMahieu, *J. Org. Chem.*, **46**, 4614 (1981).

313. L. Cassar, M. Foa, and A. Gardano, *J. Organomet. Chem.*, **121**, C55 (1976).

314. V. Galamb, M. Gopal, and H. Alper, *Organometallics*, **2**, 801 (1983).

315. K. Orito, M. Miyazawa, and H. Suginome, *Synlett*, 245 (1994).

316. J. F. Carpentier, Y. Castanet, J. Brocard, A. Mortreux, and F. Petit, *Tetrahedron Lett.*, **32**, 4705 (1991).

317. J. F. Carpentier, Y. Castanet, J. Brocard, A. Mortruex, and F. Petit, *Tetrahedron Lett.*, **33**, 2001 (1992).

318. T. Kobayashi and M. Tanaka, *J. Mol. Catal.*, **47**, 41 (1988); T. Kobayashi, F. Abe, and M. Tanaka, *J. Mol. Catal.*, **45**, 91 (1988).

319. H. Alper, K. Hashem, and J. Heveling, *Organometallics*, **1**, 775 (1982).

320. H. Arzoumanian, G. Buono, M. Choukrad, and J. F. Petrignani, *Organometallics*, **7**, 59 (1988).

321. T. Okano, T. Nakagaki, H. Konishi, and J. Kiji, *J. Mol. Catal.*, **54**, 65 (1989).

322. S. R. Adapa and C. S. N. Prasad, *J. Chem. Soc., Perkin Trans. 1*, 1706 (1989).

323. H. Urata, N. X. Hu, H. Maekawa, and T. Fuchikami, *Tetrahedron Lett.*, **32**, 4733 (1991); H. Urata, H. Maekawa, S. Takahashi, and T. Fuchikami, *J. Org. Chem.*, **56**, 4320 (1991).

324. T. Takahashi, H. Ikeda, and J. Tsuji, *Tetrahedron Lett.*, **21**, 3885 (1980).
325. J. M. Baird, J. R. Kern, and G. R. Lee, *J. Org. Chem.*, **56**, 1928 (1991).
326. Mitsubishi Petrochemical, *Jpn. Kokai Tokkyo Koho*, JP 60 38843; *Chem. Abstr.*, **103**, 53812t (1985).
327. G. Gavinato and L. Toniolo, *J. Mol. Catal.*, **69**, 283(1991); **75**, 169 (1992).
328. S. Cacchi, P. G. Ciattini, E. Morera, and G. Ortar, *Tetrahedron Lett.*, **27**, 3931 (1986).
329. R. E. Dolle, S. J. Schmidt, and L. I. Kruse, *Chem. Commun.*, 904 (1987).
330. H. Hotta, T. Suzuki, S. Miyano, and Y. Inoue, *J. Mol. Catal.*, **54**, L5 (1989).
331. T. Ohta, M. Ito, K. Inagaki, and H. Takaya, *Tetrahedron Lett.*, **34**, 1615 (1993).
332. S. Cacchi, E. Morera, and G. Ortar, *Tetrahedron Lett.*, **26**, 1109 (1985).
333. D. A. Holt, M. A. Levy, D. L. Ladd, H. J. Oh, J. M. Erb, J. I. Heaslop, and B. W. Metcalf, *J. Med. Chem.*, **33**, 937, 943 (1990).
334. S. Cacchi and A. Lupi, *Tetrahedron Lett.*, **33**, 3939 (1992).
335. A. Murai, N. Tanimoto, N. Sakamoto, and T. Masamune, *J. Am. Chem. Soc.*, **110**, 985 (1988).
336. V. V. Grushin and H. Alper, *Organometallics*, 12, 3846 (1993).
337. G. P. Roth and J. A. Thomas, *Tetrahedron Lett.*, **33**, 1959 (1992).
338. M. Uchiyama, T. Suzuki, and Y. Yamazaki, *Nippon Kagaku Kaishi (J. Chem. Soc. Jpn.)*, 236 (1982) (in Japanese).
339. M. Ochiai, K. Sumi, Y. Takaoka, M. Kunishima, Y. Nagao, M. Shiro, and E. Fujita, *Tetrahedron*, **44**, 4095 (1988).
340. T. Kitamura, I. Mihara, H. Taniguchi, and P. J. Stang, *Chem. Commun.*, 614 (1990).
341. V. V. Grushin and H. Alper, *J. Org. Chem.*, **58**, 4794 (1993).
342. K. Nagira, K. Kikukawa, F. Wada, and T. Matsuda, *J. Org. Chem.*, **45**, 2365 (1980).
343. K. Kikukawa, K. Kono, K. Nagira, F. Wada, and T. Matsuda, *Tetrahedron Lett.*, **21**, 2877 (1980); *J. Org. Chem.*, **46**, 4413 (1981).
344. K. Kikukawa, K. Kono, F. Wada, and T. Matsuda, *Chem. Lett.*, 35 (1982).
345. T. Sakakura, M. Chaisupakitsin, T. Hayashi, and M. Tanaka, *J. Organomet. Chem.*, **334**, 205 (1987).
346. K. Itoh, H. Hashimoto, M. Miura, and M. Nomura, *Chem. Lett.*, 77 (1989); *J. Mol. Catal.*, **59**, 325 (1990); **83**, 125 (1993).
347. R. A. Head and A. Ibbotson, *Tetrahedron Lett.*, **25**, 5939 (1984).
348. H. Horino, H. Sakaba, and M. Arai, *Synthesis*, 715 (1989).
349. R. Takeuchi, K. Suzuki, and N. Sato, *J. Mol. Catal.*, **66**, 277 (1991); *Synthesis*, 923 (1990).
350. H. Watanabe, Y. Hashizume, and K. Uneyama, *Tetrahedron Lett.*, **33**, 4333 (1992).
351. A. Schoenberg and R. F. Heck, *J. Org. Chem.*, **39**, 3327 (1974).
352. M. Mori, K. Chiba, and Y. Ban, *Heterocycles*, **6**, 1841 (1977).
353. M. Mori, Y. Uozumi, and Y. Ban, *Chem. Commun.*, 842 (1986); M. Mori, Y. Uozumi, M. Kimura, and Y. Ban, *Tetrahedron*, **42**, 3793 (1986).
354. K. Chiba, M. Mori, and Y. Ban, *Tetrahedron*, **41**, 387 (1980); *Chem. Commun.*, 770 (1980).
355 S. J. Brickner, J. J. Gaikema, J. T. Torrado, L. J. Greenfield. and D. A. Ulanowicz, *Tetrahedron Lett.*, **29**, 5601 (1988).
356. R. J. Perry and S. R. Turner, *J. Org. Chem.*, **56**, 6573 (1991).
357. R. J. Perry and B. D. Wilson, *J. Org. Chem.*, **58**, 7016 (1993).
358. R. J. Perry and B. D. Wilson, *J. Org. Chem.*, **57**, 2883 and 6351 (1992).

359. A. L. Meyers, A. J. Robichard, and M. J. McKennon, *Tetrahedron Lett.*, **33**, 1181 (1992).
360. P. P. Nicholas, *J. Org. Chem.*, **52**, 5266 (1987).
361. M. Tanaka, T. Kobayashi, and T. Sakakura, *Chem. Commun.*, 837 (1985). T. Kobayashi, H. Yamashita, T. Sakakura, and M. Tanaka, *J. Mol. Catal.*, **41**, 379 (1987).
362. T. Kobayashi, T. Sakakura, and M. Tanaka, *Tetrahedron Lett.*, **28**, 2721 (1987).
363. F. Ozawa, H. Soyama, T. Yamamoto, and A. Yamamoto, *Tetrahedron Lett.*, **23**, 3383 (1982); F. Ozawa, T. Sugimoto, Y. Yuasa, M. Santra, T. Yamamoto, and A. Yamamoto, *Organometallics*, **3**, 683 (1984).
364. T. Kobayashi and M. Tanaka, *J. Organomet. Chem.*, **233**, C64 (1982).
365. F. Ozawa, H. Yanagihara, and A. Yamamoto, *J. Org. Chem.*, **51**, 415 (1986).
366. R. Mutin, C. Luca, J. Thivolle-Cazat, V. Dufaud, F. Dang, and J. M. Basset, *Chem. Commun.*, 896 (1988); *J. Mol. Catal.*, **51**, L15 (1989).
367. F. Ozawa, N. Kawasaki, T. Yamamoto, and A. Yamamoto, *Chem. Lett.*, 567 (1985).
368. T. Sakakura, H. Yamashita, T. Kobayashi, T. Hayashi, and M. Tanaka, *J. Org. Chem.*, **52**, 5733 (1987); M. Tanaka, T. Kobayashi, T. Sakakura, H. Itatani, S. Danno, and K. Zushi, *J. Mol. Catal.*, **32**, 115 (1985).
369. B. Morin, A. Hirschauer, F. Hugues, D. Commereuc, and Y. Chauvin, *J. Mol. Catal.*, **34**, 317 (1986).
370. T. Son, H. Yanagihara, F. Ozawa, and A. Yamamoto, *Bull. Chem. Soc. Jpn.*, **61**, 1251 (1988).
371. F. Ozawa, H. Soyama, H. Yanagihara, I. Aoyama, H. Takino, K. Izawa, T. Yamamoto, and A. Yamamoto, *J. Am. Chem. Soc.*, **107**, 3235 (1985); F. Ozawa, N. Kawasaki, H. Okamoto, T. Yamamoto, and A. Yamamoto, *Organometallics*, **6**, 1640 (1987); L. Huang, F. Ozawa, and A. Yamamoto, *Organometallics*, **9**, 2603, 2612 (1990).
372. H. Yamashita, T. Sakakura, T. Kobayashi, and M. Tanaka, *J. Mol. Catal.*, **48**, 69 (1988).
373. A Schoenberg and R. F. Heck, *J. Am. Chem. Soc.*, **96**, 7761 (1974).
374. H. Yoshida, N. Sugita, K. Kudo, and Y. Takezaki, *Bull. Chem. Soc. Jpn.*, **49**, 1681 (1976).
375. I. Pri-Bar and O. Buchman, *J. Org. Chem.*, **53**, 624 (1988).
376. Y. Ben-David, M. Portnoy, and D. Milstein, *Chem. Commun.*, 1816 (1989).
377. V. P. Baillargeon and J. K. Stille, *J. Am. Chem. Soc.*, **105**, 7175 (1983); **108**, 452 (1986).
378. I. Pri-Bar and O. Buchman, *J. Org. Chem.*, **49**, 4009 (1984).
379. K. Kikukawa, T. Totoki, F. Wada, and T. Matsuda, *J. Organomet. Chem.*, **270**, 283 (1984).
380. Y. Tamaru, H. Ochiai, Y. Yamada, and Z. Yoshida, *Tetrahedron Lett.*, **24**, 3869 (1983).
381. Y. Uchiyama, T. Suzuki, and Y. Yamazaki, *Chem. Lett.*, 1201 (1983).
382. Y. Wakita, T. Yasunaga, M. Akita, and M. Kojima, *J. Organomet. Chem.*, **301**, C17 (1986).
383. T. Ishiyama, N. Miyaura, and A. Suzuki, *Bull. Chem. Soc. Jpn.*, **64**, 1999 (1991).
384. T. Ishiyama, H. Kizaki, N. Miyaura, and A. Suzuki, *Tetrahedron Lett.*, **34**, 7595 (1993).
385. T. Ishiyama, T. Ohe, N. Miyaura, and A. Suzuki, *Tetrahedron Lett.*, **33**, 4465 (1992).
386. T. Ishiyama, N. Miyaura, and A. Suzuki, *Tetrahedron Lett.*, **32**, 6923 (1991).

387. N. A. Bumagin, A. B. Ponomaryov, and I. P. Beletskaya, *Tetrahedron Lett.*, **26**, 4819 (1985).
388. Y. Wakita, T. Yasunaga, and M. Koyima, *J. Organomet. Chem.*, **288**, 261 (1985).
389. M. Tanaka, *Tetrahedron Lett.*, 2601 (1979).
390. Reviews: J. K. Stille, *Angew. Chem., Int. Ed. Engl.*, **25**, 508(1986); *Pure Appl. Chem.*, **57**, 1771 (1985).
391. Review: T. N. Mitchell, *Synthesis*, 803 (1992).
392. G. T. Crisp, W. J. Scott, and J. K. Stille, *J. Am. Chem. Soc.*, **106**, 7500 (1984).
393. T. Kobayashi and M. Tanaka, *J. Organomet. Chem.*, **205**, C27 (1981).
394. N. A. Bumagin, I. G. Bumagina, N. A. Kashin, and I. P. Beletskaya, *Izv. Akad. Nauk SSSR, Ser. Khim.*, **261**, 1141 (1981).
395. A. M. Echavarren and J. K. Stille, *J. Am. Chem. Soc.*, **110**, 1557 (1988).
396. K. Kikukawa, K. Kono, F. Wada, and T. Matsuda, *Chem. Lett.*, 35 (1982).
397. W. F. Goure, M. E. Wright, P. D. Davis, S. S. Labadie, and J. K. Stille, *J. Am. Chem. Soc.*, **106**, 6417 (1984).
398. W. J. Scott, G. T. Crisp, and J. K. Stille, *Org. Synth.*, **68**, 116 (1990).
399. A. C. Gyorkos, J. K. Stille, and L. S. Hegedus, *J. Am. Chem. Soc.*, **112**, 8465 (1990).
400. Y. Hatanaka and T. Hiyama, *Chem. Lett.*, 2049 (1989); Y. Hatanaka, S. Fukushima, and T. Hiyama, *Tetrahedron*, **48**, 2113 (1992).
401. S. Aoki and E. Nakamura, *Synlett*, 741 (1990).
402. R. F. Heck, *J. Am. Chem. Soc.*, **90**, 5546 (1968).
403. T. Kobayashi and M. Tanaka, *Tetrahedron Lett.*, **27**, 4745 (1986).
404. E. Negishi, Y. Zhang, I. Shimoyama, and G. Wu, *J. Am. Chem. Soc.*, **111**, 8018 (1989).
405. E. Negishi, C. Coperet, T. Sugihara, I. Shimoyama, Y. Zhang, G. Wu, and J. M. Tour, *Tetrahedron*, **50**, 425 (1994).
406. Y. Uozumi, N. Kawasaki, E. Mori, M. Mori, and M. Shibasaki, *J. Am. Chem. Soc.*, **111**, 3725 (1989).
407. I. Shimoyama, Y. Zhang, G. Wu, and E. Negishi, *Tetrahedron Lett.*, **31**, 2841 (1990).
408. E. Negishi and J. M. Tour, *Tetrahedron Lett.*, **27**, 4869 (1986).
409. J. M. Tour and E. Negishi, *J. Am. Chem. Soc.*, **107**, 8289 (1985).
410. G. Wu, I. Shimoyama, and E. Negishi, *J. Org. Chem.*, **56**, 6506 (1991).
411. S. Torii, H. Okumoto, and L. H. Xu, *Tetrahedron Lett.*, **31**, 7175 (1990).
412. E. Negishi, *Pure Appl. Chem.*, **20**, 65 (1987).
413. Z. W. An, M. Catellani, and G. P. Chiusoli, *J. Organomet. Chem.*, **371**, C51 (1989).
414. L. S. Liebeskind and M. S. South, *J. Org. Chem.*, **45**, 5426 (1980).
415. K. Kobayashi and M. Tanaka, *Chem. Commun.*, 333 (1981); *J. Chem. Soc. Jpn.*, 537 (1985) (in Japanese).
416. P. G. Ciattini, E. Morera, and G. Ortar, *Tetrahedron Lett.*, **32**, 6449 (1991).
417. Y. Tamaru, H. Ochiai, and Z. Yoshida, *Tetrahedron Lett.*, **25**, 3861 (1984).
418. E. Brocato, C. Castagnoli, M. Catellani, and G. P. Chiusoli, *Tetrahedron Lett.*, **33**, 7433 (1992).
419. V. N. Kalinin, M. V. Shostakovsky, and A. B. Ponomaryov, *Tetrahedron Lett.*, **31**, 4073 (1990).
420. P. G. Ciattini, E. Morera, G. Ortar, and S. S. Rossi, *Tetrahedron*, **47**, 6449 (1991).
421. S. Torii, H. Okumoto, and L. H. Xu, *Tetrahedron Lett.*, **32**, 237 (1991); S. Torii, H. Okumoto, L. H. Xu, M. Sadakane, M. V. Shostakovsky, A. B. Ponomaryov, and V. N. Kalinin, *Tetrahedron*, **49**, 6773 (1993).

422. V. N. Kalinin, M. V. Shostakovsky, and A. B. Ponomaryov, *Tetrahedron Lett.*, **33**, 373 (1992).
423. S. Torii, L. H. Xu, M. Sadakane, and H. Okumoto, *Synlett*, 513 (1992).
424. S. Torii, L. H. Xu, and H. Okumoto, *Synlett*, 695 (1991).
425. A. Arcadi, S. Cacchi, V. Carnicelli, and F. Marinelli, *Tetrahedron*, **50**, 437 (1994).
426. Y. Huang and H. Alper, *J. Org. Chem.*, **56**, 4534 (1991).
427. Y. Inoue, K. Ohuchi, I. F. Yen, and S. Imaizumi, *Bull. Chem. Soc. Jpn.*, **62**, 3518 (1989); Y. Inoue, M. Taniguchi, H. Hashimoto, K. Ohuchi, and S. Imaizumi, *Chem. Lett.*, 81 (1988); *Tetrahedron Lett.*, **29**, 5941 (1988).
428. M. Tanaka, *Bull. Chem. Soc. Jpn.*, **54**, 637 (1981).
429. Reviews: K. Tamao, in *Comprehensive Organic Synthesis*, Vol. 3, Pergamon Press, Oxford, 1991, p. 435; D. W. Knight, in *Comprehensive Organic Synthesis*, Vol. 3, Pergamon Press, Oxford, 1991, p. 481.
430. Review: E. Negishi, *Acc. Chem. Res.*, **15**, 340 (1982).
431. M. Kumada, *Pure Appl. Chem.*, **52**, 669 (1980); K. Tamao, K. Sumitani, and M. Kumada, *J. Am. Chem. Soc.*, **94**, 4374 (1972); *Bull. Chem. Soc. Jpn.*, **49**, 1958 (1976).
432. R. J. Corriu and J. P. Masse, *Chem. Commun.*, 144 (1972).
433. S. Murahashi, N. Yamamura, K. Yanagisawa, N. Mita, and K. Kondo, *J. Org. Chem.*, **44**, 2408 (1979); S. Murahashi, T. Naota, and Y. Tanigawa, *Org. Synth.*, **62**, 39 (1984).
434. A. Minato, K. Suzuki, K. Tamao, and M. Kumada, *Tetrahedron Lett.*, **25**, 83 (1984).
435. A. Minato, K. Tamao, T. Hayashi, K. Suzuki, and M. Kumada, *Tetrahedron Lett.*, **22**, 5319 (1981).
436. T. J. Luong and G. Linstrumelle, *Synthesis*, 738 (1982).
437. M. Yamamura, I. Moritani, and S. Murahashi, *J. Organomet. Chem.*, **91**, C39 (1975).
438. A. Sekiya and N. Ishikawa, *J. Organomet. Chem.*, **118**, 349 (1976); **125**, 281 (1977).
439. J. F. Fauvarque and A. Jutand, *Bull. Soc. Chim. Fr.*, 765 (1976).
440. H. P. Dang and G. Linstrumelle, *Tetrahedron Lett.*, 191 (1978).
441. R. Rossi and A. Capita, *Tetrahedron Lett.*, **27**, 2529 (1986).
442. D. A. Widdowson and Y. Z. Zhang, *Tetrahedron*, **42**, 2111 (1986).
443. H. Jendralla and L. J. Chen, *Synthesis*, 827 (1990).
444. K. Kondo and S. Murahashi, *Tetrahedron Lett.*, 1237 (1979).
445. T. Hayashi, M. Konishi, and M. Kumada, *Tetrahedron Lett.*, 1871 (1979).
446. T. Hayashi, M. Konishi, Y. Kobori, M. Kumada, T. Higuchi, and T. Hirotsu, *J. Am. Chem. Soc.*, **106**, 158 (1984).
447. T. Katayama and M. Umeno, *Chem. Lett.*, 2076 (1991).
448. A. Minato, K. Suzuki, and K. Tamao, *J. Am. Chem. Soc.*, **109**, 1257 (1987).
449. P. L. Castle and D. A. Widdowson, *Tetrahedron Lett.*, **27**, 6013 (1986).
450. K. Yuan and W. J. Scott, *Tetrahedron Lett.*, **30**, 4779 (1989).
451. S. Baba, and E. Negishi, *J. Am. Chem. Soc.*, **98**, 6729 (1976); E. Negishi, T. Takahashi, and S. Baba, *Org. Synth.*, **66**, 60 (1987).
452. Review, E. Erdik, *Tetrahedron*, **48**, 9577 (1992).
453. E. Negishi, T. Takahashi, S. Baba, D. E. VanHorn, and N. Okukado, *J. Am. Chem. Soc.*, **109**, 2393 (1987).
454. A. Pimm, P. Kocienski, and S. D. A. Street, *Synlett*, 886 (1992).
455. E. Negishi, H. Matsushita, and N. Okukado, *Tetrahedron Lett.*, **22**, 2715 (1981).
456. M. Kobayashi and E. Negishi, *J. Org. Chem.*, **45**, 5223 (1980); E. Negishi, L. F. Valente, and M. Kobayashi, *J. Am. Chem. Soc.*, **102**, 3298 (1980).

457. E. Negishi, H. Matsushita, M. Kobayashi, and L. Rand, *Tetrahedron Lett.*, **24**, 3823 (1983).
458. E. Negishi, A. O. King, and N. Okukado, *J. Org. Chem.*, **42**, 1821 (1977).
459. M. A.Tius, J. G. Galeno, and J. H. Zaidi, *Tetrahedron Lett.*, **29**, 6909 (1988).
460. E. Negishi, T. Takahashi, and A. O. King, *Org. Synth.*, **66**, 67 (1987).
461. J. B. Campbell, J. W. Firor, and T. W. Davenport, *Synth. Commun.*, **19**, 2265 (1989).
462. A. Pelter, M. Rowlands, and J. H. Jenkins, *Tetrahedron Lett.*, **28**, 5213 (1987); A. Pelter, M. Rowlands, and G. Clements, *Synthesis*, 51 (1987); A. Pelter, J. M. Maud, I. Jenkins, C. Sadeka, and G. Goles, *Tetrahedron Lett.*, **30**, 3461 (1989).
463. A. S. Bell, D. A. Roberts, and K. S. Ruddock, *Tetrahedron Lett.*, **29**, 5013 (1988); *Synthesis*, 843 (1987).
464. M. Gardette, N. Jabri, A. Alexakis, and J. F. Normant, *Tetrahedron*, **40**, 2741 (1984); A. Alexakis and J. M. Duffault, *Tetrahedron Lett.*, **29**, 6243 (1988); J. M. Deffault, J. Einhorn, and A. Alexakis, *Tetrahedron Lett.*, **32**, 3701 (1991).
465. F. Bjorkling, T. Norin, and R. Unelius, *Synth. Commun.*, **15**, 463, 472 (1985).
466. E. Negishi, N. Okukado, A. O. King, D. E. Van Horn, and B. I. Spiegel, *J. Am. Chem. Soc.*, **100**, 2254 (1978).
467. N. Jabri, A. Alexakis, and J. F. Normant, *Tetrahedron Lett.*, **22**, 959, 3851 (1981); **40**, 2741 (1984); *Tetrahedron*, **40**, 2741 (1984).
468. E. Negishi, Z. R. Owczarczyk, and D. R. Swanson, *Tetrahedron Lett.*, **32**, 4453 (1991); E. Negishi and K. Akiyoshi, *Chem. Lett.*, 1007 (1987).
469. T. L. Gilchrist and R. J. Summersell, *Tetrahedron Lett.*, **28**, 1469 (1987).
470. Y. Tamaru, H. Ochiai, T. Nakamura, and Z. Yoshida, *Angew. Chem., Int. Ed. Engl.*, **26**, 1157 (1987).
471. R. M. Keenan and L. I. Kruse, *Synth. Commun.*, **19**, 793 (1989).
472. Q. Y. Chen and Y. B. He, *Tetrahedron Lett.*, **28**, 2387 (1987).
473. G. P. Roth and C. Sapino, *Tetrahedron Lett.*, **32**, 4073 (1991); G. P. Roth and C. E. Fuller, *J. Org. Chem.*, **56**, 3493 (1991).
474. A. Minato, *J. Org. Chem.*, **56**, 4052 (1991).
475. J. F. Fauvarque and A. Jutand, *J. Organomet. Chem.*, **132**, C17 (1977); **177**, 273 (1979); **209**, 109 (1981).
476. F. Orsini, F. Pelizzoni, and M. Vallarino, *Synth. Commun.*, **17**, 1389 (1987); *J. Organomet. Chem.*, **367**, 375 (1989).
477. H. Yamanaka, M. An-naka, Y. Kondo, and T. Sakamoto, *Chem. Pharm. Bull.*, **33**, 4309 (1985).
478. Y. Tamaru, H. Ochiai, T. Nakamura, K. Tsubaki, and Z. Yoshida, *Tetrahedron Lett.*, **27**, 955 (1986).
479. E. Nakamura, S. Aoki, K. Sekiya, H. Oshino, and I. Kuwajima, *J. Am. Chem. Soc.*, **109**, 8056 (1987); *Tetrahedron Lett.*, **27**, 83 (1986).
480. F. Ramiandrasoa and C. Descoins, *Synth. Commun.*, **19**, 2703 (1989).
481. T. Watanabe, N. Miyaura, and A. Suzuki, *J. Organomet. Chem.*, **444**, C1 (1993).
482. T. Sakamoto, S. Nishimura, Y. Kondo, and H. Yamanaka, *Synthesis*, 485 (1988).
483. K. Asao, H. Iio, and T. Tokoroyama, *Tetrahedron Lett.*, **30**, 6401 (1989).
484. T. Hayashi, T. Hagihara, Y. Katsuro, and M. Kumada, *Bull. Chem. Soc. Jpn.*, **56**, 362 (1983).
485. H. Stadtmuller, R. Lentz, C. E. Tucker, T. Studemann, W. Dorner, and P. Knochel, *J. Am. Chem. Soc.*, **115**, 7027 (1993).
486. T. Kitazume and N. Ishikawa, *J. Am. Chem. Soc.*, **107**, 5186 (1985).
487. T. Hayashi, A. Yamamoto, M. Hojo, and Y. Ito, *Chem. Commun.*, 495 (1989).

488. S. Hyuga, Y. Chiba, N. Yamashina, S. Hara, and A. Suzuki, *Chem. Lett.*, 1757 (1987); S. Hyuga, N. Yamashina, S. Hara, and A. Suzuki, *Chem. Lett.*, 809 (1988).
489. C. J. Elsevier, H. H. Mooiweer, H. Kleijn, and P. Vermeer, *Tetrahedron Lett.*, **25**, 5571 (1984).
490. K. Ruitenberg, H. Kleijn, C. J. Elsevier, J. Meijer, P. Vermeer, *Tetrahedron Lett.*, **22**, 1451 (1981).
491. C. J. Elsevier and P. Vermeer, *J. Org. Chem.*, **50**, 3042 (1985).
492. A. M. Caporusso, L. Lardicci, and F. Da Settimo, *Tetrahedron Lett.*, **27**, 1067 (1986).
493. K. Ruitenberg, H. Kleijn, J. Meier, E. A. Oostveen, and P. Vermeer, *J. Organomet. Chem.*, **224**, 399 (1982).
494. W. de Graaf, A. Smits, J. Boersma, G. van Koten, and W. P. M. Hoekstra, *Tetrahedron*, **44**, 6699 (1988).
495. Y. Abe, M. Sato, H. Goto, R. Sugawara, E. Takahashi, and T. Kato, *Chem. Pharm. Bull.*, **31**, 1108, 4346 (1983).
496. Reviews: A. Suzuki, *Acc. Chem. Res.*, **15**, 178 (1982); *Pure Appl. Chem.*, **57**, 1749 (1985); **63**, 419 (1991); **66**, 213 (1994).
497. Review: V. Snieckus, *Chem. Rev.*, **90**, 879 (1990).
498. Review: D. S. Matteson, *Tetrahedron*, **45**, 1859 (1989).
499. N. Miyaura and A. Suzuki, *Chem. Commun.*, 866 (1979); N. Miyaura, K. Yamada, and A. Suzuki, *Tetrahedron Lett.*, 3437 (1979); N. Miyaura, T. Yanagi, and A. Suzuki, *Synth. Commun.*, **11**, 513 (1981).
500. N. Miyaura, K. Yamada, and A. Suzuki, *Tetrahedron Lett.*, 3437 (1979); N. Miyaura, K. Yamada, H. Suginome, and A. Suzuki, *J. Am. Chem. Soc.*, **107**, 972 (1985); N. Miyaura, M. Sato, and A. Suzuki, *Tetrahedron Lett.*, **27**, 3745 (1986); *Chem. Lett.*, 1329 (1986); T. Ishiyama, N. Miyaura, and A. Suzuki, *Chem. Lett.*, 25 (1987); M. Sato, N. Miyaura, and A. Suzuki, *J. Am. Chem. Soc.*, **111**, 314 (1989).
501. Y. Yoshino, N. Miyaura, and A. Suzuki, *Bull. Chem. Soc. Jpn.*, **61**, 3008 (1988); M. Sato, N. Miyaura, and A. Suzuki, *Chem. Lett.*, 1405 (1989).
502. T. Iihara, J. M. Fu, M. Bourguignon, and V. Snieckus, *Synthesis*, 184 (1989); M. A. Siddiqui and V. Snieckus, *Tetrahedron Lett.*, **31**, 1523 (1990); M. J. Sharp, W. Cheng, and V. Snieckus, *Tetrahedron Lett.*, **28**, 5093 (1987); M. J. Sharp and V. Snieckus, *Tetrahedron Lett.*, **26**, 5997 (1985).
503. M. Ishikura, M. Kamada, and M. Terashima, *Synthesis*, 936 (1984).
504. S. Gronowitz, V. Bobosik, and K. Lawitz, *Chem. Scr.*, **23**, 120 (1984); *Chem. Abstr.*, **101**, 9070 (1984).
505. A. Huth, I. Beetz, and I. Schumann, *Tetrahedron*, **45**, 6679 (1989).
505a. R. B. Miller and S. Dugar, *Organometallics*, **3**, 1261 (1984).
505b. Y. Yang, *Synth. Commun.*, **19**, 1001 (1989).
505c. M. K. Manthey, S. G. Pyne, and R. J. W. Truscott, *J. Org. Chem.*, **55**, 4581 (1990).
505d. A. L. Casalnuovo and J. C. Calabress, *J. Am. Chem. Soc.*, **112**, 4324 (1990).
506. W. J. Thompson and J. Gaudino, *J. Org. Chem.*, **49**, 5237 (1984); W. J. Thompson, J. H. Jones, P. A. Lyle, and J. E. Thies, *J. Org. Chem.*, **53**, 2052 (1988).
507. G. Merck, A. Villiger, and R. Buchecker, *Tetrahedron Lett.*, **35**, 3277 (1994).
508. W. C. Shieh and J. A. Carlson, *J. Org. Chem.*, **57**, 379 (1992).
509. N. Yasuda, L. Xavier, D. L. Rieger, Y. Li, A. E. DeDamp, and U. H. Dolling, *Tetrahedron Lett.*, **34**, 3211 (1993).
510. M. Uemura, H. Nishimura, K. Kamikawa, K. Nakayama, and Y. Hayashi, *Tetrahedron Lett.*, **35**, 1909 (1994); M. Uemura, H. Nishimura, T. Minami, and

Y. Hayashi, *J. Am. Chem. Soc.*, **113**, 5402 (1991); K. Mori and P. Puapoomchareon, *Liebigs Ann. Chem.*, 159 (1990).

511. T. I. Wallow and B. M. Novak, *J. Am. Chem. Soc.*, **113**, 7411 (1991).
512. Y. H. Kim and O. W. Webster, *J. Am. Chem. Soc.*, **112**, 4592 (1990).
513. T. Watanabe, N. Miyaura, and A. Suzuki, *Synlett*, 207 (1992).
514. S. Gronowitz, A. B. Hornfeldt, V. Kristjansson, and T. Musil, *Chem. Scr.*, **26**, 305 (1986).
515. N. M. Ali, A. McKillop, M. B. Mitchell, R. A. Rabelo, and P. J. Wallbank, *Tetrahedron*, **48**, 8117 (1992).
516. S. C. Suri and V. Nair, *Synthesis*, 695 (1990); V. Nair, D. W. Powell, and S. C. Suri, *Synth. Commun.*, **17**, 1897 (1987).
517. N. Miyaura and A. Suzuki, *J. Organomet. Chem.*, **213**, C58 (1981).
518. N. Miyaura and A. Suzuki, *Org. Synth.*, **68**, 130 (1990).
519. N. Miyaura, H. Suginome, and A. Suzuki, *Tetrahedron Lett.*, **24**, 1527 (1983).
520. W. Roush and R. Riva, *J. Org. Chem.*, **53**, 710 (1988).
521. Y. Kobayashi, T. Shimazaki, H. Taguchi, and F. Sato, *J. Org. Chem.*, **55**, 5324 (1990); *Tetrahedron Lett.*, **28**, 5849 (1987).
522. J. R. Pougny, I. Frechard-Ortuno, C. Huynh, and G. Limstrumelle, *Tetrahedron Lett.*, **30**, 6335 (1989).
523. J. Uenishi, J. M. Beau, R. W. Armstrong, and Y. Kishi, *J. Am. Chem. Soc.*, **109**, 4756 (1987).
524. M. Sato, N. Miyaura, and A. Suzuki, *Chem. Lett.*, 1405 (1989).
525. T. Ohe, N. Miyaura, and A. Suzuki, *J. Org. Chem.*, **58**, 2201 (1993); N. Miyaura, M. Ishikawa, and A. Suzuki, *Tetrahedron Lett.*, **33**, 2571 (1992).
526. N. Miyaura, T. Ishiyama, M. Ishikawa, and A. Suzuki, *Tetrahedron Lett.*, **27**, 6369 (1986); N. Miyaura, T. Ishiyama, H. Sasaki, M. Ishikawa, M. Satoh, and A. Suzuki, *J. Am. Chem. Soc.*, **111**, 314 (1989); T. Ohe, N, Miyaura, and A. Suzuki, *Synlett*, 221 (1990).
527. T. Ishiyama, N. Miyaura, and A. Suzuki, *Org. Synth.*, **71**, 89 (1993).
528. J. A. Soderquist, B. Santiago, and I. Rivera, *Tetrahedron Lett.*, **31**, 4981 (1990).
529. M. Catellani, G. P. Chiusoli, and V. Fornasari, *Gazz. Chim. Ital.*, **120**, 779 (1990).
530. P. G. Ciattini, E. Morera, and G. Ortar, *Tetrahedron Lett.*, **33**, 4815 (1992).
531. Y. Satoh, H. Serizawa, N. Miyaura, S. Hara, and A. Suzuki, *Tetrahedron Lett.*, **29**, 1811 (1988); S. Hyuga, Y. Chiba, N. Yamashina, S. Hara, and A. Suzuki, *Chem. Lett.*, 1757 (1987).
532. M. Ogima, S. Hyuga, S. Hara, and A. Suzuki, *Chem. Lett.*, 1959 (1989).
533. T. Ishiyama, N. Miyaura, and A. Suzuki, *Synlett*, 687 (1991).
534. T. Ishiyama, K. Nishijima, N. Miyaura, and A. Suzuki, *J. Am. Chem. Soc.*, **115**, 7219 (1993); Y. Hoshino, T. Ishiyama, N. Miyaura, and A. Suzuki, *Tetrahedron Lett.*, **29**, 3983 (1988); *Org. Synth.*, **71**, 89 (1993).
535. Z. Z. Song and H. N. C. Wong, *J. Org. Chem.*, **59**, 33 (1994).
536. T. Ishiyama, S. Abe, N. Miyaura, and A. Suzuki, *Chem Lett.*, 691 (1992).
537. Hyuga, Y. Chiba, N. Yamashina, S. Hara, and A. Suzuki, *Chem. Lett.*, 1757 (1987).
538. K. Takai, K. Oshima, and H. Nozaki, *Tetrahedron Lett.*, **21**, 2531 (1980); M. Sato, K. Takai, K. Oshima, and H. Nozaki, *Tetrahedron Lett.*, **22**, 1609, 2531 (1981); *Bull. Chem. Soc. Jpn.*, **57**, 108 (1984).
539. K. Asao, H. Lio, and T. Tokoroyama, *Synthesis*, 382 (1990).
540. F. Charbonnier, A. Moyano, and A. E. Greene, *J. Org. Chem.*, **52**, 2303 (1987).
541. E. J. Corey and S. W. Wright, *J. Org. Chem.*, **55**, 1670 (1990).

542. M. G. Saulnier, J. F. Kadow, M. M. Tun, D. R. Langley, and D. M. Vyas, *J. Am. Chem. Soc.*, **111**, 8320 (1989).
543. K. Hirota, Y. Isobe, and Y. Maki, *J. Chem. Soc., Perkin Trans. 1*, 2513 (1989).
544. K. Hirota, Y. Kitade, Y. Kanbe, and Y. Maki, *J. Org. Chem.*, **57**, 5268 (1992).
545. N. Okukado, D. E. Van Horn, W. L. Klima, and E. Negishi, *Tetrahedron Lett.*, 1027 (1978).
546. M. Kosugi, K. Sasazawa, Y. Shimizu, and T. Migita, *Chem. Lett.*, 301 (1977).
547. V. Farina and B. Krishnan, *J. Am. Chem. Soc.*, **113**, 9585 (1991).
548. Review. I. P. Beletzkaya, *J. Organomet. Chem.*, **250**, 551 (1983).
549. N. A. Bumagin, I. G. Bumagina, and I. P. Beletskaya, *Dokl. Akad. Nauk SSSR*, **274**, 818 (1984); *Chem. Abstr.*, **101**, 111062 (1984); A. N. Kashin, I. G. Bumagina, N. A. Bumagin, I. P. Beletzkaya, and O. A. Reutov, *Izv. Akad. Nauk SSSR, Ser. Khim.*, 479 (1980); *Chem. Abstr.*, **93**, 26019h (1980).
550. D. R. McKean, G. Parrinello, A. F. Renaldo, and J. K. Stille, *J. Org. Chem.*, **52**, 422 (1987).
551. J. K. Stille and B. L. Groh, *J. Am. Chem. Soc.*, **109**, 813 (1987).
552. C. Chan, P. B. Cox, and S. M. Roberts, *Chem. Commun.*, 971 (1988).
553. L. Crombie, M. A. Horsham, and S. R. M. Jarrett, *Tetrahedron Lett.*, **30**, 4299 (1989).
554. W. J. Scott, G. T. Crisp, and J. K. Stille, *J. Am. Chem. Soc.*, **106**, 4630 (1984); W. J. Scott and J. K. Stille, *J. Am. Chem. Soc.*, **108**, 3033 (1986); A. M. Echavarren and J. K. Stille, *J. Am. Chem. Soc.*, **109**, 5478 (1987); M. R. Pena and J. K. Stille, *Tetrahedron Lett.*, **28**, 6573 (1987).
555. Review: V. N. Kalinin, *Synthesis*, 413 (1992).
556. A. Dondoni, G. Fantin, M. Fogagnolo, A. Medici, and P. Pedrini, *Synthesis*, 693 (1987); *Tetrahedron Lett.*, **27**, 5269 (1986).
557. M. Kosugi, A. Fukiage, M. Takayanagi, H. Sano, T. Migita, and M. Satoh, *Chem. Lett.*, 1351 (1988).
558. T. R. Bailey, *Tetrahedron Lett.*, **27**, 4407 (1986).
559. G. T. Crisp, *Synth. Commun.*, **19**, 307 (1989).
560. K. Yui, Y. Aso, T. Otsubo, and F. Ogura, *Chem. Lett.*, 1179 (1988).
561. Y. Yang, S. B. Hornfeldt, and S. Gronowitz, *Synthesis*, 130 (1989); J. Malm, P. Bjork, S. Gronowitz, and A. B. Hornfeldt, *Tetrahedron Lett.*, **33**, 2199 (1992).
562. Y. Yamamoto, Y. Azuma, and H. Mitoh, *Synthesis*, 564 (1986).
563. K. Kikukawa, K. Kono, F. Wada, and T. Matsuda, *J. Org. Chem.*, **48**, 1333 (1983).
564. K. Kikukawa, H. Umekawa, and T. Matsuda, *J. Organomet. Chem.*, **311**, C44 (1986).
565. J. M. Clough, I. S. Mann, and D. A. Widdowson, *Tetrahedron Lett.*, **28**, 2645 (1987).
566. S. Gronowitz, P. Bjork, and A. B. Hornfeldt, *J. Organomet. Chem.*, **460**, 127 (1993).
567. J. Malm, P. Bjork, S. Gronowitz, and A. B. Hornfeldt, *Tetrahedron Lett.*, **33**, 2199 (1992).
568. J. Malm, P. Bjork, S. Gronowitz, and A. B. Hornfeldt, *Tetrahedron Lett.*, **35**, 3195 (1994).
569. V. Farina and G. P. Roth, *Tetrahedron Lett.*, **32**, 4243 (1991).
570. V. Farina, B. Krishnan, D. R. Marshall, and G. P. Roth, *J. Org. Chem.*, **58**, 5434 (1993).
571. P. G. Ciattini, E. Morera, and G. Ortar, *Tetrahedron Lett.*, **35**, 2405 (1994).
572. J. M. Saa, G. Martorell, and A. Garcia-Raso, *J. Org. Chem.*, **57**, 678 (1992).

573. J. K. Stille, A. M. Eschavarren, R. M. Williams, and J. A. Hendrix, *Org. Synth.*, **71**, 97 (1993).
574. E. Gomez-Bengoa and A. M. Echavarren, *J. Org. Chem.*, **56**, 3497 (1991).
575. J. M. Saa and G. Martorell, *J. Org. Chem.*, **58**, 1963 (1993).
576. E. J. Corey and I. N. Houpis, *J. Am. Chem. Soc.*, **112**, 8997 (1990).
577. V. Farina, S. R. Baker, D. A. Benigni, S. I. Hauck, and C. Sapiro, *J. Org. Chem.*, **55**, 5833 (1990).
578. S. R. Baker, G. P. Roth, and C. Sapino, *Synth. Commun.*, **20**, 2185 (1990).
579. J. K. Stille and M. Tanaka, *J. Am. Chem. Soc.*, **109**, 3785 (1987); A. Kalivretenos, J. K. Stille, and L. S. Hegedus, *J. Org. Chem.*, **56**, 2883 (1991).
580. E. Piers, R. W. Friesen, and B. A. Keay, *Tetrahedron*, **47**, 4555 (1990).
581. G. Pattenden and S. M. Thom, *Synlett*, 215 (1993).
582. K. C. Nicolaou, T. K. Chakraborty, A. D. Piscopio, N. Minowa, and P. Bertinato, *J. Am. Chem. Soc.*, **115**, 4419 (1993).
583. D. Badone, R. Cecchi, and U. Guzzi, *J. Org. Chem.*, **57**, 6321 (1992).
584. Y. Yamamoto, T. Seko, and H. Nemoto, *J. Org. Chem.*, **54**, 4734 (1989).
585. E. Dubois and J. M. Neau, *Chem. Commun.*, 1191 (1990); *Tetrahedron Lett.*, **31**, 5165 (1990).
586. R. W. Friesen and C. F. Sturino, *J. Org. Chem.*, **55**, 2572, 5808 (1990); R. W. Friesen and R. W. Loo, *J. Org. Chem.*, **56**, 4821 (1991); R. W. Friesen and A. K. Daljeet, *Tetrahedron Lett.*, **31**, 6133 (1990).
587. L. S. Liebeskind and R. W. Fengl, *J. Org. Chem.*, **55**, 5359 (1990); **58**, 3543, 3550 (1993); L. S. Liebeskind and J. Wang, *Tetrahedron Lett.*, **31**, 4293 (1990).
588. R. J. Hinkle, G. T. Poulter, and P. J. Stang, *J. Am. Chem. Soc.*, **115**, 11626 (1993).
589. J. I. Levin, *Tetrahedron Lett.*, **34**, 6211 (1993).
590. J. K. Stille and M. P. Sweet, *Organometallics*, **9**, 3189 (1990).
591. C. M. Hettrick, J. K. King, and W. J. Scott, *J. Org. Chem.*, **56**, 1489 (1991).
592. M. Kosugi, T. Sumiya, Y. Obara, M. Suzuki, H. Sano, and T. Migita, *Bull. Chem. Soc. Jpn.*, **60**, 767 (1987).
593. T. Sakamoto, Y. Kondo, A. Yasuhara, and H. Yamanaka, *Tetrahedron*, **47**, 1877 (1991).
594. R. M. Moriarty and W. R. Epa, *Tetrahedron Lett.*, **33**, 4095 (1992).
595. G. K. Cook, W. J. Hornback, C. L. Jordan, J. H. McDonald, and J. E. Munroe, *J. Org. Chem.*, **54**, 5828 (1989).
596. B. Burns, R. Grigg, P. Ratananukul, and V. Sridharan, *Tetrahedron Lett.*, **29**, 5565 (1988); B. Burns, R. Grigg, V. Sridharan, P. Stevenson, S. Sukanthini, and T. Worakun, *Tetrahedron Lett.*, **30**, 1135 (1989).
597. L. A. Paquette, A. M. Doherty, and C. M. Rayner, *J. Am. Chem. Soc.*, **114**, 3910 (1992); P. C. Astles and L. A. Paquette, *Synlett*, 444 (1992).
598. D. Milstein and J. K. Stille, *J. Am. Chem. Soc.*, **101**, 4981, 4992 (1979).
599. M. Kosugi, M. Ishiguro, Y. Negishi, H. Sano, and T. Migita, *Chem. Lett.*, 1511 (1984).
600. M. Kosugi, T. Sumiya, T. Ogata, H. Sano, and T. Migita, *Chem. Lett.*, 1225 (1984).
601. M. Pereyre, B. Bellegarde, J. Mendelsohn, and J. Valade, *J. Organomet. Chem.*, **11**, 97 (1968).
602. M. Kosugi, H. Hagihara, T. Sumiya, and T. Migita, *Bull. Chem. Soc. Jpn.*, **57**, 242 (1984); *Chem. Lett.*, 939 (1982); 839 (1983); *Chem. Commun.*, 344 (1983).
603. I. Kuwajima and H. Urabe, *J. Am. Chem. Soc.*, **104**, 6831 (1982).
604. T. Kobayashi, T. Sakakura, and M. Tanaka, *Tetrahedron Lett.*, **26**, 3463 (1985).

605. M. Kosugi, M. Koshiba, A. Atoh, H. Sano, and T. Migita, *Bull. Chem. Soc. Jpn.*, **59**, 677 (1986).
606. A. Capita, R. Rossi, and B. Scamuzzi, *Tetrahedron Lett.*, **30**, 2699 (1989).
607. C. Juxiang and G. T. Crisp, *Synth. Commun.*, **22**, 683 (1992).
608. M. Kosugi, M. Kaneyama, and T. Migita, *Chem. Lett.*, 927 (1983).
609. M. Kosugi, K. Shimizu, A. Ohtani, and T. Migita, *Chem. Lett.*, 829 (1981).
610. H. Azarian, S. S. Dua, C. Eaborn, D. R. M. Walton, and A. Pidcock, *J. Organomet. Chem.*, **117**, C55 (1976); **215**, 49 (1981).
611. A. N. Kashin, I. G. Bumagina, N. A. Bumagin, V. N. Bakunin, and I. P. Beletskaya, *J. Org. Chem. USSR*, **17**, 905 (1981).
612. V. Farina and S. I. Hauk, *J. Org. Chem.*, **56**, 4317 (1991).
613. T. R. Kelly, O. Li, and V. Bhushan, *Tetrahedron Lett.*, **31**, 161 (1990).
614. R. Grigg, A. Teasdale, and V. Sridharan, *Tetrahedron Lett.*, **32**, 3859 (1991).
615. M. Mori, N. Kaneta, and M. Shibasaki, *J. Org. Chem.*, **56**, 3486 (1991).
616. G. Stork and R. C. A. Isaacs, *J. Am. Chem. Soc.*, **112**, 7399 (1990).
617. H. Yamashita, T. Kobayashi, T. Hayashi, and M. Tanaka, *Chem. Lett.*, 761 (1991).
618. S. Aoki, T. Fujimura, E. Nakamura, and I. Kuwajima, *J. Am. Chem. Soc.*, **110**, 3296 (1988).
619. Review, Y. Hatanaka and T. Hiyama, *Synlett*, 845 (1991).
620. Y. Hatanaka and T. Hiyama, *J. Org. Chem.*, **53**, 918 (1988).
621. Y. Hatanaka, and T. Hiyama, *J. Org. Chem.*, **54**, 268 (1989); *Tetrahedron Lett.*, **31**, 2719 (1990); Y. Hatanaka, S. Fukushima, and T. Hiyama, *Chem. Lett.*, 1711 (1989).
622. K. Tamao, K. Kobayashi, and Y. Ito, *Tetrahedron Lett.*, **30**, 6051 (1989).
623. Y. Hatanaka, Y. Ebuna, and T. Hiyama, *J. Am. Chem. Soc.*, **113**, 7075 (1991).
624. Y. Hatanaka and T. Hiyama, *J. Am. Chem. Soc.*, **112**, 7793 (1990).
625. H. Matsumoto, S. Nagashima, K. Yoshihiro, and Y. Nagai, *J. Organomet. Chem.*, **85**, C1 (1975); H. Matsumoto, K. Yoshihiro, S. Nagashima, H. Watanabe, and Y. Nagai, *J. Organomet. Chem.*, **128**, 409 (1977); H. Matsumoto, S. Nagashima, T. Sato, and Y. Nagai, *Angew. Chem.*, **17**, 279 (1978).
626. C. Eaborn, R. W. Griffiths, and A. Pidcock, *J. Organomet. Chem.*, **225**, 331 (1982).
627. Y. Hatanaka and T. Hiyama, *Tetrahedron Lett.*, **28**, 4715 (1987).
628. B. M. Trost and J. Yoshida, *Tetrahedron Lett.*, **24**, 4895 (1983).
629. J. D. Rich and T. E. Krafft, *Organometallics*, **9**, 2040 (1990).
630. J. D. Rich, *J. Am. Chem. Soc.*, **111**, 5886 (1989).
631. T. E. Krafft, J. D. Rich, and P. J. MaDermott, *J. Org. Chem.*, **55**, 5430 (1990).
632. D. Azarian, S. S. Dua, C. Eaborn, and D. D. H. Walton, *J. Organomet. Chem.*, **117**, C55 (1976).
633. H. Matsumoto, M. Kasahara, I. Matsubara, M. Takahashi, T. Arai, M. Hasegawa, T. Nakano, and Y. Nagai, *J. Organomet. Chem.*, **250**, 99 (1983); *Chem. Lett.*, 613 (1982).
634. E. J. Corey and H. Kigoshi, *Tetrahedron Lett.*, **32**, 5025 (1991).
635. J. Ichikawa, T. Minami, T. Sonoda, and H. Kobayashi, *Tetrahedron Lett.*, **33**, 3779 (1992).
636. T. Tsuda, T. Yoshida, and T. Saegusa, *J. Org. Chem.*, **53**, 607 (1988).
637. N. Shimizu, T. Kitamura, K. Watanabe, T. Yamaguchi, H. Shigyo, and T. Ohta, *Tetrahedron Lett.*, **34**, 3421 (1993).
638. Y. Xu and J. Zhang, *Tetrahedron Lett.*, **26**, 4771 (1985); *Synthesis*, 778 (1984); Y. Xu, and Z. Li, *Synthesis*, 240 (1986); Y. Xu, Z. Li, J. Xia, H. Guo, and Y. Huang, *Synthesis*, 377 (1989).

639. Y. Xu, H. Wei, J. Zhang, and G. Huang, *Tetrahedron Lett.*, **30**, 949 (1989); **29**, 1955 (1988).
640. H. Lei, M. S. Stoakes, and A. W. Schwacher, *Synthesis*, 1255 (1992).
641. T. Hirao, T. Masunaga, Y. Oshiro, and T. Agawa, *Tetrahedron Lett.*, **21**, 3595 (1980); *Synthesis*, 56 (1981); T. Hirao, T. Masunaga, N. Yamada, Y. Ohshiro, and T. Agawa, *Bull. Chem. Soc. Jpn.*, **55**, 909 (1982).
642. K. S. Petrakis and T. L. Nagabhushan, *J. Am. Chem. Soc.*, **109**, 2831 (1987).
643. D. A. Holt and J. M. Erb, *Tetrahedron Lett.*, **30**, 5393 (1989).
644. Y. Xu, A. Li, J. Xia, H. Guo, and Y. Huang, *Synthesis*, 781 (1984); 691 (1986).
645. L. Kurz, G. Lee, D. Morgans, Jr., M. J. Waldyke, and T. Ward, *Tetrahedron Lett.*, **31**, 6321 (1990).
646. Y. Uozumi, A. Tahahashi, S. Y. Lee, and T. Hayashi, *J. Org. Chem.*, **58**, 1945 (1993).
647. S. E. Tunney and J. K. Stille, *J. Org. Chem.*, **52**, 748 (1987).
648. R. J. Hinkle, P. J. Stang, and M. H. Kowalski, *J. Org. Chem.*, **55**, 5033 (1990); M. H. Kowalski, R. J. Hinkle, and P. J. Stang, *J. Org. Chem.*, **54**, 2783 (1989).
649. M. Uno, K. Seto, and S. Takahashi, *Chem. Commun.*, 932 (1984); M. Uno, H. Seto, W. Ueda, M. Masuda, and S. Takahashi, *Synthesis*, 506 (1985); M. Uno, T. Takahashi, and S. Takahashi, *J. Chem. Soc. Perkin Trans. 1*, 647 (1990).
650. T. Sakamoto, E. Katoh, Y. Kondo, and H. Yamanaka, *Chem. Pharm. Bull.*, **36**, 1664 (1988); *Heterocycles*, **27**, 1353 (1988); T. Sakamoto, Y. Kondo, T. Suginome, S. Ohta, and H. Yamanaka, *Synthesis*, 552 (1992).
651. M. Uno, K. Seto, M. Masuda, W. Ueda, and S. Takahashi, *Tetrahedron Lett.*, **26**, 1553 (1985).
652. M. A. Ciufolini, H. B. Qi, and M. E. Browne, *Tetrahedron Lett.*, **28**, 171 (1987); *J. Org. Chem.*, **53**, 4149 (1988).
653. G. Fournet, G. Balme, and J. Gore, *Tetrahedron*, **46**, 7763 (1990).
654. E. Piers and P. C. Marais, *J. Org. Chem.*, **55**, 3454 (1990); **58**, 11 (1993).
655. J. Y. Legros and J. C. Fiaud, *Tetrahedron Lett.*, **33**, 2509 (1992).
656. K. Takagi, T. Okamoto, Y. Sakakibara, A. Ohno, S. Oka, and N. Hayama, *Chem. Lett.*, 471 (1973); *Bull. Chem. Soc. Jpn.*, **49**, 3177 (1976).
657. Y. Akita, M. Shimazaki, and A. Ohta, *Synthesis*, 974 (1981).
658. A. Sekiya and N. Ishikawa, *Chem. Lett.*, 277 (1975).
659. K. Takagi and Y. Sakakibara, *Chem. Lett.*, 1957 (1989).
660. E. Piers and F. Fleming, *Chem. Commun.*, 756 (1989).
661. K. Yamamura and S. Murahashi, *Tetrahedron Lett.*, 4429 (1977).
662. J. R. Dalton and S. L. Regen, *J. Org. Chem.*, **44**, 4443 (1979).
663. N. Chatani and T. Hanafusa, *J. Org. Chem.*, **51**, 4714 (1986).
664. W. Oppolzer and D. A. Roberts, *Helv. Chim. Acta*, **63**, 1703 (1980).
665. A. Sabel, J. Smidt, R. Jira, and H. Prigge, *Chem. Ber.*, **102**, 2939 (1969).
666. P. Henry, *Acc. Chem. Res.*, **6**, 16 (1973).
667. H. C. Volger, *Recl. Trav. Chim. Pays-Bas*, **87**, 501 (1968).
668. E. W. Stern, M. L. Spector, and H. P. Leftin, *J. Catal.*, **6**, 152 (1966).
669. C. F. Kohll and van Helden, *Recl. Trav. Chim. Pays-Bas*, **87**, 481 (1968).
670. P. M. Henry, *J. Am. Chem. Soc.*, **94**, 7311 (1972).
671. M. Tamura and T. Yasui, *Kogyo Kagaku Zasshi, J. Ind. Chem.* (in Japanese) **72**, 572 (1969).
672. M. Yamaji, Y. Fujiwara, T. Imanaka, and S. Teranishi, *Bull. Chem. Soc. Jpn.*, **43**, 2659 (1970); M. Yamaji, Y. Fujiwara, R. Asano, and S. Teranishi, *Bull. Chem. Soc. Jpn.*, **46**, 90 (1973).
673. M. Julia and C. Blasioli, *Bull. Soc. Chim. Fr.*, 1941 (1976).

674. J. Tsuji and H. Yasuda, *Synth. Commun.*, 103 (1978).
675. M. Kosugi, T. Shimizu, and T. Migita, *Chem. Lett.*, 13 (1978); *Bull. Chem. Soc. Jpn.*, **53**, 1385 (1980).
676. H. Y. Cristan, B. Chabanol, A. Cheme, and H. Christol, *Synthesis*, 892 (1981).
677. M. Foa, R. Santi, and F. Garavaglia, *J. Organomet. Chem.*, **206**, C29 (1981).
678. S. Murahashi, M.Yamamura, K. Yanagihara, N. Mita, and K. Kondo, *J. Org. Chem.*, **44**, 2408 (1979).
679. D. L. Boger and J. S. Panek, *Tetrahedron Lett.*, **25**, 3175 (1984).
680. R. A. Egli, *Helv. Chim. Acta,* **51**, 2090 (1968).
681. T. R. Bosin, M. G. Raymond, and A. R. Buckpitt, *Tetrahedron Lett.*, 4699 (1973).
682. Review, S. Ram and R. E. Ehrenkaufer, *Synthesis*, 91 (1988).
683. N. A. Cortese and R. F. Heck, *J. Org. Chem.*, **42**, 3491 (1977).
684. P. Helquist, *Tetrahedron Lett.*, 1913 (1978).
685. Y. Akita and A. Ohta, *Heterocycles*, **16**, 1325 (1981).
686. M. K. Anwer and A. F. Spatola, *Tetrahedron Lett.*, **26**, 1381 (1985); M. K. Anwer, D. B. Sherman, J. G. Roney, and A. F. Spatola, *J. Org. Chem.*, **54**, 1284 (1989).
687. R. G. Pews, J. E. Hunter, and R. M. Wehmeyer, *Tetrahedron Lett.*, **32**, 7191 (1991).
688. M. Numazawa, K. Kimura, M. Ogata, and M. Nagaoka, *J. Org. Chem.*, **50**, 5421 (1985).
689. S. Cacchi, E. Morera, and G. Ortar, *Org. Synth.*, **68**, 138 (1990); *Tetrahedron Lett.*, **25**, 4821 (1984).
690. R. Dolk, S. J. Schmidt, and L. I. Kruse, *Tetrahedron Lett.*, **29**, 1581 (1988).
691. S. Cacchi, P. G. Ciattini, E. Morera, and G. Ortar, *Tetrahedron Lett.*, **27**, 5541 (1986).
692. G. A. Peterson, F. A. Kunng, J. S. McCallum, and W. D. Wulff, *Tetrahedron Lett.*, **28**, 1381 (1987).
693. Q. Y. Chen and Y. B. He, *Synthesis*, 896 (1988).
694. W. Cabri, S. De Bernardinis, F. Francalanci, and S. Penco, *J. Chem. Soc., Perkin Trans. 1*, 428 (1990); W. Cabri, S. De Bernardinis, F. Francalanci, S. Penco, and R. Santi, *J. Org. Chem.*, **55**, 350 (1990).
695. B. J. Hussey, R. A. W. Johnstone, and I. D. Entwistle, *Tetrahedron*, **38**, 3775 (1982).
696. A. Zask and P. Helquist, *J. Org. Chem.*, **43**, 1619 (1978).
697. Y. Ben-David, M. Gozin, M. Portnoy, and D. Milstein, *J. Mol. Catal.*, **73**, 173 (1992).
698. Y. Tamaru, Y. Yamamoto, Y. Yamada, and Z. Yoshida, *Tetrahedron Lett.*, 1401 (1979).
699. J. L. Fabre, M. Julia, and J. N. Verpeaux, *Tetrahedron Lett.*, **23**, 2469 (1982); **24**, 4311 (1983).
700. H. Pri-Bar and O. Buchman, *J. Org. Chem.*, **51**, 734 (1986).
701. F. R. S. Clark, R. O. C. Norman, and C. B. Thomas, *J. Chem. Soc., Perkin Trans. 1*, 121 (1975).
702. P. Bamfield and P. M. Quan, *Synthesis*, 537 (1978).
703. R. Grigg, P. Stevenson, and T. Worakun, *Tetrahedron*, **44**, 2049 (1988); *Chem. Commun.*, 1073 (1984); 971 (1985).
704. Mitsubishi Chemical, *Jpn. Pat. Kokai*, 61-137838, 64-13036, 2-53742..
705. A. Jutand and A. Mosleh, *Synlett*, 568 (1993).
706. S. Torii, H. Tanaka, and K. Morisaki, *Tetrahedron Lett.*, **26**, 1655 (1985).
707. A. Jutand, S. Negri, and A. Mosleh, *Chem. Commun.*, 1729 (1992).
708. M. Uchiyama, T. Suzuki, and Y. Yamazaki, *Chem. Lett.*, 1165 (1983).

709. M. Miura, H. Hashimnoto, K. Itoh, and M. Nomura, *Chem. Lett.*, 459 (1990).
710. D. E. Ames and A. Opalko, *Tetrahedron*, **40**, 1919 (1984); *Synthesis*, 234 (1983).
711. G. Bringmann, J. R. Jansen, and H. P. Rink, *Angew. Chem., Int. Ed. Engl.*, **25**, 913 (1986); G. Bringmann, J. R. Jansen, H. Reuscher, M. Rubenacker, K. Peters, and H. G. von Schnering, *Tetrahedron Lett.*, **30**, 5249 (1989); **31**, 643 (1990).
712. P. P. Deshpande and O. R. Martin, *Tetrahedron Lett.*, **31**, 6313 (1990).
713. T. Matsumoto, T. Hosoya, and K. Suzuki, *J. Am. Chem. Soc.*, **114**, 3568 (1992); T. Hosoya, E. Takashiro, T. Matsumoto, and K. Suzuki, *J. Am. Chem. Soc.*, **116**, 1004 (1994).
714. J. E. Rice and Z. W. Cai, *Tetrahedron Lett.*, **33**, 1675 (1992); *J. Org. Chem.*, **58**, 1415 (1993).
715. T. Kuroda and F. Suzuki, *Tetrahedron Lett.*, **32**, 6915 (1991).
716. A. P. Kozikowski and D. Ma, *Tetrahedron Lett.*, **32**, 3317 (1991).
717. G. Dyker, *Angew. Chem., Int. Ed. Engl.*, **31**, 1023 (1992); *J. Org. Chem.*, **58**, 6426 (1993).
718. C. M. Andersson and A. Hallberg, *J. Org. Chem.*, **53**, 4257 (1988).
719. Y. Tohda, K. Sonogashira, N. Hagihara, and Y. Kobayashi, *Synthesis*, 777 (1977).
720. M. W. Logue and K. Teng, *J. Org. Chem.*, **47**, 2549 (1982).
721. T. N. Mitchell and K. Kwetkat, *Synthesis*, 1001 (1990).
722. K. Hartke, H. D. Gerber, and U. Roesrath, *Tetrahedron Lett.*, **30**, 1073 (1989).
723. M. Kadokura, T. Mitsudo, and Y. Watanabe, *Chem. Commun.*, 252 (1986).
724. T. Mitsudo, M. Kadokura, and Y. Watanabe, *J. Org. Chem.*, **26**, 5143 (1985).
725. E. Negishi, V. Bagheri, S. Chatterjee, F. T. Lue, J. A. Miller, and A. T. Stoll, *Tetrahedron Lett.*, **24**, 5181 (1983).
726. R. A. Grey, *J. Org. Chem.*, **49**, 2288 (1984).
727. T. Sato, T. Itoh, and T. Fujisawa, *Chem. Lett.*, 1559 (1982).
728. Y. Tamaru , H. Ochiai, T. Nakamura, K. Tsubaki, and Z. Yoshida, *Tetrahedron Lett.*, **26**, 5529 (1985); *Org. Synth.*, **67**, 98 (1988).
729. T. Sato, K. Naruse, M. Enokiya, and T. Fujisawa, *Chem. Lett.*, 1135 (1981).
730. M. Iyoda, M. Sakaitani, H. Otsuka, and M. Oda, *Chem. Lett.*, 127 (1985); *Tetrahedron Lett.*, **26**, 4777 (1985).
731. N. Jabri, A. Alexakis, and J. F. Normant, *Tetrahedron Lett.*, **24**, 5081 (1983); *Tetrahedron*, **42**, 1369 (1986).
732. J. Yamada and Y. Yamamoto, *Chem. Commun.*, 1302 (1987).
733. M. Kosugi, Y. Shimizu, and T. Migita, *Chem. Lett.*, 1423 (1977).
734. D. Milstein and J. K. Stille, *J. Org. Chem.*, **44**, 1613 (1979); *J. Am. Chem. Soc.*, **100**, 3636 (1978).
735. A. F. Renaldo, J. W. Labadie, and J. K. Stille, *Org. Synth.*, **67**, 86 (1988).
736. J. E. Baldwin, R. M. Adlington, and S. H. Ramcharitar, *Chem. Commun.*, 940 (1991).
737. J. W. Labadie, D. Tueting, and J. K. Stille, *J. Org. Chem.*, **48**, 4634 (1983); *J. Am. Chem. Soc.*, **105**, 6129 (1983).
738. J. P. Verlhac, E. Chanson, B. Jousseaume, and J. P. Quintard, *Tetrahedron Lett.*, **26**, 6075 (1985).
739. M. Kosugi, H. Naka, S. Harada, H. Sano, and T. Migita, *Chem. Lett.*, 1371 (1987).
740. H. B. Kwon, B. H. McKee, and J. K. Stille, *J. Org. Chem.*, **55**, 3114 (1990).
741. J. A. Soderquist and W. W. Leong, *Tetrahedron Lett.*, **24**, 2361 (1983).
742. L. Balas, B. Jousseaume, H. A. Shin, J. B. Verlhac, and F. Willian, *Organometallics*, 10, 366 (1991); B. Jeusseaume, H. A. Kwon, J. B. Verlhac, F. Denat, and J. Dubac, *Synlett*, 117 (1993).

743. M. Isaku and E. Nakamura, *J. Am. Chem. Soc.*, **112**, 7428 (1990).
744. J. Tsuji, K. Ohno, and T. Kajimoto, *Tetrahedron Lett.*, 4565 (1965).
745. F. Guibe, P. Four, and H. Riviere, *Chem. Commun.*, 432 (1980); *J. Org. Chem.*, **46**, 4439 (1981).
746. T. Fukuyama, S. C. Lin, and L. Li, *J. Am. Chem. Soc.*, **112**, 7050 (1990).
747. D. A. Evans and H. P. Ng, *Tetrahedron Lett.*, **34**, 2229 (1993).
748. N. A. Bumagin, Yu. V. Gulevich, and I. P. Beletakaya, *J. Organomet. Chem.*, **282**, 421 (1985).
749. A. Ricci, A. Degl'Innoceti, S. Chimichi, M. Fiorenea, G. Rossini and H. J. Bestmann, *J. Org. Chem.*, **50**, 130 (1985).
750. K. Yamamoto, A. Hayashi, S. Suzuki, and J. Tsuji, *Organometallics*, **6**, 974 (1987).
751. S. Aoki, T. Fujimura, E. Nakamura, and I. Kuwajima, *Tetrahedron Lett.*, **30**, 6541 (1989); *J. Org. Chem.*, **56**, 2809 (1991).
752. S. S. Labadie, *J. Org. Chem.*, **54**, 2496 (1989).
753. Review, J. Tsuji and K. Ohno, *Synthesis*, 157 (1969).
754. J. W. Verbicky, B. A. Dellacoletta, and L. Williams, *Tetrahedron Lett.*, **23**, 371 (1982).
755. J. A. Miller, J. A. Nelson, and M. P. Byrne, *J. Org. Chem.*, **58**, 18 (1993).
756. E. Mossetig and R. Mozingo, *Org. React.*, **4**, 362 (1948).
757. T. Nozoe and T. Kinugasa, *J. Chem. Soc. Jpn.*, **59**, 772 (1938) (in Japanese).
758. J. G. Burr, Jr, *J. Am. Chem. Soc.*, **73**, 3502 (1951).
759. G. P. Chiusoli and G. Agnes, *Chim. Ind. (Milan)*, **46**, 548 (1964).
760. F. Henin and J. P. Pete, *Tetahedron Lett.*, **24**, 4687 (1983).
761. F. Henin, J. Muzart, and J. P. Pete, *Tetrahedron Lett.*, **27**, 6339 (1986).
762. K. Hori, M. Ando, N. Takaishi, and Y. Inamoto, *Tetrahedron Lett.*, **28**, 5883 (1987).
763. J. H. Simpson and J. K. Stille, *J. Org. Chem.*, **50**, 1759 (1985).
764. B. K. Bhatt, S. D. Shin, J. R. Falck, and C. Mioskowski, *Tetrahedron Lett.*, **33**, 4885 (1992).
765. H. Nagashima, K. Sato, and J. Tsuji, *Tetrahedron Lett.*, **23**, 3085 (1982).
766. H. Nagashima, K. Sato, and J. Tsuji, *Tetrahedron*, **41**, 5645 (1985).
767. G. S. Jones and W. J. Scott, *J. Am. Chem. Soc.*, **114**, 1491 (1992).
768. H. Urata, H. Maekawa, S. Takahashi, and T. Fuchikami, *J. Org. Chem.*, **56**, 4320 (1991); H. Urata, N. X. Hu, H. Maekawa, and T. Fuchikami, *Tetrahedron Lett.*, **32**, 4733 (1991).
769. H. Urata, O. Kosugekawa, Y. Ishii, H. Yugari, and T. Fuchikami, *Tetrahedron Lett.*, **30**, 4403 (1989).
770. H. Urata, Y. Ishii, and T. Fuchikami, *Tetrahedron Lett.*, **30**, 4407 (1989)
771. T. Kobayashi and M. Tanaka, *J. Organomet. Chem.*, **205**, C27 (1981).
772. J. Tsuji, K. Sato, and H. Nagashima, *Chem. Lett.*, 1169 (1981); *Tetrahedron*, **41**, 393 (1985).
773. H. Nagashima, K. Sato, and J. Tsuji, *Chem. Lett.*, 1605 (1981).
774. J. Tsuji, K. Sato, and H. Nagashima, *Tetrahedron Lett.*, **23**, 893 (1982); *Tetrahedron*, **41**, 5003 (1985).
775. M. Mori, N. Kanda, I. Oda, and Y. Ban, *Tetrahedron*, **41**, 5465 (1985); M. Mori, Y. Kubo, and Y. Ban, *Tetrahedron*, **44**, 4321 (1988).
776. D. P. Curran and C. T. Chang, *Tetrahedron Lett.*, **31**, 933 (1990).
777. M. Mori, N. Kanda, and Y. Ban, *Chem. Commun.*, 1375 (1986).
778. Z. Y. Yang and D. J. Burton, *J. Org. Chem.*, **57**, 4676 (1992).
779. T. Ishihara, M. Kuroboshi, and Y. Okada, *Chem. Lett.*, 1895 (1986).

780. Q. Y. Chen, Z. Y. Yang, C. X. Zhao, and Z. M. Qiu, *J. Chem. Soc., Perkin Trans. 1*, 563 (1988).
781. W. Qiu and D. J. Burton, *J. Org. Chem.*, **58**, 419 (1993).
782. T. Fuchikami, Y. Shibata, and H. Urata, *Chem. Lett.*, 521 (1987).
783. H. Urata, H. Yugari, and T. Fuchikami, *Chem. Lett.*, 833 (1987).
784. M. Kosugi, T. Ohya, and T. Migita, *Bull. Chem. Soc. Jpn.*, **56**, 3539 (1983).
785. Y. C. Shen and Y. F. Zhou, *Tetrahedron Lett.*, **32**, 513 (1991); *Synth. Commun.*, **22**, 657 (1992).

2 Reactions of Allylic Compounds via π-Allylpalladium Complexes Catalyzed by Pd(0)

2.1 Reaction Patterns and Various Allylic Compounds Used for Catalytic Reactions

Application of π-allylpalladium chemistry to organic synthesis has made remarkable progress[1]. As described in Chapter 3, Section 3, π-allylpalladium complexes react with soft carbon nucleophiles such as malonates, β-keto esters, and enamines in DMSO to form carbon–carbon bonds[2, 3]. The characteristic feature of this reaction is that whereas organometallic reagents are considered to be nucleophilic and react with electrophiles, typically carbonyl compounds, π-allylpalladium complexes are electrophilic and react with nucleophiles such as active methylene compounds, and Pd(0) is formed after the reaction.

When the π-allylpalladium complexes are formed by the reaction of alkenes with $PdCl_2$ and react with nucleophiles, the whole reaction constitutes the stoichiometric functionalization of alkenes[4,5].

In addition, a catalytic version of π-allylpalladium chemistry has been developed[6,7]. Formation of the π-allylpalladium complexes by the oxidative addition of various allylic compounds to Pd(0) and subsequent reaction of the complex with soft carbon nucleophiles are the basis of catalytic allylation. After the reaction, Pd(0) is reformed, and undergoes oxidative addition to the allylic compounds again, making the reaction catalytic. In addition to the soft carbon nucleophiles, hard carbon nucleophiles of organometallic compounds of main group metals are allylated with π-allylpalladium complexes. The reaction proceeds via transmetallation. These catalytic reactions are treated in this chapter.

In addition to the catalytic allylation of carbon nucleophiles, several other catalytic transformations of allylic compounds are known as illustrated. Sometimes these reactions are competitive with each other, and the chemo-selectivity depends on reactants and reaction conditions.

Mainly allylic esters are used as the substrates for the catalytic reactions. In addition, the allylic compounds shown are known to react with Pd(0) to form π-allylpalladium complexes. Even allylic nitro compounds[8,9] and sul-fones[10–12] are used for the allylation. The reactivities of these allylic compounds are very different.

Acetate,   Carbonate,   Carbamate,   Oxirane,   Phenyl ether,   Alcohol,

Phosphate,   Chloride,   Nitro,   Sulfone,   Amine and ammonium salts,   Vinylcyclo-propane

Allylic acetates are widely used. The oxidative addition of allylic acetates to Pd(0) is reversible, and their reaction must be carried out in the presence of bases. An important improvement in π-allylpalladium chemistry has been achieved by the introduction of allylic carbonates. Carbonates are highly reactive. More importantly, their reactions can be carried out under neutral conditions[13,14]. Also reactions of allylic carbamates[14], allyl aryl ethers[6,15], and vinyl epoxides[16,17] proceed under neutral conditions without addition of bases.

2.2 Allylation Reactions

2.2.1 Stereochemistry

The stereochemistry of the Pd-catalyzed allylation of nucleophiles has been studied extensively[5,18–20]. In the first step, π-allylpalladium complex formation by the attack of Pd(0) on an allylic part proceeds by inversion (*anti* attack). Then subsequent reaction of soft carbon nucleophiles, N- and O-nucleophiles proceeds by inversion to give **1**. Thus overall retention is observed. On the other hand, the reaction of hard carbon nucleophiles of organometallic compounds proceeds via transmetallation, which affords **2** by retention, and reductive elimination affords the final product **3**. Thus the overall inversion is observed in this case[21,22].

Convincing evidence for oxidative addition by inversion has been presented by the reaction of chiral (*S*)-(*E*)-3-acetoxy-1-phenyl-1-butene (**4**) with Pd(0)(dppe), followed by the treatment with NaBF$_4$ to give optically active the π-allylpalladium complex (1*R*,2*S*,3*S*) **5** with 81% stereoselectivity[19].

Furthermore, the catalytic allylation of malonate with optically active (*S*)-(*E*)-3-acetoxy-1-phenyl-1-butene (**4**) yields the (*S*)-(*E*)-malonates **7** and **8** in a ratio of 92 : 8. Thus overall retention is observed in the catalytic reaction[23]. The intermediate complex **6** is formed by inversion. Then in the catalytic reaction of (*S*)-(*Z*)-3-acetoxy-1-phenyl-1-butene (**9**) with malonate, the oxidative addition generates the complex **10**, which has the sterically disfavored *anti* form. Then the π–σ–π rearrangement (rotation) of the complex **10** moves the Pd from front to the rear side to give the favored *syn* complex **6**, which has the same configuration as that from the (*S*)-(*E*)-acetate **4**. Finally the (*S*)-(*E*)-malonates **7** and **8** are obtained in a ratio of 90 : 10. Thus the reaction of (*Z*)-acetate **9** proceeds by inversion, π–σ–π rearrangement and inversion of configuration accompanied by *Z* to *E* isomerization[24].

The reaction of phenylzinc reagent proceeds with opposite stereochemistry, namely by retention of configuration at the final step via transmetallation. Both the (S)-(E)- and (R)-(Z)-allylic acetates **4** and **9** afford the (R)-(E)-phenylated product **11** by overall inversion[23].

Based on the above-mentioned stereochemistry of the allylation reactions, nucleophiles have been classified into Nu^1 (overall retention group) and Nu^2 (overall inversion group) by the following experiments with the cyclic *exo*- and *endo*-acetates **12** and **13**[25]. No Pd-catalyzed reaction takes place with the *exo*-allylic acetate **12**, because attack of Pd(0) from the rear side to form π-allylpalladium is sterically difficult. On the other hand, smooth π-allylpalladium complex formation should take place with the *endo*-allylic acetate **13**. The Nu^1-type nucleophiles must attack the π-allylic ligand from the *endo* side **14**, namely *trans* to the *exo*-oriented Pd, but this is difficult. On the other hand, the attack of the Nu^2-type nucleophiles is directed to the Pd, and subsequent reductive elimination affords the *exo* products **15**. Thus the allylation reaction of **13** takes place with the Nu^2 nucleophiles (PhZnCl, formate, indenide anion) and no reaction with Nu^1 nucleophiles (malonate, secondary amines, LiP(S)Ph$_2$, cyclopentadienide anion).

Many examples of stereospecific allylation consistent with the above mechanism have been reported. As one example, the regioselective and highly diastereoselective allylation of the lactone **17** with the optically active allylic phosphate **16** proceeded with no appreciable racemization of the allylic part to give the lactones **18** and **19**, and the reaction has been used for the synthesis of a polypropionate chain[26].

However, the stereospecificity mentioned above is not retained in some catalytic reactions, depending on the structure of allylic part and the reaction conditions. In such a case, a π-allylpalladium intermediate isomerizes prior to the reaction with a nucleophile. The isomerization during the allylation occurs by the rear side attack of acetate anion on the π-allylpalladium acetate. The isomerization has also been explained by the attack of Pd(0) species on π-allylpalladium[27,28]. In the following example, transfer from C—O chirality to C—C is possible by the reaction of optically active allylic esters with carbon nucleophiles. Chirality transfer in the following cyclization of optically active allylic carbonate **20** (61% *ee*) to give **23** has been studied under various conditions, and different degrees of chirality transfer were observed. The optical purity of **23** was determined after converting **23** into the lactone **24**[27]. Formation of the complex **21** from **20** by inversion and its cyclization gives **23** by inversion and then (3*R*)-**24**. For this efficient chirality transfer, carbonate as a leaving group is better than acetate. Using Pd$_2$(dba)$_3$ (10 mol%) and TMPP in DMSO, nearly complete chirality transfer was observed. The degree of chirality transfer depends on the concentration of Pd catalyst; higher concentrations give lower optical yields. The explanation is that Pd(0) is a nucleophile and its attack on the π-allylpalladium intermediate **21** from the rear side causes racemization to give **22**. Cyclization of **22** should give **25** and (3*S*)-**26**.

As supporting evidence, rapid isomerization of the *cis*- and *trans*-π-allylpalladium complexes **27** and **28** is catalyzed by Pd(Ph$_3$P)$_4$ in THF even at −15 °C to give a 45 : 55 equilibrium mixture from either **27** or **28**[29–31]. Actually, in the intramolecular reaction of soft nucleophiles of **29** and **30**, a *trans–cis* mixture (**31** and **32**) (1 : 1) was obtained from *trans*-allylic acetate **29**. On the other hand, the corresponding *cis* compound **30** afforded only the *cis* product **32** by retention without isomerization[32].

In some cases, π-allylpalladium complex formation by retention (*syn* attack) has been observed. The reaction of the cyclic allylic chloride **33** with Pd(0) affords the π-allylpalladium chlorides **34** and **35** by retention or inversion depending on the solvents and Pd species. For example, retention is observed in benzene, THF, or dichloromethane with $Pd_2(dba)_3$. However, the complex formation proceeds by inversion in these solvents with $Pd(Ph_3P)_4$, whereas in MeCN and DMSO it is always inversion[33]. The *syn* attack in this case may be due to coordination of Pd to chlorine in **33**, because Pd is halophilic. The definite *syn* attack in complex formation has been observed using stereochemically biased substrates. The reaction of the *exo*-allylic diphenylphosphinoacetate **36** with phenylzinc proceeds smoothly to give **37**. The reaction can be explained by complex formation by a *syn* mechanism[31]. However, these *syn* attacks are exceptional, and normally *anti* attack dominates.

	trans	cis
	34	**35**
$Pd_2(dba)_3$	100	0
$Pd(Ph_3P)_4$	0	100

2.2.2 Allylation of Soft Carbon Nucleophiles

π-Allylpalladium cations can be regarded as 'soft' electrophiles, and react most smoothly with 'soft' nucleophiles. Compounds which have the electron-withdrawing groups (EWGs) shown are allylated. Typically, active methylene or methine compounds which are activated by two EWGs, such as carbonyl, sulfonyl, cyano, and nitro, are allylated by the Pd-catalyzed reactions of allylic compounds. Aryl groups (arylacetonitrile, phenylacetate)[14], olefinic bonds (β,γ-unsaturated sulfones, ketones)[11], and imino groups[34–36] are also activating groups. Methoxy(phenylthio)acetonitrile as a precursor of an ester is also reactive for allylation[37]. Nitroalkanes alone are allylated without being activated by other EWGs[38,39]. Both stoichiometric and catalytic allylations of nucleophiles via π-allylpalladium intermediates are called Tsuji–Trost reactions.

$$\begin{array}{cccccccc}
<\!\!\begin{array}{l}CO_2R\\CO_2R\end{array} & <\!\!\begin{array}{l}COR\\CO_2R\end{array} & <\!\!\begin{array}{l}CN\\CO_2R\end{array} & <\!\!\begin{array}{l}CO_2R\\SO_2Ph\end{array}, & <\!\!\begin{array}{l}COR\\COR\end{array}, & <\!\!\begin{array}{l}NO_2\\CO_2R\end{array}, & <\!\!\begin{array}{l}NC\\CO_2R\end{array}
\end{array}$$

$$RCH_2NO_2 \quad , \quad \underset{CN}{}\!\!<\!\!\begin{array}{l}CO_2R\\SPh\end{array}, \quad <\!\!\begin{array}{l}SO_2Ph\\SO_2Ph\end{array}, \quad <\!\!\begin{array}{l}N=\!\!<\\PO(OEt)_2\end{array}, \quad <\!\!\begin{array}{l}N=\!\!<\\CO_2R\end{array},$$

PhCH₂CN, PhCH₂CO₂R, PhCH₂SO₂R,

MX = SiMe₃
SnR₃
BR₂, Li

Interestingly, the allylation of a stabilized carbon nucleophile has been found to be reversible. Complete isomerization of dimethyl methylmalonate, involving bis-allylic C—C bond cleavage, from a secondary carbon **38** to a primary carbon **39** was observed by treatment with a Pd catalyst for 24 h. The C—C bond cleavage of a monoallylic system proceeds slowly[40].

2.2.2.1 Allylation under basic conditions. Allylation can be carried out under basic conditions with allylic acetates and phosphates, and under neutral conditions with carbonates and vinyloxiranes. The allylations under neutral conditions are treated separately in Section 2.2.2.2 from those under basic conditions. However, in some cases, allylations of the same substrates are carried out under both basic and neutral conditions to give similar results. These reactions are treated together in this section for convenience. Allylic acetates are widely used for Pd-catalyzed allylation in the presence of bases; tertiary amines or NaH are commonly used[6,7,41]. As a base, basic alumina or KF on alumina is conveniently used, because it is easy to remove by filtration after the reaction[42]. Allyl phosphates are more reactive than acetates. The allylation with **40** proceeds stepwise. At first allylic phosphate reacts with malonate and then allylic acetate reacts with amine to give **41**[43].

The intramolecular allylation of soft carbon nucleophiles with allylic acetates as a good cyclization method has been extensively applied to syntheses of various three, four, five and six-membered rings, and medium and macrocyclic compounds[44]. Only a few typical examples of the cyclizations are treated among numerous applications.

Examples of four-membered ring formation are rare. The cyclization of the cyclic allylic acetate **42** afforded a 2 : 1 mixture of the four-membered ring compound **43** and the six-membered ring compound **44**[45].

Intramolecular bis-allylation with a diester of 1,4-butenediol derivatives is a good synthetic method for vinylcyclopropanes[46]. Derivatives of chrysanthemic acid (**47a** and **47b**) were prepared by stepwise bis-allylation via **46** using the diacetate of the 1,4-butenediol derivative **45**. The starting material **49** for the bicyclo[5.1.0]octene ring **50** is easily prepared by the Pd-catalyzed selective reaction of the allylic chloride part in the 1-acetoxy-4-chloro-2-alkene system **48**[47]. As a ligand, dppe is the most suitable. Elimination takes place with Ph₃P. On prolonged reaction, the vinylcyclopropane **50** undergoes ring opening via the π-allylpalladium intermediate **51** to afford the dienes **52** and **53**. The reaction has been applied to the stereoselective synthesis of the side chain of glaucasterol (**55**). Cyclization of the allylic dichlorobenzoate **54** proceeded without racemization. A complete chirality transfer from C-22 to C-24 in a steroid side-chain took place[48]. The bicyclo[3,3,0]oct-6-en-2-one **58** was prepared via the cyclopropanation of **56** to give **57** and its rearrangement[49].

On the other hand, the Pd-catalyzed ring opening of the vinylcyclopropane **59** bearing two EWGs takes place to give the π-allylpalladium intermediate **60**, which reacts with amines[50] and carbon nucleophiles to afford **61**[51]. The (1,3-butadienyl)cyclopropane **62** substituted by two EWGs forms the π-allyl-palladium complex **63** by ring opening, and recyclizes to form the five-membered ring compound **64**[52]. This is another example of the cleavage of allylic C—C bonds. In these ring-opening reactions, the formation of a stabilized carbanion is a driving force of the π-allylpalladium intermediate formation. The optically active cyclopentene **66** was prepared by the asymmetric vinylcy-clopropane–cyclopentene rearrangement using the chiral sulfoxide **65** as a chiral source with 89% stereoselectivity[53].

Five-membered rings are prepared most easily by the intramolecular allylation of malonates or phenylsulfonylacetates. Formation of a spiro ring in **68** from the allylic acetate **67** is an example[54]. Changing the nucleophile from malonate to a β-keto ester reveals the problem of C versus O alkylation in the case of five-membered ring formation. Another problem is the selectivity between five- or seven-membered rings [54a]. In the cyclization of the allyl phenyl ether **69** (Pd-catalyzed reaction of allyl phenyl ethers proceeds under neutral conditions, see Section 2.2.2.2), the cyclopentanonecarboxylate **71** is obtained by the usual C-alkylation as a main product and the cycloheptenonecarboxylate **72** as a minor product when Bu_3P, Ph_3P, or dppe is used as a ligand in MeCN. When $(PhO)_3P$ is used, exclusively O-allylation takes place to give the furan **70**[15]. This is a rare example of the Pd-catalyzed O-allylation reaction. The O-allylation product **70** can be rearranged to the cyclopentanonecarboxylate **71** by further reaction with Pd–dppe or Ph_3P[15,55].

The ligand effect seems to depend on the substrates. Treatment of the prostaglandin precursor **73** with $Pd(Ph_3P)_4$ produces only the *O*-allylated product **74**. The use of dppe effects a [1,3] rearrangement to produce the cyclopentanone **75**[55]. Usually a five-membered ring, rather than seven-membered, is predominantly formed. The exceptionally exclusive formation of seven-membered ring compound **77** from **76** is explained by the inductive effect of an oxygen adjacent to the allyl system in the intermediate complex[56].

No *O*-allylation is observed in formation of the six-membered ring compound **79** by intramolecular allylation of the β-keto ester **78**[15,57]. Intramolecular allylation is useful for lactone formation. On the other hand, exclusive formation of the eight-membered ring lactone **81** from **80** may be in part derived from the preference for the nucleophile to attack the less substituted terminus of the allyl system[58].

Intramolecular allylation offers a useful synthetic method for macrocyclic compounds. An application to the synthesis of humulene (**83**) by the cyclization of **82** is the first example[59]. In the synthesis of the 16-membered ring keto lactone **85**, the sulfone group in **84** is necessary for the allylation, and after the cyclization it is removed to generate the double bond in **85**[60]. The allylation proceeds smoothly in the presence of an ethoxy group at the center carbon of the allyl group in **84**. After the allylation, the ethoxy group is hydrolyzed to generate a ketone. Another example is the 14-membered ring formation applied to the synthesis of cembranolide isolobophytilide (**88**). The cyclization of **86** exhibits high diastereoselectivity at the nucleophilic center since the macrocycle **87** was obtained as a single diastereomer[61]. In these macrocyclizations, use of BSA as a base is important. 2-Ethoxyallyl acetate (**89**) is an acetonylation reagent, although it is less reactive than simple allylic acetates. This functionality was employed for cyclopentenone annulation to prepare **90**[62] and for the synthesis of pyrenophorin[63]. Another acetonylation reagent is 2-chloroallyl acetate[64].

Intermolecular allylation is a useful synthetic method. The intermediate **92** of the Torgov steroid synthesis can be prepared by the selective C-allylation of cyclopentanone carboxylate with the allylic acetate **91**[65]. Dihydropyranyl acetate (**93**) undergoes regio- and stereoselective reaction at the anomeric carbon with net retention to give **94**[66]. Two different allylic groups (**95** and **97**) can be introduced stepwise into the bis(arylsulfonyl)methane **96**, and both arylsulfonyl groups are removed by treatment with sodium amalgam as shown by the synthesis of the pheromone **98**[67].

The regiochemistry of Pd-catalyzed allylation needs further consideration. Generally in the Pd-catalyzed allylation of carbon nucleophiles with unsymmetrically substituted allylic acetates, the attack at the less substituted side of the allylic systems is observed with high regioselectivity. However, it has been proposed that at low temperature and short reaction times the reaction is under kinetic control, but at elevated temperature and longer reaction times the reaction is under thermodynamic control. It was demonstrated that allylic malonate rearranges kinetically to its thermodynamically more stable regioisomer in the presence of Pd(0) catalyst[40]. In addition, the regioselectivity is influenced considerably by the leaving groups, nucleophiles, and ligands. Studies carried out on the regioselectivity revealed that allylation mainly at the more substituted side is frequently observed[12]. For example, in allylation in refluxing THF with the allylic acetates **99** and **100**, mainly substitution at the tertiary carbon to give **101** takes place and the more common products **102**, substituted at the primary carbon, are minor products. Taking advantage of this regioselectivity, 3,3-dimethyl-1-pentene-4-nitrile (**103**), an important compound for pyrethroid synthesis, can be prepared by the allylation of cyanoacetate with either **99** or **100** with high regioselectivity.

Chemoselective C-alkylation of the highly acidic and enolic triacetic acid lactone **104** (pK_a = 4.94) and tetronic acid (pK_a = 3.76) is possible by use of DBU[68]. No O-alkylation takes place. The same compound **105** is obtained by the regioslective allylation of copper-protected methyl 3,5-dioxohexanoate[69]. It is known that base-catalyzed alkylation of nitro compounds affords O-alkylation products, and the smooth Pd-catalyzed C-allylation of nitroalkanes[38,39], nitroacetate[70], and phenylsulfonylnitromethane[71] is possible. Chemoselective C-allylation of nitroethane (**106**) or the nitroacetate **107** has been applied to the synthesis of the skeleton of the ergoline alkaloid **108**[70].

Simple esters cannot be allylated with allyl acetates, but the Schiff base **109** derived from α-amino acid esters such as glycine or alanine is allylated with allyl acetate. In this way, the α-allyl-α-amino acid **110** can be prepared after hydrolysis[34]. The α-allyl-α-aminophosphonate **112** is prepared by allylation of the Schiff base **111** of diethyl aminomethylphosphonates. [35,36]. Asymmetric synthesis in this reaction using the (+)-*N,N*-dicyclohexylsulfamoylisobornyl alcohol ester of glycine and DIOP as a chiral ligand achieved 99% *ee*[72].

Allylation of simple ketones is not easy. However, their metal enolates can be allylated. Ketones are allylated as the Li enoate **113**[73] or after treatment of their potassium enolates with the triethylborane **114**[74]. Ketone enolates of boron and zinc can be allylated[75]. Tin enolate **115** is also used for regioselective monoallylation of ketones with allylic acetates, and used for annulation to give **116**[76]. The *O*-allylisourea **117**, formed by the reaction of allyl alcohol with dicyclohexylcarbodiimide, is used for the allylation of ketones using dppe as a ligand[77]. No allylation of simple esters with allylic acetates takes place. However, in an economical synthetic route to phytone, the Li enolate **118**, generated from β, γ-unsaturated 4-methyl-3-pentenoate with LDA, undergoes Pd-catalyzed allylation with allylic chlorides at room temperature in toluene to give the α-allylated ester **119** in a good yield[78].

Enamines derived from ketones are allylated[79]. The intramolecular asymmetric allylation (chirality transfer) of cyclohexanone via its *S*-proline allyl ester enamine **120** proceeds to give α-allylcyclohexanone (**121**) with 98% *ee*[80,81]. Low *ee* was observed in intermolecular allylation. Similarly, the asymmetric allylation of imines and hydrazones of aldehydes and ketones has been carried out[82].

The allyl-substituted cyclopentadiene **122** was prepared by the reaction of cyclopentadiene anion with allylic acetates[83]. Allyl chloride reacts with carbon nucleophiles without Pd catalyst, but sometimes Pd catalyst accelerates the reaction of allylic chlorides and gives higher selectivity. As an example, allylation of the anion of 6,6-dimethylfulvene **123** with allyl chloride proceeded regioselectively at the methyl group, yielding **124**[84]. The uncatalyzed reaction was not selective.

Diacetates of 1,4-butenediol derivatives are useful for double allylation to give cyclic compounds. 1,4-Diacetoxy-2-butene (**126**) reacts with the cyclohexanone enamine **125** to give bicyclo[4.3.1]decenone (**127**) and vinylbicyclo[3.2.1]octanone (**128**)[85,86]. The reaction of the 3-ketoglutarate **130** with *cis*-cyclopentene-3,5-diacetate (**129**) affords the furan derivative **131**[87]. The *C*- and *O*-allylations of ambident lithium [(phenylsulfonyl)methylene]nitronate (**132**) with **129** give isoxazoline-2-oxide **133**, which is converted into *cis*-3-hydroxy-4-cyanocyclopentene (**134**)[88]. Similarly, chiral *cis*-3-amino-4-hydroxycyclopentene was prepared by the cyclization of *N*-tosylcarbamate[89].

The 1,3-allylic diacetate **135** can be used for the formation of the methylenecyclopentane **137** with the dianionic compound **136**[86]. The cyclohexanone-2-carboxylate **138** itself undergoes a similar annulation with the 1,3-allylic diacetate **135** to form the methylenecyclohexane derivative **139**[90]. The reaction was applied as a key step in the synthesis of huperzin A[91]. On the other hand, *C*- and *O*-allylations of simple *β*-diketones or *β*-keto esters take place, yielding a dihydropyran **140**[92].

The allylic geminal diacetate **141** undergoes the monoallylation of malonates to give **142** and the two regioisomers **143** and **144**[93,94]. The dimethylacetal **145** or *ortho* esters of aromatic and α,β-unsaturated carbonyl compounds react with trimethylsilyl cyanide to give the methyl ether of cyanohydrin[95].

The α-isocyanocarboxylate **146** is allylated smoothly and the product **147** is converted into an amino acid[96]. The electronic effect of substituents on the regioselectivity of the allylation has been studied. The α-acetoxy-β, γ-unsaturated nitrile **148** and esters react with nucleophiles regioselectively at the γ-carbon, yielding the α, β-unsaturated nitriles **149**[97]. Similarly, the reaction of the acetoxyallylic phosphonate **150** with carbon nucleophiles or secondary amines affords conjugated vinylic phosphonate regioselectively[98]. The reaction of the allylic acetate **151**, substituted by *p*-nitrophenyl and *p*-anisyl groups at both sides, with triacetic acid lactone proceeds regioselectively, and the carbon nucleophile is introduced at the anisyl-substituted carbon to give **152**. Only a small amount of the isomer **153** was formed (**152** : **153** = 97 : 3)[99]. These results indicate that the allylic substitution takes place at the carbon with a higher electron density.

2-(Trimethylsilylmethyl)allyl acetate (**154**), which has a silyl group at the allylic position, undergoes the [3 + 2] cycloaddition reaction with electron deficient alkenes to give methylenecyclopentane derivatives. After the π-allylpalladium complex formation, elimination of the TMS group is facilitated by the proximal positive charge, and a dipolar intermediate **155** [the Pd complex of trimethylenemethane (TMM)] is generated as a reactive intermediate. (For other preparative methods for Pd–TMM complex, see Sections 2.2.2.2 and 7.4.) The cyclization of the intermediate **155** with an electron-deficient double bond proceeds by Michael addition to the double bond and Pd-catalyzed intramolecular allylation of the carbanion formed by the Michael addition to give the five-membered ring compound **156**[100,101]. Triisopropyl phosphite is a particularly good ligand for the cycloaddition. Various cyclic compounds are prepared by this reaction. Cyclohexenone (and also cycloheptenone) is not a good substrate, but the 3-siloxy-1-phenylsulfonylcyclohexene **157** serves as an excellent substrate. The highly diastereoselective methylenecyclopenteneannulation provides a versatile adduct **158** that allows simple conversion to cyclopentenone and cyclohexenone and other useful compounds[102]. The cycloaddition has been utilized for syntheses of various cyclic natural products. Addition to an activated norbornene is a key reaction of albene synthesis[103]. The adduct **159** to cyclopentenone is useful for the syntheses of loganin aglycone and loganin[104]. Similarly, a precursor to pentalenene was synthesized by the cycloaddition of a cyclopentenone derivative as a key step[105]. Cyclohexenone is a poor partner, but addition to the 2-cyclohexenone-2-carboxylate derivative **160** proceeds smoothly in a high yield and is used as a key step in the synthesis of the kempane diterpene **161**[106].

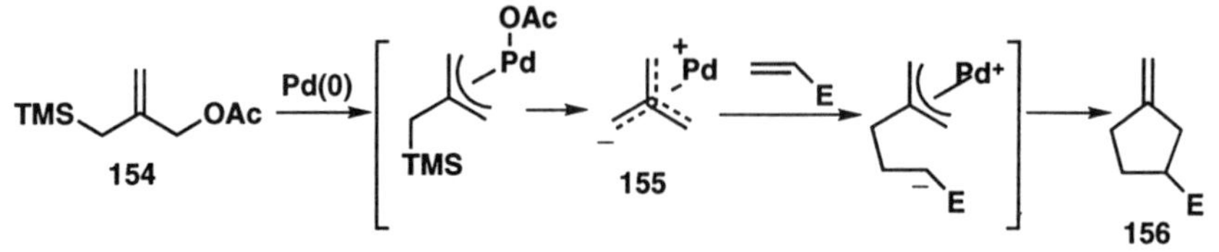
OAc
Pd
TMS
OAc
Pd(0)
154
TMS
Pd
E
155
Pd+
E
E
156

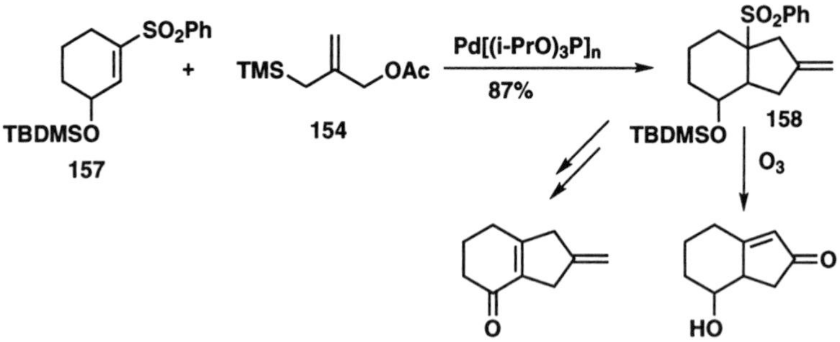
SO2Ph
+
TMS
OAc
Pd[(i-PrO)3P]n
87%
SO2Ph
TBDMSO
157
154
TBDMSO
158
O3
O
O
HO

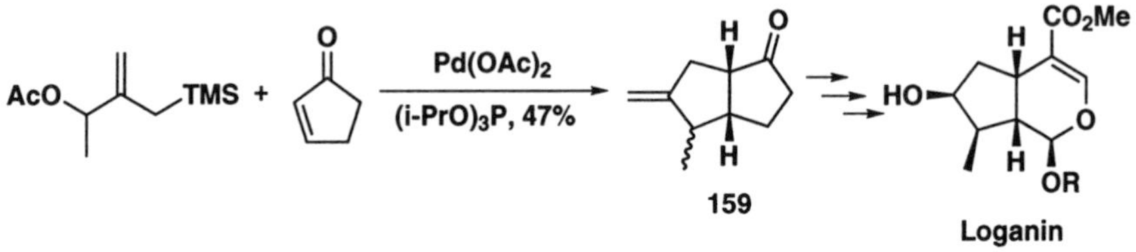
AcO
TMS +
O
Pd(OAc)2
(i-PrO)3P, 47%
H
O
H
159
CO2Me
H
HO
O
H
OR
Loganin

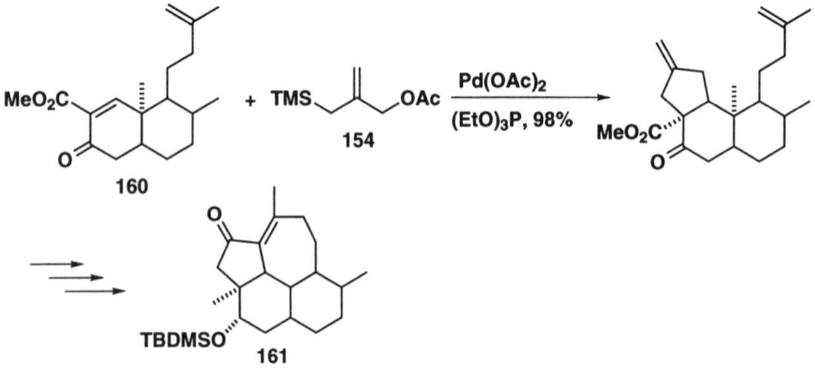
MeO2C
O
160
+ TMS
OAc
154
Pd(OAc)2
(EtO)3P, 98%
MeO2C
O
O
TBDMSO
161

A key intermediate, **163**, which possesses all but one chiral center of (+)-brefeldin, has been prepared by the enantiocontrolled cycloaddition of the chiral α,β-unsaturated ester **162** to **154**[107]. Synthesis of phyllocladane skeleton **165** has been carried out by the Pd-catalyzed cycloaddition of the unsaturated diester **164** and cobalt-catalyzed cycloaddition of alkynes as key reactions[108]. Intramolecular cycloaddition to the vinylsulfone in **166** proceeds smoothly to give a mixture of the *trans* and *cis* isomers in a ratio of 2.4 : 1[109]. Diastereocontrolled cycloaddition of the hindered vinylsulfone **167** affords a single stereoisomeric adduct, **168**, which is used for the synthesis of the spirocarbocyclic ring of ginkgolide[110].

The cyclooctenonecarboxylate **169** undergoes smooth cycloaddition with **154** to give **170** in a good yield, but no reaction takes place with cyclooctenone (**171**)[111]. Cyclohexenone behaves similarly. Cycloheptatrienone (**172**) undergoes [6 + 3] cycloaddition to form the nine-membered carbocycle **173**[112].

Chemoselectivity in the cycloaddition of 2-methylenecycloheptenone (**174**) changes on addition of In(acac)₃. The allylic carbonate **175** reacts with the ketone **174** in the presence of In(acac)₃ to give the methylenetetrahydrofuran **176,** and the methylenecyclopentane **177** is obtained in its absence[113]. The cycloaddition of ynones to produce the methylenetetrahydrofuran proceeds smoothly only in the presence of In(acac)₃ (10 mol%)[114].

Aldehydes take part in the cycloaddition to give the methylenetetrahydrofuran **178** by the co-catalysis of Pd and Sn compounds[115]. A similar product **180** is obtained by the reaction of the allyl acetate **179**, which has a tributyltin group instead of a TMS group, with aldehydes[116]. The pyrrolidine derivative **182** is formed by the addition of the tosylimine **181** to **154**[117].

The cyclic 2,4-dienoate **184**, formed by the Pd-catalyzed cyclization of the 1,6-enyne **183**, reacted with **154** to form the azulene derivative **185**[118]. The 3-methylenepyrrolidine **188** is formed by the reaction of the Zn reagent **186** with the chiral imine **187** with high diastereomeric excess. The structure of the allylic ethers is important for obtaining high diastereoselectivity[119].

The allylic esters **189** and **191** conjugated with cyclopropane undergo regioselective reactions without opening the cyclopropane ring. The soft carbon nucleophiles are introduced at the terminal carbon to give **190**, and phenylation with phenylzinc chloride takes place on the cyclopropane ring to form **192**[120].

2-Aza-π-allylpalladium is formed from the Schiff base **193** and reacts with malonate to give a derivative of aspartic acid **194** after hydrolysis of the product[121].

Allylic alcohols are less reactive allylating agents, and few examples of allylation with allylic alcohols are known[7,122]. However, smooth allylation of acetylacetone (**195**) using Pd(acac)$_2$ and Ph$_3$P was reported[7]. Allylation of carbon nucleophiles with the allylic alcohol **196** activated by treatment of allylic alkoxide with Ph$_3$B is possible[123]. Intramolecular allylation of the silyl ether **197** with allylic alcohol proceeds in CCl$_4$ as a solvent. Ph$_3$PCl$^+$ CCl$_3^-$ is generated from CCl$_4$ and Ph$_3$P, transposing the allylic alcohol into an oxyphosphonium group **198,** and liberating Cl$^-$ anion, which deprotects the silyl ether to generate *in situ* the alkoxy nucleophile **199** to give the tetrahydrofuran derivative **200**[124].

Allylation with allyl borates takes place smoothly under neutral conditions. Allylic alcohols are also used for allylation in the presence of boron oxide by *in situ* formation of allylic borates[125]. Similarly, arsenic oxide is used for allylation with allylic alcohols[126]. In addition, it was claimed that the allyl alkyl ethers **201**, which are inert by themselves, can be used for the allylation in the presence of boron oxide[127].

Allylic nitro compounds form π-allylpalladium complexes by displacement of the nitro group and react with nucleophiles, and allylation with the tertiary nitro compound **202** takes place at the more substituted side without rearrangement to give **203**[8,9,128].

The alkyl allyl sulfonate **205** was prepared by rearrangement of the alkyl allyl sulfite **204**[129]. Allyl sulfones are normally employed as carbon nucleophiles for alkylation, but allyl sulfones react with Pd(0) to form π-allylpalladium complexes, and are displaced with nucleophiles[10–12]. In the reaction of substituted allylic compounds, usually a nucleophile is introduced at the less substituted side. However, substitution at the more substituted side of allyl sulfones is observed to some extent depending on the ligands and leaving groups. The primary prenyl sulfone **206** affords the primary and tertiary substituted malonates **207** and **208** in a ratio of 66 : 34 using dppe as a ligand. On the other hand, the isomeric sulfone **209** yields the tertiary malonate as a main product (**207** : **208** = 23 : 77) using Ph₃P[12]. The pheromone **213** of the Monarch butterfly was prepared by combination of this dual reactivity of allylic sulfones. The first step is the Michael addition of methyl vinyl ketone to the methallyl sulfone **210** to give **211**. Then the allylic sulfone group in **212** is displaced with malonate via a π-allylpalladium intermediate[10].

Allylic amine is a less reactive leaving group[7], but the allylic ammonium salts **214** (quaternary ammonium salts) can be used for allylation[130,131]. Allylic sulfonium salts are also used for the allylation[130]. The allylic nitrile in the cyclic aminonitrile **215** can be displaced probably via π-allylic complex formation. The possibility of the formation of the dihydropyridinium salts **216** and subsequent conjugate addition are less likely[132].

Asymmetric allylation of carbon nucleophiles has been carried out extensively using Pd catalysts coordinated by various chiral phosphine ligands and even with nitrogen ligands, and *ee* > 90% has been achieved in several cases. However, in most cases, a high *ee* has been achieved only with the 1,3-diaryl-substituted allylic compounds **217**, and the synthetic usefulness of the reaction is limited. Therefore, only references are cited[24,133].

2.2.2.2 Allylation under neutral conditions. Reactions which proceed under neutral conditions are highly desirable. Allylation with allylic acetates and phosphates is carried out under basic conditions. Almost no reaction of these allylic compounds takes place in the absence of bases. The useful allylation under neutral conditions is possible with some allylic compounds. Among them, allylic carbonates **218** are the most reactive and their reactions proceed under neutral conditions[13,14,134]. In the mechanism shown, the oxidative addition of the allyl carbonates **218** is followed by decarboxylation as an irreversible process to afford the π-allylpalladium alkoxide **219**, and the generated alkoxide is sufficiently basic to pick up a proton from active methylene compounds, yielding **220**. This *in situ* formation of the alkoxide, which is a poor nucleophile, is the reason why the reaction of allyl carbonates can be carried out without addition of bases from outside. In addition, the formation of π-allylpalladium complexes from allylic carbonates involving decarboxylation is irreversible. On the other hand, the complex formation from allylic acetates is reversible. This is another advantage of allylic carbonates over allylic acetates. Allylic carbamates behave similarly and can be used for the allylation under neutral conditions[135]. The chemoselective reaction of the allylic carbonate part in the allylic compound **221** without attacking the allylic acetate part under neutral conditions clearly shows the higher reactivity of allylic carbonates than allylic acetates[14]. Similarly, the nitroacetate **223** undergoes smooth chemoselective *C*-allylation with the allylic carbonate **222** under neutral conditions[135a]. The allylation of carbon and nitrogen nucleophiles with the allylic carbonates **224** or acetates proceeds under milder conditions in an aqueous solution using water-soluble TMSPP as the ligand. The catalyst stays in aqueous phase and can be recycled[136].

Since allylation with allylic carbonates proceeds under mild neutral conditions, neutral allylation has a wide application to alkylation of labile compounds which are sensitive to acids or bases. As a typical example, successful *C*-allylation of the rather sensitive molecule of ascorbic acid (**225**) to give **226** is possible only with allyl carbonate[137]. Similarly, Meldrum's acid is allylated smoothly[138]. Pd-catalyzed reaction of carbon nucleophiles with isopropyl 2-methylene-3,5-dioxahexylcarbonate (**227**)[139] followed by hydrolysis is a good method for acetonylation of carbon nucleophiles.

Pd-catalyzed reaction of allyl acetate substituted by a TMS group with nucleophiles gives alkenylsilanes regioselectively[140–142]. Furthermore, the α-substituted γ-trimethylsilylated allyl carbonate **228** reacts with carbon nucleophiles regioselectively at the carbon opposite to the silyl group, affording the alkenylsilane **229**. The whole process means that regioselective nucleophilic substitution exclusively at the more substituted side of the allylic carbonate **230** can be achieved indirectly by these reactions[142]. Similarly, nitrogen nucleophiles react regioselectively[143].

Allylation of the alkoxymalonitrile **231** followed by hydrolysis affords acyl cyanide, which is converted into the amide **232**. Hence the reagent **231** can be used as an acyl anion equivalent[144]. Methoxy(phenylthio)acetonitrile is allylated with allylic carbonates or vinyloxiranes. After allylation, they are converted into esters or lactones. The intramolecular version using **233** has been applied to the synthesis of the macrolide **234**[37]. The β,γ-unsaturated nitrile **235** is prepared by the reaction of allylic carbonate with trimethylsilyl cyanide[145].

Allylation of the 10-carborane **236** (pK_a = 18–22) with diallyl carbonate is possible under neutral conditions to give **237**[146]. Allylation and rearrangement of the trialkylalkynylborane **238** affords the trisubstituted alkene **239** stereoselectively[147].

The π-allylpalladium complexes **241** formed from the allyl carbonates **240** bearing an anion-stabilizing EWG are converted into the Pd complexes of TMM (trimethylenemethane) as reactive, dipolar intermediates **242** by intramolecular deprotonation with the alkoxide anion, and undergo [3 + 2] cycloaddition to give five-membered ring compounds **244** by Michael addition to an electron-deficient double bond and subsequent intramolecular allylation of the generated carbanion **243**. This cycloaddition proceeds under neutral conditions, yielding the functionalized methylenecyclopentanes **244**[148]. The synthetic method for the reagent **240** has been improved and the reagent has been used for the preparation of the pentalenolactone intermediate **245**[149].The asymmetric cycloaddition of **240** with acrylate using chiral ferrocenyl phosphine afforded the adduct **246** in an optical yield of 78%[150]. As a related method, treatment of 2-chloromethylallyl phenyl sulfone with Pd(0) generates a TMM complex under basic conditions which is used for pentalenolactone synthesis[151]. The oxa-TMM complex **248** is generated from 5,5-dimethyl-4-methylene-1,3-dioxolan-2-one (**247**). Although interesting Pd-catalyzed reations are expected from the cyclic carbonate **247**, it has low nucleophilicity. Its cycloaddition to phenyl isocyanate gives oxazolidinone[152].

The vinylcyclopropane **249**, activated by two EWGs, is easily prepared by the Pd-catalyzed reaction of the biscarbonate of 1,4-butenediol and generates the dipolar allylpalladium species **250** by the ring opening under neutral conditions, which undergoes cycloaddition with acrylate to give **251**[153]. The activated vinylcyclopropane **249** also undergoes cycloaddition with phenyl isocyanate via the zwitterionic π-allylpalladium intermediate **250** to give the γ-lactam **252**. 2,2-Diphenyl-5,5-bis(methoxycarbonyl)-3-vinylcyclopentanone was obtained in 85% yield by the reaction of diphenylketene[154].

Diphenylketene (**253**) reacts with allyl carbonate or acetate to give the α-allylated ester **255** at 0 °C in DMF. The reaction proceeds via the intermediate **254** formed by the insertion of the C=C bond of the ketene into π-allylpalladium, followed by reductive elimination. Depending on the reaction conditions, the decarbonylation and elimination of β-hydrogen take place in benzene at 25 °C to afford the conjugated diene **256**[155].

Another useful allylating reagent under neutral conditions is the vinyloxirane **257** (vinyl epoxide). The carbon–oxygen bond in **257** is easily cleaved with Pd(0) by oxidative addition to form the π-allylpalladium complex **258**. This cleavage generates the alkoxide, which picks up a proton from nucleophiles, yielding **259,** and hence the allylation reaction with vinyl epoxide can be carried out under neutral conditions. In addition, the reaction shows high regioselectivity, namely 1,4-addition rather than 1,2-addition is the main path[16,17]. The 1,4-adducts **260** are allylic alcohols and further Pd-catalyzed reaction with a nucleophile is possible. In the presence of vinyloxirane and allylic acetate in the same molecule **261**, only the vinyloxirane moiety reacts with acetoacetate chemoselectively under neutral conditions to give the 1,4-adduct **262** as the main product and the 1,2-adduct **263** as the minor product. No reaction takes place with the allylic acetate moiety[16].

The reaction of vinyloxiranes with malonate proceeds regio- and stereoselectively. The reaction has been utilized for the introduction of a 15-hydroxy group in a steroid related to oogoniol (**265**)[156]. The oxirane **264** is the β-form and the attack of Pd(0) takes place from the α-side by inversion. Then the nucleophile comes from the β-side. Thus overall reaction is *syn*-S_N2' type. In the intramolecular reaction, the stereochemical information is transmitted to the newly formed stereogenic center. Thus the formation of the six-membered ring lactone **267** from **266** proceeded with overall retention of the stereochemistry, and was employed to control the stereochemistry of C-15 in the prostaglandin **268**[157]. The method has also been employed to create the butenolide of cardenolides[158]. However, the cyclization of **269** gave the six-membered ring compound **270** by 1,2-addition, rather than the normal 1,4-addition to form an eight-membered ring. On the other hand, a nine-membered ring by 1,4-addition was formed rather than a seven-membered ring[159].

Intramolecular 1,4-addition is useful for macrolide synthesis. An unusual molecule of punctaporonin B (**272**) has been synthesized by this 1,4-addition of **271**[160]. Cyclization to form the seventeen-membered ring macrolide **273** was carried out at 0.1–0.5 M concentration[161]. The choice of ligands seems to be important in the macrocyclization. The 26-membered ring model **274** for a synthesis of the ring system of tetrin A was obtained in 92% yield by using triisopropyl phosphite as a ligand[162].

Vinyloxiranes react with CO_2 under severe conditions without a catalyst, but the Pd-catalyzed reaction proceeds at room temperature and 1 atm to give cyclic carbonates in good yields[163]. The carboxylate nucleophile **276** is generated *in situ* by capturing the oxygen-leaving group in the Pd-catalyzed reaction of the vinyloxirane **275** with CO_2 to afford the cyclic carbonate **277**[164]. 1,2-Addition, rather than the normal 1, 4-addition, is observed by the tethering of the carboxylate anion to the oxygen leaving group. This reaction offers a method for the *cis*-hydroxylation of epoxides, and was used for the syntheses of (+)-citreoviral[165] and (−)-*exo*-brevicomin (**278**)[166]. Similarly, vinyloxiranes react with aldehydes to form the cyclic acetal **279** of aldehydes[167].

279

The four-membered vinyloxetane **280** is cleaved with Pd(0) and used for allylation; a homoallylic alcohol unit can be introduced into the keto ester **281** as a nucleophile with this reagent to form **282**[168].

Allyl aryl ethers are used for allylation under basic conditions[6], but they can be cleaved under neutral conditions. Formation of the five-membered ring compound **284** based on the cyclization of **283** has been applied to the syntheses of methyl jasmonate (**285**)[15], and sarkomycin[169]. The trisannulation reagent **286** for steroid synthesis undergoes Pd-catalyzed cyclization and aldol condensation to afford CD rings **287** of steroids with a functionalized 18-methyl group[170]. The 3-vinylcyclopentanonecarboxylate **289**, formed from **288**, is useful for the synthesis of 18-hydroxyestrone (**290**)[171].

Phenyl 4,6-di-*O*-benzyl-2,3-dideoxy-D-*erythro*-hex-2-enopyranoside (**291**) reacts regiospecifically with malonate or acetoacetate using dppb as a ligand at the anomeric center and stereoselectively to give the α- and β-*C*-glycopyranoside **292** (α : β = 75 : 25)[172]. On the other hand, the facile regioselective *C*-glycosylation of trifluoroacetylglucal proceeds under basic conditions. No reaction takes place with the corresponding acetate[173].

2.2.3 Allylation of Other Nucleophiles

Some nucleophiles other than carbon nucleophiles are allylated. Amines are good nucleophiles. Diethylamine is allylated with allyl alcohol[7]. Allylamines are formed by the reaction of allyl alcohol with ammonia by using dppb as a ligand. Di- and triallylamines are produced commercially from allyl alcohol and ammonia[174].

Intramolecular amination with allylic acetates is used for the synthesis of cyclic alkaloids[175]. Cyclization of **293** affords the six-membered ring compound **294** rather than a four-membered ring. The reaction is particularly useful for 1-azaspirocyclization, which is suitable for the synthesis of perhy-drohistrionicotoxin (**295**)[176,177].

Allylic chlorides react with amines without a Pd catalyst, but the reaction is accelerated with such a catalyst. For example, in the Pd-catalyzed reaction of the allylic chloroacetate **296** with amine at 25 °C, only the allylic chloride is displaced without attacking the allylic acetate moiety. The uncatalyzed reaction of **296** proceeds at 80 °C[178]. The secondary amine **297** is obtained without significant formation of a tertiary amine when the reaction is carried out in an aqueous solution using water-soluble TMSPP as a ligand[179]. Macrocyclic amines are prepared by intramolecular allylation of amines with allylic acetates. The 21-membered cyclic amine skeleton of the spermidine alkaloid inandenin-12-one (**298**) has been prepared in high yields using dppb as a ligand. The ratio of dppb to Pd(Ph₃P)₄ should be < 1 for successful cyclization. Neutralization, hydrogenation, and acetamide hydrolysis complete the synthesis of inandenin-12-one[180].

Imidazole can be *N*-allylated. The *N*-glycosylimidazole **299** is prepared by regiospecific amination at the anomeric center with retention of configuration. Phenoxy is a good leaving group in this reaction[181]. Heterocyclic amines such as the purine base **300** are easily allylated[182].

The reaction of the vinylcyclopropanedicarboxylate **301** with amines affords an allylic amine via the π-allylpalladium complex **302**[50]. Similarly, three-membered ring *N*-tosyl-2-(1,3-butadienyl)aziridine (**303**) and the four-membered ring azetidine **304** can be rearranged to the five- and six-membered ring unsaturated cyclic amines[183].

Preparation of primary allylamines by the selective monoallylation of ammonia is not possible and they are prepared by indirect methods. The monoallylation of Li and Na amides of di-*t*-butoxycarbonyl (Boc) (**305**), followed by hydrolysis, affords a primary allylamine (**306**)[184].

The reaction of an azide[185,186] or a trimethylsilylazide[187] followed by the treatment with Ph₃P is another preparative method for the primary allylic amine **307**.

Regioselective 1,4-azidohydroxylation to give **309** takes place by the reaction of the vinyloxirane **308** with sodium azide[188]. The reaction of the cyclopentadiene monoepoxide **310** with sodium azide or purine base offers a good synthetic method for the carbocyclic nucleoside **311**[189–191]

The cyclic carbamate (oxazolidin-2-one) **313** is formed by the reaction of phenyl isocyanate (**312**) with vinyloxirane[192]. Nitrogen serves as a nucleophile and attacks the carbon vicinal to the oxygen exclusively. The thermodynamically less stable *Z*-isomer **315** was obtained as a major product (10 : 1) by the reaction of 2-methoxy-1-naphthyl isocyanate (**314**) with a vinyloxirane[193]. Formation of the oxazolidinone **316** by this method is used for the *cis*-hydroxyamination as shown by a synthesis of acosamine (**317**)[192]. 2,3-Dihydroxyalkenes react similarly[194]. The reaction of cyclic carbonates with aryl isocyanates affords cyclic carbamates[195].

Sodium *p*-toluenesulfonamide (**319**) reacts with the allylic lactone **318** to give an allylic tosylamide with retention of configuration[196].

Carbamates are allylated in the presence of strong bases in DMSO or HMPA[197]. Phthalimide (**320**) and succimide are allylated with the allyl-isourea **321** at room temperature or the allylic acetate **322** at 100 °C[198,199]. Di-*t*-butyl iminodicarbonate is used as a nitrogen nucleophile[200].

The secondary allylic methylamine **324** can be prepared by the allylation of *N*-methylhydroxylamine (**323**), followed by hydrogenolysis[201]. Monoallylation of hydroxylamine, which leads to primary allylamines, is achieved using the *N,O*-bis-Boc-protected hydroxylamine **326**. N^6-Hydroxylysine (**328**) was synthesized by chemoselective reaction of (*Z*)-4-acetoxy-2-butenyl methyl carbonate (**325**) with two different nucleophiles: first with *N,O*-Boc-protected hydroxylamine (**326**) under neutral conditions and then with methyl (diphenylmethyleneamino)acetate (**327**) in the presence of BSA[202]. The primary allylic amine **331** is prepared by the highly selective monoallylation of 4,4′-dimethoxybenzhydrylamine (**329**). Deprotection of the allylated secondary amine **330** with 80% formic acid affords the primary allylamine **331**. The reaction was applied to the total synthesis of gabaculine **332**[203].

2-Pyridone (**333**) is allylated with allylic carbonates on the nitrogen atom rather than on the oxygen atom, but 2-thiopyridone (**334**) is allylated on the sulfur atom[204].

Alcohols are poor *O*-nucleophiles, and the Pd-catalyzed allylation of alcohols to form allyl alkyl ethers is somewhat sluggish. As one method, the allyl alkyl carbonates **335** are decarboxylated to give alkyl allyl ethers in the absence of other nucleophiles[205]. Interestingly, secondary alcohols in methyl α-D-glucopyranoside (**336**) can be allylated by the reaction of an excess of allyl ethyl or methyl carbonate, rather than the expected allyl ethyl or methyl ether formation[206]. The allyloxycarbonyl group was introduced unexpectedly to the highly hindered C-12α alcohol in the steroid **337** by its reaction with allyl ethyl carbonate using dppb as a ligand. Surprisingly, no decarboxylation took place[207].

Complete chirality transfer has been observed in the intramolecular allylation of an alcohol with the activated allylic ester of 2,6-dichlorobenzoic acid **338** to give the 2-substituted tetrahydrofuran **339**[208].

Silyl ethers serve as precursors of nucleophiles and liberate a nucleophilic alkoxide by desilylation with a chloride anion generated from CCl_4 under the reaction conditions described before[124]. Rapid intramolecular stereoselective reaction of an alcohol with a vinyloxirane has been observed in dichloromethane when an alkoxide is generated by desilylation of the silyl ether **340** with TBAF. The *cis*- and *trans*-pyranopyran systems **341** and **342** can be prepared selectively from the *trans*- and *cis*-epoxides **340**, respectively. The reaction is applicable to the preparation of 1,2-diol systems[209]. The method is useful for the enantioselective synthesis of the AB ring fragment of gambiertoxin[210]. Similarly, tributyltin alkoxides as nucleophiles are used for the preparation of allyl alkyl ethers[211].

| from | trans epoxide | 99 | : | 1 |
| from | cis epoxide | 2 | : | 98 |

Carboxylate anions are better nucleophiles for allylation. The monoepoxide of cyclopentadiene **343** is attacked by AcOH regio- and stereoselectively via π-allylpalladium complex formation to give the *cis*-3,5-disubstituted cyclopentene **344**[212]. The attacks of both the Pd and the acetoxy anion proceed by inversion (overall retention) to give the *cis* product.

Phenols are highly reactive *O*-nucleophiles and allylated easily with allylic carbonates under neutral conditions. EWGs on phenols favor the reaction[213]. Allylic acetates are used for the allylation of phenol in the presence of KF-alumina as a base[214].

Carbon–phosphorus bonds are formed by the allylation of various phosphorus compounds. The allyldiphenylphosphine sulfide **346** is formed by the reaction of allylic acetates with lithium diphenylthiophosphide **345**[215].

The Pd-catalyzed reaction of an allylic alcohol, phosphine and an aldehyde in boiling dioxane gives conjugated dienes. Pd-catalyzed reaction of an allylic alcohol with a phosphine affords a phosphonium salt, which reacts with an aldehyde to form a conjugated diene by a Wittig-type reaction[216]. This Wittig-type reaction is interesting because it can be carried out as a one-pot reaction without using a base. A modification of this Wittig-type reaction is the Pd-catalyzed reaction of an allylic alcohol, phenyl isocyanate, an aldehyde and Bu₃P in refluxing MeCN. An allylic carbamate is formed *in situ*, which reacts with Bu₃P and the aldehyde via a phosphonium salt to form a conjugated diene. All-*trans*-β-carotene (**349**) has been prepared by the reaction of the allylic alcohol **347** and the dialdehyde **348**[217]. Allylic isoureas also are used[218]. As an alternative method, the allylic phosphonium salt **350** is formed by the Pd-catalyzed reaction of Ph₃P with geranyl acetate in the presence of NaBr, and converted *in situ* to the ylide by treatment with BuLi. The conjugated diene **351** was obtained by the reaction of benzaldehyde[219]. Similarly, allylic nitro compounds are used for stereoselective (*E*)-alkene formation[220].

The dimethyl allylphosphonate **352** is prepared by the reaction of allylic acetates with trimethyl phosphite[221].

Various *S*-nucleophiles are allylated. Allylic acetates or carbonates react with thiols or trimethylsilyl sulfide (**353**) to give the allylic sulfide **354**[222]. Allyl sulfides are prepared by Pd-catalyzed allylic rearrangement of the dithiocarbonate **355** with elimination of COS under mild conditions. The benzyl alkyl sulfide **357** can be prepared from the dithiocarbonate **356** at 65 °C[223,224]. The allyl aryl sufide **359** is prepared by the reaction of an allylic carbonate with the aromatic thiol **358** by use of dppb under neutral conditions[225]. The *O*-allyl phosphoro- or phosphonothionate **360** undergoes the thiono–thiolo allylic rearrangement (from *O*-allyl to *S*-allyl rearrangement) to afford **361** and **362** at 130 °C[226].

The reaction of the allylic acetate **363** with the sodium arylsulfinate **364** affords the allyl aryl sulfone **365**[227,227a]. The allylic nitro compound **366** undergoes denitrosulfonylation with sodium phenylsulfinate (**367**) without allylic rearrangement to give the allyl phenyl sulfone **368**[228,229]. The β,γ-epoxy nitro compound **369** reacts with **367** to give the allylic sulfone **372**. In this reaction, the allylic nitro compound **370** is formed, which reacts with Pd(0) to afford the π-allylpalladium intermediate **371**[230]. Allylic *p*-toluenesulfinate–sulfone rearrangement of the chiral sulfinate **373** proceeds to give the sulfone **374** with a chirality transfer of 91.8%[231]. Asymmetric induction of 87% was also achieved in the Pd-catalyzed rearrangement of **375** using DIOP as a chiral ligand[232].

Allylic phenyl selenides are obtained by the reaction of allylic acetates with diphenyldiselenide and SmI_2[233].

2.3 Carbonylation

Carbonylation of various allylic compounds in alcohols gives β,γ-unsaturated esters, but allylic compounds are less reactive than aryl or alkenyl halides. Allylic halides can be carbonylated in alcohols under pressure. Carbonylation of allyl chloride proceeds smoothly in THF to give the 4-chlorobutyl ester **376** in a good yield[234,235]. Smooth carbonylation of allyl chloride proceeds at 1 atm under mild conditions in two phases using a phase-transfer agent and water-soluble DPMSPP as a ligand[236], or by use of EtONa and dppe[237]. In the presence of a double bond, the carbonylation of the allylic chloride **377** is followed by intramolecular insertion of the double bond into the Pd–acyl bond, affording the cyclopentenoneacetate derivative **378**[238].

Allylic carbonates are most reactive. Their carbonylation proceeds under mild conditions, namely at 50 °C under 1–20 atm of CO. Facile exchange of CO_2 with CO takes place[239]. The carbonylation of 2,7-octadienyl methyl carbonate (379) in MeOH affords the 3,8-nonadienoate 380 as expected, but carbonylation in AcOH produces the cyclized acid 381 and the bicyclic ketones 382 and 383 by the insertion of the internal alkene into π-allylpalladium before CO insertion[240] (see Section 2.11). The alkylidenesuccinate 385 is prepared in good yields by the carbonylation of the allylic carbonate 384 obtained by DABCO-mediated addition of aldehydes to acrylate. The $E : Z$ ratios are different depending on the substrates[241].

It is known that π-allylpalladium acetate is converted into allyl acetate by reductive elimination when it is treated with CO[242,243]. For this reason, the carbonylation of allylic acetates themselves is difficult. The allylic acetate 386 is carbonylated in the presence of NaBr (20–50 mol%) under severe conditions, probably via allylic bromides[244]. However, the carbonylation of 5-phenyl-2,4-pentadienyl acetate (387) was carried out in the presence of Et_3N without using NaBr at 100 °C to yield methyl 6-phenyl-3,5-hexadienoate (388)[245]. The dicarbonylation of 1,4-diacetoxy-2-butene to form the 3-hexenedioate also proceeds by using tetrabutylphosphonium chloride as a ligand in 49% yield[246].

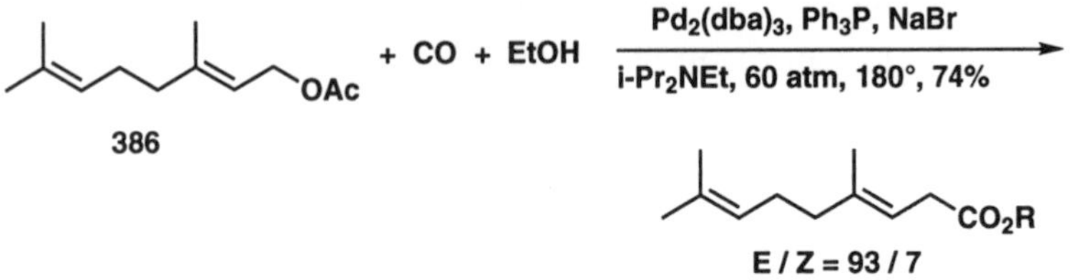

+ CO + EtOH
Pd2(dba)3, Ph3P, NaBr
i-Pr2NEt, 60 atm, 180°, 74%
OAc
386
CO2R
E / Z = 93 / 7

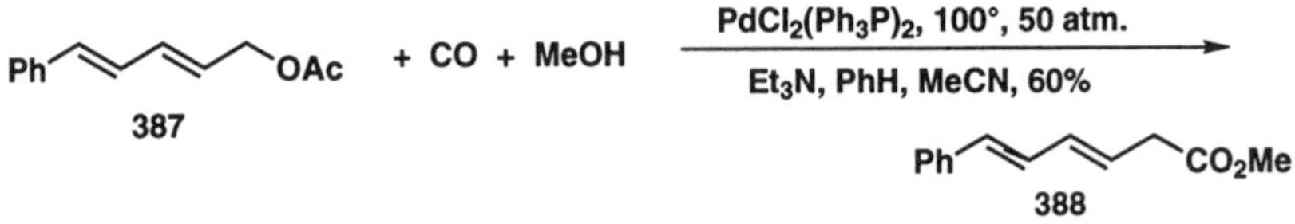

Ph
OAc
+ CO + MeOH
PdCl2(Ph3P)2, 100°, 50 atm.
Et3N, PhH, MeCN, 60%
387
Ph
CO2Me
388

Allylic phosphates are used for carbonylation in the presence of amines under pressure. Carbonylation of diethyl neryl phosphate (**389**) affords ethyl homonerate (**390**), maintaining the geometric integrity of the double bond[244]. The carbonylation of allyl phosphate in the presence of the imine **392** affords the β-lactam **393**. The reaction may be explained by the formation of the ketene **391** from the acyl phosphate, and its stereoselective [2 + 2] cycloaddition to the imine **392** to give the β-lactam **393**[247].

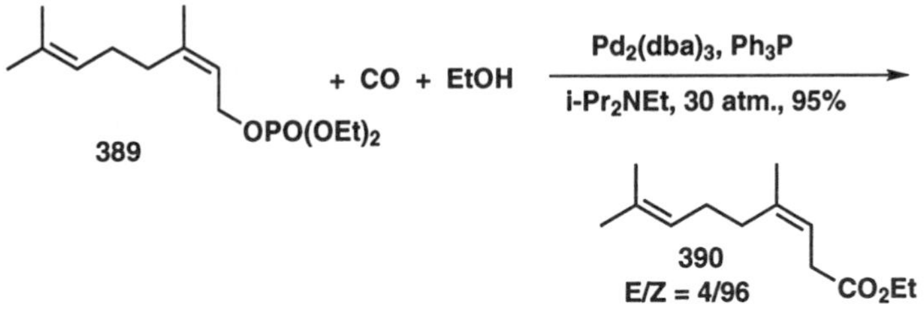

+ CO + EtOH
Pd2(dba)3, Ph3P
i-Pr2NEt, 30 atm., 95%
OPO(OEt)2
389
390
E/Z = 4/96
CO2Et

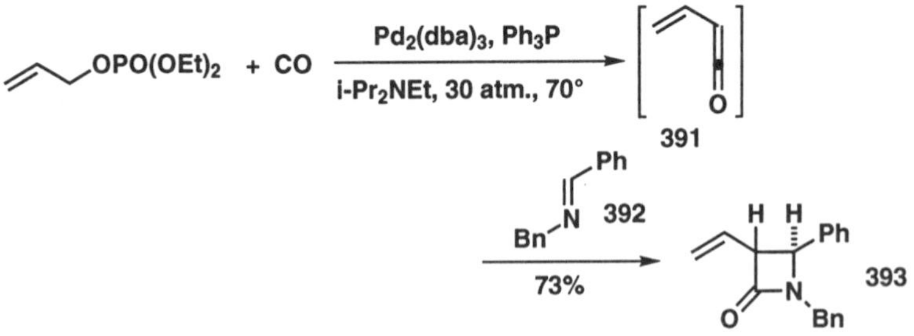

OPO(OEt)2 + CO
Pd2(dba)3, Ph3P
i-Pr2NEt, 30 atm., 70°
O
391
Ph
N 392
Bn
73%
H H
Ph
N
O
Bn
393

Carbonylation of allylic alcohols requires severe conditions[248]. The carbonylation of allylic alcohols proceeds smoothly in the presence of LiCl and Ti(IV) isopropoxide[249]. The allylic methyl ether **394** can be carbonylated with the use of $PdCl_2$[250] or π-allylpalladium coordinated by BF_4, PF_6, and chloride anions, or with $PdCl_2$–PPh_3 as catalysts at 100 °C and 30 atm in toluene[251].

Allylamines are not easily cleaved with Pd catalysts, but the carbonylation of the allylic amine **395** proceeds at 110 °C to give the β,γ-unsaturated amide **396** by using dppp as a ligand[252]. Decarboxylation–carbonylation of allyl diethylcarbamate under severe conditions (100 °C, 80 atm) affords β,γ-unsaturated amides[253]. The 3-vinylaziridine **397** is converted into the α-vinyl-β-lactam **398** under mild conditions[254].

The 3,7-octadienoate **399** is obtained by the carbonylation of allyl chloride and butadiene[255].

Ketones can be prepared by trapping (transmetallation) the acyl palladium intermediate **402** with organometallic reagents. The allylic chloride **400** is carbonylated to give the mixed diallylic ketone **403** in the presence of allyltributylstannane (**401**) in moderate yields[256]. Alkenyl- and arylstannanes are also used for ketone synthesis from allylic chlorides[257,258]. Total syntheses of dendrolasin (**404**)[258] and manoalide[259] have been carried out employing this reaction. Similarly, formation of the ketone **406** takes place with the alkylzinc reagent **405**[260].

The β,γ-unsaturated aldehyde **407** is prepared in good yields by the carbonylation of an allylic chloride under mild conditions using tributyltin hydride as a hydride source[261]. Aldehydes are obtained in moderate yields by the reaction of CO and H_2[262].

Unusual cyclocarbonylation of allylic acetates proceeds in the presence of acetic anhydride and an amine to afford acetates of phenol derivatives. The cinnamyl acetate derivative **408** undergoes carbonylation and Friedel–Crafts-type cyclization to form the α-naphthyl acetate **410** under severe conditions[263,264]. The reaction proceeds at 140–170 °C under 50–70 atm of CO in the presence of acetic anhydride and Et_3N. Addition of acetic anhydride is essential for the cyclization. The key step seems to be the Friedel–Crafts-type cyclization of an acylpalladium complex as shown by **409**. When MeOH is added instead of acetic anhydride, β,γ-unsaturated esters such as **388** are formed as described before. From a naphthalein derivative, a phenanthrene ring is formed rather than an anthracene ring[265]. The reaction can be applied to the syntheses of acetoxybenzofuran, acetoxyindole, and the acetoxyben-zothiophene **411** from corresponding heterocycles[266]. Even the 2,4-dienyl acetate **387** undergoes a similar cyclization, affording the phenyl acetate derivatives **412**. The commonly observed five-membered ring formation does not take place in this case[245].

2.4 Reactions of Hard Nucleophiles via Transmetallation

Various allylic compounds react with hard carbon nucleophiles of organome-tallic compounds of Zn, B, Al, Sn, and Si via transmetallation. Tables sum-marizing allylic leaving groups and alkenyl and aryl metallic reagents used in the coupling and regioselectivity have been published[267]. Two kinds of cross-coupling of allyl groups are known, the reaction with hard carbon nucleophiles and the formation of allylic metal compounds from bimetallic compounds, and umpolung of π-allylpalladium complexes can be achieved.

2.4.1 Cross-Coupling of Allylic Groups with Hard Carbon Nucleophiles

1,4-Dienes and allylarenes can be prepared by the Pd-catalyzed coupling of allylic compounds with hard carbon nucleophiles derived from alkenyl and aryl compounds of main group metals. Allylic compounds with various leaving groups can be used. Some of them are unreactive with soft nucleophiles, but react with hard carbon nucleophiles. The order of the reactivity of allylic leaving groups in the coupling of alkenylaluminums was found to be Cl > OAc > OAlMe$_2$ > OPO(OEt)$_2$ > OSiMe$_3$[268]. In contrast, the reverse order OPh > > OAc, Cl, was found in the coupling of alkenylboranes. Grignard reagents react with allylic alcohols and the allyl silyl ethers **413** and **414** using dppf as a suitable ligand[269]. The allylic sulfide **415** reacts with Grignard reagents in refluxing diethyl ether[270].

Perfluoroalkylzinc iodides, prepared *in situ* from iodides and ultrasonically dispersed Zn, are coupled with allylic halides via an allylic rearrangement[271]. The Pd-catalyzed homocoupling of allylic acetate in the presence of Zn to give a mixture of regioisomers **416** and **417** may proceed via *in situ* formation of allylzinc species[272,273].

Phenylisoprene (**419**) is obtained in a good yield by the coupling of phenyl-zinc chloride with isoprenyl chloride (**418**). Coupling of allylic chlorides with alkenylzinc or aluminum reagents affords 1,4-dienes of terpenoid origin[274]. The allylic lactone **420** reacts with phenylzinc chloride, yielding the coupled product with inversion of the stereochemistry of allylic carbon[21]. Organozinc reagents undergo γ-attack of the α,β-unsaturated acetal and orthoester **421**. By this reaction, the overall conjugate addition to form a β-substituted aldehyde and the ester **422** is possible[275]. The reaction of allyl acetate with the Reformatsky reagent **423** proceeds smoothly at room temperature to give the 4-pentenoate **424**[276].

Organoboranes are reactive compounds for cross-coupling[277]. The synthesis of humulene (**83**) by the intramolecular cross-coupling of allylic bromide with alkenylborane is an example[278]. The reaction of vinylborane with vinyloxirane (**425**) affords the homoallylic alcohol **426** by 1,2-addition as main products and the allylic alcohol **427** by 1,4-addition as a minor product[279]. Two phenyl groups in sodium tetraphenylborate (**428**) are used for the coupling with allylic acetate[280] or allyl chloride[33,281].

It was claimed that the *Z*-form of the allylic acetate **430** was retained in homoallylic ketone **431** obtained by reaction with the potassium enolate of 3-vinylcyclopentanone (**429**), after treatment with triethylborane[282]. Usually this is not possible. The reaction of a (*Z*)-allylic chloride with an alkenylaluminum reagent to give 1,4-dienes proceeds with retention of the stereochemistry to a considerable extent when it is carried out at −70 °C[283].

Organoaluminum reagents are used for coupling with allylic chlorides. 1,4-Dienes of terpenoid origin can be prepared by the coupling of allylic chlorides with dimethyl(alkenyl)aluminum, and applied to the completely stereo- and regiospecific synthesis of α-farnesene (**432**)[274]. Organoaluminum reagents undergo γ-attack of the α,β-unsaturated acetal **433**, and the overall conjugate addition to form the β-substituted aldehyde **434** is possible[275]. The alkenylaluminum **436**, formed *in situ* from **435**, reacts intramolecularly with an allylic aluminum alkoxide as a leaving group in the presence of ZnCl₂[284].

Organotin reagents are extensively used for coupling[285,286]. The cross-coupling reactions of allylic halides and triflates with alkyl-, alkenyl-, and arylstannanes and also allylstannanes proceed smoothly[257,258,281,287]. The cross-coupling of an allylic bromide with the allylstannane **437** takes place to give **438** without undergoing homocoupling[281].

The coupling of the allylic halide **439** with the alkenyltin reagent **440** leads to the 1,4-diene **441** in a high yield[257,258,288]. The reaction proceeds with retention of the double bond stereochemistry in the alkenyltin reagent. Inversion of configuration of the cyclic allylic sp^3 carbon center in **439** was observed as expected. The allylic halides usually undergo regioselective coupling at the primary allylic carbon. The coupling of the 1-ethoxyvinylstannane (**442**) with the 2-ethoxyallyl phosphate **443** produced 2,4-diethoxy-1,4-pentadiene (**444**) as a precursor of a β-diketone[289]. The coupling of the allylic bromide **445** with an aryltin reagent yields the allylated arene **446**[258,287].

Under standard conditions for the coupling of allylic halides with tin reagents [Pd(dba)$_2$, Ph$_3$P, THF, 50 °C], most allylic acetates fail to react. Only cinnamyl and allyl acetate, which do not undergo elimination to form a diene, can be used[290,291]. Further studies revealed that simple allylic acetates react with a variety of aryl- and alkenyltin reagents using DMF as a solvent, adding 3 equiv. of LiCl for the substrate, and using Pd(dba)$_2$ in the absence of a ligand[292]. Addition of Ph$_3$P stops the reaction. LiCl assists the transmetallation by displacing π-allylpalladium acetate with chloride. Geranyl acetate (**447**) is phenylated smoothly with phenyltin to give **448**. Vinyloxiranes are more reactive than allylic acetates. The 4-phenylated allylic alcohol **450** can be prepared by the 1,4-addition of phenyltrimethyltin to the vinyloxirane **449**[293,294].

Simple ketones cannot be allylated with Pd catalysts, but tin enolates of ketones are allylated with allylic acetates[76]. Most conveniently, an enol acetate as a ketone enolate equivalent can be allylated with allyl carbonates by the bimetallic catalyses of Pd(0) and tin methoxide[13,295]. The enol acetate **451** derived from an unsymmetrical ketone gives a single isomer **452** regiospecifically. The reaction can be explained by the following mechanism. The first step is the formation of the tin enolate **454** by the reaction of the enol acetate **453** with the tin methoxide **455**, which is a known reaction[296]. The transmetallation of the tin enolate **454** with π-allylpalladium methoxide (**456**), formed from allyl methyl carbonate, gives the π-allylpalladium enolate **457**, which undergoes reductive elimination, yielding the allylated ketone **458**. At the same time, both the tin methoxide **455** and Pd(0) are regenerated and hence the whole reaction proceeds by the bimetallic catalyses of Pd and Sn.

The β, γ-unsaturated ketones **460a** and **460b** are prepared by the coupling of the acylstannane **459** with an allylic chloride[297].

Some organosilicon compounds undergo transmetallation. The allylic cyanide **461** was prepared by the reaction of an allylic carbonate with trimethylsilyl cyanide[298]. The *ortho* esters and acetals of the α, β-unsaturated carbonyl compounds **462** undergo cyanation with trimethylsilyl cyanide[95].

Coupling of allyl chloride with the (*Z*)-alkenylpentafluorosilicate **463** using $Pd(OAc)_2$ as a catalyst at room temperature gives a 1,4-diene in good yields. The reaction has been applied to the synthesis of recifeiolide[299].

Silyl enol ethers are other ketone or aldehyde enolate equivalents and react with allyl carbonate to give allyl ketones or aldehydes[13,300]. The transmetallation of the π-allylpalladium methoxide, formed from allyl alkyl carbonate, with the silyl enol ether **464** forms the palladium enolate **465**, which undergoes reductive elimination to afford the allyl ketone or aldehyde **466**. For this reaction, neither fluoride anion nor a Lewis acid is necessary for the activation of silyl enol ethers. The reaction also proceeds with metallic Pd supported on silica by a special method[301]. The ketene silyl acetal **467** derived from esters or lactones also reacts with allyl carbonates, affording allylated esters or lactones by using dppe as a ligand[302].

Allylic acetates react with ketene silyl acetals. In this reaction, in addition to the allylated ester **468**, the cyclopropane derivative **469**, which is formed by the use of bidentate ligands, is obtained[303]. Formation of a cyclopropane derivative **471** has been observed by the stoichiometric reaction of the π-allylpalladium complex **470** with Li enolates of carboxylates[304,305] (see Chapter 3, Section 3.2). Also enolates of the ketone *N*-isobutyryloxazolidinone and iso-propyl phenylsulfone react with the π-allylpalladium complex **470** to form the cyclopropane **471**[306]. The most efficient cyclopropanation to form **473** is the reaction of allyl bromide with the ketene silyl acetal **472** catalyzed by π-allyl-palladium chloride in the presence of thallium acetate. TMEDA as a ligand is essential for high chemoselectivity[307].

468, 53%   **469, 38%**

471

472   **473**

2.4.2 Formation of Allylic Metal Compounds

Allylic metal compounds useful for further transformations can be prepared by Pd-catalyzed reactions of allylic compounds with bimetallic reagents. By this transformation, umpolung of nucleophilic π-allylpalladium complexes to elec-trophilic allylmetal species can be accomplished. Transfer of an allyl moiety from Pd to Sn is a typical umpolung.

The allylstannane **474** is prepared by the reaction of allylic acetates or phos-phates with tributyltin chloride and SmI$_2$[286,308] or electroreduction[309]. Bu$_3$SnAlEt$_2$ prepared *in situ* is used for the preparation of the allylstannane **475**. These reactions correspond to inversion of an allyl cation to an allyl anion[310, 311]. The reaction has been applied to the reductive cyclization of the alkenyl bromide in **476** with the allylic acetate to yield **477**[312]. Intramolecular coupling of the allylic acetate in **478** with aryl bromide proceeds using Bu$_3$SnAlEt$_2$ (**479**) by *in situ* formation of the allylstannane **480** and its reaction with the aryl bromide via transmetallation. (Another mechanistic possibility is the formation of an arylstannane and its coupling with allylic acetate)[312]. The reaction of $(R_3Sn)_2$ with allylic acetates affords allystannanes[313,314]. In the intermolecular reaction of the bromoindole **481** with allyl acetate and $(Bu_3Sn)_2$, the allylstannane **482** is formed first by the Pd-catalyzed transmetallation. Further transmetallation with indolylpalladium bromide affords the allylated indole **483**[315].

The allylstannane species **484** is generated *in situ* by the umpolung of a π-allylpalladium complex when an allyl ester or alcohol is treated with $SnCl_2$ in the presence of Pd catalyst, and reacts with aldehydes[316,317]. A similar reaction takes place by use of Et_2Zn[318] or Zn, or by electroreduction instead of using $SnCl_2$[319,320]. As another route to umpolung, an Sm species is formed *in situ* from the allylic acetate **485** and SmI_2, and reacts with ketone intramolecularly to give the alcohol **486**[321].

Allylsilane is prepared by the reaction of allyl chloride or acetate with $(Me_3Si)_2$[322–324]. However, the elimination to form conjugated dienes, rather than the silylation, takes place with substituted allylic compounds. On the other hand, the substituted allylic silane **487** can be prepared smoothly by using $Pd(dba)_2$ in the presence of an excess of LiCl[325]. 1,1-Dichlorophenyl-2,2,2-trimethyldisilane (**488**) is a good silylating agent when dppf is used as a ligand to give the allylic silane **489** in good yields[326]. Inversion of stereochemistry is observed as expected. Geranylsilane (**491**) is prepared by the reaction of geranyl phosphate (**490**) with $PhMe_2SiAlEt_2$[327]. $(Me_3Si)_3Al$ is used for the same purpose[328].

2.5 1,4-Elimination to Form Conjugated Dienes

When allylic compounds are treated with Pd(0) catalyst in the absence of any nucleophile, 1,4-elimination is a sole reaction path, as shown by **492**, and conjugated dienes are formed as a mixture of *E* and *Z* isomers[329]. From terminal allylic compounds, terminal conjugated dienes are formed. The reaction has been applied to the syntheses of a pheromone, 12-acetoxy-1,3-dodecadiene (**493**)[330], ambergris fragrance[331], and aklavinone[332]. Selective elimination of the acetate of the cyanohydrin **494** derived from 2-nonenal is a key reaction for the formation of the 1,3-diene unit in pellitorine (**495**)[333]. Facile aromatization occurs by bis-elimination of the 1,4-diacetoxy-2-cyclohexene **496**[334].

1,4-Elimination of geranyl acetate using $Pd(OAc)_2$ and Ph_3P is not regioselective[329]. Interestingly, the addition of an equivalent amount of propargylzinc bromide not only accelerates the elimination, but also gives high regioselectivity. A 75 : 25 mixture of *trans*- and *cis*-ocimene was obtained from geranyl acetate without producing any myrcene (**497**), while neryl acetate gave pure myrcene (**497**). An exclusive *anti*-1,4-elimination is observed[335]. Allylic amines are difficult to cleave, but they can be cleaved by using a cationic Pd complex and dppb as a ligand. As another example of regioselective 1,4-elimination, diethyl(7-hydroxylgeranyl)amine (**498**) is converted into the myrcene-type conjugated diene **499** selectively and the reaction is applied to the industrial production of a fragrant compound (kovanol)[336]. 1,4-Elimination of allylic amines is also possible by conversion into amine salts using AcOH as a solvent[337]. Allylic chlorination of diprenyl ether (**500**), followed by Pd-catalyzed elimination cleanly affords the diene **501**, from which citral (**502**) is synthesized by Claisen and Cope rearrangements[338].

Allylic sulfones undergo Pd-catalyzed elimination. The synthesis of an α,β-unsaturated ketone by the elimination of the allylic sulfone group in **503** is an example[339].

Elimination proceeds with high selectivity in cyclic systems, and is particularly useful for the selective preparation of both homoannular and heteroannular conjugated dienes **504** and **505** in decalin systems. These two conjugated cyclic dienes exist in many natural products, and their regioselective preparation is highly desirable. In particular, the selective formation of the homoannular diene **504** is considered to be difficult, because acid-catalyzed elimination affords only the heteroannular diene **505**. By the Pd method using allylic carbonates under mild conditions, it is possible to prepare both the homo- and heteroannular conjugated dienes at will. The regiospecificity is dependent on the stereochemistry of the allylic carbonates in the decalin systems. The heteroannular diene **508** is obtained selectively from the β-oriented 3-allylic carbonate **506**. The π-allylpalladium intermediate, in which Pd is α-oriented, is formed by inversion of the stereochemistry, and rapid elimination takes place via the α-oriented angular C-5 σ-allylpalladium methoxide **507** to give the heteroannular conjugated diene **508**, rather than via the C-3 σ-allylpalladium methoxide **509**. This selectivity may be explained by different rates of elimination. The elimination of H—Pd—OAc from the tertiary carbon of **507** seems to be faster than that from **509**. On the other hand, β-oriented π-allylpalladium methoxide is formed from 3-α-allylic carbonate **510**, and the homoannular diene **512** is obtained as a major product via the β-oriented 3-σ-allylpalladium methoxide **511**[340]. Formation of β-oriented angular C-5 σ-allylpalladium compound **513** is sterically unfavorable.

In steroid systems, the homoannular diene in ring A and the heteroannular diene in AB rings are generated. The allylic 3α-carbonate **514** affords the homoannular conjugated diene **515** as a main product and a small amount of the heteroannular diene **516**. On the other hand, the heteroannular conjugated diene **516** is obtained exclusively from 3β-carbonates **517**. The elimination reaction proceeds smoothly at room temperature.

The reaction has been applied successfully to the synthesis of a precursor of provitamin D, **520**, which has a homoannular conjugated diene in the B ring[340]. Treatment of the 7α-carbonate **518** with Pd catalyst at 40 °C afforded the 5,7-diene **520** regioselectively in good yield. No heteroannular diene **521** was detected. In the intermediate complex **519**, the β-oriented 7σ-allylpalladium undergoes facile *syn*-elimination of 8β-hydrogen to afford **520** exclusively.

Only the heteroannular diene **523** is formed by treatment of both α- and β-allylic carbonates **522** and **524** in a hydrindan system with a Pd catalyst. No homoannular diene is formed.

The preparative method for the Pd(0) catalyst active in these regioselective eliminations under mild conditions is crucial. The very active catalyst is prepared by mixing equimolar amounts of Pd(OAc)$_2$ or Pd(acac)$_2$ and pure *n*-

Bu$_3$P. A rapid redox reaction takes place to yield the active Pd(0) species and tributylphosphine oxide. The Pd(0) thus generated is a phosphine-free catalyst[341]. Severe reaction conditions are necessary, or no reaction takes place, when Pd$_2$(dba)$_3$ is used in the elimination reaction of cyclic allylic compounds with an excess of *n*-Bu$_3$P[342].

$$Pd(OAc)_2 + Bu_3P \longrightarrow Pd(0) + Bu_3PO + Ac_2O$$

The optically active 1,4-cyclohexenediol monoacetate **525**, prepared by hydrolysis of the *meso*-diacetate with lipase, was converted into the optically pure cyclohexenone **526** by an elimination reaction in the presence of ammonium formate. Optically active carvone (**527**) was prepared from **526**[343].

The substituted vinyl-β-lactone **528** undergoes Pd-catalyzed ring opening to form the 2,4-dienecarboxylate **529** in aprotic solvents. The conjugated diene **530** is formed by decarboxylation in DMF[344]. 3-Acetoxy-2-methyl-4-pentenoic acid(**531**) undergoes the Pd-catalyzed decarboxylation–elimination to form a conjugated diene[345]. This elimination offers regioselective generation of conjugated dienes and the new double bond has mainly an *E* configuration. The reaction has been used for the construction of the polyene system in vitamin A derivative **532**[346]. Another route to diene formation via π-allyl-palladium is the reaction of the dicarbonate of the 2-butene-1,4-diol derivative **533**[347]. The reaction involves umpolung of a π-allylpalladium intermediate, and may be related to the enone formation described in Section 2.10.3.

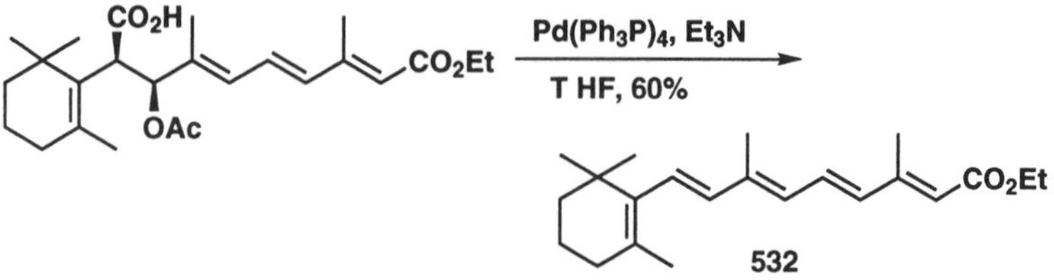

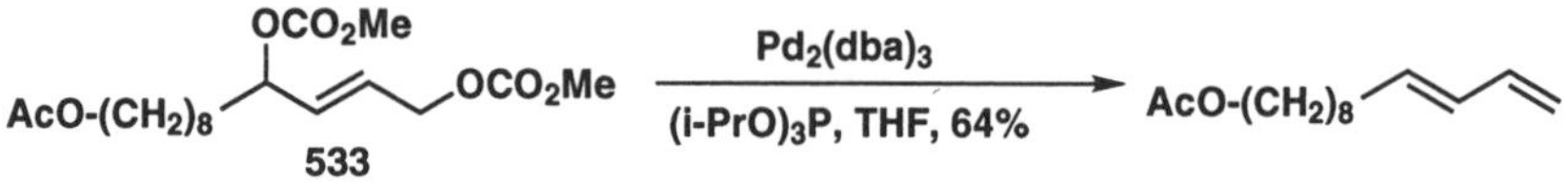

532

533

The vinyloxiranes **534** and **537** are converted into dienes by the elimination of β-hydrogen via π-allylpalladium complexes. Depending on the structures, different hydrogens are eliminated, giving either an allylic alcohol **538** or an unconjugated ketone **536** via a 1-hydroxy-1,3-diene **535**[348]. The vinyloxirane **539**, prepared from glucose, is converted into a cyclopentenone derivative **541** via rearrangement to form the unsaturated aldehyde **540**[349]. As a related elimination reaction, the epoxy ketone **542** is converted into the β-diketone **543** under severe conditions[350]. The Pd-catalyzed ring opening of cyclopentadiene-1,4-epiperoxide (**544**) affords 4-hydroxy-2-cyclopentenone (**545**) as a major product[351].

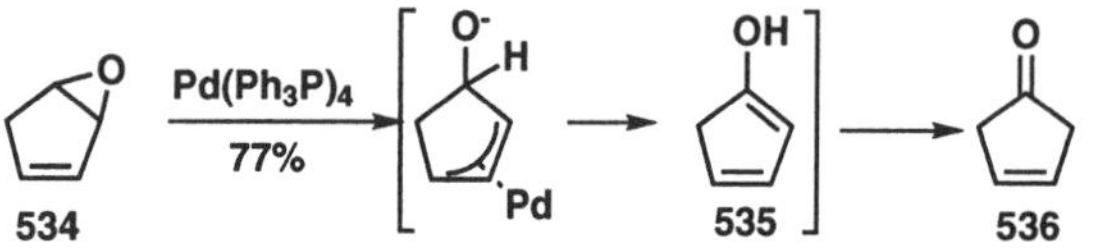

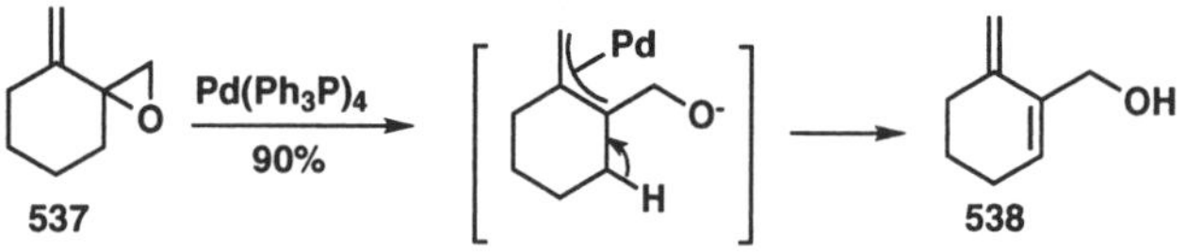

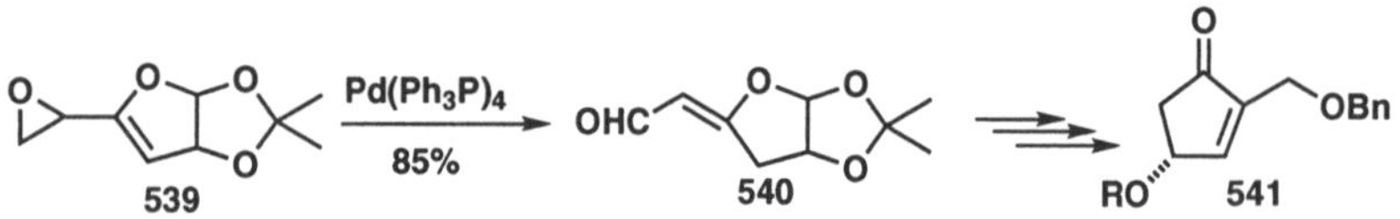

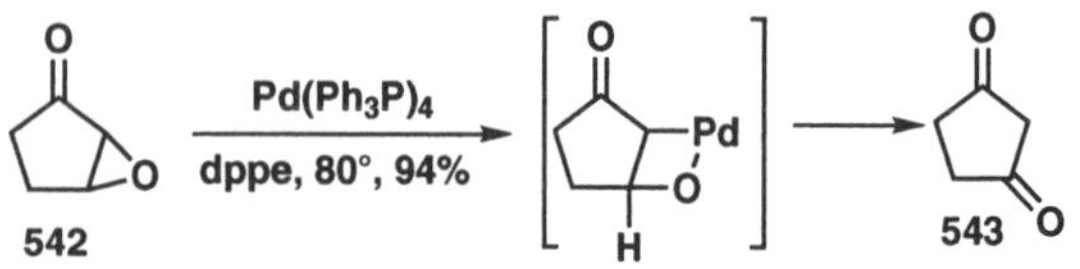

534 77% **535** **536**

537 90% **538**

539 85% **540** **541**

542 dppe, 80°, 94% **543**

The steroidal 4β-acetoxy-5β,6β-epoxy-2-en-1-one system **546** was converted at room temperature into the 6β-hydroxy-2,4-dien-1-one **547** by reductive elimination of the vicinal oxygen function, and the reaction has been applied to the synthesis of withanolide[352].

2.6 Preparation of α,β-Unsaturated Carbonyl Compounds by the Reactions of Silyl Enol Ethers and Enol Acetates with Allyl Carbonates

New synthetic methods for α,β-unsaturated ketones, aldehydes, and esters by the dehydrogenation of the corresponding carbonyl compounds have been developed based on π-allylpalladium enolate chemistry. The reactions of the silyl enol ether **548** with an allyl carbonate generates the π-allylpalladium enolate **549** via transmetallation. The intermediate complex undergoes elimination of β-hydrogen to form the α, β-unsaturated carbonyl compound **550**[13,353]. The allyl group is a hydrogen acceptor and is converted into propylene. The enone formation is competitive with the allylation which proceeds by reductive coupling (see Section 2.4.1). Chemoselective enone formation can be achieved by the reaction of silyl enol ethers with allyl carbonates in boiling MeCN. The ratio of Pd to phosphine ligand is important. The lower ratio or even the absence of the ligand favors enone formation[354]. The reaction also proceeds in the liquid phase using Pd metal supported on silica by a special method[355].

The reaction can be applied to the synthesis of α, β-unsaturated esters and lactones by treatment of the ketene silyl acetal **551** with an allyl carbonate in boiling MeCN[356]. The preparation of the α, β-unsaturated lactone **552** by this method has been used in the total synthesis of lauthisan[357].

Another preparative method for the enone **554** is the reaction of the enol acetate **553** with allyl methyl carbonate using a bimetallic catalyst of Pd and Tin methoxide[354,358]. The enone formation is competitive with the allylation reaction (see Section 2.4.1). MeCN as a solvent and a low Pd to ligand ratio favor enone formation. Two regioisomeric steroidal dienones, **558** and **559,** are prepared regioselectively from the respective dienol acetates **556** and **557** formed from the steroidal α, β-unsaturated ketone **555**. Enone formation from both silyl enol ethers and enol acetates proceeds via π-allylpalladium enolates as common intermediates.

2.7 Oxidation of Alcohols via Allyl Carbonates

Smooth oxidation of alcohols under mild neutral conditions without using inorganic oxidizing agents is possible via π-allylpalladium alkoxides. The reaction of the allyl alkyl carbonate **560** with a Pd(0) catalyst generates the π-allylpalladium alkoxide **561**. This intermediate complex undergoes elimination of β-hydrogen to give the carbonyl compound **562** in the absence of nucleophiles, offering a good method for the oxidation of alcohols[359]. The treatment of alkyl allyl carbonates **560** in boiling MeCN with Pd(OAc)$_2$ gives ketones **562** selectively. This is a clean oxidation reaction, because byproducts of the reaction are CO$_2$ and propylene. The reaction is applied to secondary alcohols and both primary and secondary allylic or benzylic alcohols. The oxidation of primary alcohols is too slow to be practical.

The slow oxidation of primary alcohols, particularly MeOH, is utilized for the oxidation of allylic or secondary alcohols with allyl methyl carbonate without forming carbonates of the alcohols to be oxidized. Allyl methyl carbonate (**564**) forms π-allylpalladium methoxide, then exchange of the methoxide with a secondary or allylic alcohol **563** present in the reaction medium takes place to form the π-allylpalladium alkoxide **565**, which undergoes elimination of β-hydrogen to give the ketone or aldehyde **566**. The lactol **567** was oxidized selectively with diallyl carbonate to the lactone **568** without attacking the secondary alcohol in the synthesis of echinosporin[360].

2.8 Hydrogenolysis of Allylic Compounds

The Pd-catalyzed hydrogenolysis of allylic compounds by various hydrides gives alkenes. From terminal allylic compounds, either 1-alkenes or 2-alkenes are formed depending on the hydride sources [360a].

At first, the hydrogenolysis was carried out with ammonium formate[361,362]. Later it was found that many metal hydrides and reducing agents such as $LiAlH_4$,[363], various borohydrides[364–367], hydrosilanes (polymethylhydrosiloxane, PMHS)[368], tin hydride[369–372], butylzinc chloride[335], SmI_2[373], and *N*-propyl-1,4-dihydronicotinamide[374], and electrolysis[375] can be used for the hydrogenolysis of allylic compounds. These reducing agents, except formic acid, give mainly 2-alkenes from terminal allylic compounds. Formic acid is a unique reagent for hydrogenolysis. The first report on the use of formic acid was an attempted preparation of allylic formates by Pd(0)-catalyzed transesterification between allylic acetates and formic acid, but a mixture of alkenes was obtained by the hydrogenolysis, instead of transesterification[376]. More useful terminal alkenes can be prepared smoothly by using ammonium formate[361]. The regioselectivity observed in the hydrogenolysis of the geranyl derivatives **569** with representative hydrides to give 1-alkene **570a** and 2-alkene **570b** is shown in the following table; only formate affords the 1-alkene **570a** with high regioselectivity.

			ratio(%)		
leaving group	catalysts	hydrides	570b	570a	Ref.
X = OCO$_2$Me	Pd$_2$(dba)$_3$ - PBu$_3$	HCO$_2$H	0	100	362
X = OAc	Pd(Ph$_3$P)$_4$	PMHS	52	48	368
X = OAc	Pd(Ph$_3$P)$_4$	BuZnCl	96	4	335
X = OPh	Pd(Ph$_3$P)$_4$	LiBHEt$_3$	98	2	367
X = OAc	Pd(Ph$_3$P)$_4$	SmI$_2$ -H$^+$	93	7	373

Treatment of various terminal allylic compounds with ammonium formate affords 1-alkenes as the main products. High regioselectivity is observed by use of Bu$_3$P as a ligand and the reaction proceeds more smoothly by use of formic acid and Et$_3$N, since triethylammonium formate is more soluble than ammonium formate in organic solvents[362]. Needless to say, 1-alkenes are more synthetically valuable than 2-alkenes, and the regioselective formation of 1-alkenes has a high synthetic value. The Pd-catalyzed chemoselective allylation of the α-methoxycarbonyl lactone **571** with 4-acetoxy-2-butenyl carbonate under neutral conditions, followed by hydrogenolysis of the allylic acetate moiety with triethylammonium formate to give α-(3-butenyl)lactone, offers a convenient method for the 3-butenylation of active methylene compounds[377]. Hydrogenolysis of allylic chlorides with sodium formate is carried out in a heptane–water mixture using a water-soluble phosphine[378].

The formation of 1-and 2-alkenes can be understood by the following mechanism. In the presence of formate anion, the π-allylpalladium complex **572** is converted into the π-allylpalladium formate **573**. The most interesting feature is the attack of the hydride from formate to the more substituted side of the σ-allylic system by the cyclic mechanism shown by **574** to form the 1-alkene **575**[367]. The decarboxylation and hydride transfer should be a concerted process. On the other hand, other metal hydrides form the Pd hydride **576** by direct transmetallation, and the hydride is transferred to the less hindered side of the σ-allyl system by subsequent reductive elimination to give the 2-alkenes **577**. In this mechanism of formate reduction, the π-allylpalladium formate **573** is an intermediate, which can be formed directly from the allylic formates **578**, and hence allylic formates can be used more conveniently for hydrogenolysis without using triethylammonium formate. Formation of the π-allylpalladium formate **573** as an intermediate in the hydrogenolysis was confirmed by an NMR spectrum[379].

In addition to the preparation of 1-alkenes, the hydrogenolysis of allylic compounds with formate is used for the protection and deprotection of carboxylic acids, alcohols, and amines as allyl derivatives (see Section 2.9).

Various terminal allylic compounds are converted into 1-alkenes at room temperature[362]. Regioselective hydrogenolysis with formate is used for the formation of an *exo*-methylene group from cyclic allylic compounds by the formal *anti* thermodynamic isomerization of internal double bonds to the exocyclic position[380]. Selective conversion of myrtenyl formate (**579**) into β-pinene is an example. The allylic sulfone **580** and the allylic nitro compound

581 are prepared from cyclic ketones. Treatment of these allylic compounds with triethylammonium formate generates *exo*-methylene compounds. The preparation of **582** by regioselective hydrogenolysis of the allylic acetate has been applied to the syntheses of gomisin A and schizandrin[381].

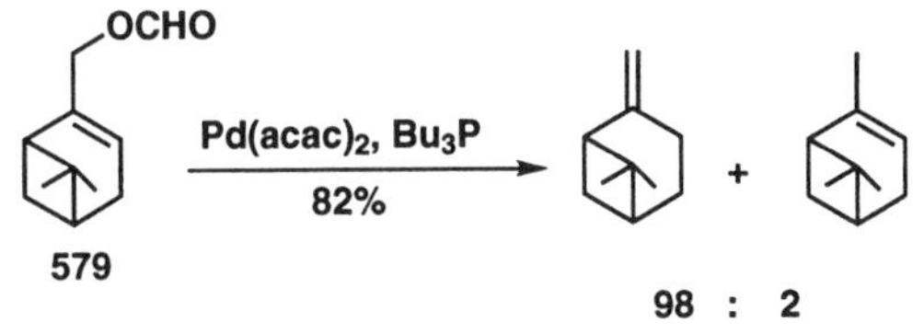
OCHO
Pd(acac)2, Bu3P
82%
+
579
98 : 2

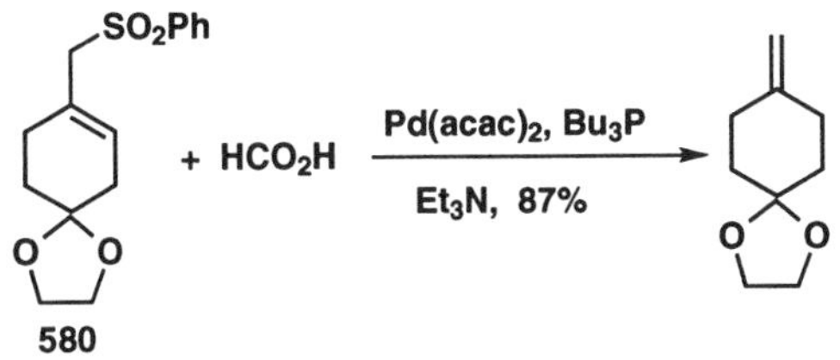
SO2Ph
+ HCO2H
Pd(acac)2, Bu3P
Et3N, 87%
580

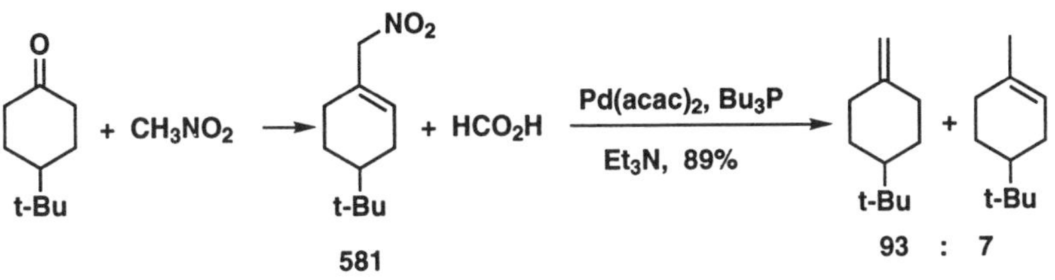
O
+ CH3NO2
t-Bu
NO2
+ HCO2H
t-Bu
581
Pd(acac)2, Bu3P
Et3N, 89%
+
t-Bu
t-Bu
93 : 7

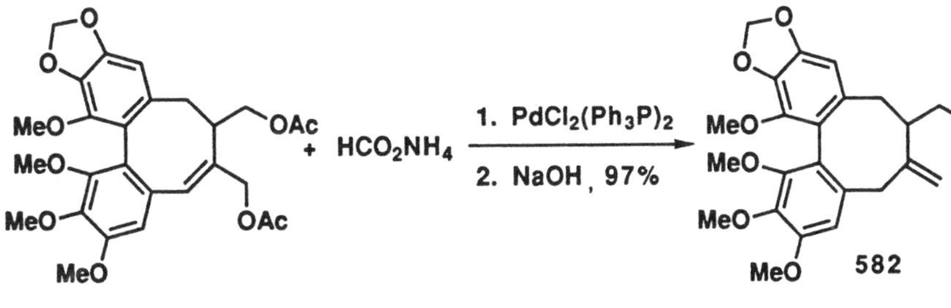
MeO
MeO
MeO
MeO
OAc
OAc
+ HCO2NH4
1. PdCl2(Ph3P)2
2. NaOH, 97%
MeO
MeO
MeO
MeO
OH
582

As a further application of the reaction, the conversion of an endocyclic double bond to an *exo*-methylene is possible[382]. The epoxidation of an *endo*-alkene followed by diethylaluminum amide-mediated isomerization affords the allylic alcohol **583** with an *exo* double bond[383]. The hydroxy group is eliminated selectively by Pd-catalyzed hydrogenolysis after converting it into allylic formate, yielding the *exo*-methylene compound **584**. The conversion of carvone (**585**) into 1,3-disiloxy-4-methylenecyclohexane (**586**) is an example[382].

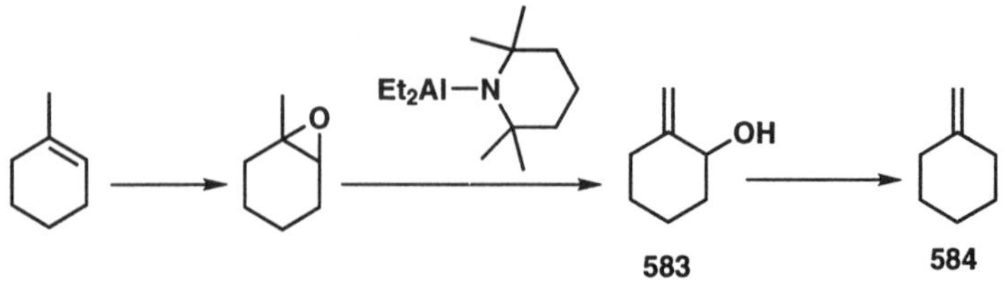
Et2Al—N
O
OH
583
584

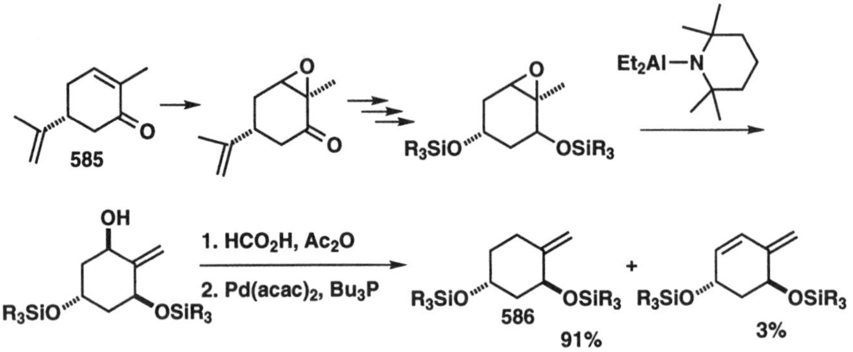
O
O
585
O
O
R3SiO
OSiR3
Et2Al—N
OH
R3SiO
OSiR3
1. HCO2H, Ac2O
2. Pd(acac)2, Bu3P
R3SiO
OSiR3
586
91%
+
R3SiO
OSiR3
3%

Cyclic ketones are converted into vinyl and isopropenyl groups by the following sequence of reactions. The tertiary allylic alcohol, formed by the reaction of 4-*t*-butylcyclohexanone with isopropenyllithium, is converted into the carbonate **587** (formates of tertiary allylic alcohols cannot be prepared) and their reaction with triethlyammonium formate at room temperature generates the isopropenyl group **588** without forming the isopropylidene group **590**. Also the primary allylic formate **589** is converted regioselectively into the isopropenyl group **588**[384]. This reaction is useful for the regio- and stereoselective preparation of the C-17 β-isopropenyl steroid **592** via the allylic carbonate **591** formed from C-17 keto steroids. Generation of the correct stereochemistry at C-20 of steroid side-chains has been achieved by hydroboration of **592**[385]. Quantitative and completely regioselective removal of the nitro group from the allylic nitro system **593** has been applied to the synthesis of kainic acid[386].

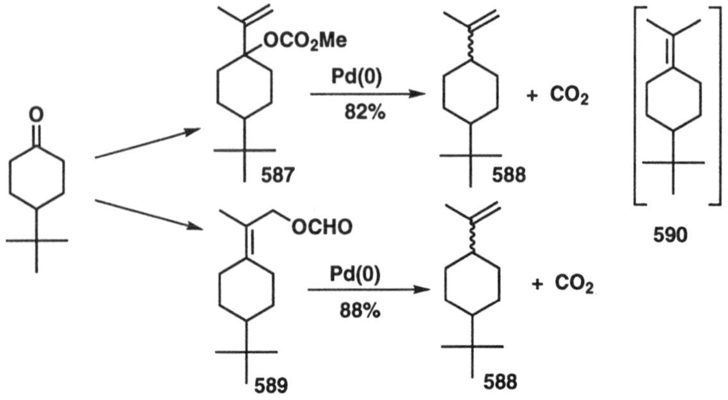
O
OCO2Me
Pd(0)
82%
+ CO2
587
588
590
OCHO
Pd(0)
88%
+ CO2
589
588

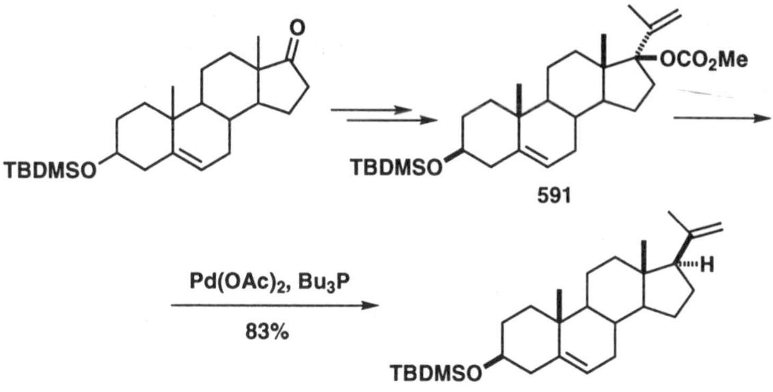

591

Pd(OAc)$_2$, Bu$_3$P
83%

592

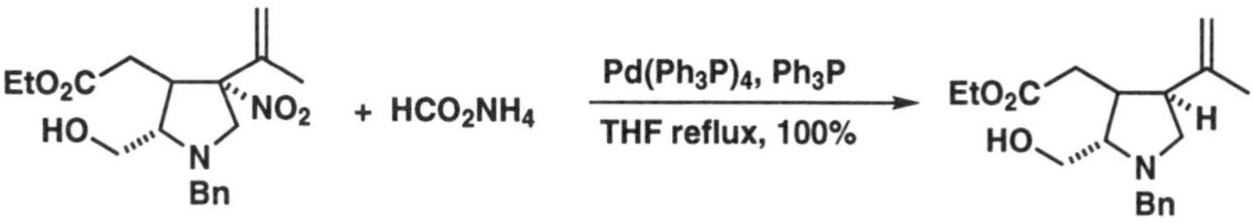

593

+ HCO$_2$NH$_4$

Pd(Ph$_3$P)$_4$, Ph$_3$P
THF reflux, 100%

Asymmetric hydrogenolysis of allylic esters with formic acid with satisfactory *ee* was observed[387]. Geranyl methyl carbonate (**594**) was reduced to **570** with formic acid using 1,8-bis(dimethylamino)naphthalene as a base and MOP-Phen as the best chiral ligand, achieving 85% *ee*.

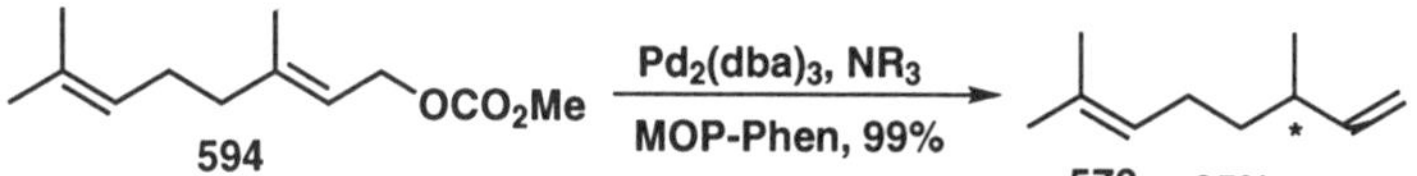

594

Pd$_2$(dba)$_3$, NR$_3$
MOP-Phen, 99%

570   85% ee

NR$_3$ =

(R)-MOP-Phen =

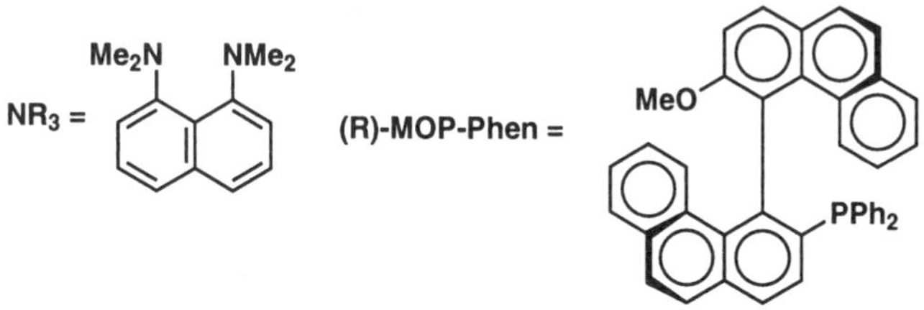

Furthermore, the regioselective hydrogenolysis can be extended to internal allylic systems. In this case, clean differentiation of a tertiary carbon from a secondary carbon in an allylic system is a problem. The regioselectivity in the hydrogenolysis of unsymmetrically substituted internal allylic compounds depends on the nature and size of the substituents. The less substituted alkene **596** was obtained from **595** as the main product, but the selectivity was only

77%. On the other hand, the expected alkene **598** was regioselectively formed from the allylic carbonate **597**[388]. In these reactions, the hydride from formate preferentially attacks the tertiary carbon rather than the secondary carbon.

The regioselective hydrogenolysis of an internal allylic system with triethyl-ammonium formate can be applied to cyclic allylic systems remarkably well, and the stereospecific generation of either a *cis* or *trans* ring junction in hydrindane and decalin systems is possible depending on the stereochemistry of allylic formates[389]. The hydrogenolysis of an allylic formate in hydrindane proceeds regioselectively; the hydride attacks the more hindered angular carbon. In addition, the reaction is stereospecific. Formation of the π-allylpalladium formate takes place with inversion of stereochemistry, and the subsequent decarboxylation–hydride transfer to the angular carbon takes place from the same side of the Pd with retention. Thus overall inversion is observed. The reaction of the 5β-formate **599** generates a *trans* junction in **601** via the α-oriented σ-palladium formate **600**, and a *cis* junction in **604** is formed by the reaction of the 5α-formate **602** via **603**.

The same regioselective and stereospecific reactions are observed in decalin systems. The 3β-formate **605** is converted into the α-oriented σ-allylpalladium complex **606**, and the hydride transfer generates the *trans*-decalin **607**, while the *cis* junction in **610** is generated from the 3α-formate **608** by attack of the hydride from the β-side (**609**). An active catalyst for the reaction is prepared by mixing Pd(OAc)$_2$ and Bu$_3$P in a 1 : 1 ratio; with this catalyst the reaction proceeds at room temperature. The reaction proceeded in boiling dioxane when a catalyst prepared from Pd(OAc)$_2$ and Bu$_3$P in a 1 : 4 ratio was used[390].

The method has been successfully applied to steroids. The *trans* AB ring junction in **612** can be generated cleanly from the β-allylic formate **611**; the *cis* junction in **614** is formed by the hydrogenolysis of the α-allylic formate **613**.

Stereocontrolled construction of a natural configuration at C-20 **619** in steroid side-chains is an important problem in steroid synthesis. In addition, the preparation of the unnatural epimer **617** has increasing importance because of its interesting biological activity. Both epimers can be prepared cleanly from a common intermediate **615** by palladium-catalyzed hydrogenolysis with ammonium formate. The (*E*)- and (*Z*)-C-20 allylic alcohols are prepared by stereoselective reaction of the C-20 keto steroid **615** with (*E*)- and (*Z*)-alkenyllithium reagents, and the tertiary allylic alcohols are converted into the carbonates **616** and **618**. The reaction of the (*E*)-allylic carbonate **616** with triethylammonium formate proceeded at room temperature, affording regioselectively and stereospecifically the unnatural configuration at C-20 **617**. On the other hand, the natural configuration at C-20 **619** was generated from the (*Z*)-allylic carbonate **618**[391].

R₃Si = t-BuMe₂Si

The regioselective and stereospecific construction of C-20 stereochemistry is explained by the following mechanism. The Pd(0) species attacks the (*E*)-β-carbonate **616** from the α-side by inversion to form the π-allylpalladium species **620**, which has a stable *syn* structure[392]. Then concerted decarboxylation–hydride transfer as in **621** takes place from the α-side to give the unnatural configuration in **617**. On the other hand, the π-allylpalladium complex **622** formed from the (Z)-allylic carbonate **618** has an unstable *anti* structure owing to a large repulsion between the methyl group and the side-chain (the stereochemical terms *syn* and *anti* are related to the middle hydrogen on C-2 in a π-allyl system)[392]. Thus the σ-allylpalladium compound **623** rotates to form the more stable *syn* structure **624** before decarboxylation. By this rotation, the Pd also moves from the α- to the β-side. Then the decarboxylation–hydride transfer of **625** takes place from the β-side. In this way, the natural configuration in **619** is generated. In these selective hydrogenolyses, the active catalyst is prepared by mixing $Pd(OAc)_2$ or $Pd(acac)_2$ and Bu_3P in a 1 : 1 ratio.

Hydride attacks regioselectively at the Si-substituted carbon in the hydrogenolysis of the silylated allylic carbonate **626** with formate, affording the allylic silane **627**[142].

TMS $\diagdown\diagup\diagdown\diagup$ Ph + HCO$_2$H $\xrightarrow[\text{Et}_3\text{N, 81\%}]{\text{Pd}_2(\text{dba})_3, \text{Bu}_3\text{P}}$ TMS $\diagdown\diagup\diagdown\diagup$ Ph

OCO$_2$Me

626   **627**

Highly regio- and stereoselective 4α-deuteration in steroids is possible by the hydrogenolysis of the cyclic allylic β-carbonate **628** with NaBD$_4$; the extent of 6α-deuteration is only 3%[393].

The Pd-catalyzed hydrogenolysis of vinyloxiranes with formate affords homoallyl alcohols, rather than allylic alcohols regioselectively. The reaction is stereospecific and proceeds by inversion of the stereochemistry of the C—O bond[394,395]. The stereochemistry of the products is controlled by the geometry of the alkene group in vinyloxiranes. The stereoselective formation of stereoisomers of the *syn* hydroxy group in **630** and the *anti* in **632** from the (*E*)-epoxide **629** and the (*Z*)-epoxide **631** respectively is an example.

E- and *Z*-isomers of the vinylepoxides **633** and **635** give isomeric products. Based on this selective hydrogenolysis, the (*S*)- and (*R*)-tetrahydrogeranylacetones **634** and **636** have been prepared stereospecifically from vinyloxiranes which contain (*E*)- and (*Z*)-alkene groups (**633** and **635**). (−)-Serricornin and (−)-nupharamine have been synthesized applying the regio- and stereoselective hydrogenolysis of vinyloxiranes as a key reaction[396,397]. These stereoselective hydrogenolyses can be explained as follows. The π-allylpalladium complex **638** is formed by inversion from the (*E*)-vinylepoxide **637** and converted into **639** by regioselectively attacking the tertiary carbon rather than the secondary carbon from the rear side. On the other hand, the *Z*-isomer **640** generates the complex **641**, which has a sterically unstable *anti* form, and hence the complex **641** rearranges to the stable *syn* form **643** by π–σ–π rearrangement via rotation as shown by **642**. Finally, **644** is formed by hydrogenolysis from the front side.

The vinylcylopropanes **645** and **647** bearing two EWGs (ketone and ester) form π-allylpalladium intermediates owing to the generation of a stabilized carbanion by ring cleavage, and undergo regioselective and stereoselective hydrogenolysis[398]. The cyclopropane **645** attached to the *trans* double bond gives **646** with inversion of stereochemistry, and the *cis* compound **647** undergoes π–σ–π rearrangement before hydrogenolysis to give **648**, which has an opposite stereochemistry to that of **646**. This stereoselective reaction has been applied to the syntheses of clavukerin A and isoclavukerin A[399].

Allylic O, S, and Se bonds are cleaved with various borohydrides such as $NaBH_4$, $LiBHEt_3$, and $NaBH_3CN$[364–367]. Allylic nitro compounds are prepared by the reaction of ketones with nitromethane[400,401], and the nitro group in **649** can be eliminated by hydrogenolysis. Formate and $NaBH_4$ give the different regioisomers **650** and **651**[402]. Selective allylic deoxygenation of acetoxy groups in unsaturated carbohydrates has been carried out using Ph_2SiH_2–$ZnCl_2$[403] or $NaBH_4$[404]. Unexpectedly, the reduction of **652** with Ph_2SiD_2 proceeds with retention of stereochemistry to give **653**. This may be due to the presence of acetoxy groups.

Allylic amines can be cleaved. Hydrogenolysis of allylic amines of different stereochemistry with NaBH$_3$CN was applied to the preparation of both diastereoisomers **655** and **657** of cyclopentenylglycine from the cyclic amines **654** and **656** of different stereochemistry[405].

Desulfonylation of equally substituted allylic sulfones with NaBH$_4$ and LiBHEt$_3$ usually yields a mixture of regioisomeric alkenes[406,407]. However, the regioselective attack of the less substituted side of the unsymmetrically substituted allylic system with LiEt$_3$BH has been utilized for the removal of the allylic sulfone group in synthesis of the polyprenoid **658**[408].

Tributyltin hydride is used for hydrogenolysis of allylic esters[369–372].

2.9 Allyl Group as a Protecting Group

It is widely recognized that an allyl group is a useful protecting group for acids, amines, and alcohols. Facile formation of a π-allylpalladium complex from allylic compounds can be used for their deprotection[409]. The deprotection can be achieved by two methods: hydrogenolysis with various hydrides or allyl group transfer to other nucleophiles.

The first report on the Pd-catalyzed deprotection of an allyl group was the treatment of allyl esters with ammonium formate under almost neutral conditions; allyl cinnamate (**659**) was converted into cinnamic acid without reducing the double bond[361]. Only carbon dioxide and propylene are formed by the deprotection. The allylic ester **660** is easily prepared by esterification of acids or by the tin compound-catalyzed transesterification of a methyl ester with allyl or prenyl alcohol, and the allyl or prenyl ester can be deprotected with triethyl-ammonium formate[410]. In the synthesis of various prostaglandins, the methyl ester must be hydrolyzed to give the free carboxylic acid by using an enzyme, because prostaglandins are not stable under acidic or basic conditions, necessary for the usual hydrolysis. However, the allyl ester of the prostaglandins **661** can be deprotected smoothly to give the free acid **662** by treatment with triethylammonium formate at 40 °C under neutral conditions without using the enzyme[411]. The allyl phosphate **663** can be deprotected with $BuNH_2$ and formic acid, [412,413] or tributyltin hydride[414].

Another method for deallylation of allyl esters is the transfer of the allyl group to reactive nucleophiles. Amines such as morpholine are used[415-417]. Potassium salts of higher carboxylic acids are used as an accepter of the allyl group[418]. The method is applied to the protection and deprotection of the acid function in rather unstable β-lactam **664**[419,420].

Allylic anchoring groups **665** and **666** based on 4-bromocrotonic acid[421] and *cis*-1,4-butenediol[422,423] are used as protecting groups of terminal amino acids in solid phase synthesis of peptides by attaching to polymers (polystyrene, cellulose). For their deprotection, dimedone, morpholine[421], tin hydride[423], *N*-hydroxybenzotriazole[422] are used. But morpholine seems to be the best deprotecting agent[424]. 4-Trimethylsilyl-2-butenyl ester **667** can be deprotected by the elimination reaction, giving butadiene and the silyl ester, which is hydrolyzed easily to the free acid[425].

Allylamines are difficult to cleave with Pd catalysts. Therefore, amines are protected as carbamates, but not as allylamines. Also, allyl ethers used for the protection of alcohols cannot be cleaved smoothly, hence alcohols are protected as carbonates. In other words, amines and alcohols are protected by an allyloxycarbonyl (AOC or Alloc) group.

The amino group of the amino acids **668**, protected as allyl carbamates, can be cleaved by Pd-catalyzed treatment with formic acid[135], dimedone (**669**)[415,426], *N*-hydroxysuccimide[427], tin hydride[428,429], and silylamine[429] under mild conditions. These methods are applied to the protection of amino acids used in peptide synthesis without racemization. *p*-Nitrocinnamyloxycarbonyl is used as an acid-stable protecting group of amino groups, and is removed by cinnamyl group transfer to *N,N*-dimethylbarbituric acid (DMBA) (**670**). As an allyl group acceptor, DMBA is said to be better than dimedone (**669**)[430]. *N*-allyloxycarbonyl is a good protecting group for glucosamine derivatives[431].

The protected nucleoside-3-phosphoramidite monomer units such as **671** are used in the solid-phase oligonucleotide synthesis. In the 60mer synthesis, 104 allylic protective groups are removed in almost 100% overall yield by the single Pd-catalyze reaction with formic acid and BuNH₂[432]. N,O-protection of uridine derivatives was carried out under phase-transfer conditions[433].

671

Deprotection of allylic esters, carbamates, and carbonates by transferring the allyl group to Et_2NH can be carried out smoothly in aqueous media employing water-soluble TMSPP as a ligand. In this way, the catalyst can be recycled[434]. 2-Thiobenzoic acid (**672**) as an S nucleophile is a good acceptor of the allyl group[435]. Surprisingly, it has been found that deprotective trans-allylation from allylamines to DMBA (**670**) under mild conditions is possible in CH_2Cl_2. Thus, the allyl group can be used for the protection of amino groups. Based on this reaction, diallylamine (**673**) can be regarded as a convenient NH_3 equivalent and used for the preparation of primary amines by alkylation and deallylation[436].

Another protecting group of amines is 1-isopropylallyloxycarbonyl, which can be deprotected by decarboxylation and a β-elimination reaction of the (π-1-isopropylallyl)palladium intermediate under neutral conditions, generating CO_2 and 4-methyl-1,3-pentadiene. The method can be applied to the amino acid **674** and peptides without racemization[437].

Alcohols and phenols are protected as allylic ethers or allylic carbonates. Allyl phenyl ethers as protected phenols are cleaved with $HSnBu_3$[429] and $NaBH_4$[437a]. Allyl ethers are used for the protection of alcohols in sugar molecules, but it is difficult to cleave them with Pd catalysts. They are cleaved by isomerization to enol ethers and subsequent hydrolysis[438]. The anomeric allyl ether groups in the carbohydrate **675** can be deprotected by catalysis with $PdCl_2$–AcONa–AcOH[439] or $Pd(Ph_3P)_4$ in AcOH[440]. Also the anomeric allyl ether **676** is cleaved by oxidation to the ketone with $PdCl_2$–CuCl–O_2, followed by photolysis[441]. $PdCl_2$ and CuCl (1 equiv. each) in DMF are used for the deprotection of anomeric allyl ethers[442]. Interestingly, allyl ether formation takes place on treatment of a sugar with an excess of allyl methyl carbonate in the presence of a Pd(0) catalyst[206]. Allyl carbonates as the protecting group of alcohols are easily deprotected with ammonium formate[413,443] or $HSnBu_3$[370]. Allyl carbonates are deprotected by allyl transfer to Et_2NH using TMSPP in aqueous media[434].

$$R-O-\overset{\overset{\displaystyle O}{\|}}{C}-O\diagup\diagdown + HCO_2H \xrightarrow[\substack{Et_3N \\ (HSnBu_3)}]{Pd(0),\ Ph_3P} ROH + 2\ CO_2 + \diagup\diagdown$$

A method for protecting ketones and aldehydes is the formation of oximes, but sometimes further protection of the oximes is required. For this purpose, the oximes can be protected as allyl ethers. The oxime ethers can be cleaved with triethylammonium formate in boiling dioxane[444]. The allyl ether of oximes is cleaved under mild conditions without attacking the acetal group in **677**.

2.10 Reactions of Allyl β-Keto Carboxylates and Related Compounds

Needless to say, β-keto esters are important compounds in organic synthesis. The usefulness of β-keto esters has been enormously expanded based on Pd-catalyzed reactions of allyl β-keto carboxylates **678**. Allylic carbon–oxygen bond cleavage and facile decarboxylation take place by treatment of allyl β-keto carboxylates with Pd(0) catalysts, forming the π-allylpalladium enolates **679**, which undergo various transformations depending on the reaction conditions[13]. Similarly, derivatives of allyl acetate which have other EWGs, such as malonates, nitroacetates, cyanoacetates, and sulfonylacetates, all undergo Pd-catalyzed decarboxylation and further transformations. In addition to allyl β-keto carboxylates, allyl enol carbonates **680**, which can be regarded as structural isomers of allyl β-keto esters, undergo similar reactions via the formation of the π-allylpalladium enolates **679**. It should be pointed out that similar π-allylpalladium enolates are formed by the Pd-catalyzed reactions of silyl enol ethers and enol acetates with allyl carbonates (see Sections 2.4 and 2.6).

2.10.1 Decarboxylation and Allylation

The reductive coupling of the π-allylpalladium enolates **679** gives the allylated ketones. This reaction is also possible thermally, as the Carroll reaction, which proceeds by heating to 200 °C. On the other hand, the Pd-catalyzed Carroll-type reaction can be carried out under mild conditions even at room temperature via the π-allylpalladium enolate **682**[445,446]. The Pd-catalyzed reaction is different from the thermal reaction in the mechanism and is more versatile than the thermal reaction, which is explained by the [3,3] sigmatropic rearrangement of the enolate form **681**. For example, the thermal Carroll rearrangement of the α,α-disubstituted keto ester **683** is not possible, because there is no possibility of enolization. However, the smooth rearrangement of the β-keto ester **683** takes place with a Pd catalyst via the π-allylpalladium enolate to give the allyl ketone **684**.

Geranyl acetoacetate (**685**) is converted into geranylacetone (**686**). On the other hand, a mixture of *E*- and *Z*-isomers of **688** is obtained from neryl acetoacetate (**687**). The decarboxylation and allylation of the allyl malonate or cyanoacetate **689** affords the α-allylated acetate or nitrile[447]. The trifluoromethyl ketone **691** is prepared from cinnamyl 4,4,4-trifluoroacetoacetate (**690**)[448].

α,α-Diallylation of ketones with two different allyl groups to give **694** is possible as a one-pot reaction by combining the allylation of the allyl β-keto ester **692** with the allylic carbonate **693**, followed by decarboxylative allylation. The α, β-diallylated cyclic ketone **697** is prepared by intramolecular allylation of **695** to generate **696**, which undergoes decarboxylative allylation to give **697**. These chemoselective diallylation reactions are based on the fact that the allylation of β-keto esters with allylic carbonates is faster than the decarboxylation–allylation of allyl β-keto esters[449].

The decarboxylation–allylation of allyl enol carbonates proceeds smoothly[450]. The isomeric enol carbonates **699** and **701** of the enone **698** undergo regiospecific allylation, giving the regioisomers **700** and **702** selectively.

A silyl enol ether derived from an allyl β-keto carboxylate undergoes Pd-catalyzed intramolecular decarboxylation–allylation to afford the enol ether **703**, offering a synthetic method for thermodynamically stable silyl enol ethers[451]. Although it is not decarboxylation–allylation, an interesting method for regioselective tin enolate formation can be mentioned here. The Pd-catalyzed hydrostannolysis of an α-disubstituted allyl β-keto ester yields the corresponding tributylstannyl β-keto carboxylate **704**, which on heating undergoes facile decarboxylation, yielding regioselectively the tin enolate **705**[452].

2.10.2 Decarboxylation and Elimination (Enone Synthesis)

When the reaction of allyl β-keto carboxylates is carried out in boiling MeCN using dppe or Ph₃P, the decarboxylation is followed by the elimination of β-hydrogen from the intermediate enolate **706**, affording the α, β-unsaturated ketone **707**[453]. For chemoselective enone formation, the ratio of Pd to Ph₃P is important. A ratio of 1–1.5 : 1 (Ph₃P : Pd) is essential for the chemoselective formation of the enones. This means that coordinative unsaturation favors the elimination. The allylation becomes the main reaction path by increasing the ratio of Ph₃P. The allyl group is the proton acceptor in the elimination. As supporting evidence, the enone **709** and 1-phenylpropylene (**710**) were obtained in equal amounts by the reaction of the cinnamyl β-keto ester **708** as the allylic component. In addition to the allylation, the protonation is competitive with the elimination, yielding saturated ketones in a considerable amount depending on the reaction conditions and substrates. The presence of an α-hydrogen in β-keto esters favors the protonation.

708   82%   81%
709   710

As one application of enone formation, α-substituted cyclopentenones can be prepared. Dieckmann condensation of diallyl adipate (**711**), followed by alkylation gives the α-substituted cyclopentanonecarboxylate **712**, which undergoes facile Pd-catalyzed decarboxylation and elimination of β-hydrogen in boiling MeCN by keeping the Pd : Ph_3P ratio at 1 : 1–1.5, yielding the α-alkylcyclopentenone **713**. When the ratio is higher than 2, the decarboxylation-allylation takes place. This method has been applied to the synthesis of 2-methylcyclopentenone (**713**, R = Me), which is a useful intermediate for cyclopentanoid synthesis[454]. Production of α-(2-pentynyl)-2-cyclopentenone (**714**) based on this method is a key step in the commercial production of methyl jasmonate (**715**)[455].

711   712   713

714   715

Enone formation–aromatization has been used for the synthesis of 7-hydro-xyalkavinone (**716**)[456]. The isoflavone **717** was prepared by the elimination[457]. The unsaturated β-keto allyl esters **718** and **719**, obtained in two steps from myrcene, were subjected to enone formation. The reaction can be carried out even at room temperature using dinitriles such as adiponitrile (**720**) or 1,6-dicyanohexane as a solvent and a weak ligand to give the pseudo-ionone isomers **721** and **722** without giving an allylated product[458].

716

Allyl enol carbonates derived from ketones and aldehydes undergo Pd-catalyzed decarboxylation–elimination, and are used for the preparation of α, β-unsaturated ketones and aldehydes. The reaction is regiospecific. The regioisomeric enol carbonates **724** and **726**, prepared from **723**, are converted into two isomeric enones, **725** and **727**, selectively. The saturated aldehyde **728** can be converted into the α,β-unsaturated aldehyde **730** via the enol carbonate **729**[459].

The allyl cyanoacetate **731** can be converted into an α, β-unsaturated nitrile by the decarboxylation–elimination reaction[460], but allyl malonates cannot be converted into unsaturated esters, the protonation and allylation products being formed instead.

2.10.3 Decarboxylation and Deacetoxylation (Preparation of α-Methylene Compounds)

An α-acetoxymethyl group can be introduced into allyl β-keto carboxylates by the treatment with formaldehyde followed by acetylation, and used as a good leaving group. When allyl β-keto esters with the acetoxymethyl group at the α-position are treated at room temperature with Pd(0) catalyst, decarboxylation is followed by deacetoxylation to give the exomethylene ketone **732**[461]. The allyl group is an acceptor of the acetoxy group, and allyl acetate (**733**) is formed, indicating that the acetoxy group is eliminated as allyl acetate more easily than β-hydrogen. The elimination reaction proceeds even at room temperature under mild, neutral conditions. Hence the method offers a good synthetic method for this important functional group under extremely mild conditions.

The reaction can be applied to allyl malonates. Alkylation of diallyl malonate (**734**) with bromoacetate and acetoxymethylation afford the mixed triester **735**. Treatment of the triester **735** with Pd catalyst affords allyl ethyl itaconate (**736**). In a similar way, α-methylene lactone and the lactam **737** can be prepared[462].

2.10.4 Aldol Condensation and Michael Addition

The decarboxylation of allyl β-keto carboxylates generates π-allylpalladium enolates. Aldol condensation and Michael addition are typical reactions for metal enolates. Actually Pd enolates undergo intramolecular aldol condensation and Michael addition. When an aldehyde group is present in the allyl β-keto ester **738**, intramolecular aldol condensation takes place yielding the cyclic aldol **739** as a main product[463]. At the same time, the diketone **740** is formed as a minor product by β-elimination. This is Pd-catalyzed aldol condensation under neutral conditions. The reaction proceeds even in the presence of water, showing that the Pd enolate is not decomposed with water. The spiro-aldol **742** is obtained from **741**. Allyl acetates with other EWGs such as allyl malonate, cyanoacetate **743**, and sulfonylacetate undergo similar aldol-type cyclizations[464].

The Pd enolates also undergo intramolecular Michael addition when an enone of suitable size is present in the allyl β-keto ester **744**[465]. The main product is the saturated ketone **745**, but the unsaturated ketone **746** and ally-lated product **747** are also obtained as byproducts. The Pd-catalyzed Michael addition takes place with the allyl malonate **748**, cyanoacetates, and sulfony-lacetates[464].

2.10.5 Decarboxylation and Hydrogenolysis

β-Keto esters and malonates are useful compounds in organic synthesis. After alkylation, they are hydrolyzed and decarboxylated to give alkylated ketones or acids. However, severe conditions are required for the hydrolysis, strongly acidic or basic conditions and high temperatures being required. On the other hand, allyl β-keto carboxylates and allyl malonates, after alkylation, can be decarboxylated under extremely mild conditions, namely at room temperature and under neutral conditions, by palladium-catalyzed hydrogenolysis with ammonium formate. Other products are CO_2 and propylene. The reaction can be carried out without attacking acid- or base-sensitive functional groups. The acid-sensitive THP ether group in **749** is not cleaved, and no retro-Michael is observed in the decarboxylation of **750**[466]. The decarboxylation of **751** has been used in the total synthesis of glycinoeclepin[467]. The α-keto acid **753** can be prepared from the diallyl ketosuccinate **752** by chemoselective decarboxylation of only the allyl β-keto carboxylate; the allyl α-keto carboxylate is hydrolyzed without decarboxylation[468]. The α-fluoro ketone **755** is prepared by hydrogenolysis of the allyl α-fluoro-β-keto ester **754** with formate under mild conditions[469]. Another method for removal of the allyl ester group in **756** involves decarboxylation and allyl transfer to secondary amines such as morpholine[470].

Hydrogenolysis of the diallyl alkylmalonate **757** with formic acid in boiling dioxane affords the monocarboxylic acid **758**. Allyl ethyl malonates are converted into ethyl carboxylates[471]. The malonic allyl ester *N*-allylimide **759** undergoes smooth deallylation in refluxing dioxane to give the simple imide **760**[472]. The allyl cyanoacetate **761** undergoes smooth decarboxylation to give the nitrile **762**[473]. The diallyl α-methylmalonate **763**, attached to a β-lactam ring, undergoes palladium-catalyzed stereoselective decarboxylation and hydrogenolysis using an excess of formic acid without amine to give the mono acid **764**, which has the desired β-oriented methyl group with high selectivity. Protection of the amide nitrogen with TBDMS is essential, otherwise no stereoselectivity is observed[474].

2.11 Reactions of Alkenes and Alkynes

1-Hexyne and allyl carbonate undergo dimerization–allylation to give (Z)-5-allyl-4-butyl-1,4-undecadien-6-yne (**765**) in 85% yield[475].

The intermolecular insertion of alkenes into π-allylpalladium is unknown, except with norbornadiene[476]. On the other hand, the intramolecular insertion of alkene group in **766** into π-allylpalladium proceeds smoothly to give the

765

cyclic compound **767** after β-elimination. The reaction can be regarded as a metallo–ene reaction[477]. Complete chirality transfer was observed in the cyclization of **768**. The alkene insertion takes place from the same side of the Pd to give **769**[478]. It is important to use AcOH as a solvent in this reaction; no reaction takes place in its absence. However, the reaction proceeds satisfactorily in MeCN, provided that the substrate is an appropriate derivative containing the 4-substituted-4-hydroxy-2,7-octadienyl acetate moiety[479].

767

The five- and also six-membered ring compound **770** is formed by this reaction[480]. However, the six-membered ring compound **771** is formed by the reaction of 2,7-octadienyl carbonate containing an allylic silane moiety by preferential elimination of the silyl group rather than β-hydrogen in the final step by *endo*-cyclization in MeCN using TMPP as a ligand, without forming a five-membered ring by *exo*-cyclization[481]. After the alkene insertion into the π-allylpalladium, even a strained four-membered ring compound **773** is formed by the subsequent Heck-type alkene insertion, because the intermediate **772** contains neopentylpalladium, which cannot undergo β-elimination. In the presence of formic acid, the neopentylpalladium intermediate **772** is trapped with hydride to form a methyl group **774**[482].

Similarly to alkenes, alkynes also insert. In the reaction of **775** carried out under a CO atmosphere in AcOH, sequential insertions of alkyne, CO, alkene, and CO take place in this order, yielding the keto ester **776**[483]. However, the same reaction carried out in THF in the presence of LiCl affords the ketone **777**, but not the keto ester[484]. The tricyclic terpenoid hirsutene (**779**) has been synthesized via the Pd-catalyzed metallo–ene carbonylation reaction of **778** with 85% diastereoselectivity as the key reaction[485]. Kainic acid and allo-kainic acid (**783**) have been synthesized by the intramolecular insertion of an alkene in **780**, followed by carbonylation to give **781** and **782**[486].

The reaction of the allylic acetate with a diene system **784** affords the poly-fused ring system **785** by three repeated alkene insertions[487]. An even more strained molecule of the [5.5.5.5] fenestrane **788** has been constructed by a one-pot reaction in a satisfactory yield by the Pd-catalyzed carbonylation–cyclization of **786** without undergoing elimination of β-hydrogen in the σ-alkylpalladium intermediate **787** owing to unfavorable stereochemistry for *syn* elimination[488].

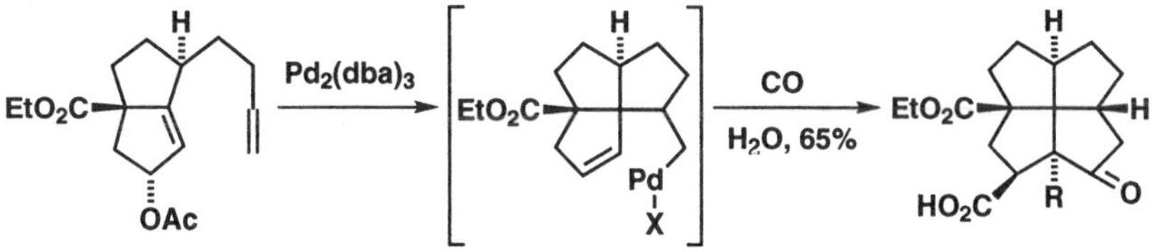

786 **787** **788**

The intramolecular insertion of a conjugated diene into π-allylpalladium, initially formed in **789**, generates another π-allyl complex **790**, which is trapped with acetate anion to give a new allylic acetate **791**. No further reaction of the allylic acetate with alkene takes place[489].

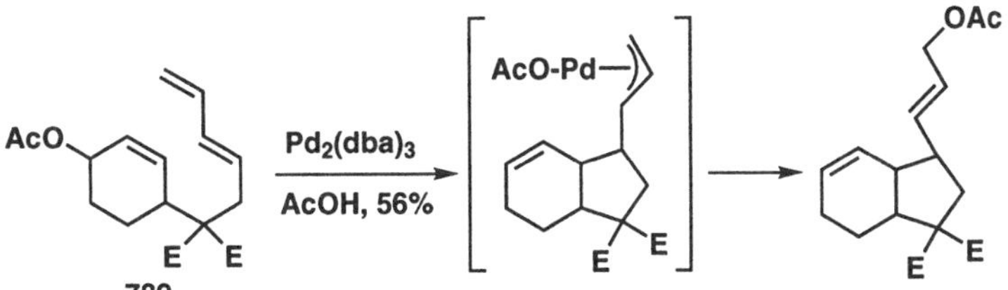

789 **790** **791**

The 1,5-hexadien-3-ol derivatives **792** and **794** are cyclized to form the cyclopentadiene derivatives **793** and **795** by insertion of an alkene into π-allylpalladium formed from allylic alcohols in the presence of trifluoroacetic acid (10 mol%) in AcOH[490].

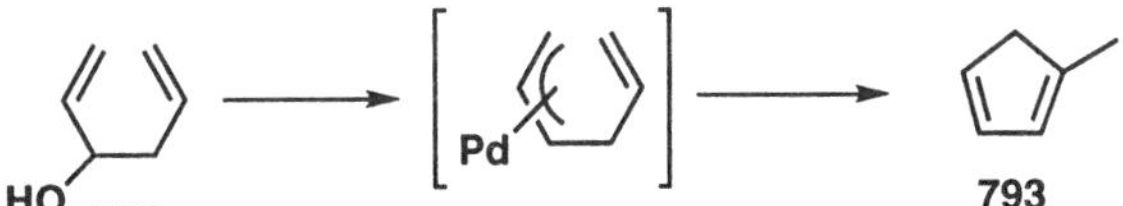

792 **793**

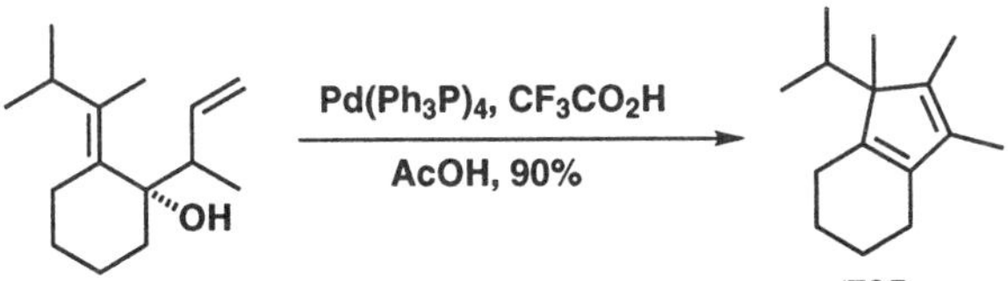

794 **795**

2.12 Allylic Rearrangement, Claisen Rearrangement, and Related Reactions Catalyzed by Pd(0) and Pd(II)

The rearrangements of various allylic compounds catalyzed by both Pd(II) and Pd(0) are treated in this section[491]. Related reactions such as the Carroll rearrangement are treated in Section 2.10.1 and the Pd(II)-catalyzed Cope rearrangement is treated in Chapter 5, Section 3.

Allylic ester rearrangement is catalyzed by both Pd(II) and Pd(0) compounds, but their catalyses are different mechanistically. Allylic rearrangement of allylic acetates takes place by the use of $Pd(OAc)_2$–Ph_3P [Pd(0)–phosphine] as a catalyst[492,493]. An equilibrium mixture of **796** and **797** in a ratio of 1.9 : 1.0 was obtained[494]. The Pd(0)–Ph_3P-catalyzed rearrangement is explained by π-allylpalladium complex formation[495].

The stereoselective allylic rearrangement of the allylic alcohol **798** catalyzed by $PdCl_2(MeCN)_2$ and Ph_3P under Mitsunobu inversion conditions is explained as proceeding via a π-allylpalladium intermediate[496]. The smooth rearrangement of the allylic *p*-tolylsulfone **799** via a π-allylpalladium intermediate is catalyzed by a Pd(0) catalyst[497].

The Pd(0)-catalyzed rearrangement of the *N*-allylenamine **800** in CF_3CO_2H affords the δ,ε-unsaturated imine **801**, which is hydrolyzed to give the γ, δ-unsaturated aldehyde **802**[498]. The vinyloxaspirohexane **803** undergoes rearrangement–ring expansion to give the cyclopentanone **804** in the presence of 1 equiv. of *p*-nitrophenol[499].

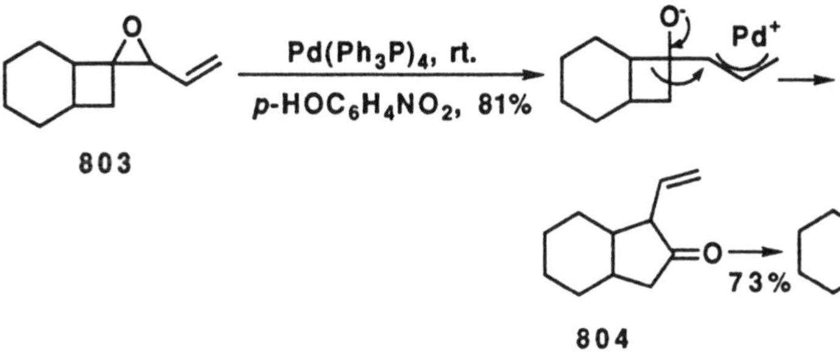

803

804

The rearrangement of allylic esters is catalyzed efficiently by PdCl$_2$(MeCN)$_2$[500]. The allylic rearrangement of **805** and **807** in prostaglandin synthesis is catalyzed efficiently by PdCl$_2$(MeCN)$_2$[501–503]. The reaction goes in one direction irreversibly, yielding the thermodynamically stable products **806** and **808**, possibly for steric reasons. In addition, complete transfer of the chirality of the carbon–oxygen bond is observed. The rearrangements of the *E*- and *Z*-isomers **805** and **807** generate opposite stereochemistry after the rearrangement. A minor product **809** was formed by π–σ–π-rearrangement via rotation. A similar chirality transfer in the cholesterol side-chain has been carried out[504].

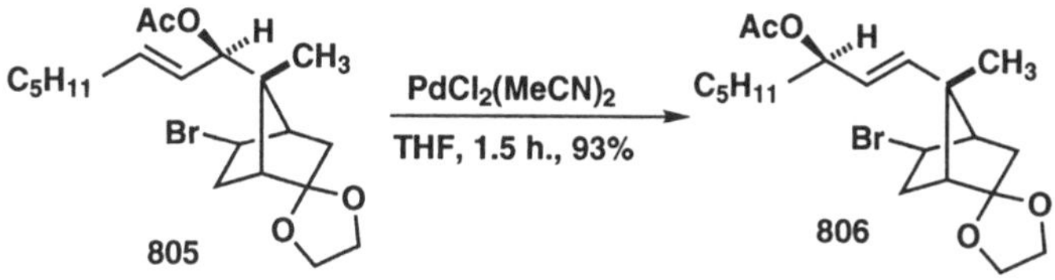

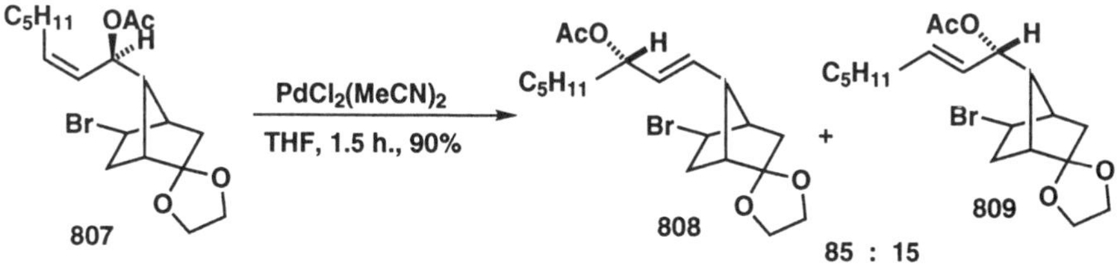

805   **806**

807   **808**   **809**

85 : 15

The mechanism of the rearrangement catalyzed by Pd(II), typically by PdCl$_2$(RCN)$_2$, is explained by the oxypalladation of an alkene to form **810** as an intermediate, or cyclization-induced rearrangement. As a limitation, no rearrangement takes place when the allylic ester **812** is substituted at the C-2 position of the allyl group, while a smooth rearrangement of **811** takes place[500].

The (*E,Z*)-3-acetoxy-1,4-diene **813** is isomerized rapidly at room temperature with a Pd(II) catalyst regio- and stereoselectively to the conjugated (*E,Z*)-1-acetoxy-2,4-diene **814**[505]. On the other hand, (*E,E*)-**815** is obtained via a π-allylpalladium intermediate from the (*E,Z*)-diene **813** when Pd(Ph₃P)₄ is used as a catalyst. The 1,5-diene-3,4-diol diacetate **816** is isomerized to the more stable conjugated diene **817** with complete transfer of chirality[506]. Cyanohydrin acetates of α,β-unsaturated aldehydes rearrange to give γ-acetoxy-α,β-unsaturated nitrile[507]. The optically active cyanohydrin acetate **818** undergoes Pd(II)-catalyzed stereospecific chirality transfer to give the α,β-unsaturated nitrile **819**[508].

Conversion of *S*-allylthioimidates into *N*-allylthioamides is catalyzed by Pd(II). 2-Allylthiopyridine (**820**) is converted into the less stable 1-allyl-2-thio-pyridone **821** owing to Pd complex formation[509]. Claisen rearrangement of 2-(allylthio)pyrimidin-4-(3*H*)-one (**822**) affords the *N*-1-allylation product **823** as the main product rather than the *N*-3-allylation product **824**[510]. The smooth rearrangement of the allylic thionobenzoate **825** to the allyl thiolo-benzoate **826** is catalyzed by both PdCl$_2$(PhCN)$_2$ and Pd(Ph$_3$P)$_4$ by different mechanisms[511].

Allyl imidates are converted into allylamides[512]. Both Pd(II) and Pd(0) catalyze the rearrangement under mild conditions. Pd(II) catalysis via a cyclic carbonium ion intermediate is characterized by exclusive [3,3] regioselectivity and high stereoselectivity. On the other hand, Pd(0) catalyst gives both the [3,3] and [1,3] rearrangement products via the formation of a π-allylpalladium complex. The rearrangement of the optically active allyl imidate **827** catalyzed by Pd(II) at 25 °C gives the (*E*)- and (*Z*)-alkene isomers **828** and **829** in a ratio of 78 : 22; the *E*- and *Z*-isomers were of *S* and *R* chirality, respectively[513]. Complete chirality transfer is observed in the Pd(II)-catalyzed rearrangement of the allyl trichloroacetimidate **830** from O to N at room temperature[514,515]. The allyl methyl-*N*-aryldithiocarboimidate **831** undergoes rearrangement in boiling dioxane to form the *N*-allyl-*N*-phenyldithiocarbamate **832**[516].

The PdCl$_2$(MeCN)$_2$-catalyzed Claisen rearrangement of allyl vinyl ethers proceeds smoothly even at room temperature. The yields are highly dependent on the substituents. The Pd(0) complex is inactive[517]. By rearrangement of **833**, the *syn* product **834** is obtained with high diastereoselectivity. In contrast to the thermal Claisen rearrangement, the Pd(II)-catalyzed Claisen rearrangement is always *threo*-selective irrespective of the geometry of allylic alkenes[518]. The Pd(II)-catalyzed reaction of an allylic alcohol with the ketene acetal **835** at room temperature generates the *ortho* ester **836**, which undergoes Claisen rearrangement via **837** in boiling xylene with a catalytic amount of PdCl$_2$(Ph$_3$P)$_2$ to give the γ,δ-unsaturated ester **838**[519].

2.13 Reactions of 2,3-Alkadienyl Derivatives

The 2,3-alkadienyl esters **839** are reactive compounds toward Pd catalysts and form the α-alkylidene-π-allylpalladium complexes **840**, which react further to give two kinds of products, namely the 1,2- and 1,4-diene derivatives **841** and **842**, depending on the reactants.

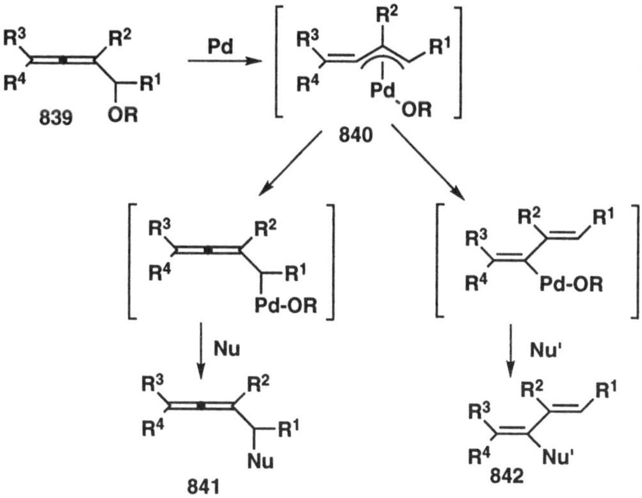

The reaction of 2,3-butadienyl acetate (**843**) with soft carbon nucleophiles such as dimethyl malonate gives dimethyl 2,3-butadienylmalonate (**844**)[520]. On the other hand, the reaction of the 2,3-butadienyl phosphate **845** with hard carbon nucleophiles such as Mg and Zn reagents affords the 2-alkyl-1,3-butadiene **846**[520,521]. The 3-methoxy-1,3-butadiene **848** is obtained by the reaction of the 2-methoxy-2,3-butadienyl carbonate **847** with organozinc reagent.

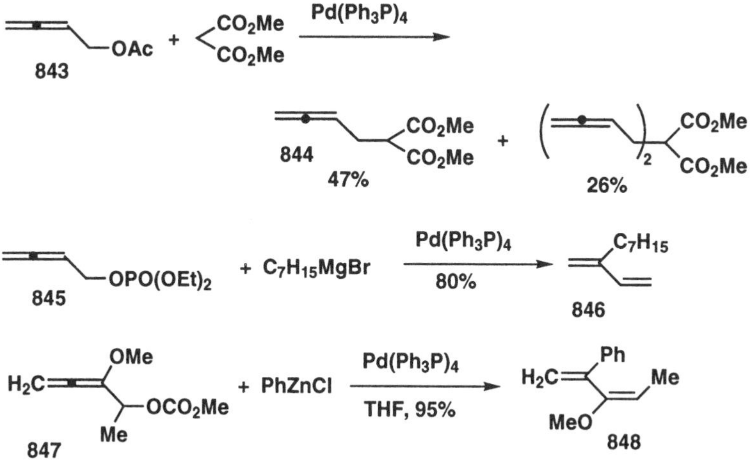

The 3-alkyl-1,3-butadiene-2-carboxylate (2-vinylacrylate) **850** is obtained in a high yield by the carbonylation of the 2-alkyl-2,3-butadienyl carbonate **849** under mild conditions (room temperature, 1 atm)[522]. The corresponding acids are obtained in moderate yields by the carbonylation of 2,3-alkadienyl alcohols under severe conditions (100 °C, 20 atm) using a cationic Pd catalyst and *p*-TsOH[523].

The 2,3-alkadienyl acetate **851** reacts with terminal alkynes to give the 2-alkynyl-1,3-diene derivative **852** without using CuI and a base. In the absence of other reactants, the terminal alkyne **853** is formed by an unusual elimination as an intermediate, which reacts further with **851** to give the dimer **854**. Hydrogenolysis of **851** with formic acid affords the 2, 4-diene **855**[524].

2.14 References

1. Reviews: S. Godleski, in *Comprehensive Organic Synthesis*, Vol. 4, Pergamon Press, Oxford, 1991, p. 585. C. G. Frost, J. Howarth, and J. M. J. Williasm, *Tetrahedron, Asymm.*, **3**, 1089 (1992).
2. J. Tsuji, H. Takahashi, and M. Morikawa, *Tetrahedron Lett.*, 4387 (1965).
3. J. Tsuji, *Acc. Chem. Res.*, **2**, 144 (1969).
4. B. M. Trost and T. J. Fullerton, *J. Am. Chem. Soc.*, **95**, 292 (1973).
5. D. J. Collins, W. R. Jackson, and R. N. Timms, *Tetrahedron Lett.*, 495 (1976); *Aust. J. Chem.*, **30**, 2167 (1977).
6. G. Hata, K. Takahashi, and A. Miyake, *Chem. Commun.*, 1392 (1970); *Bull. Chem. Soc. Jpn.*, **45**, 230 (1972).
7. K. E. Atkins, W. E. Walker, and R. M. Manyik, *Tetrahedron Lett.*, 3821 (1970).
8. R. Tamura and L. S. Hegedus, *J. Am. Chem. Soc.*, **104**, 3727 (1982); R. Tamura, Y. Kai, M. Kakihana, K. Hayashi, M. Tsuji, T. Nakamura, and D. Oda, *J. Org. Chem.*, **51**, 4375 (1986).
9. N. Ono, I. Hamamoto, and A. Kaji, *Chem. Commun.*, 821 (1982); *Synthesis*, 950 (1985).

10. B. M. Trost, N. R. Schmuff, and M. J. Miller, *J. Am. Chem. Soc.*, **102**, 5979 (1980); B. M. Trost and C. A. Merlic, *J. Org. Chem.*, **55**, 1127 (1290).
11. H. Kotake, T. Yamamoto, and H. Kinoshita, *Chem. Lett.*, 1331 (1982).
12. T. Cuvigny, M. Julia, and C. Rolando, *J. Organomet. Chem.*, **285**, 395 (1985).
13. Reviews, J. Tsuji and I. Minami, *Acc. Chem. Res.*, **20**, 140 (1987); J. Tsuji, *Tetrahedron*, **42**, 4361 (1986).
14. J. Tsuji, I. Shimizu, I. Minami, and Y. Ohashi, *Tetrahedron Lett.*, **23**, 4809 (1982); J. Tsuji, I. Shimizu, I. Minami, Y. Ohashi, T. Sugiura, and K. Takahashi, *J. Org. Chem.*, **50**, 1523 (1985).
15. J. Tsuji, Y. Kobayashi, H. Kataoka, and T. Takahashi, *Tetrahedron Lett.*, **21**, 1475 (1980).
16. J. Tsuji, H. Kataoka, and Y. Kobayashi, *Tetrahedron Lett.*, **22**, 2575 (1981).
17. B. M. Trost and G. A. Molander, *J. Am. Chem. Soc.*, **103**, 5969 (1981).
18. B. M. Trost and L. Weber, *J. Am. Chem. Soc.*, **97**, 1611 (1975); B. M. Trost, L. Weber, P. E. Strege, T. J. Fullerton, and T. Dietsche, *J. Am. Chem. Soc.*, **100**, 3416 (1978); B. M. Trost and T. R. Verhoeven, *J. Am. Chem. Soc.*, **100**, 3435 (1978); **102**, 4730 (1980). *J. Org. Chem.*, **41**, 3215 (1976).
19. T. Hayashi, T. Hagihara, M. Konishi, and M. Kumada, *J. Am. Chem. Soc.*, **105**, 7767 (1983); T. Hayashi, M. Konishi, and M. Kumada, *Chem. Commun.*, 107 (1984).
20. B. Akermark, J. E. Backvall, A. Lowenborg, and K. Zetterberg, *J. Organomet. Chem.*, **166**, C33 (1979); B. Akermark and A. Jutand, *J. Organomet. Chem.*, **217**, C41 (1981).
21. H. Matsushita and E. Negishi, *Chem. Commun.*, 160 (1982).
22. J. S. Temple, M. Riediker, and J. Schwartz, *J. Am. Chem. Soc.*, **104**, 1310 (1982).
23. T. Hayashi, A. Yamamoto, and T. Hagihara, *J. Org. Chem.*, **51**, 723 (1986).
24. B. Bosnich and P. B. Mackenzie, *Pure Appl. Chem.*, **54**, 189 (1982).
25. J. C. Fiaud and J. Y. Legros, *J. Org. Chem.*, **52**, 1907 (1987).
26. F. E. Ziegler, A. Kneisley, J. K. Thottathil, and R. T. Wester, *J. Am. Chem. Soc.*, **110**, 5434 (1988); F. E. Ziegler, W. T. Cain, A. Kneisley, E. P. Stirchak, and R. T. Wester, *J. Am. Chem. Soc.*, **110**, 5442 (1988).
27. T. Takahashi, Y. Jimbo, K. Kitamura, and J. Tsuji, *Tetrahedron Lett.*, **25**, 5921 (1984).
28. P. R. Auburn, P. B. Mackenzie, and B. Bosnich, *J. Am. Chem. Soc.*, **107**, 2033, 2046 (1985).
29. S. Ogoshi and H. Kurosawa, *Chem. Lett.*, 1745 (1990); H. Kurosawa, S. Ogoshi, N. Chatani, Y. Kawasaki, and S. Murai, *Organometallics*, **12**, 2869 (1993).
30. K. L. Granberg and J. E. Bäckvall, *J. Am. Chem. Soc.*, **114**, 6858 (1992).
31. I. Stary, J. Zajicek, and P. Kocosky, *Tetrahedron*, **48**, 7229 (1992); *J. Am. Chem. Soc.*, **111**, 4981 (1989).
32. J. E. Bäckvall, J. O. Vagberg, and K. L. Granberg, *Tetrahedron Lett.*, **30**, 617 (1989).
33. H. Kurosawa, S. Ogoshi, Y. Kawasaki, S. Murai, M. Miyoshi, and I. Ikeda, *J. Am. Chem. Soc.*, **112**, 2813 (1990); H. Kurosawa, H. Kajimaru, S. Ogoshi, H. Yoneda, K. Miki, N. Kasai, S. Murai, and I. Ikeda, *J. Am. Chem. Soc.*, **114**, 8417 (1992).
34. J. P. Genet, S. Juge, S. Achi, S. Mallart, J. R. Montes, and G. Levif, *Tetrahedron*, **44**, 5263 (1988); D. Ferroud, J. P. Genet, and R. Kiolle, *Tetrahedron Lett.*, **27**, 23 (1986).
35. J. P. Genet, J. Uziel and S. Juge, *Tetrahedron Lett.*, **29**, 4559 (1988).
36. J. P. Genet, J. Uziel, M. Port, A. M. Touzin, S. Roland, S. Thorimbert, and S. Tanier, *Tetrahedron Lett.*, **33**, 77 (1992).
37. B. M. Trost and J. R. Granja, *J. Am. Chem. Soc.*, **113**, 1044 (1991).
38. H. Piotrowska and W. Sas, *Tetrahedron*, **38**, 1321 (1982).

39. P. A. Wade, S. D. Morrow, and S. A. Hardinger, *J. Org. Chem.*, **47**, 365 (1982).
40. Y. I. M. Nilsson, P. G. Andersson, and J. E. Bäckvall, *J. Am. Chem. Soc.*, **115**, 6609 (1993).
41. Review, B. M. Trost, *Acc. Chem. Res.*, **13**, 385 (1980).
42. D. Ferroud, J. P. Genet, and J. Muzart, *Tetrahedron Lett.*, **25**, 4379 (1984).
43. Y. Tanigawa, K. Nishimura, A. Kawasaki, and S. Murahashi, *Tetrahedron Lett.*, **23**, 5549 (1982).
44. Review, B. M. Trost, *Angew. Chem., Int. Ed. Engl.*, **28**, 1173 (1989).
45. B. M. Trost and T. R. Verhoeven, *J. Am. Chem. Soc.*, **99**, 3867 (1977); **102**, 4743 (1980).
46. J. P. Genet, F. Piau, and J. Ficini, *Tetrahedron Lett.*, **21**, 3183 (1980); J. P. Genet and F. Piau, *J. Org. Chem.*, **46**, 2414 (1981); J. P. Genet, M. Balabane, and F. Charbonnier, *Tetrahedron Lett.*, **23**, 5027 (1982); F. Colobert and J. P. Genet, *Tetrahedron Lett.*, **26**, 2779 (1985).
47. J. E. Bäckvall, J. O. Vagberg, C. Zercher, J. P. Genet, and A. Denis, *J. Org. Chem.*, **52**, 5430 (1987).
48. J. P. Genet and J. M. Gaudin, *Tetrahedron*, **43**, 5315 (1987).
49. S. Hashimoto, T. Shinoda, and S. Ikegami, *Tetrahedron Lett.*, **27**, 2885 (1986).
50. G. P. Chiusoli, M. Costa, L. Pallini, and G. Terenghi, *Transition Met. Chem.*, **6**, 317 (1981); **7**, 304 (1982).
51. K. Burgess, *J. Org. Chem.*, **52**, 2046 (1987); *Tetrahedron Lett.*, **26**, 3049 (1995).
52. Y. Morizawa, K. Oshima, and H. Nozaki, *Tetrahedron Lett.*, **23**, 2871 (1982); *Isr. J. Chem.*, **24**, 149 (1984).
53. K. Hiroi and Y. Arinaga, *Tetrahedron Lett.*, **35**, 153 (1994).
54. S. A. Godleski and R. S. Valpey, *J. Org. Chem.*, **47**, 381 (1982).
54a. J. Tsuji, *Pure Appl. Chem.*, **54** 197 (1982).
55. B. M. Trost, T. A. Runge, and L. N. Jungheim, *J. Am. Chem. Soc.*, **102**, 2840, 7910 (1980); **103**, 2485, 7559 (1981).
56. B. M. Trost and T. A. Runge, *J. Am. Chem. Soc.*, **103**, 2485, 7559 (1981).
57. K. Yamamato and J. Tsuji, *Tetrahedron Lett.*, **23**, 3089 (1982).
58. B. M. Trost and T. R. Verhoeven, *J. Am. Chem. Soc.*, **101**, 1595 (1979).
59. Y. Kitagawa, A. Itoh, S. Hashimoto, H. Yamamoto, and H. Nozaki, *J. Am. Chem. Soc.*, **99**, 3865 (1977).
60. B. M. Trost and S. J. Brickner, *J. Am. Chem. Soc.*, **105**, 568 (1983).
61. J. A. Marshall, R. C. Andrews, and L. Lebroda, *J. Org. Chem.*, **52**, 2378 (1987).
62. B. M. Trost and D. P. Curran, *J. Am. Chem. Soc.*, **102**, 5699 (1980).
63. B. M. Trost and W. Gowland, *J. Org. Chem.*, **44**, 3448 (1979).
64. P. F. Schuda and B. Berstein, *Synth. Commun.*, **14**, 293 (1984).
65. S. G. Will, P. Magriotis, E. R. Marinelli, J. Dolan, and F. Johnson, *J. Org. Chem.*, **50**, 5432 (1985).
66. L. V. Dunkerton and A. J. Serino, *J. Org. Chem.*, **47**, 2812 (1982).
67. D. Ferroud, J. M. Gaudin, and J. P. Genet, *Tetrahedron Lett.*, **27**, 845 (1986).
68. M. Moreno-Manas, J. Ribas, and A. Virgili, *J. Org. Chem.*, **53**, 5328 (1988); M. Moreno-Manas, M. Prat, J. Ribas, and A. Virgili, *Tetrahedron Lett.*, **29**, 581 (1988).
69. J. Marquet, M. Moreno-Manas, and M. Prat, *Tetrahedron Lett.*, **30**, 3105 (1989).
70. J. P. Genet and D. Ferroud, *Tetrahedron Lett.*, **25**, 3579 (1984); J. P. Genet and S. Grisoni, *Tetrahedron Lett.*, **27**, 4165 (1986); **29**, 4543 (1988).
71. P. A. Wade, H. R. Hinney, N. V. Amin, P. D. Vail, S. D. Morrow, S. A. Hardinger, and M. S. D. Saft, *J. Org. Chem.*, **46**, 765 (1981).
72. J. P. Genet, N. Kopola, S. Juge, J. Ruiz-Montes, O. A. C. Antunes, and S. Tanier, *Tetrahedron Lett.*, **31**, 3133 (1990).

73. J. C. Fiaud and J. L. Malleron, *Chem. Commun.*, 1159 (1981).
74. E. Negishi, H. Matsushita, S. Chatterjee, and R. A. John, *J. Org. Chem.*, **47**, 3188 (1982); E. Negishi, F. T. Luo, A. J. Pecora, and A. Silveira, *J. Org. Chem.*, **48**, 2427 (1983).
75. E. Negishi and R. A. John, *J. Org. Chem.*, **48**, 4098 (1983).
76. B. M. Trost and E. Keinan, *Tetrahedron Lett.*, **21**, 2591 (1980); B. M. Trost and C. R. Self, *J. Org. Chem.*, **49**, 468 (1984).
77. Y. Inoue, M. Toyofuku, and H. Hashimoto, *Chem. Lett.*, 1227 (1984).
78. G. Mignani, F. Grass, M. Aufrand, and D. Morel, *Tetrahedron Lett.*, **30**, 2383 (1989).
79. H. Onoue, I. Moritani, and S. Murahashi, *J. Org. Chem.*, **55**, 1127 (1990).
80. K. Hiroi, J. Abe, K. Suya, S. Sato, and T. Koyama, *J. Org. Chem.*, **59**, 203 (1994).
81. K. Hiroi, J. Abe, K. Suya, and S. Sato, *Tetrahedron Lett.*, **30**, 1543 (1989); *Chem. Commun.*, 469 (1986); K. Hiroi and J. Abe, *Tetrahedron Lett.*, **31**, 3623 (1990); K. Hiroi, T. Koyama, and K. Anzai, *Chem. Lett.*, 235 (1990).
82. K. Hiroi, M. Haraguchi, Y. Masuda, and J. Abe, *Chem. Lett.*, 2409 (1992).
83. J. C. Fiaud and J. L. Malleron, *Tetrahedron Lett.*, **21**, 4437 (1980).
84. J. E. Nystrom, J. O. Vagberg, and B. C. Soderberg, *Tetrahedron Lett.*, **32**, 5247 (1991).
85. H. Onoue, I. Moritani, and S. Murahashi, *Tetrahedron Lett.*, 121 (1973).
86. X. Lu and Y. Huang, *Tetrahedron Lett.*, **27**, 1615 (1986); **29**, 5663 (1988).
87. M. Mori, S. Nukui, and M. Shibasaki, *Chem. Lett.*, 1797 (1991).
88. B. M. Trost, L. Li, and S. D. Guile, *J. Am. Chem. Soc.*, **114**, 8745 (1992).
89. B. M. Trost and D. L. Van Vranken, *Angew. Chem., Int. Ed. Engl.*, **31**, 228 (1992).
90. D. Gravel, S. Benoit, S. Kamanovic, and H. Sivaramakrishnan, *Tetrahedron Lett.*, **33**, 1407 (1992).
91. G. Campiani, L. Q. Sun, A. P. Kozikowski, P. Aagaard, and M. McKinney, *J. Org. Chem.*, **58**, 7660 (1993).
92. Y. Huang and X. Lu, *Tetrahedron Lett.*, **28**, 6219 (1987).
93. X. Lu and Y. Huang, *J. Organomet. Chem.*, **268**, 185 (1984).
94. B. M. Trost and J. Vercauteren, *Tetrahedron Lett.*, **26**, 131 (1985).
95. T. Mukaiyama, T. Soga, and H. Takenoshita, *Chem. Lett.*, 997 (1989), 229 (1990).
96. Y. Ito, M. Sawamura, M. Matsuoka, Y. Matsumoto, and T. Hayashi, *Tetrahedron Lett.*, **28**, 4849 (1987).
97. J. Tsuji, H. Ueno, Y. Kobayashi, and H. Okumoto, *Tetrahedron Lett.*, **22**, 2573 (1981).
98. J. Zhu and X. Lu, *Tetrahedron Lett.*, **28**, 1897 (1987); *Chem. Commun.*, 1318 (1987).
99. M. Moreno-Manas and J. Ribas, *Tetrahedron Lett.*, **30**, 3109 (1989); M. Prat, J. Ribas, and M. Moreno-Manas, *Tetrahedron*, **48**, 1695 (1992).
100. Reviews, B. M. Trost, *Angew. Chem.*, **25**, 1 (1986); *Pure Appl. Chem.*, **60**, 1615 (1988); D. M. T. Chan, in *Comprehensive Organic Synthesis*, Vol. 5, Pergamon Press, Oxford, 1991, p. 289.
101. B. M. Trost and D. M. T. Chan, *J. Am. Chem. Soc.*, **101**, 6429, 6432 (1979); **102**, 6359 (1980); **105**, 2315, 2326 (1983).
102. B. M. Trost, P. Seoane, S. Mignani, and M. Acemoglu, *J. Am. Chem. Soc.*, **111**, 7487 (1989).
103. B. M. Trost and P. Renaut, *J. Am. Chem. Soc.*, **104**, 6668 (1982).
104. B. M. Trost and T. N. Nanninga, *J. Am. Chem. Soc.*, **107**, 1293 (1985).
105. R. Baker and R. B. Keen, *J. Organomet. Chem.*, **285**, 419 (1985).

106. L. A. Paquette, D. R. Sauer, D. G. Cleary, M. A. Kinsella, C. M. Blackwell, and L. G. Anderson, *J. Am. Chem. Soc.*, **114**, 7375 (1992).
107. B. M. Trost, J. Lynch, P. Renant, and D. H. Steinman, *J. Am. Chem. Soc.*, **118**, 284 (1986).
108 J. P. Gotteland and M. Malacria, *Tetrahedron Lett.*, **30**, 2541 (1989); *Synlett*, 667 (1990); *J. Org. Chem.*, **58**, 4298 (1993).
109. B. M. Trost and T. A. Grese, *J. Org. Chem.*, **57**, 686 (1992).
110. B. M. Trost and M. Acemoglu, *Tetrahedron Lett.*, **30**, 1495 (1989).
111. D. G. Cleary and L. A. Paquette, *Synth. Commun.*, **17**, 497 (1987).
112. B. M. Trost and P. R. Seoane, *J. Am. Chem. Soc.*, **109**, 615 (1987); B. M. Trost and M. C. Matelich, *J. Am. Chem. Soc.*, **113**, 9007 (1991).
113. B. M. Trost, S. Sharma, and T. Schmidt, *J. Am. Chem. Soc.*, **114**, 7903 (1992).
114. B. M. Trost, S. Sharma, and T. Schmidt, *Tetrahedron Lett.*, **34**, 7183 (1993).
115. B. M. Trost and S. A. King, *Tetrahedron Lett.*, **27**, 5971 (1986); B. M. Trost, S. A. King, and T. Schmidt, *J. Am. Chem. Soc.*, **111**, 5902 (1989).
116. B. M. Trost and P. J. Bonk, *J. Am. Chem. Soc.*, **107**, 1778, 8277 (1985).
117. B. M. Trost and C. M. Marrs, *J. Am. Chem. Soc.*, **115**, 6636 (1993).
118. B. M. Trost and D. T. MacPherson, *J. Am. Chem. Soc.*, **109**, 3483 (1987).
119. T. A. J. van der Heide, J. L. van der Baan, V. de Kimpe, F. Bickelhaupt, and G. W. Klumpp, *Tetrahedron Lett.*, **34**, 3309 (1993).
120. A. Stolle, J. Salaun, and A. De Meijere, *Tetrahedron Lett.*, **31**, 4593 (1990); *Synlett*, 327 (1991); A. Stolle, J. Ollivier, P. P. Piras, J. Salaun, and A. de Meijere, *J. Am. Chem. Soc.*, **114**, 4051 (1992); G. McGaffin, S. Michalski, A. Stolle, S. Brase, J. Salaun, and A. de Meijere, *Synlett*, 558 (1992).
121. M. J. O'Donnell, X. Yang, and M. Li, *Tetrahedron Lett.*, **31**, 5135 (1990).
122. D. E. Bergbreiter and D. A. Weatherford, *Chem. Commun.*, 883 (1989).
123. I. Stary, I. G. Stera, and P. Kocovsky, *Tetrahedron Lett.*, **34**, 179 (1993); *Tetrahedron*, **50**, 529 (1994).
124. S. A. Stanton, S. W. Felman, C. S. Parkhurst, and S. A. Godleski, *J. Am. Chem. Soc.*, **105**, 1964 (1983).
125. X. Lu, X. Jiang, and X. Tao, *J. Organomet. Chem.*, **344**, 109 (1988).
126. X. Lu, L. Lu, and J. Sun, *J. Mol. Catal.*, **41**, 245 (1987).
127. X. Lu and X. Jiang, *J. Organomet. Chem.*, **359**, 139 (1989).
128. Review: R. Tamura, A. Kamimura, and N. Ono, *Synthesis*, 423 (1991).
129. Y. Tamaru, K. Nagano, T. Bando, and Z. Yoshida, *J. Org. Chem.*, **55**, 1823 (1990).
130. T. Hirao, N. Yamada, Y. Ohshiro, and T. Agawa, *J. Organomet. Chem.*, **236**, 409 (1982).
131. R. Hunter and C. D. Simin, *Tetrahedron Lett.*, **29**, 2257 (1988).
132. F. Guibe, D. S. Grierson, and H. P. Husson, *Tetrahedron Lett.*, **23**, 5055 (1982).
133. P. R. Auburn, P. B. Mackenzie, and B. Bosnich, *J. Am. Chem. Soc.*, **107**, 2033, 2046 (1985); T. Hayashi, A. Yamamoto, T. Hagihara, and Y, Ito, *Tetrahedron Lett.*, **27**, 191 (1986); *Chem. Lett.*, 177 (1987); *J. Am. Chem. Soc.*, **111**, 6301 (1989); *Pure Appl. Chem.*, **60**, 7 (1988); M. Yamaguchi, T. Shima, T. Yamagishi, and M. Hida, *Tetrahedron Lett.*, **31**, 5049 (1990); *Tetrahedron Asymmetry*, **2**, 663 (1991); J. C. Fiaud and J. Y. Legros, *J. Org. Chem.*, **55**, 4840 (1990); Y. Oda, T. Minami, T. Sasaki, Y. Umezu, and M. Yamaguchi, *Tetrahedron Lett.*, **31**, 3905, 5049 (1990); Y. Okada, T. Minami, Y. Sasaki, Y. Umezu, and M. Yamaguchi, *Tetrahedron Lett.*, **31**, 3905 (1990); Y. Okada, T. Minami, Y. Umezu, S. Nishikawa, R. Mori, and Y. Nakayama, *Tetrahedron Asym.*, **2**, 667 (1991); U. Leutenegger, G. Umbricht, C. Fahrni, P. von Matt, and A. Pfaltz, *Tetrahedron*, **48**, 2143 (1992);

Helv. Chim. Acta, **74**, 232 (1991); *Acc. Chem. Res.*, **26**, 339 (1993); B. M. Trost, D. L. van Vranken, and C. Bingel, *J. Am. Chem. Soc.*, **114**, 9327 (1992); B. M. Trost and D. J. Murphy, *Organometallics*, **4**, 1143 (1985); J. Sprinz and G. Helmchen, *Tetrahedron Lett.*, **34**, 1769 (1993); J. Sprinz, M. Kiefer, G. Helmchen, M. Reggelin, G. Huttner, O. Walter, and L. Zsolnai, *Tetrahedron Lett.*, **35**, 1523 (1994); G.J. Dawson, C. G. Frost, C. J. Martin, J. M. J. Williams, and S. J. Coote, *Tetrahedron Lett.*, **34**, 2015, 3149, 7793 (1993); O. Reiser, *Angew. Chem., Int. Ed. Engl.*, **32**, 547 (1993); A. Togni, *Tetrahedron Asym.*, **2**, 683 (1991); M. Bovens, A. Togni, and L. M. Venanzi, *J. Organomet. Chem.*, **451**, C28 (1993); H. Kubota, M. Nakajima, and K. Koga, *Tetrahedron Lett.*, **34**, 8135 (1993); Review: G. Consiglio and R. M. Waymouth, *Chem. Rev.*, **89**, 257 (1989).

134. M. Safi and D. Sinou, *Tetrahedron Lett.*, **32**, 2025 (1991).

135. I. Minami, Y. Ohashi, I. Shimizu, and J. Tsuji, *Tetrahedron Lett.*, **26**, 2449 (1985); J. P. Genet and D. Ferroud, *Tetrahedron Lett.*, **25**, 3579 (1984).

135a. J. P. Genet and D. Ferroud, *Tetrahedron Lett.*, **25**, 3579 (1984).

136. E. Blart, J. P. Genet, M. Safi, M. Savignac, and D. Sinou, *Synlett*, 715 (1992); *Tetrahedron*, **50**, 505 (1994).

137. M. Moreno-Manas, R. Pleixats, and M. Villarroya, *J. Org. Chem.*, **55**, 4925 (1990).

138. E. Bernocchi, S. Cacchi, E. Morera, and G. Ortar, *Synlett*, 161 (1992).

139. I. Ikeda, X. P. Gu, T. Okuhara, and M. Okahara, *Synthesis*, 535 (1988); 32 (1990).

140. T. Hirao, J. Enda, Y. Ohshiro, and T. Agawa, *Tetrahedron Lett.*, **22**, 3079 (1981).

141. B. M. Trost and C. R. Self, *J. Am. Chem. Soc.*, **105**, 5942 (1983).

142. J. Tsuji, M. Yuhara, M. Minato, H. Yamada, F. Sato, and Y. Kobayashi, *Tetrahedron Lett.*, **29**, 343 (1988).

143. H. Inami, T. Ito, H. Urabe, and F. Sato, *Tetrahedron Lett.*, **34**, 5919 (1993).

144. H. Nemoto, Y. Kubota, and Y. Yamamoto, *J. Org. Chem.*, **55**, 4515 (1990).

145. Y. Tsuji, N. Yamada, and S. Tanaka, *J. Org. Chem.*, **58**, 16 (1993).

146. H. Nemoto, F. G. Rong, and Y. Yamamoto, *J. Org. Chem.*, **55**, 6065 (1990).

147. C. Yan, L. N. Sheng, and D. M. Zhi, *Tetrahedron Lett.*, **31**, 2405 (1990).

148. I. Shimizu, Y. Ohashi, and J. Tsuji, *Tetrahedron Lett.*, **25**, 5157 (1984).

149. P. Breuilles and D. Uguen, *Tetrahedron Lett.*, **28**, 6053 (1987).

150. A. Yamamoto, Y. Ito, and T. Hayashi, *Tetrahedron Lett.*, **30**, 375 (1989).

151. P. Breuilles and D. Uguen, *Tetrahedron Lett.*, **29**, 201 (1988).

152. K. Ohe, H. Matsuda, T. Ishihara, S. Ogoshi, N. Chatani, and S. Murai, *J. Org. Chem.*, **58**, 1173 (1993).

153. I. Shimizu, Y. Ohashi, and J. Tsuji, *Tetrahedron Lett.*, **26**, 3825 (1985).

154. K. Yamamoto, T. Ishida, and J. Tsuji, *Chem. Lett.*, 1157 (1987).

155. T. Mitsudo, M. Kadokura, and Y. Watanabe, *Chem. Commun.*, 1539 (1986); *J. Org. Chem.*, **52**, 1695 (1987).

156. T. Takahashi, A. Ootake, and J. Tsuji, *Tetrahedron Lett.*, **25**, 1921 (1984); *Tetrahedron*, **41**, 5747 (1985).

157. T. Takahashi, H. Kataoka, and J. Tsuji, *J. Am. Chem. Soc.*, **105**, 147 (1983); T. Takahashi, M. Miyazawa, H. Ueno, and J. Tsuji, *Tetrahedron Lett.*, **27**, 3881 (1986).

158. J. Wicha and M. M. Kabat, *Chem. Commun.*, 985 (1983).

159. B. M. Trost, B. A. Vos, C. M. Brezezowski, and D. P. Martina, *Tetrahedron Lett.*, **33**, 717 (1992).

160. A. S. Kende, I. Kaldor, and R. Aslanian, *J. Am. Chem. Soc.*, **110**, 6265 (1988).

161. B. M. Trost and R. W. Warner, *J. Am. Chem. Soc.*, **105**, 5940 (1983).

162. B. M. Trost, J. T. Hane, and P. Metz, *Tetrahedron Lett.*, **27**, 5695 (1986).

163. T. Fujinami, T. Suzuki, M. Kamiya, S. Fukuzawa, and S. Sakai, *Chem. Lett.*, 199 (1985).
164. B. M. Trost and S. R. Angle, *J. Am. Chem. Soc.*, **107**, 6123 (1985).
165. B. M. Trost, J. K. Lynch, and S. R. Angle, *Tetrahedron Lett.*, **28**, 375 (1987).
166. S. Wershofen and H. D. Scharf, *Synthesis*, 854 (1988).
167. S. Suzuki, Y. Fujita, Y. Kobayashi, and F. Sato, *Tetrahedron Lett.*, **27**, 69 (1986).
168. R. C. Larock and S. K. Stolz-Dunn, *Tetrahedron Lett.*, **30**, 3487 (1989).
169. Y. Kobayashi and J. Tsuji, *Tetrahedron Lett.*, **22**, 4295 (1981).
170. J. Tsuji, Y. Kobayashi, H. Kataoka, and T. Takahashi, *Tetrahedron Lett.*, **21**, 3393 (1980).
171. J. Tsuji, H. Okumoto, Y. Kobayashi, and T. Takahashi, *Tetrahedron Lett.*, **22**, 1357 (1981).
172. M. Brakta, P. Lhoste, and D. Sinou, *J. Org. Chem.*, **54**, 1890 (1989).
173. T. V. Rajanbabu, *J. Org. Chem.*, **50**, 3642 (1985).
174. N. Nagato, *Jpn. Pat. Kokai*, Hei-1-165555, 1-153660.
175. B. M. Trost and J. P. Genet, *J. Am. Chem. Soc.*, **98**, 8516 (1976); J. P. Genet, M. Balabane, J. E. Backvall, and J. E. Nystrom, *Tetrahedron Lett.*, **24**, 2745 (1983).
176. S. A. Godleski, J. D. Meinhart, D. J. Miller, and S. V. Wallendael, *Tetrahedron Lett.*, **22**, 2247 (1981); S. A. Godleski, D. J. Heathcock, J. D. Meinhart, and S. V. Wallendael, *J. Org. Chem.*, **48**, 2101 (1983).
177. W. Carruthers and S. A. Cumming, *J. Chem. Soc., Perkin Trans. 1*, 2386 (1983); *Chem. Commun.*, 360 (1983).
178. J. P. Genet, M. Balabane, J. E. Bäckvall, and J. E. Nystrom, *Tetrahedron Lett.*, **24**, 2745 (1983); J. E. Nystrom, T. Rein, and J. E. Backvall, *Org. Synth.*, **67**, 105 (1988).
179. J. P. Genet, E. Blart, and M. Savignac, *Synlett*, 715 (1992).
180. B. M. Trost and J. Cossy, *J. Am. Chem. Soc.*, **104**, 6881 (1982).
181. V. Bolitt, B. Chaquir, and D. Sinou, *Tetrahedron Lett.*, **33**, 2481 (1992).
182. L. Gundersen, T. Bennehe, and K. Undheim, *Tetrahedron Lett.*, **33**, 1085 (1992).
183. K. Fugami, Y. Morizawa, K. Oshima, and H. Nozaki, *Tetrahedron Lett.*, **26**, 857 (1985).
184. R. D. Connell, T. Rein, B. Akermark, and P. Helquist, *J. Org. Chem.*, **53**, 3845 (1988).
185. S. Murahashi, Y. Taniguchi, Y. Imada, and Y. Tanigawa, *J. Org. Chem.*, **54**, 3292 (1989); S. Murahashi, Y. Tanigawa, Y. Imada, and Y. Taniguchi, *Tetrahedron Lett.*, **27**, 227 (1986).
186. P. Aufranc, J. Ollivier, A. Stolle, C. Bremer, M. Es-Sayed, A. de Meijere, and J. Salaun, *Tetrahedron Lett.*, **34**, 4191 (1993).
187. M. Safi, R. Fahrnag, and D. Sinou, *Tetrahedron Lett.*, **31**, 527 (1990).
188. A. Tenaglia and B. Waegell, *Tetrahedron Lett.*, **29**, 4851 (1988).
189. B. M. Trost, G. H. Kuo, and T. Benneche, *J. Am. Chem. Soc.*, **110**, 621 (1988).
190. D. R. Deardorff, M. J. Shulman, and J. E. Sheppeck, *Tetrahedron Lett.*, **30**, 6625 (1989).
191. M. R. Peel, D. D. Sternbach, and M. R. Johnson, *J. Org. Chem.*, **56**, 4990 (1991).
192. B. M. Trost and A. R. Sudhakar, *J. Am. Chem. Soc.*, **109**, 3792 (1987).
193. B. M. Trost and A. R. Sudhakar, *J. Am. Chem. Soc.*, **110**, 7933 (1988).
194. D. Xu and K. B. Sharpless, *Tetrahedron Lett.*, **34**, 951 (1993).
195. T. Bando, H. Harayama, Y. Fukazawa, M. Shiro, K. Fugami, S. Tanaka, and Y. Tamaru, *J. Org. Chem.*, **59**, 1465 (1994).
196. S. E. Bystrom, R. Aslanian, and J. E. Bäckvall, *Tetrahedron Lett.*, **26**, 1749 (1985).
197. M. Takagi and K. Yamamoto, *Chem. Lett.*, 2123 (1989).

198. Y. Inoue, M. Taguchi, M. Toyofuku, and H. Hashimoto, *Bull. Chem. Soc. Jpn.*, **57**, 3021 (1984).
199. C. R. Johnson, P. A. Ple, L. Su, M. J. Heeg, and J. P. Adams, *Synlett*, 388 (1992).
200. R. D. Connel, T. Rein, B. Akermark, and P. Helquist, *J. Org. Chem.*, **53**, 3845 (1988).
201. S. Murahashi, Y. Imada, Y. Taniguchi, and Y. Kodera, *Tetrahedron Lett.*, **29**, 2973 (1988).
202. J. P. Genet, S. Thorimbert, and A. M. Touzin, *Tetrahedron Lett.*, **34**, 1159 (1993); G. P. Genet, S. Thorimbert, S. Mallart, and N. Kardos, *Synthesis*, 321 (1993).
203. B. M. Trost and E. Keinan, *J. Org. Chem.*, **44**, 3451 (1979).
204. M. Moreno-Manas, R. Pleixats, and M. Villarroya, *Tetrahedron*, **49**, 1457, 1465 (1993).
205. F. Guibe and Y. Saint M'Leux, *Tetrahedron Lett.*, **22**, 3591 (1981).
206. R. Lakhmiri, P. Lhoste, and D. Sinou, *Synth. Commun.*, **20**, 1551 (1990); R. Lakhmiri, P. Lhoste, P. Boullanger, and D. Sinou, *J. Chem. Res.* (S), 342 (1990).
207. A. P. Davis, B. J. Dorgan, and E. R. Mageean, *Chem. Commun.*, 492 (1993).
208. G. Stork and J. M. Poirier, *J. Am. Chem. Soc.*, **105**, 1073 (1983).
209. T. Suzuki, O. Sato, and M. Hirama, *Tetrahedron Lett.*, **31**, 4747 (1990);
210. T. Suzuki, O. Sato, H. Hirama, Y. Yamamoto, M. Murata, T. Yasumoto, and N. Harada, *Tetrahedron Lett.*, **32**, 4505 (1991).
211. E. Keinan, M.Sahai, Z. Roth, A. Nudelman, and H. Herzig, *J. Org. Chem.*, **50**, 3558 (1985); B. M. Trost and A. Tenaglia, *Tetrahedron Lett.*, **29**, 2931 (1988).
212. D. R. Deardorff and D. C. Myles, *Org. Synth.*, **67**, 114 (1988); D. R. Deardorff, D. C. Myles, and K. D. MacFerrin, *Tetrahedron Lett.*, **26**, 5615 (1985).
213. C. Goux, P. Lhoste, and D. Sinou, *Synlett*, 725 (1992).
214. J. Muzart, J. P. Genet, and A. Denis, *J. Organomet. Chem.*, **326**, C23 (1987).
215. J. C. Fiaud, *Chem. Commun.*, 1055 (1983).
216. M. Moreno-Manas and A. Trius, *Tetrahedron Lett.*, 3109 (1981); *Bull. Chem. Soc. Jpn.*, **56**, 2154 (1983); M. Moreno-Manas, R. M. Ortuno, M. Prat, and M. A. Galan, *Synth. Commun.*, **16**, 1003 (1986).
217. N. Okukado, O. Uchikawa, and Y. Nakamura, *Chem. Lett.*, 1449 (1988).
218. Y. Inoue, M. Toyofuku, and H. Hashimoto, *Bull. Chem. Soc. Jpn.*, **59**, 1279 (1979).
219. Y. Tsukahara, H. Kinoshita, K. Inomata, and H. Kotake, *Bull. Chem. Soc. Jpn.*, **57**, 3013 (1984).
220. R. Tamura, M. Kato, K. Saegusa, M. Kakihara, and D. Oda, *J. Org. Chem.*, **52**, 4121 (1987).
221. R. Malet, M. Moreno-Manas, and R. Pleixats, *Synth. Commun.*, **22**, 2219 (1992).
222. B. M. Trost and T. S. Scanlan, *Tetrahedron Lett.*, **27**, 4141 (1986).
223. P. R. Auburn, J. Whelan, and B. Bosnich, *Chem. Commun.*, 146 (1986).
224. X. Lu and Z. Ni, *Synthesis*, 66 (1987).
225. C. Goux, P. Lhoste, and D. Sinou, *Tetrahedron Lett.*, **33**, 8099 (1992).
226. Y. Yamada, K. Mukai, H. Yoshioka, Y. Tamaru, and Z. Yoshida, *Tetrahedron Lett.*, 5015 (1979); Y. Tamaru, Z. Yoshida, Y. Yamada, K. Mukai, and H. Yoshioka, *J. Org. Chem.*, **48**, 1293 (1983).
227. K. Inomata, T. Yamamoto, and H. Kotake, *Chem. Lett.*, 1357 (1981).
227a. Review: U. M. Dzhemilev and R. V. Kunakova, *J. Organomet. Chem.*, **455**, 1 (1993); U. M. Dzhemilev, R. V. Kunakova, and R. L. Gaisin, *Izv. Akad. Nauk SSR, Ser. Khim.*, 696 (1981); 157 (1983).
228. R. Tamura, K. Hayashi, M. Kakihana, M. Tsuji, and D. Oda, *Tetrahedron Lett.*, **26**, 851 (1985); *Chem. Lett.*, 229 (1985).

229. N. Ono, I. Hamamoto, T. Yanai, and A. Kaji, *Chem. Commun.*, 523 (1985).
230. R. Tamura, M. Kato, K. Saegusa, D. Oda, T. Egawa, and T. Yamamoto, *J. Org. Chem.*, **52**, 1640 (1987).
231. K. Hiroi, R. Kitayama, and S. Sato, *Chem. Commun.*, 303 (1984); *Chem. Lett.*, 929 (1984); K. Hiroi, M. Yamamoto, Y. Kurihara, and H. Yonezawa, *Tetrahedron Lett.*, **31**, 2619 (1990).
232. K. Hiroi and K. Makino, *Chem. Lett.*, 617 (1986).
233. S. Fukuzawa, T. Fujinami, and S. Sakai, *Chem. Lett.*, 927 (1990).
234. J. Tsuji, M. Morikawa, and J. Kiji, *Tetrahedron Lett.*, 1811 (1963); *J. Am. Chem. Soc.*, **86**, 4350 (1964).
235. W. T. Dent, R. Long, and G. H. Whitefield, *J. Chem. Soc.*, 1988 (1964).
236. J. Kiji, T. Okano, W. Nishiumi, and H. Konishi, *Chem. Lett.*, 957 (1988); T. Okano, N. Okabe, and J. Kiji, *Bull. Chem. Soc. Jpn.*, **65**, 2589 (1992).
237. J. Kiji, T. Okano, H. Konishi, and W. Nishiumi, *Chem. Lett.*, 1873 (1989).
238. E. Negishi, G. Wu, and J. M. Tour, *Tetrahedron Lett.*, **29**, 6745 (1989).
239. J. Tsuji, K. Sato, and H. Okumoto, *Tetrahedron Lett.*, **23**, 5189 (1982); *J. Org. Chem.*, **49**, 1341 (1984).
240. K. Yamamoto, M. Terakado, K. Murai, M. Murayama, J. Tsuji, K. Takahashi, and K. Mikami, *Chem. Lett.*, 955 (1989).
241. K. Yamamoto, R. Deguchi, and J. Tsuji, *Bull. Chem. Soc. Jpn.*, **58**, 3397 (1985); S. Z. Wang, K. Yamamoto, H. Yamada, and T. Takahashi, *Tetrahedron*, **48**, 2333 (1992).
242. Y. Takahashi, K. Tsujiyama, S. Sakai, and Y. Ishii, *Tetrahedron Lett.*, 1913 (1970).
243. T. Yamamoto, O. Saito, and A. Yamamoto, *J. Am. Chem. Soc.*, **103**, 5601 (1981).
244. S. Murahashi, Y. Imada, Y. Taniguchi, and S. Higashiura, *Tetrahedron Lett.*, **29**, 4945 (1988); *J. Org. Chem.*, **58**, 1538 (1993).
245. Y. Ishii, C. Gao, M. Iwasaki, and M. Hidai, *Chem. Commun.*, 695 (1991); *J. Org. Chem.*, **58**, 6818 (1993).
246. S. Duprat, H. Deweerdt, J. Jenck, and P. Kalck, *J. Mol. Catal.*, **80**, L9 (1993).
247. S. Torii, H. Okumoto, M. Sadakane, A. K. M. Abdul Hai, and H. Tanaka, *Tetrahedron Lett.*, **34**, 6553 (1993); *J. Org. Chem.*, **59**, 3040 (1994).
248. J. Tsuji, J. Kiji, S. Imamura, and M. Morikawa, *J. Am. Chem. Soc.*, **86**, 4350 (1964).
249. K. Itoh, N. Hamaguchi, M. Miura, and M. Nomura, *J. Mol. Catal.*, **75**, 117 (1992).
250. S. Imamura and J. Tsuji, *Tetrahedron*, **25**, 4187 (1969).
251. M. C. Bonnet, D. Neibecker, B. Stitou, and I. Tkatchenko, *J. Organomet. Chem.*, **366**, C9 (1989); D. Neibecker, J. Poirier, and I. Takatchenko, *J. Org. Chem.*, **54**, 2459 (1989); M.C. Bonnet, J. Coombes, B. Manzano, D. Neibecker, and I. Tkatchenko, *J. Mol. Catal.*, **52**, 263 (1989).
252. S. Murahashi, Y. Imada, and K. Nishimura, *Chem. Commun.*, 1578 (1988); *Tetrahedron*, **50**, 453 (1994).
253. S. Z. Wang, H. Takeda, and K. Yamamoto, *Synlett*, 323 (1992).
254. G. W. Spears, K. Nakanishi, and Y. Ohfune, *Synlett*, 91 (1991).
255. D. Medema, R.Van Helden, and C. F. Kohll, *Inorg. Chim. Acta*, **3**, 255 (1969).
256. J. H. Merrifield, J. P. Godschalx, and J. K. Stille, *Organometallics*, **3**, 1108 (1984).
257. F. K. Sheffy and J. K. Stille, *J. Am. Chem. Soc.*, **105**, 7173 (1983).
258 F. K. Sheffy, J. Godschalx, and J. K. Stille, *J. Am. Chem. Soc.*, **106**, 4833 (1984).
259. S. Katsumura, S. Fujiwara, and S. Isoe, *Tetrahedron Lett.*, **28**, 1191 (1987); **29**, 1173 (1988).

260. Y. Tamaru, K. Yasui, H. Takanabe, S. Tanaka, and K. Fugami, *Angew. Chem., Int. Ed. Engl.*, **31**, 645 (1992).
261. V. P. Baillargeon and J. K. Stille, *J. Am. Chem. Soc.*, **105**, 7175 (1983); **108**, 452 (1986).
262. A. Kasahara, T. Izumi, and H. Yanai, *Chem. Ind. (London)*, 898 (1983).
263. Y. Ishii and M. Hidai, *J. Organomet. Chem.*, **428**, 289 (1992).
264. H. Matsuzaka, Y. Hiroe, M. Iwasaki, Y. Ishii, Y. Koyasu, and M. Hidai, *Chem. Lett.*, 377 (1988); *Chem. Commun.*, 575 (1987); *J. Org. Chem.*, **53**, 3832 (1988).
265. M. Iwasaki, H. Matsuzaka, Y. Hiroe, Y. Ishii, Y. Koyasu, and M. Hidai, *Chem. Lett.*, 1159 (1988).
266. M. Iwasaki, J. Li, Y. Kobayashi, H. Matsuzaka, Y. Ishii, and M. Hidai, *Tetrahedron Lett.*, **30**, 95 (1989); *J. Org. Chem.*, **56**, 1922 (1991).
267. Review, K. Tamao, in *Comprehensive Organic Synthesis*, Vol. 3, Pergamon Press, Oxford, 1991, p. 468.
268. E. Negishi, S. Chatterjee, and H. Matsushita, *Tetrahedron Lett.*, **22**, 3737 (1981).
269. T. Hayashi, M. Konishi, and M. Kumada, *J. Organomet. Chem.*, **186**, C1 (1980); T. Hayashi, M. Konishi, K. Yokota, and M. Kumada, *Chem. Commun.*, 313 (1981); *J. Organomet. Chem.*, **285**, 359 (1985).
270. H. Okamura and H. Takei, *Tetrahedron Lett.*, 3425 (1979).
271. T. Kitazume and N. Ishikawa, *Chem. Lett.*, 137 (1982).
272. Review, E. Erdik, *Tetrahedron*, **48**, 9577 (1992).
273. S. Sasaoka, T. Yamamoto, H. Kinoshita, K. Inomata, and H. Kotake, *Chem. Lett.*, 315 (1985).
274. E. Negishi and H. Matsushita, *Org. Synth., Coll. Vol.* **7**, 245 (1990); *J. Am. Chem. Soc.*, **103**, 2882 (1981).
275. S. Chatterjee and E. Negishi, *J. Org. Chem.*, **50**, 3406 (1985).
276. G. P. Boldrini, M. Mengoli, E. Tagliavini, C. Trombini, and A. Umani-Ronchi, *Tetrahedron Lett.*, **27**, 4223 (1986).
277. N. Miyaura, T. Yano, and A. Suzuki, *Tetrahedron Lett.*, **21**, 2865 (1980).
278. N. Miyaura, H. Suginome, and A. Suzuki, *Tetrahedron Lett.*, **25**, 761 (1984).
279. N. Miyaura, Y. Tanabe, H. Suginome, and A. Suzuki, *J. Organomet. Chem.*, **233**, C13 (1982).
280. J. Y. Legros and J. C. Fiaud, *Tetrahedron Lett.*, **31**, 7453 (1990).
281. J. Godschalx and J. K. Stille, *Tetrahedron Lett.*, **21**, 2599 (1980); **24**, 1905 (1983).
282. E. T. Luo and E. Negishi, *Tetrahedron Lett.*, **26**, 2177 (1985); *J. Org. Chem.*, **50**, 4762 (1985); E. Negishi and R. A. John, *J. Org. Chem.*, **48**, 4098 (1983).
283. M. W. Hutzinger and A. C. Oehlschlager, *J. Org. Chem.*, **56**, 2918 (1991).
284. S. Chatterjee and E. Negishi, *J. Organomet. Chem.*, **285**, C1 (1985).
285. Review, J. K. Stille, *Angew. Chem., Int. Ed. Engl.*, **25**, 508 (1986).
286. Review, T. N. Mitchell, *Synthesis*, 803 (1992).
287. N. A. Bumagin, I. G. Bumagina, A. N. Kashin, and I. P. Beletskaya, *Zh. Obshch. Khim.*, **52**, 714 (1982).
288. V. Farina, S. R. Baker, D. A. Benigni, and C. Sapino, Jr, *Tetrahedron Lett.*, **29**, 5739 (1988).
289. M. Kosugi, K. Ohashi, K. Akazawa, T. Kawazoe, H. Sano, and T. Migita, *Chem. Lett.*, 1237 (1987).
290. B. M. Trost and E. Keinan, *Tetrahedron Lett.*, **21**, 2591, 2595 (1980).
291. I. P. Beletskaya, A. N. Kasatkin, S. A. Lebedev, and N. A. Bumagin, *Izv. Akad. Nauk SSSR, Ser. Khim.*, 2414 (1981).
292. L. Del Valle, J. K. Stille, and L. S. Hegedus, *J. Org. Chem.*, **55**, 3019 (1990).

293. A. M. Echavarren, D. R. Tueting, and J. K. Stille, *Tetrahedron Lett.*, **28**, 5627 (1987); *J. Am. Chem. Soc.*, **110**, 4039 (1988).
294. D. R. Tueting, A. M. Eschavarren, and J. K. Stille, *Tetrahedron*, **45**, 979 (1989).
295. J. Tsuji, I. Minami, and I. Shimizu, *Tetrahedron Lett.*, **24**, 4713 (1983).
296. M. Pereyre, B. Bellegrade, J. Mendelsohn, and J. Valade, *J. Organomet. Chem.*, **11**, 97 (1968).
297. M. Kosugi, H. Naka, S. Harada, H. Sano, and T. Migita, *Chem. Lett.*, 1371 (1987).
298. Y. Tsuji, N. Yamada, and S. Tanaka, *J. Org. Chem.*, **58**, 16 (1993).
299. J. Yoshida, K. Tamao, M. Takahashi, and M. Kumada, *Tetrahedron Lett.*, 2161 (1978); K. Tamao, T. Kakui, and M. Kumada, *Tetrahedron Lett.*, 619 (1978).
300. J. Tsuji, K. Takahashi, I. Minami, and I. Shimizu, *Tetrahedron Lett.*, **25**, 4783 (1984).
301. T. Baba, K. Nakano, S. Nishiyama, S. Tsuruya, and M. Masai, *Chem. Commun.*, 348 (1990).
302. J. Tsuji, I. Minami, and I. Shimizu, *Chem. Lett.*, 1325 (1983).
303. C. Carfagna, L. Mariani, A. Musco, G. Sallese, and R. Santi, *J. Org. Chem.*, **56**, 3924 (1991); C. Carfagna, R. Galarini, A. Musco, and R. Santi, *J. Mol. Catal.*, **72**, 19 (1992).
304. L. S. Hegedus, W. H. Darlington, and C. E. Russel, *J. Org. Chem.*, **45**, 5193 (1980).
305. H. M. R. Hoffmann, A. R. Otte, and A. Wilde, *Angew. Chem., Int. Ed. Engl.*, **31**, 234 (1992).
306. A. Wilde, A. R. Otte, and H. M. R. Hoffmann, *Chem. Commun.*, 615 (1993).
307. M. Formica, A. Musco, and R. Pontelloni, *J. Mol. Catal.*, **84**, 239   (1993).
308. T. Tabuchi, J. Inanaga, and M. Yamaguchi, *Tetrahedron Lett.*, **28**, 215 (1987).
309. J. Yoshida, H. Funahashi, H. Iwasaki, and N. Kawabata, *Tetrahedron Lett.*, **27**, 4469 (1986).
310. S. Matsubara, K.Wakamatsu, Y. Morizawa, N. Tsuboniwa, K. Oshima, and H. Nozaki, *Bull. Chem. Soc. Jpn.*, **58**, 1196 (1985).
311. B. M. Trost and J. W. Herndon, *J. Am. Chem. Soc.*, **106**, 6835 (1984).
312. B. M. Trost and R. Walchli, *J. Am. Chem. Soc.*, **109**, 3487 (1987).
313. N. A. Bumagin, A. N. Kasatkin, and I. P. Beletskaya, *Izv. Akad. Nauk SSSR, Ser. Khim.*, 636 (1984); *Chem. Abstr.*, **101**, 91109 (1984).
314. B. M. Trost and K. M. Pietrusiewicz, *Tetrahedron Lett.*, **26**, 4039 (1985).
315. Y. Yokoyama, S. Ito, Y. Takahashi, and Y. Murakami, *Tetrahedron Lett.*, **26**, 6457 (1985).
316. Y. Masuyama, R. Hayashi, K. Otake, and Y. Kurusu, *Chem. Commun.*, 44 (1988).
317. J. Takahara, Y. Masuyama, and Y. Kurusu, *J. Am. Chem. Soc.*, **110**, 4473 (1988); **114**, 2577 (1992).
318. K. Yasui, Y. Goto, T. Yajima, Y. Taniseki, K. Fugami, A. Tanaka, and Y. Tamaru, *Tetrahedron Lett.*, **34**, 7619 (1993).
319. Y. Masuyama, N. Kinugawa, and Y. Kurusu, *J. Org. Chem.*, **52**, 3702 (1987).
320. P. Zhang, W. Zhang, T. Zhang, Z. Wang, and W. Zhou, *Chem. Commun.*, 491 (1991).
321. T. Tabuchi, J. Inanaga, and M. Yamaguchi, *Tetrahedron Lett.*, **27**, 1195 (1986).
322. H. Matsumoto, T. Yako, S. Nagashima, T. Motegi, and Y. Nagai, *J. Organomet. Chem.*, **148**, 97 (1978).
323. R. Calas, J. Dunogues, G. Deleris, and N. Duffaurt, *J. Organomet. Chem.*, **225**, 117 (1982).

324. H. Urata, H. Suzuki, Y. Morooka , and T. Ikawa, *Bull. Chem. Soc. Jpn.*, **57**, 607 (1984).
325. Y. Tsuji, S. Kajita, S. Isobe, and M. Funato, *J. Org. Chem.*, **58**, 3607 (1993).
326. Y. Matsumoto, A. Ohno, and T. Hayashi, *Organometallics*, **12**, 4051 (1993).
327. Y. Okuda, M. Sato, K. Oshima, and H. Nozaki, *Tetrahedron Lett.*, **24**, 2015 (1983).
328. B. M. Trost, J. Yoshida, and M. Lautens, *J. Am. Chem. Soc.*, **105**, 4494 (1983).
329. J. Tsuji, T. Yamakawa, M. Kaito, and T. Mandai, *Tetrahedron Lett.*, 2075 (1978).
330. T. Mandai, H. Yasuda, M. Kaito, J. Tsuji, R. Yamaoka, and H. Fukami, *Tetrahedron*, **35**, 309 (1979).
331. I. C. Coste-Maniere, J. P. Zahra, and B. Waegell, *Tetrahedron Lett.*, **29**, 1017 (1988).
332. F. M. Hauser and D. Mal, *J. Am. Chem. Soc.*, **106**, 1098 (1984).
333. T. Mandai, J. Gotoh, J. Otera, and M. Kawada, *Chem. Lett.*, 313 (1980).
334. B. M. Trost, T. R. Verhoeven, and J. M. Fortunak, *Tetrahedron Lett.*, 2301 (1979).
335. H. Matsushita and E. Negishi, *J. Org. Chem.*, **47**, 4161 (1982).
336. H. Kumobayashi, S. Mitsuhashi, S. Akutagawa, and S. Otsuka, *Chem. Lett.*, 157 (1986).
337. A. J. Chalk, V. Wertheimer, and S. A. Magennis, *J. Mol. Catal.*, **19**, 189 (1983).
338. S. Suzuki, Y. Fujita, and T. Nishida, *Tetrahedron Lett.*, **24**, 5737 (1983).
339. K. Inomata, Y. Murata, H. Kato, Y. Tsukahara, H. Kinoshita, and H. Kotake, *Chem. Lett.*, 931 (1985).
340. T. Mandai, T. Matsumoto, Y. Nakao, H. Teramoto, M. Kawada, and J. Tsuji, *Tetrahedron Lett.*, **33**, 2549 (1992).
341. T. Mandai, T. Matsumoto, J. Tsuji, and S. Saito, *Tetrahedron Lett.*, **34**, 2513 (1993).
342. T. Takahashi, N. Nakagawa, T. Minoshima, H. Yamada, and J. Tsuji, *Tetrahedron Lett.*, **31**, 4041 (1990).
343. S. Takano, M. Moriya, and K. Ogasawara, *Synlett*, 601 (1993); S. Takano, Y. Higashi, T. Kamikubo, M. Moriya, and K. Ogasawara, *Chem. Commun.*, 788 (1993).
344. A. F. Noels, J. J. Herman, and P. Teyssie, *J. Org. Chem.*, **41**, 2527 (1976).
345. B. M. Trost and J. M. Fortunak, *J. Am. Chem. Soc.*, **102**, 2841 (1980).
346. B. M. Trost and J. M. Fortunak, *Tetrahedron Lett.*, **22**, 3459 (1981).
347. B. M. Trost and G. B. Tometzki, *J. Org. Chem.*, **53**, 915 (1988).
348. M. Suzuki, Y. Oda, and R. Noyori, *J. Am. Chem. Soc.*, **101**, 1623 (1979).
349. S. Achab, J. P. Cosson, and B. C. Das, *Chem. Commun.*, 1040 (1984).
350. M. Suzuki, A. Watanabe, and R. Noyori, *J. Am. Chem. Soc.*, **102**, 2095 (1980).
351. M. Suzuki, Y. Oda, and R. Noyori, *Tetrahedron Lett.*, **22**, 4413 (1981).
352. E. Keinan and M. Sahai, *J. Org. Chem.*, **48**, 2550 (1983).
353. J. Tsuji, I. Minami, and I. Shimizu, *Tetrahedron Lett.*, **24**, 5635 (1983).
354. I. Minami, K. Takahashi, I. Shimizu, T. Kimura, and J. Tsuji, *Tetrahedron*, **42**, 2971 (1986).
355. T. Baba, K. Nakano, S. Nishiyama, S. Tsuruya, and M. Masai, *Chem. Commun.*, 1697 (1989).
356. J. Tsuji, K. Takahashi, I. Minami, and I. Shimizu, *Tetrahedron Lett.*, **25**, 4783 (1984).
357. K. Tsushima and A. Murai, *Chem. Lett.*, 761 (1990).
358. J. Tsuji, I. Minami, and I. Shimizu, *Tetrahedron Lett.*, **24**, 5639 (1983).
359. J. Tsuji, I. Minami, and I. Shimizu, *Tetrahedron Lett.*, **25**, 279 (1984); *Tetrahedron*, **43**, 3903 (1987).

360. A. S. Smith, G. A. Sulikpwski, and K. Fujimoto, *J. Am. Chem. Soc.*, **111**, 8039 (1989).
360a. Review: T. Tsuji and T. Mandai, Synthesis (1996), in press.
361. J. Tsuji and T. Yamakawa, *Tetrahedron Lett.*, 613 (1979).
362. J. Tsuji, I. Minami, and I. Shimizu, *Synthesis*, 623 (1986); *Chem. Lett.*, 1017 (1984).
363. D. N. Jones and S. D. Knox, *Chem. Commun.*, 165 (1975).
364. H. Kotake, T. Yamamoto, and H. Kinoshita, *Chem. Lett.*, 1331 (1982).
365. M. Mohri, H. Kinoshita, K. Inomata, and H. Kotake, *Chem. Lett.*, 451 (1985).
366. R. O. Hutchins, K. Learn, and R. P. Fulton, *Tetrahedron Lett.*, **21**, 27 (1980).
367. R. O. Hutchins and K. Learn, *J. Org. Chem.*, **47**, 4380 (1982).
368. E. Keinan and N. Greenspoon, *J. Org. Chem.*, **48**, 3545 (1983); **53**, 3723 (1988).
369. E. Keinan and N. Greenspoon, *Tetrahedron Lett.*, **23**, 241 (1982).
370. F. Guibe and Y. Saint M'Leux, *Tetrahedron Lett.*, **22**, 3591 (1981).
371. F. Guibe, Y. T. Xian, A. M. Zigna, and G. Balavoine, *Tetrahedron Lett.*, **26**, 3559 (1985).
372. F. Guibe, H. X. Xian, and G. Balavoine, *J. Organomet. Chem.*, **306**, 257 and 267 (1986).
373. T. Tabuchi, J. Inanaga, and M. Yamaguchi, *Tetrahedron Lett.*, **27**, 601 (1986).
374. K. Nakamura, A. Ohno, and S. Oka, *Tetrahedron Lett.*, **24**, 3335 (1983).
375. S. Torii, H. Tanaka, T. Katoh, and K. Morisaki, *Tetrahedron Lett.*, **25**, 3207 (1984).
376. H. Hey and H. J. Arpe, *Angew. Chem., Int. Ed. Engl.*, **12**, 928 (1973).
377. T. Takahashi, M. Miyazawa, H. Ueno, and J. Tsuji, *Tetrahedron Lett.*, **27**, 3881 (1986).
378. T. Okano, Y. Moriyama, H. Konishi, and J. Kiji, *Chem. Lett.*, 1463 (1986).
379. M. Oshima, I. Shimizu, A. Yamamoto, and F. Ozawa, *Organometallics*, **10**, 1221 (1991).
380. T. Mandai, T. Matsumoto, and J. Tsuji, *Synlett*, 113 (1993).
381. C. Mukaiyama, H. Mitsuhashi, and T. Wakamatsu, *Tetrahedron Lett.*, **33**, 4165 (1992).
382. T. Mandai and J. Tsuji, unpublished results.
383. A. Yasuda, S. Tanaka, K. Oshima, H. Yamamoto, and H. Nozaki, *J. Am. Chem. Soc.*, **96**, 6513 (1974).
384. T. Mandai, T. Murakami, T. Suzuki, M. Fujita, M. Kawada, and J. Tsuji, *Tetrahedron Lett.*, **33**, 2987 (1992).
385. M. M. Midland and Y. C, Kwon, *J. Am. Chem. Soc.*, **105**, 3725 (1983).
386. A. Barco, S. Benetti, and G. Spalluto, *J. Org. Chem.*, **57**, 6279 (1992).
387. T. Hayashi, H. Iwamura, M. Naito, Y. Matsumoto, Y. Uozumi, M. Miki, and K. Yanagi, *J. Am. Chem. Soc.*, **116**, 775 (1994); *Synthesis*, 526 (1994).
388. T. Mandai and J. Tsuji, unpublished results.
389. T. Mandai, T. Matsumoto, M. Kawada, and J. Tsuji, *J. Org. Chem.*, **57**, 1326 (1992); *Tetrahedron*, **49**, 5483 (1993).
390. L. F. Tietze and P. S. V. Subba Rao, *Synlett*, 291 (1993).
391. T. Mandai, T. Matsumoto, M. Kawada, and J. Tsuji, *J. Org. Chem.*, **57**, 6090 (1992); *Tetrahedron*, **50**, 475 (1994).
392. F. A. Cotton and G. Wilkinson, *Advanced Inorganic Chemistry*, 5th Ed. Wiley–Interscience, New York, 1988, p. 77.
393. M. H. Rabinowitz, *Tetrahedron Lett.*, **32**, 6081 (1991).
394. I. Shimizu, M. Oshima, M. Nisar, and J. Tsuji, *Chem. Lett.*, 1775 (1986).
395. M. Oshima, H. Yamazaki, I. Shimizu, M. Nisar, and J. Tsuji, *J. Am. Chem. Soc.*, **111**, 6280 (1989).

396. I. Shimizu, K. Hayashi, and M. Oshima, *Tetrahedron Lett.*, **31**, 4757 (1990).
397. I. Shimizu and H. Yamazaki, *Chem. Lett.*, 777 (1990).
398. I. Shimizu and F. Aida, *Chem. Lett.*, 601 (1988).
399. I. Shimizu and T. Ishikawa, *Tetrahedron Lett.*, **35**, 1905 (1994).
400. D. H. R. Barton, I. Fernandez, C. S. Richard, and S. Z. Zard, *Tetrahedron*, **43**, 551 (1987).
401. R. Tamura, M. Sato, and D. Oda, *J. Org. Chem.*, **51**, 4368 (1986).
402. N. Ono, I. Hamamoto, A. Kamimura, and A. Kaji, *J. Org. Chem.*, **51**, 3734 (1986).
403. N. Greenspoon and E. Keinan, *J. Org. Chem.*, **53**, 3723 (1988).
404. M. E. Jung and J. D. Trifunovich, *Tetrahedron Lett.*, **33**, 2921 (1992).
405. A. Bourgeois-Cury, D. Doan, and J. Gore, *Tetrahedron Lett.*, **33**, 1277 (1992).
406. K. Inomata, S. Igarashi, M. Mohri, T. Yamamoto, H. Kinoshita, and H. Kotake, *Chem. Lett.*, 707 (1987).
407. M. Mohri, H. Kinoshita, K. Inomata, H. Kotake, H. Takagaki, and K. Yamazaki, *Chem. Lett.*, 1177 (1986).
408. D. Eren and E. Keinan, *J. Am. Chem. Soc.*, **110**, 4356 (1988).
409. Review: C. J. Salomon, E. G. Mata, and O. A. Mascaretti, *Tetrahedron*, **49**, 3714 (1993).
410. J. Otera, T. Yano, A. Kawabata, and H. Nozaki, *Tetrahedron Lett.*, **27**, 2383 (1986).
411. N. Ono, M. Tsuboi, S. Okamoto, T. Tanami, and F. Sato, *Chem. Lett.*, 2095 (1992).
412. Y. Hayakawa, S. Wakabayashi, T. Nobori, and R. Noyori, *Tetrahedron Lett.*, **28**, 2259 (1987); Y. Hayakawa, M. Uchiyama, H. Kato, and R. Noyori, *Tetrahedron Lett.*, **26**, 6505.
413. Y. Hayakawa, H. Kato, M. Uchiyama, H. Kajino, and R. Noyori, *J. Org. Chem.*, **51**, 2400 (1986).
414. H. X. Zhang, F. Guibe, and G. Balavoine, *Tetrahedron Lett.*, **29**, 623 (1988).
415. H. Kunz and C. Unverzagt, *Angew. Chem., Int. Ed. Engl.*, **23**, 436 (1984).
416. H. Kunz and H. Waldman, *Angew. Chem., Int. Ed. Engl.*, **23**, 71 (1984).
417. R. Deziel, *Tetrahedron Lett.*, **28**, 4371 (1987).
418. P. D. Jeffery and S. W. McCombie, *J. Org. Chem.*, **47**, 587 (1982).
419. D. Habich, P. Naab, and K. Metzger, *Tetrahedron Lett.*, **24**, 2559 (1983).
420. D. Gala, M. Steinman, and R. S. Jaret, *J. Org. Chem.*, **51**, 4488 (1986).
421. H. Kunz and R. Dombo, *Angew. Chem., Int. Ed. Engl.*, **27**, 711 (1988).
422. B. Blankemeyer-Menge and R. Frank, *Tetrahedron Lett.*, **29**, 5871 (1988).
423. F. Guibe, O. Dangles, G. Balavoine, and A. Loffet, *Tetrahedron Lett.*, **30**, 2641 (1989).
424. P. L. Williams, G. Jou, F. Albericio, and E. Giralt, *Tetrahedron Lett.*, **32**, 4207 (1991).
425. H. Mastalerz, *J. Org. Chem.*, **49**, 4092 (1984).
426. H. Kunz and C. Unverzagt, *Angew. Chem., Int. Ed. Engl.*, **26**, 294 (1987).
427. H. Kinoshita, K. Inomata, T. Kameda, and H. Kotake, *Chem. Lett.*, 515 (1985).
428. O. Dangles, F. Guibe, G. Balavoine, S. Lavielle, and A. Marquet, *J. Org. Chem.*, **52**, 4984 (1987)
429. F. Guibe, O. Dangles, and G. Balavoine, *Tetrahedron Lett.*, **27**, 2365 (1986).
430. H. Kunz and J. Marz, *Angew. Chem., Int. Ed. Engl.*, **27**, 1375 (1988).
431. P. Boullanger, J. Banoub, and G. Descotes, *Can. J. Chem.*, **65**, 1343 (1987); **68**, 828 (1990).
432. Y. Hayakawa, S. Yakabayashi, H. Kato, and R. Noyori, *J. Am. Chem. Soc.*, **112**, 1691 (1990).

433. M. Sekine, *J. Org. Chem.*, **54**, 232 (1989).
434. J. P. Genet, E. Blart, M. Savignac, S. Lemeune, and J. M. Paris, *Tetrahedron Lett.*, **34**, 4189 (1993); *Tetrahedron*, **50**, 497 (1994).
435. J. P. Genet, E. Blart, M. Savignac, S. Lemeune, S. Lemaire-Audoire, and J. M. Bernard, *Synlett*, 680 (1993).
436. F. Garro-Helion, A. Merzouk, and F. Guibe, *J. Org. Chem.*, **58**, 6109 (1993).
437. I. Minami, M. Yuhara, and J. Tsuji, *Tetrahedron Lett.*, **28**, 2737 (1987).
437a.R. Beugelmans, S. Bourdet, A. Bigot, and J. Zhu, *Tetrahedron Lett.*, **35**, 4349 (1994).
438. E. J. Corey and W. J. Suggs, *J. Org. Chem.*, **38**, 3224 (1973).
439. T. Ogawa, S. Nakabayashi, and T. Kitazima, *Carbohydr. Res.*, **114**, 225 (1983).
440. K. Nakayama, K. Uoto, K. Higashi, T. Soga, and T. Kusama, *Chem. Pharm. Bull.*, **40**, 1718 (1992).
441. J. Luning, U. Moller, N. Debski, and P. Welzel, *Tetrahedron Lett.*, **34**, 5871 (1993).
442. H. B. Mereyala and S. Guntha, *Tetrahedron Lett.*, **34**, 6929 (1993).
443. M. Kloosterman, J. H. Boom, P. Chatelard, P. Boullanger, and G. Descotes, *Tetrahedron Lett.*, **26**, 5045 (1985).
444. T. Yamada, K. Goto, Y. Mitsuda, and J. Tsuji, *Tetrahedron Lett.*, **28**, 4557 (1987).
445. I. Shimizu, T. Yamada, and J. Tsuji, *Tetrahedron Lett.*, **21**, 3199 (1980).
446. T. Tsuda, Y. Chujo, S. Nishi, K. Tawara, and T. Saegusa, *J. Am. Chem. Soc.*, **102**, 6381 (1980).
447. J. Tsuji, T. Yamada, I. Minami, M. Yuhara, M. Nisar, and I. Shimizu, *J. Org. Chem.*, **52**, 2988 (1987).
448. I. Shimizu, H. Ishii, and A. Tanaka, *Chem. Lett.*, 1127 (1989).
449. I. Shimizu, Y. Ohashi, and J. Tsuji, *Tetrahedron Lett.*, **24**, 3865 (1983).
450. J. Tsuji, I. Minami, and I. Shimizu, *Tetrahedron Lett.*, **24**, 1793 (1983).
451. J. Tsuji, Y. Ohashi, and I. Minami, *Tetrahedron Lett.*, **28**, 2397 (1987).
452. F. Guibe, Y. T. Xian, A. H. Zigna, and G. Balavoine, *Tetrahedron Lett.*, **26**, 3559 (1985).
453. I. Shimizu and J. Tsuji, *J. Am. Chem. Soc.*, **104**, 5844 (1982); I. Minami, M. Nisar, M. Yuhara, I. Shimizu, and J. Tsuji, *Synthesis*, 992 (1987).
454. J. Tsuji, M. Nisar, I. Shimizu, and I. Minami, *Synthesis*, 1009 (1984).
455. H. Kataoka, T. Yamada, K. Goto, and J. Tsuji, *Tetrahedron*, **43**, 4107 (1987).
456. G. A. Kraus and L. Chen, *J. Org. Chem.*, **56**, 5098 (1991).
457. D. M. X. Donnelly, J. P. Finet, and B. A. Rattigan, *J. Chem. Soc., Perkin Trans. 1*, 1729 (1993).
458. C. Mercier, G. Mignani, M. Aufrand, and G. Allmang, *Tetrahedron Lett.*, **32**, 1433 (1991).
459. I. Shimizu, I. Minami, and J. Tsuji, *Tetrahedron Lett.*, **24**, 1797 (1983).
460. I. Minami, M. Yuhara, I. Shimizu, and J. Tsuji, *Chem. Commun.*, 118 (1986).
461. J. Tsuji, M. Nisar, and I. Minami, *Tetrahedron Lett.*, **27**, 3483 (1986).
462. J. Tsuji, M. Nisar, and I. Minami, *Chem. Lett.*, 23 (1987).
463. J. Nokami, T. Mandai, H. Watanabe, H. Ohyama, and J. Tsuji, *J. Am. Chem. Soc.*, **111**, 4126 (1989).
464. J. Nikami, H. Watanabe, and J. Tsuji, unpublished results.
465. J. Nokami, H. Watanabe, T. Mandai, M. Kawada, and J. Tsuji, *Tetrahedron Lett.*, **30**, 4829 (1989).
466. J. Tsuji, M. Nisar, and I. Shimizu, *J. Org. Chem.*, **50**, 3416 (1985).
467. A. Miura, N. Tanimoto, N. Sakamoto, and T. Masamune, *J. Am. Chem. Soc.*, **110**, 1985 (1988).
468. I. Shimizu, T. Makuta, and M. Oshima, *Chem. Lett.*, 1457 (1989).
469. I. Shimizu and H. Ishii, *Chem. Lett.*, 577 (1989); *Tetrahedron*, **50**, 487 (1994).

470. G. Casy, A. G. Sutherland, R. J. K. Taylor, and P. G. Urben, *Synthesis*, 767 (1989).
471. T. Mandai, M. Imaji, H. Takada, M. Kawada, J. Nokami, and J. Tsuji, *J. Org. Chem.*, **54**, 5395 (1989).
472. T. Koch and M. Hesse, *Synthesis*, 931 (1992); T. Koch, V. I. Ognyanov, and M. Hesse, *Helv. Chim. Acta*, **75**, 62 (1992).
473. J. Nokami, M. Ohkura, Y. Danoh, and Y. Sakamoto, *Tetrahedron Lett.*, **32**, 2409 (1991).
474. T. Murayama, A. Yoshida, T. Kobayashi, and T. Miura, *Tetrahedron Lett.*, **35**, 2271 (1994).
475. Y. Inoue, K. Ohuchi, T. Kawamata, J. Ishiyama, and S. Imaizumi, *Chem. Lett.*, 835 (1991).
476. R. C. Larock, K. Takagi, J. P. Burkhart, and S. S. Hershberger, *Tetrahedron*, **42**, 3759 (1986).
477. Reviews: W. Oppolzer, *Pure Appl. Chem.*, **60**, 39 (1988); **62**, 1941 (1990); *Chimia*, **42**, 212 (1988); *Angew. Chem., Int. Ed. Engl.*, **28**, 38 (1989); *Acc. Chem. Res.*, **23**, 34 (1990); in *Comprehensive Organic Synthesis*, Vol. 5, Pergamon Press, Oxford, 1991, p. 29.
478. W. Oppolzer, J. M. Gaudin, and T. N. Birkinshaw, *Tetrahedron Lett.*, **29**, 4705 (1988).
479. E. Negishi, S. Iyer, and C. J. Rousset, *Tetrahedron Lett.*, **30**, 291 (1989).
480. W. Oppolzer, R. E. Swenson, and J. M. Gaudin, *Terahedron Lett.*, **29**, 5529 (1988).
481. M. Terakado, M. Miyazawa. and K. Yamamoto, *Synlett*, 134 (1994).
482. R. Grigg, V. Sridharan, and S. Sukirthalingam, *Tetrahedron Lett.*, **32**, 3855 (1991).
483. W. Oppolzer, T. H. Keller, M. B. Zurita, and C. Stone, *Tetrahedron Lett.*, **30**, 5883 (1989); W. Oppolzer, J. Z. Xu, and C. Stone, *Helv. Chim. Acta*, **74**, 465 (1991).
484. N. C. Ihle and C. H. Heathcock, *J. Org. Chem.*, **58**, 560 (1993).
485. W. Oppolzer and C. Robyr, *Tetrahedron*, **50**, 415 (1994).
486. S. E. Yoo, S. H. Lee, K. Y. Yi, and N. Jeong, *Tetrahedron Lett.*, **31**, 6877 (1990).
487. W. Oppolzer and R. J. DeVita, *J. Org. Chem.*, **56**, 6256 (1991).
488. R. Keese, R. Guidetti-Grept, and B. Herzog, *Tetrahedron Lett.*, **33**, 1207 (1992).
489. B. M. Trost and J. I. Luengo, *J. Am. Chem. Soc.*, **110**, 8239 (1988).
490. T. Zair, C. Santelli-Rouvier, and M. Santelli, *Tetrahedron Lett.*, **32**, 4501 (1991); *Tetrahedron*, **49**, 3313 (1993).
491. Reviews, R. P. Lutz, *Chem. Rev.*, **84**, 205 (1984); L. E. Overman, *Angew. Chem., Int. Ed. Engl.*, **23**, 579 (1984).
492. W. E. Walker, R. M. Manyik, K. E. Atkins, and M. L. Farmer, *Tetrahedron Lett.*, 3817 (1970).
493. D. Rose and H. Lepper, *J. Organomet. Chem.*, **49**, 473 (1973).
494. J. Tsuji, K. Tsuruoka, and K. Yamamoto, *Bull. Chem. Soc. Jpn.*, **49**, 1701 (1976).
495. E. Curzon, B. T. Golding, C. Pierpoint, and B. W. Waters, *J. Organomet. Chem.*, **262**, 263 (1984).
496. S. Lumin and J. R. Falck, *Tetrahedron Lett.*, **33**, 2091 (1992).
497. K. Inomata, Y. Murata, H. Kato, Y. Tsukahara, H. Kinoshita, and H. Kotake, *Chem. Lett.*, 931 (1985)
498. S. Murahashi, Y. Makabe, and K. Kunita, *Tetrahedron Lett.*, **26**, 5563 (1985).
499. S. Kim, K. H. Uh, S. Lee, and J. H. Park, *Tetrahedron Lett.*, **32**, 3395 (1991).
500. L. E. Overman and F. M. Knoll, *Tetrahedron Lett.*, 321 (1979).
501. P. A. Grieco, T. Takigawa, B. L. Bongers, and H. Tanaka, *J. Am. Chem. Soc.*, **102**, 7587 (1980).

502. U. Hoffmann, C. O. Messe, M. Hecker, and V. Ullrich, *Tetrahedron Lett.*, **28**, 5655 (1987).
503. S. J. Danishefsky, M. P. Cabal, and K. Chow, *J. Am. Chem. Soc.*, **111**, 3456 (1989).
504. S. Takatsuto, M. Oshiguro, and N. Ikekawa, *Chem. Commun.*, 258 (1982).
505. B. T. Golding and C. Pierpoint, *Chem. Commun.*, 1030 (1981).
506. S. Saito, S. Hamano, H. Moriyama, K. Okuda, and T. Moriwaki, *Tetrahedron Lett.*, **29**, 1157 (1988).
507. T. Mandai, S. Hashio, J. Goto, and M. Kawada, *Tetrahedron Lett.*, **22**, 2187 (1981); T. Mandai, K. Hara, M. Kawada, and J. Nokami, *Tetrahedron Lett.*, **24**, 1517 (1983).
508. H. Abe, H. Nitta, A. Mori, and S. Inoue, *Chem. Lett.*, 2443 (1992).
509. Y, Tamaru, M. Kagotani, and Z. Yoshida, *J. Org. Chem.*, **45**, 5221 (1980); *Tetrahedron Lett.*, **22**, 4245 (1981); *J. Org. Chem.*, **50**, 764 (1985).
510. M. Mizutani, Y. Sanemitsu, Y. Tamaru, and Z. Yoshida, *J. Org. Chem.*, **50**, 764 (1985).
511. P. R. Auburn, J. Whelan, and B. Bosnich, *Organometallics*, **5**, 1533 (1986).
512. T. Ikariya, Y. Ishikawa, K. Hirai, and S. Yoshikawa, *Chem. Lett.*, 1815 (1982).
513. T. G. Schenck and B. Bosnich, *J. Am. Chem. Soc.*, **107**, 2058 (1985).
514. M. Mehmandoust, Y. Petit, and M. Larcheveque, *Tetrahedron Lett.*, **33**, 4313 (1992).
515. J. Genda, A. C. Helland, B. Ernst, and D. Bellus, *Synthesis*, 729 (1993).
516. J. Garin, E. Melendez, F. L. Merchan, T. Tejero, S. Uriel, and J. Ayestaran, *Synthesis*, 147 (1991); 4489 (1988).
517. J. L. van der Baan and F. Bickelhaupt, *Tetrahedron Lett.*, **27**, 6267 (1986).
518. K. Mikami, T. Takahashi, and T. Nakai, *Tetrahedron Lett.*, **28**, 5879 (1987).
519. M. Oshima, M. Murakami, and T. Mukaiyama, *Chem. Lett.*, 1535 (1984).
520. D. Djahanbuni, B. Cazes, and J. Gore, *Tetrahedron Lett.*, **25**, 203 (1984); *Tetrahedron*, **43**, 3441, 3453 (1987); B. Cazes, D. Djahanbini, J. Gore, J. P. Genet, and J. M. Gaudin, *Synthesis*, 983 (1988).
521. H. Kleijn, H. Wertmijze, J. Meijer, and P. Vermeer, *Recl. Trav. Chim. Pays-Bas*, **102**, 378 (1983).
522. J. Nokami, A. Maihara, and J. Tsuji, *Tetrahedron Lett.*, **31**, 5629 (1990).
523. M. E. Piotti and H. Alper, *J. Org. Chem.*, **59**, 1956 (1994).
524. Z. Ni and A. Padwa, *Synlett*, 869 (1992).

3 Reactions of Conjugated Dienes Catalyzed by Pd(0)

Several types of Pd-catalyzed or -promoted reactions of conjugated dienes via π-allylpalladium complexes are known. The Pd(II)-promoted oxidative difunctionalization reactions of conjugated dienes with various nucleophiles is treated in Chapter 3, Section 4, and Pd(0)-catalyzed addition reactions of conjugated dienes to aryl and alkenyl halides in this chapter, Section 1.1.1. Other Pd(0)-catalyzed reactions of conjugated dienes are treated in this section.

3.1 Dimerization and Telomerization of Conjugated Dienes and Related Reactions

3.1.1 General Reaction Patterns

Pd-catalyzed reactions of butadiene are different from those catalyzed by other transition metal complexes. Unlike Ni(0) catalysts, neither the well known cyclodimerization nor cyclotrimerization to form COD or CDT[1,2] takes place with Pd(0) catalysts. Pd(0) complexes catalyze two important reactions of conjugated dienes[3,4]. The first type is linear dimerization. The most characteristic and useful reaction of butadiene catalyzed by Pd(0) is dimerization with incorporation of nucleophiles. The bis-π-allylpalladium complex **3** is believed to be an intermediate of 1,3,7-octatriene (**7**) and telomers **5** and **6**[5,6]. The complex **3** is the resonance form of 2,5-divinylpalladacyclopentane (**1**) and pallada-3,7-cyclononadiene (**2**) formed by the oxidative cyclization of butadiene. The second reaction characteristic of Pd is the co-cyclization of butadiene with C=O bonds of aldehydes[7–9] and CO_2[10] and C=N bonds of Schiff bases[11] and isocyanate[12] to form the six-membered heterocyclic compounds **9** with two vinyl groups. The cyclization is explained by the insertion of these unsaturated bonds into the complex **1** to generate **8** and its reductive elimination to give **9**.

An active catalytic species in the dimerization reaction is Pd(0) complex, which forms the bis-π-allylpalladium complex 3. The formation of 1,3,7-octatriene (7) is understood by the elimination of β-hydrogen from the intermediate complex 1 to give 4 and its reductive elimination. In telomer formation, a nucleophile reacts with butadiene to form the dimeric telomers in which the nucleophile is introduced mainly at the terminal position to form the 1-substituted 2,7-octadiene 5. As a minor product, the isomeric 3-substituted 1,7-octadiene 6 is formed[13,14]. The dimerization carried out in MeOD produces 1-methoxy-6-deuterio-2,7-octadiene (10) as a main product[15]. This result suggests that the telomers are formed by the 1,6- and 3,6-additions of MeO and D to the intermediate complexes 1 and 2.

For the dimerization reaction, Pd(Ph$_3$P)$_4$ or better Pd(Ph$_3$P)$_2$ coordinated by benzoquinone is used as a catalyst[15]. Most conveniently, Pd(OAc)$_2$ and Ph$_3$P are used as precursors of the catalyst. Pd(OAc)$_2$ is reduced *in situ* to Pd(0), which is an active species. When PdCl$_2$(Ph$_3$P)$_2$ is used, bases such as PhONa, Et$_3$N, and AcOK are added to remove Cl$^-$ ion, which deactivates the catalyst. In the following sections, the dimerization reactions with different nucleophiles are discussed. Although addition of conjugated dienes to nucleophiles in a 2 : 1 ratio mostly takes place, sometimes addition in a 1 : 1 ratio is observed, depending on the reaction conditions, particularly the ligands.

3.1.2 Dimerization and Related Reactions

Pd-catalyzed reactions of conjugated dienes afford linear dimers and no cyclization takes place. As an exception, the formation of vinylcyclohexene (11) by the cyclodimerization of butadiene using a ligand-free Pd catalyst has been reported[16]. Butadiene is converted into 1,3,7-octatriene (7) when heated at 100–120 °C in aprotic solvents, such as benzene and THF[15]. The dimerization proceeds smoothly in isopropyl alcohol. The formation of 1,3,7-octatriene involves hydrogen migration during the reaction. The faster rate in isopropyl alcohol in comparison with aprotic solvents suggests the participation of proton from the alcohol. The reaction carried out in deuterated isopropyl alcohol, (CH$_3$)$_2$CHOD, gave 1,3,7-octatriene (12), deuterated at C-6[13]. The addition of proton from isopropyl alcohol to C-6 and elimination of the proton from C-4 result in an increased rate of hydrogen migration from C-4 to C-6.

As another route, formation of 1,3,7-octatriene (7) proceeds at higher temperature in the absence of nucleophiles by Pd-catalyzed elimination of acetic acid or phenol via a π-allylpalladium complex from their telomers[14,17].

11

$$Me_2CH\text{-}OD \longrightarrow$$ **12**

$$\text{OPh} \xrightarrow{Pd(0)} \text{7} + PhOH$$

The linear dimerization of substituted conjugated dienes is difficult, but the Pd-catalyzed intramolecular dimerization reaction of the 1,3,9,11-tetraene **13** gives the 3-propenylidene-4-allylpiperidine derivative **14**, which has the 1,3,7-octatriene system. The corresponding 1,3,8,10-tetraene also affords the 3-propenylindene-4-allylcyclopentane derivative[18].

$$ArSO_2N \xrightarrow[Ph_3P, 72\%]{Pd(OAc)_2} ArSO_2N$$

13   **14**

Dimerization is the main path. However, trimerization to form 1,3,6,10-dodecatetraene (**15**) takes place with certain Pd complexes in the absence of a phosphine ligand. The reaction in benzene at 50 °C using π-allylpalladium acetate as a catalyst yielded 1,3,6,10-dodecatetraene (**15**) with a selectivity of 79% at a conversion of 30% based on butadiene in 22 h[19,20]. 1,3,7-Octatriene (**7**) is dimerized to 1,5,7,10,15-hexadecapentaene (**16**) with 70% selectivity by using bis-π-allylpalladium. On the other hand, 9-allyl-1,4,6,12-tridecatetraene (**17**) is formed as the main product when Ph_3P is added in a 1:1 ratio[21].

$$\xrightarrow{p\text{-}C_3H_5PdOAc}$$ **15**

$$\xrightarrow{p\text{-}C_3H_5PdCl}$$ **16**

7

$$\xrightarrow[Ph_3P]{p\text{-}C_3H_5PdCl,}$$ **17**

The dimerization of isoprene is possible, but the reaction of isoprene is slower than that of butadiene. Dimerization or telomerization of isoprene, if carried out regioselectively to give a tail-to-head dimer **18** or a head-to-tail dimer **19** would be useful for terpene synthesis. So far few reports on the successful regioselective dimerization of isoprene have appeared, and it remains an unsolved problem. The selectivity is dependent on the ligands, solvents and temperature. For example, when PdBr$_2$, dppe, and sodium phenoxide were used, the tail-to-tail dimer **21** was obtained almost exclusively when the (phenol : isoprene) ratio was 1 : 30, and the head-to-head dimer **20** was obtained when the ratio was 1 : 1.5[22]. The tail-to-tail dimer **21** was obtained in the presence of *m*-methoxybenzaldehyde[23] or CO$_2$[24]. The head-to-tail 3,7-dimethyl-1,3(*E*)-7-octatriene (**19**) was obtained regioselectively by the use of cationic Pd catalysts[25].

Few examples of the dimerization of substituted 1,3-dienes are known. An attempted synthesis of 2-trimethylsilylbutadiene by the Pd-catalyzed elimination of 2-trimethylsilyl-3-acetoxy-1-butene afforded unexpectedly 3-trimethylsilyl-1,3,7-octatriene (**22**), formed by the Pd-catalyzed dimerization of trimethylsilylbutadiene by an unknown mechanism[26]. 1,2,4-Hexatriene undergoes interesting cyclodimerization to give 1-methylene-2-propadienylcyclohexane (**23**) as the major product in 81% yield using *i*-Pr$_3$P as a ligand. The reaction can be understood by the tail-to-tail coupling of its butadiene part to form a bis-π-allylpalladium complex and reductive coupling between C-2 and C-7[27].

The dimerization and addition of butadiene to allyldimethylamine takes place to afford 6-allyl-2,7-octadienyldimethylamine (**24**) by the mechanism shown. The triene **24** is a useful starting material for some natural products.

The Cope rearrangement of **24** gives 2,6,10-undecatrienyldimethylamine[28]. Sativene (**25**)[29] and diquinane (**26**) have been synthesized by applying three different palladium-catalyzed reactions [oxidative cyclization of the 1,5-diene with Pd(OAc)$_2$, intramolecular allylation of a β-keto ester with allylic carbonate, and oxidation of terminal alkene to methyl ketone] using allyloctadienyl-dimethylamine (**24**) as a building block[30].

3.1.3 With Water, Alcohols, and Phenols

The formation of 2,6-octadienol (**27**) by the reaction of 1,3-butadiene with water has attracted attention as a novel method for the commercial production of *n*-octanol, which has a considerable market. However, the reaction of water under the usual conditions is very sluggish. The addition of CO$_2$ facilitates the telomerization of water and 2,6-octadienol (**27**) is obtained as a major product[31]. In the absence of CO$_2$, only 1,3,7-octatriene (**7**) is formed. Probably octadienyl carbonate is formed, which is easily hydrolyzed to give **27**. A commercial process for the production of 2,6-octadienol (**27**) has been developed, which is operated in two phases consisting of water and hexane in the presence of carbonate salts. The water-soluble sulfonated phosphonium salt **28** is used as a ligand[32]. The reaction proceeds in the aqueous phase and the product goes into the organic layer. The catalyst always stays in the aqueous phase, and hence 2,6-octadienol (**27**) can be separated easily from the catalyst. 2,6-Octadienol (**27**) is used for the commercial production of *n*-octanol by hydrogenation, and 1,9-nonanediol (**29**) by isomerization, hydroformylation, and reduction.

Phenol is a reactive substrate and reacts with butadiene smoothly to give octadienyl phenyl ether (**30**) in a high yield[14].

Primary alcohols react easily to form ethers[13]. The higher the class of the alcohol, the lower is the reactivity[33]. Secondary alcohols have low reactivity. For example, isopropyl alcohol is used as a good solvent for 1,3,7-octatriene formation without reacting with butadiene[13]. However, the reaction of isopropyl alcohol with butadiene to give octadienyl isopropyl ether under different conditions was reported[33]. Reaction of methanol at 70 °C using Pd(Ph$_3$P)$_2$(maleic anhydride) as a catalyst gives 1-methoxy-2,7-octadiene (**31**) (85%), 3-methoxy-1,7-octadiene (**32**) (5%), and 1,3,7-octatriene (3%). When dppe is used with PdCl$_2$ and sodium phenoxide, 1 : 1 products [1-methoxy-2-butene (**33**) and 3-methoxy-1-butene] and the 1 : 2 product [1-methoxy-2,7-octadiene (**31**)] were obtained in equal amounts[34]. α-Hydroxy esters reacted smoothly[35]. The reaction of methanol catalyzed by cationic Pd complexes affords higher oligomers **34**, and the tetramer is the main product[36].

The reaction of isoprene with MeOH catalyzed by Pd(acac)$_2$ and Ph$_3$P is not regioselective, giving a mixture of isomers[37]. However, 1-methoxy-2,6-dimethyl-2,7-octadiene (**35**), the head-to-tail dimer, was obtained in 80% yield, accompanied by the tail-to-tail dimer (15%) using π-allylpalladium chloride and Bu$_3$P. On heating, **35** was converted into 2,6-dimethyl-1,3,7-octatriene (**36**) by an elimination reaction[38].

N-Octadienylation, rather than O-octadienylation, of aldehyde oximes takes place to give the nitrone **37** as an intermediate, which undergoes 1,3-dipolar addition to butadiene, yielding the isoxazolidine **38**[39].

3.1.4 With Carboxylic Acids

Carboxylic acids react with butadiene as alkali metal carboxylates. A mixture of isomeric 1- and 3-acetoxyoctadienes (**39** and **40**) is formed by the reaction of acetic acid[13]. The reaction is very slow in acetic acid alone. It is accelerated by forming acetate by the addition of a base[40]. Addition of an equal amount of triethylamine achieved complete conversion at 80 °C after 2 h. AcONa or AcOK also can be used as a base. Trimethylolpropane phosphite (TMPP) completely eliminates the formation of 1,3,7-octatriene, and the acetoxyoctadienes **39** and **40** are obtained in 81% and 9% yields by using *N,N,N',N'*-tetramethyl-1,3-diaminobutane at 50 °C in a 2 h reaction. These two isomers undergo Pd-catalyzed allylic rearrangement with each other.

Formic acid behaves differently. The expected octadienyl formate is not formed. The reaction of butadiene carried out in formic acid and triethylamine affords 1,7-octadiene (**41**) as the major product and 1,6-octadiene as a minor product[41–43]. Formic acid is a hydride source. It is known that the Pd hydride formed from palladium formate attacks the substituted side of π-allylpalladium to form the terminal alkene[44] (see Section 2.8). The reductive dimerization of isoprene in formic acid in the presence of Et$_3$N using tri(*o*-tolyl)phosphine at room temperature afforded a mixture of dimers in 87% yield, which contained 71% of the head-to-tail dimers **42a** and **42b**. The mixture was treated with concentrated HCl to give an easily separable chloro derivative **43**. By this means, α- and β-citronellol (**44** and **45**) were prepared[45].

3.1.5 With Ammonia and Amines

Direct addition of ammonia to olefinic bonds would be an attractive method for amine synthesis, if it could be carried out smoothly. Like water, ammonia reacts with butadiene only under particular reaction conditions. Almost no reaction takes place with pure ammonia in organic solvents. The presence of water accelerates the reaction considerably, The reaction of aqueous ammonia (28%) with butadiene in MeCN in the presence of Pd(OAc)$_2$ and Ph$_3$P at 80 °C for 10 h gives tri-2,7-octadienylamine (**47**) as the main product, accompanied by a small amount of di-2,7-octadienylamine (**46**)[46,47]. Isomeric branched triamines are also formed as byproducts. The reaction proceeds stepwise, but primary amines are more reactive than ammonia and secondary amines are more reactive than primary amines. Hence the main product is the tertiary amine even when the reaction is stopped before completion. Reaction of isoprene with ammonia gives a tail-to-tail coupled amine as the main product[48].

Both aromatic and aliphatic amines react with butadiene to give the tertiary octadienylamines **48**[13]. Amines with higher basicity shows higher reactivity. Electron-donating substituents on aniline have an accelerating effect[49]. Aminoethanol reacts to give *N*-octadienylaminoethanol preferentially, showing that the amine is more reactive than a primary alcohol[50,51]. When dppe is used, 1 : 1 adducts are obtained. No reaction takes places with amides. However, sulfonamides react with butadiene smoothly to give dimers **49** and **50**[52].

The reaction of isoprene with Et_2NH in the presence of triethylammonium iodide (10 mol%) gives the 1 : 1 adducts **51** and **52** with high selectivity[53]. The reaction of isoprene with ammonia or diethylamine affords the tail-to-tail dimer **53** when $Pd(acac)_2$ and $(BuO)_3P$ are used as the catalyst. The head-to-head dimer **54** is obtained with $Pd(acac)_2$, BF_3, and tricyclohexylphosphine[54].

The reaction of butadiene with the phenylhydrazone of acetone using Pd(Ph₃P)₄ affords the *C*-octadienylated products **55** and **56** and *N*-octadienylated product **57**[55]. *N*-Methylhydrazones are *N*-octadienylated selectively [56].

55
26%

56
20%

57
62%

3.1.6 With Enamines and Carbon Nucleophiles

Enamines as nucleophiles react with butadiene, and α-octadienyl ketones or aldehydes are obtained after hydrolysis[57]. This is a good way of introducing an octadienyl group at the α-position of ketones or aldehydes, because butadiene does not react with ketones or aldehydes directly. The reaction of the pyrrolidine enamine of cyclohexanone gives, after hydrolysis, 2-(2,7-octadienyl)cyclohexanone (**58**) as the main product, accompanied by a small amount of 2,6-di(2,7-octadienyl)cyclohexanone. The reaction of the optically active enamine **59** with butadiene gave 2-(2,7-octadienyl)cyclohexanone (**60**) in 72% *ee*[58].

58

59

60   **72%ee**

Active methylene or methine compounds, to which two EWGs such as carbonyl, alkoxycarbonyl, formyl, cyano, nitro, and sulfonyl groups are attached, react with butadiene smoothly and their acidic hydrogens are displaced with the 2,7-octadienyl group to give mono- and disubstituted compounds[59]. 3-Substituted 1,7-octadienes are obtained as minor products. The reaction is carried out with a β-keto ester, β-diketone, malonate, α-formyl ketones, α-cyano and α-nitro esters, cyanoacetamide, and phenylsulfonylacetate. Di(octadienyl)malonate (**61**) obtained by this reaction is converted into an interesting fatty acid **62**, which has an acid function at the center of the 17-carbon chain. α-Octylglycine (**63**) is synthesized by the Pd-catalyzed reaction of butadiene with diethyl acetamidomalonate, followed by hydrolysis, decarboxylation, and hydrogenation[60].

Asymmetric dimerization with cyclopentanone-2-carboxylate using BPPM as a chiral ligand gave the telomer in 41% *ee*[58].

Simple ketones and esters are inert. On the other hand, nitroalkanes react smoothly in *t*-butyl alcohol as a solvent with butadiene, and their acidic hydrogens are displaced with the octadienyl group. From nitromethane, three products, **64**, **65**, and **66**, are formed, accompanied by 3-substituted 1,7-octadiene as a minor product. Hydrogenation of **65** affords a fatty amine **67** which has a primary amino function at the center of the long linear chain[46,61].

When a bidentate phosphine is used as a ligand for the reaction of β-keto esters or β-diketones, no dimerization takes place. Only a 2-butenyl group is introduced to give **68**[49,62]. Substituted dienes such as isoprene, 1,3-cyclohexadiene, and ocimene react with carbon nucleophiles to give a mixture of possible regio- and stereoisomers of 1 : 1 adducts when dppp is used as a ligand[63,64].

The reactivity of substituted dienes such as piperylene is very low. However, the intramolecular reactions of the 1,3,8,10-undecatetraene **69** [di(2,4-pentadie-nyl)malonate] with enamines and carbon, oxygen, and nitrogen nucleophiles proceed smoothly to give the five-membered ring compound **70**, in which the nucleophiles are introduced at the terminal carbon. Similarly the 2,4-hexadie-nyl-3,5-hexadienylmalonate (**71**) reacts with nucleophiles to give the corresponding six-membered ring compound **72**[65].

3.1.7 With B, Si, and Sn Compounds

Hydroboration of conjugated dienes proceeds without a catalyst to give 1,2-adducts. However, the less reactive catecholborane reacts with isoprene with catalysis by Pd(Ph$_3$P)$_4$, yielding the 1,4-adduct **73**[66].

The reaction of hydrosilanes with butadiene is different from other reactions. Different products are obtained depending on the structure of the hydrosilanes and the reaction conditions. Trimethylsilane and other trialkylsilanes react to give the 1 : 2 adduct, namely the 1-trialkylsilyl-2,6-octadiene **74**, in high yields[67–69]. Unlike other telomers which have the 2,7-octadienyl chain, the 2,6-octadienyl chain is formed by hydrosilylation. On the other hand, the 1 : 1 adduct **75** (1-trichlorosilyl-2-butene)is formed selectively with trichlorosilane, which is more reactive than trialkylsilanes[69]. The reaction gives the *Z* form stereoselectively[70]. A mixture of the 1 : 1 and 1 : 2 adducts (83.5 and 5.2%) is obtained with dichloromethylsilane[69]. The reaction of chlorosilanes proceeds even in the absence of a phosphine ligand. The reaction of trichlorosilane and dichlorosilane with isoprene proceeds regioselectively and stereoselectively to give (*Z*)-1-trichlorosilyl-2-methyl-2-butene (**76**). No reaction of trialkylsilanes with isoprene takes place[69,71–73].

Hydrosilylation of 1-vinyl-1-cyclohexene (**77**) proceeds stereoselectively to give the (*Z*)-1-ethylidene-2-silylcyclohexane **78**, which is converted into (*Z*)-2-ethylidenecyclohexanol (**79**)[74]. Hydrosilylation of cyclopentadiene affords the 3-silylated 1-cyclopentene **80**, which is an allylic silane and used for further transformations[75,75a]. Cyclization of the 1,3,8,10-undecatetraene system in the di(2,4-pentadienyl)malonate **69** via hydrosilylation gives the cyclopentane derivative **81**, which corresponds to 2,6-octadienylsilane[18,76].

Disilanes add to conjugated dienes by splitting their Si—Si bond. 1,1,2,2-Tetramethyl-1,2-disilacyclopentane (**82**) reacts with butadiene at 100 °C to give 1,1,5,5-tetramethyl-1,5-disilacyclotrideca-7,11-diene (**83**) in 83% yield[77]. The six-membered carbodisilanes undergo a similar reaction to give 14-membered compounds.

Unstrained difluorotetramethyldisilane (**84**) gives the 1 : 1 adduct **85** as the main product and the 1 : 2 adduct **86** as a minor product[78,79]. On the other hand, the dimerization and double silylation of conjugated dienes with $(Me_3Si)_2$ catalyzed by $PdCl_2(PhCN)_2$ take place at 90 °C[80]. $Pd(dba)_2$ without phosphine is an active catalyst for the reaction, which proceeds in DMF to give **87** at room temperature[81]. A five-membered ring is formed by the application of the reaction to the di-(2,4-pentadienyl)malonate (**69**)[82].

1, 4-Arylsilylation of conjugated dienes to give **88** takes place at 80 °C by the reaction of a diene, disilane, and benzoyl chloride, which undergoes facile decarbonylation at 80 °C[83].

Pd-catalyzed hydrostannation of dienes with $HSnBu_3$ affords the (Z)-2-alkenylstannane **89** with high regio- and stereoselectivities[84]. Dimerization–double stannation of butadiene with $(Bu_3Sn)_2$ to give the 1,8-distannyl-2,6-octa-diene **90** is catalyzed by $Pd_2(dba)_3$, while $Pd(Ph_3P)_4$ is not suitable[85]. A five-membered ring is formed by the reaction of the di(2,4-pentadienyl)malonate **69** with $(Me_3Sn)_2$[86].

3.1.8 Carbonylation

Carbonylation of butadiene gives two different products depending on the catalytic species. When $PdCl_2$ is used in ethanol, ethyl 3-pentenoate (**91**) is obtained[87,88]. Further carbonylation of 3-pentenoate catalyzed by cobalt carbonyl affords adipate **92**[89]. 3-Pentenoate is also obtained in the presence of acid. On the other hand, with catalysis by $Pd(OAc)_2$ and Ph_3P, methyl 3,8-nonadienoate (**93**) is obtained by dimerization–carbonylation[90,91]. The presence of chloride ion firmly attached to Pd makes the difference. The reaction is slow, and higher catalytic activity was observed by using $Pd(OAc)_2$, $(i\text{-}Pr)_3P$, and maleic anhydride[92]. Carbonylation of isoprene with either $PdCl_2$ or $Pd(OAc)_2$ and Ph_3P gives only the 4-methyl-3-pentenoate **94**[93].

$$\text{(isopropenyl)} + CO + \text{t-BuOH} \xrightarrow[\text{30 atm., 120°, 50\%}]{\text{Pd(OAc)}_2, \text{Ph}_3\text{P}} \text{CO}_2\text{-t-Bu}$$

94

3.1.9 Co-cyclization with Aldehydes, Isocyanates, and Imines

Butadiene reacts with hetero double bonds such as C=N and C=O bonds to give six-membered rings with two vinyl groups. A typical reaction is the formation of the 2-substituted 3,6-divinyltetrahydropyran **95** by reaction with an aldehyde[7–9]. Both aliphatic and aromatic aldehydes including formaldehyde take part in the reaction. In this reaction, the unsaturated acyclic alcohol **96** is also formed. The selectivity to the pyran **95** and the alcohol **96** can be controlled by the Pd : Ph₃P ratio. When the ratio is higher than three, the pyran **95** is formed exclusively by reductive coupling. On the other hand, with a lower Pd : Ph₃P ratio, the unsaturated alcohol **96** is formed as the main product[7]. Formation of the unsaturated alcohol **96** is significant, because it is formed by the reaction of π-allylpalladium intermediate with an electrophile (carbonyl group). So far there is no other example of this type of the reaction. For example, reaction of benzaldehyde gives 1-phenyl-2-vinyl-4, 6-heptadien-1-ol (**96**) and 2-phenyl-3,6-divinyltetrahydropyran (**95**). Asymmetric reaction of formaldehyde using NORPhOS as a chiral ligand gave divinylpyran in 30% *ee*[94].

$$\text{(butadiene)} + \text{PhCHO} \xrightarrow[\text{PhONa, rt., 72\%}]{\text{PdCl}_2(\text{Ph}_3\text{P})_2} \mathbf{95} + \mathbf{96}$$

95   93 : 7   **96**

Another bond which takes part in the co-cyclization is C=N bond. The C=N bond, rather than the CO bond, of phenyl isocyanate reacts with butadiene to give 3-ethylidene-1-phenyl-6-vinyl-2-piperidone (**97**) in 75% yield. Double bond migration to the conjugated position takes place. The reaction with isoprene gives the 3,6-diisopropenylpiperidines **98** and **99** without double bond migration[12]. The C=N bond in Schiff bases of aromatic aldehydes is also involved in the cyclization to give the 3,6-divinylpiperidine **100**[11]. Pd(NO₃)₂ and Ph₃P are used for the generation of an active catalyst.

$$\text{(butadiene)} + \text{Ph-N=C=O} \xrightarrow[\text{PhH, 75\%}]{\text{Pd(Ph}_3\text{P})_2\text{MA}} \left[\text{intermediate} \right] \longrightarrow \mathbf{97}$$

3.1.10 With Carbon Dioxide

CO_2 is another molecule which reacts with conjugated dienes[10,95,96]. CO_2 undergoes cyclization with butadiene to give the five- and six-membered lactones 101, 102, and 103, accompanied by the carboxylic esters 104 and 105[97,98]. Alkylphosphines such as tricyclohexyl- and triisopropylphosphine are recommended as ligands. MeCN is a good solvent[99].

The reaction of isoprene with CO_2 in the presence of organotin ethoxide and DBU by the use of dicyclohexyl(β-pyridylethyl)phosphine (106) affords the isomeric esters 107 and 108 by head-to-tail and tail-to-tail dimerizations. Tin ethoxide forms tin carbonate, which seems to be an effective carrier of CO_2[100].

3.1.11 With SO₂, CS₂, and S

Pd-catalyzed reactions of sulfur-containing nucleophiles, SO_2, CS_2 and S, were summarized in a review[101]. When 3-sulfolene (**109**) is treated with butadiene at 100 °C, *cis*- and *trans*-2,5-divinylsulfolanes (**110**) are obtained. Under these conditions, 3-sulfolene (**109**) decomposes to butadiene and SO_2, which react to give divinylsulfolane (**110**)[102]. Divinylsulfolane (**110**) can be obtained directly by the reaction of SO_2 with butadiene[103].

An SO species is generated by the reaction of SO_2 with CS_2 in the presence of a Pd catalyst and reacts with butadiene to afford the 1, 4-disulfoxide **111**[104].

The reaction of butadiene with sulfur gives the disulfide **112**, cyclic sulfide **113**, and macrocyclic mono- and trisulfides **114** and **115**[105].

3.1.12 With Other Nucleophiles

The allyl aryl sulfones **116**, **117**, and **118** can be prepared by 1 : 1 and 1 : 2 additions of arylsulfinic acid to butadiene[101,106,107]. The reaction of myrcene (**119**) with *p*-tolylsulfinic acid affords the 1 : 1 adduct **120** as the main product[107]. In these reactions, the sulfone group is introduced mainly at the more hindered side of the allylic system to afford **117** and **120**. The adduct **121** to 1,3-cyclohexadiene is obtained in a good yield by use of $(PhO)_3P$ as a ligand[108,109].

3.2 Miscellaneous Reactions

1-Phenyl-1,4-hexadiene (**122**) is obtained as a major product by the codimerization of butadiene and styrene in the presence of a Lewis acid[110]. Pd(0)-catalyzed addition reaction of butadiene and allene (1 : 2) proceeds at 120 °C to give a 3 : 1 mixture of *trans*- and *cis*-2-methyl-3-methylene-1,5,7-octatriene (**123**)[111].

3.3 Synthetic Applications of Butadiene Telomers

Various butadiene telomers obtained by Pd-catalyzed reactions have one functional group at one end and a terminal and an internal double bond, and they are useful as good building blocks for natural products such as steroids and macrolides by modification of these functional groups[112]. Some typical examples are shown.

3-Acetoxy-1,7-octadiene (40) is converted into 1,7-octadien-3-one (124) by hydrolysis and oxidation. The most useful application of this enone 124 is bisannulation to form two fused six-membered ketones[113]. The Michael addition of 2-methyl-1,3-cyclopentanedione (125) to 124 and asymmetric aldol condensation using (S)-phenylalanine afford the optically active diketone 126. The terminal alkene is oxidized with $PdCl_2$–$CuCl_2$–O_2 to give the methyl ketone 127 in 77% yield. Finally, reduction of the double bond and aldol condensation produce the important intermediate 128 of steroid synthesis in optically pure form[114].

The method was applied to the synthesis of (+)-19-nortestosterone by the following sequence of reactions. Michael addition of the bisannulation reagent 124 to the optically active keto ester 129 and decarboxylation afforded 130, and subsequent aldol condensation gave 131. Selective Pd-catalyzed oxidation of the terminal double bond afforded the diketone 132 in 78% yield. Reduction of the double bond and aldol condensation gave (+)-19-nortestosterone (133)[114].

The trisannulation reagent 7-acetoxy-1,11-dodecadien-3-one (134) is derived from the bisannulation reagent 124 in four steps. This reagent is a synthetic equivalent of 1-dodecene-3,7,11-trione, and the two ketone groups of the trione are masked as an acetoxy and a terminal alkene. The synthesis of optically active D-homo-19-norandrosta-4-en-3-one (135) by the trisannulation reaction using **134** was carried out. Michael addition and asymmetric aldol condensation mediated by L-alanine are key steps. The PdCl$_2$-catalyzed oxidation of the terminal double bond, reduction, and aldol condensation afforded **135** in an optically pure form[115].

The telomer **40** is converted into lipoic acid (**136**) by modification of three functional groups, as shown[116].

The telomer **137**, obtained by the reaction of butadiene with malonate, is a suitable compound for the syntheses of naturally occurring dodecanoic acid derivatives, such as queen substance (**138**)[117], one of the royal jelly acids (**139**)[118], and pellitorine (**140**)[119].

The telomer obtained from the nitromethane **65** is a good building block for civetonedicarboxylic acid. The nitro group was converted into a ketone, and the terminal alkenes into carboxylic acids. The acyloin condensation of protected dimethyl civetonedicarboxylate (**141**) afforded the 17-membered acyloin **142**, which was modified to introduce a triple bond **143**. Finally, the triple bond was reduced to give civetone (**144**)[120].

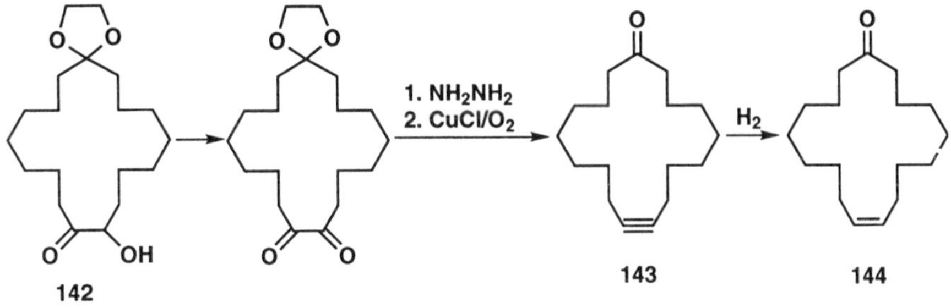

142   143   144

The telomer **145** of nitroethane was used for the synthesis of recifeiolide (**148**)[121]. The nitro group was converted into a hydroxy group via the ketone and the terminal double bond was converted into iodide to give **146**. The ester **147** of phenythioacetic acid was prepared and its intramolecular alkylation afforded the 12-membered lactone, which was converted into recifeiolide (**148**).

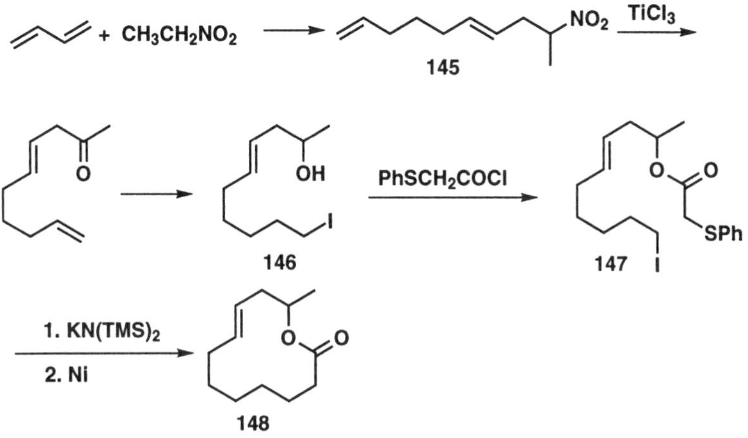

145   146   147   148

The 3,8-nonadienoate **91**, obtained by dimerization–carbonylation, has been converted into several natural products. The synthesis of brevicomin is described in Chapter 3, Section 2.3. Another royal jelly acid [2-decenedioic acid (**149**)] was prepared by cobalt carbonyl-catalyzed carbonylation of the terminal double bond, followed by isomerization of the double bond to the conjugated position to afford **149**[122]. Hexadecane-2,15-dione (**150**) can be prepared by Pd-catalyzed oxidation of the terminal double bond, hydrogenation of the internal double bond, and coupling by Kolbe electrolysis. Aldol condensation mediated by an organoaluminum reagent gave the unsaturated cyclic ketone **151** in 65% yield. Finally, the reduction of **151** afforded muscone (**152**)[123]. *n*-Octanol is produced commercially as described before[32].

3.4 References

1. Review: R. Baker, *Chem. Rev.*, **73**, 487 (1973).
2. Review: M. F. Semmelhack, *Org. React.*, **19**, 115 (1972).
3. Review: J. Tsuji, *Adv. Organomet. Chem.*, **17**, 141(1979).
4. J. Tsuji, *Acc. Chem. Res.*, **6**, 8 (1973).
5. R. Benn, P. W. Jolly, R. Mynott, B. Raspel, G. Schenker, K. P. Schick, and G. Schroth, *Organometallics*, **4**, 1945 (1985); P. W. Jolly, R. Mynott, B. Raspel, and K. P. Schick, *Organometallics*, **5**, 473 (1986).
6. A. Behr, G. V. Ilsemann, W. Keim, C. Kruger, and Y. H. Tsay, *Organometallics*, **5**, 514 (1986).
7. K. Ohno, T.Mitsuyasu, and J. Tsuji, *Tetrahedron Lett.*, 67 (1971); *Tetrahedron*, **28**, 3705 (1972).
8. P. Haynes, *Tetrahedron Lett.*, 3687 (1970).
9. R. M. Manyik, W. E. Walker, K. E. Atkins, and E. S. Hammack, *Tetrahedron Lett.*, 3813 (1970).
10. Y. Sasaki, Y. Inoue, and H. Hashimoto, *Chem. Commun.*, 605 (1976); *Bull. Chem. Soc. Jpn.*, **51**, 2375 (1978).
11. J. Kiji, K. Yamamoto, H. Tomita, and J. Furukawa, *Chem. Commun.*, 506 (1974).
12. K. Ohno and J. Tsuji, *Chem. Commun.*, 247 (1971).
13. S. Takahashi, T. Shibano, and N. Hagihara, *Tetrahedron Lett.*, 2451 (1967); *Bull. Chem. Soc. Jpn.*, **41**, 454 (1968).
14. E. J. Smutny, *J. Am. Chem. Soc.*, **89**, 6793 (1967).
15. S. Takahashi, H. Yamazaki, and N. Hagihara, *Bull. Chem. Soc. Jpn.*, **41**, 254 (1968).
16. A. Gollaszewski and J. Schwartz, *Organometallics*, **4**, 415 (1985).

17. T. Mandai, H. Yasuda, M. Kaito, J. Tsuji, R. Yamaoka, and H. Fukami, *Tetrahedron*, **35**, 309 (1979).
18. J. M. Takacs, J. Zhu, and S. Chandramouli, *J. Am. Chem. Soc.*, **114**, 773 (1992).
19. D. Medema and R.van Helden, *Recl. Trav. Chim. Pays-Bas*, **90**, 304, 324 (1971); *Inorg. Chim. Acta*, **3**, 255 (1969).
20. G. Wilke, B. Bogdanovic, P. Hardt, P. Heimbach, W. Keim, M. Kröner, W. Oberkirch, K. Tanaka, G.Steinrücke, D. Walter, and H. Zimmermann, *Angew. Chem., Int. Ed. Engl.*, **5**, 151 (1966).
21. W. Keim and H. Chung, *J. Org. Chem.*, **37**, 947 (1972).
22. K. Takahashi, G. Hata, and A. Miyake, *Bull. Chem. Soc. Jpn.*, **46**, 600 (1973).
23. M. Anteunis and A. De Smet, *Synthesis*, 800 (1974).
24. A. Musco, *J. Mol. Catal.*, **1**, 443 (1976).
25. D. Neibecker, M. Touma, and I. Takatchenko, *Synthesis*, 1023 (1984).
26. B. M. Trost and S. Mignani, *J. Org. Chem.*, **51**, 3435 (1986).
27. H. Siegel, H. Hopf, A. Germer, and P. Binger, *Chem. Ber.*, **111**, 3112 (1978).
28. C. Moberg, *Tetrahedron Lett.*, **22**, 4827 (1981); T. Antonsson and C. Moberg, *Organometallics*, **4**, 1083 (1985).
29. T. Antonsson, C. Malmberg, and C. Moberg, *Tetrahedron Lett.*, **29**, 5973 (1988).
30. C. Moberg, K. Nordstrom, and P. Helquist, *Synthesis*, 685 (1992).
31. K. E. Atkins, W. E Walker, and R. M. Manyik, *Chem. Commun.*, 330 (1971).
32. N. Yoshimura, Y. Tokitoh, M. Matsumoto, and M. Tamura, *J. Chem. Soc. Jpn. Chem. Ind.*, 119 (1993) (in Japanese); Kuraray, *Jpn. Pat. Kokai*, 85988 (1989); *HS. Pat.*, 4 927 960.
33. J. Beger and H. Reichel, *J. Prakt. Chem.*, **315**, 1067 (1973).
34. D. Commereuc and Y. Chauvin, *Bull. Soc. Chim. Fr.*, 652 (1974).
35. R. Kluter, M. Bernd, and H. Singer, *J. Organomet. Chem.*, **137**, 309 (1977).
36. P. Grenouillet, D. Neibecker, J. Poirier, and I. Tkatchenko, *Angew. Chem., Int. Ed. Engl.*, **21**, 767 (1982).
37. J. Beger, C. Duschek, and H. Reichel, *J. Prakt. Chem.*, **315**, 1077 (1973).
38. H. Yagi, E. Tanaka, H. Ishiwatari, M. Hidai, and Y. Uchida, *Synthesis*, 334 (1977).
39. R. Baker and M. S. Nobb, *Tetrahedron Lett.*, 3759 (1977).
40. W. E. Walker, R. M. Manyik, K. E. Atkins, and M. L. Farmer, *Tetrahedron Lett.*, 3817 (1970).
41. S. Gardner and D. Wright, *Tetrahedron Lett.*, 163 (1972).
42. P. Riffia, G. Gregorio, F. Conti, G. F. Pregaglia, and R. Ugo, *J. Organomet. Chem.*, **55**, 405 (1973).
43. C. U. Pitman, R. M. Hanes, and J. J. Yang, *J. Mol. Catal.*, **15**, 377 (1982).
44. J. Tsuji and T. Yamakawa, *Tetrahedron Lett.*, 613 (1979).
45. J. P. Neilan, R. M. Laine, N. Cortese, and R. F. Heck, *J. Org. Chem.*, **41**, 3455 (1976).
46. T. Mitsuyasu, M. Hara, and J. Tsuji, *Chem. Commun.*, 345 (1971).
47. J. Tsuji and M. Takahashi, *J. Mol. Catal.*, **10**, 107 (1981).
48. W. Keim and M. Roper, *J. Org. Chem.*, **46**, 3702 (1981).
49. K. Takahashi, A. Miyake and G. Hata, *Bull. Chem. Soc. Jpn.*, **45**, 1183 (1972).
50. J. D. Umpleby, *Helv. Chim. Acta*, **61**, 2243 (1978).
51. A. Groult and A. Guy, *Tetrahedron*, **39**, 1543 (1983).
52. U. M. Dzhemilev, M. M. Sirazova, and R. V. Kunakova, *Izv. Akad. Nauk SSSR, Ser. Khim.*, **372**, 2766 (1985); *Chem. Abstr.*, **103**, 37154z (1985).
53. R. W. Armbruster, M. M. Morgan, J. L. Schmidt, C. M. Lau, R. M. Riley, D. L. Zabrovski, and H. A. Dieck, *Organometallics*, **5**, 234 (1986).

54. W. Keim, K. R. Kurtz, and M. Roper, *J. Mol. Catal.*, **20**, 129, 139 (1983).
55. R. Baker, M. S. Nobbs, and D. T. Robinson, *Chem. Commun.*, 723 (1975).
56. R. Baker, M. S. Nobbs, and P. M. Winton, *J. Organomet. Chem.*, **137**, C43 (1977).
57. J. Tsuji, *Bull. Chem. Soc. Jpn.*, **46**, 1896 (1973).
58. W. Keim, A. Koehnes, T. Roethel, and D. Enders, *J. Organomet. Chem.*, **382**, 295 (1990).
59. K. Takahashi, A. Miyake, and G. Hata, *Chem. Ind. (London)*, 488 (1971); G. Hata, K. Takahashi, and A. Miyake, *J. Org. Chem.*, **36**, 2116 (1971).
60. J. P. Haudegond, Y. Chauvin, and D. Commereuc, *J. Org. Chem.*, **44**, 3063 (1979).
61. T. Mitsuyasu and J. Tsuji, *Tetrahedron*, **30**, 831 (1974).
62. P. W. Jolly and N. Kokel, *Synthesis*, 771 (1990).
63. B. M. Trost and L. Zhu, *Tetrahedron Lett.*, **33**, 1831 (1992).
64. T. Cuvigny, M. Julia, and C. Rolando, *J. Organomet. Chem.*, **285**, 395 (1985).
65. J. M. Takacs and J. Zhu, *J. Org. Chem.*, **54**, 5193 (1989); *Tetrahedron Lett.*, **31**, 1117 (1990); J. M. Takacs and S. V. Chandramouli, *J. Org. Chem.*, **58**, 7315 (1993).
66. M. Satoh, Y. Nomoto, N. Miyaura, and A. Suzuki, *Tetrahedron Lett.*, **30**, 3789 (1989).
67. S. Takahashi, T. Shibano, and N. Hagihara, *Chem. Commun.*, 161 (1969).
68. M. Hara, K. Ohno, and J. Tsuji, *Chem. Commun.*, 247 (1971).
69. J. Tsuji, M. Hara, and K. Ohno, *Tetrahedron*, **30**, 2143 (1974).
70. M. Kira, T. Hino, and H. Sakurai, *Tetrahedron Lett.*, **30**, 1099 (1989).
71. I. Ojima, *J. Organomet. Chem.*, **134**, C1 (1977).
72. I. Ojima and M. Kumagai, *J. Organomet. Chem.*, **157**, 359 (1978).
73. V. Vaisarova, J. Schraml, and J. Hetflejs, *Collect. Czech. Chem. Commun.*, **43**, 265 (1978); *Chem. Abstr.*, **89**, 43556r (1978).
74. T. Hayashi, S. Hengrasmee, and Y. Matsumoto, *Chem. Lett.*, 1377 (1990).
75. I. Ojima, N. Kumagai, and Y. Miyazawa, *Tetrahedron Lett.*, 1385 (1977).
75a. S. Kobayashi and K. Nishio, *Synthesis*, 457 (1994).
76. J. M. Takacs and S. Chandramouli, *Organometallics*, **9**, 2877 (1990).
77. H. Sakurai, Y. Kamiyama, and Y. Nakadaira, *Chem. Lett.*, 887 (1975).
78. K. Tamao, S. Okazaki, and M. Kumada, *J. Organomet. Chem.*, **146**, 87 (1978).
79. H. Matsumoto, K. Shono, A. Wada, I. Matsubara, H. Watanabe, and Y. Nagai, *J. Organomet. Chem.*, **199**, 185 (1980).
80. H. Sakurai, Y. Eriyama, Y. Kamiyama, and Y. Nakadaira, *J. Organomet. Chem.*, **264**, 229 (1984).
81. Y. Obora, Y. Tsuji, and T. Kawamura, *Organometallics*, **12**, 2853 (1993).
82. K. Tamao, G. Sun, S. Takahashi, and Y. Ito, Abstract, Ann. meeting of Chem. Soc. Jpn., 4K110 (1994).
83. Y. Obora, Y. Tsuji, and T. Kawamura, *J. Am. Chem. Soc.*, **115**, 10414 (1993).
84. H. Miyake and K. Yamamura, *Chem. Lett.*, 507 (1992).
85. Y. Tsuji and T. Kakehi, *Chem. Commun.*, 1000 (1992).
86. Y. Obora, Y. Tsuji, T. Kakehi, M. Kobayashi, Y. Shinkai, M. Ebihara, and T. Kawamura, *J. Chem. Soc., Perkin 1*, 599 (1995).
87. S. Hosaka and J. Tsuji, *Tetrahedron Lett.*, **27**, 3821 (1971).
88. S. Brewis and P. R. Hughes, *Chem. Commun.*, 157 (1965).
89. E. Drent and J. van Goch, *Eur. Pat. Appl.*, EP 284 170 (1989); *Chem. Abstr.*, **110**, 39519 (1989).
90. J. Tsuji, Y. Mori, and M. Hara, *Tetrahedron*, **28**, 3721 (1972).
91. W. E. Billups, W. E. Walker, and T. C. Shield, *Chem. Commun.*, 1067 (1971).
92. J. Kiji, T. Okano, K. Odagiri, N. Ueshima, and H. Konishi, *J. Mol. Catal.*, **18**, 109 (1982).

93. J. Tsuji and H. Yasuda, *Bull. Chem. Soc. Jpn.*, **50**, 553 (1977).
94. W. Keim, W. Meltzow, A. Koehues, and T. Roethel, *Chem. Commun.*, 1151 (1989).
95. Review, A. Behr, *Angew. Chem., Int. Ed. Engl.*, **27**, 661 (1988).
96. Review, P. Braunstein, *Chem. Rev.*, **88**, 747 (1988).
97. A. Musco, C. Perego, and V. Tartiari, *Inorg. Chim. Acta*, **28**, L147 (1978); *J. Chem. Soc., Perkin Trans. 1*, 693 (1980).
98. P. Braunstein, D. Matt, and D. Nobel, *J. Am. Chem. Soc.*, **110**, 3207 (1988).
99. A. Behr, K. D. Juszak, and W. Keim, *Synthesis*, 574 (1983); A. Behr and K. D. Juszak, *J. Organomet. Chem.*, **255**, 263 (1983).
100. H. Hoberg and M. Minato, *J. Organomet. Chem.*, **406**, C25 (1991).
101. Review: U. M. Dzhemilev and R. V. Kunakova, *J. Organomet. Chem.*, **455**, 1 (1993).
102. U. M. Dzhemilev, R. V. Kunakova, and F. V. Sharipova, *Izv. Akad. Nauk SSSR, Ser. Khim.*, 1822 (1979); 1833 (1980).
103. U. M. Dzhemilev, R. V. Kunakova, and Yu. T. Struchkov, *Dokl. Akad. Nauk SSSR*, **250**, 105 (1980).
104. U. M. Dzhemilev, R. V. Kunakova, and V. V. Fomenko, *Izv. Akad. Nauk SSSR, Ser. Khim.*, 399 (1989).
105. U. M. Dzhemilev, R. V. Kunakova, and N. Z. Baibulatova, *Dokl. Akad. Nauk SSSR*, **286**, 643 (1986).
106. U. M. Dzhemilev, R. V. Kunakova, R. I. Gaisin, E. V. Vasil'eva, and G. A. Tolstikov, *Zh. Org. Khim.*, **14**, 2223 (1978); *Chem. Abstr.*, **90**, 54606d (1979); U. M. Dzhemilev, R. V. Kunakova, and R. I. Gaisin, *Izv. Akad. Nauk SSSR, Ser. Khim.*, 2337 (1983); *Chem. Abstr.*, **100**, 51176c (1984).
107. M. Julia, M. Nel, and L. Saussine, *J. Organomet. Chem.*, **181**, C17 (1979); M. Julia, M. Nel, A. Righini, and D. Uguen, *J. Organomet. Chem.*, **235**, 113 (1982).
108. O. S. Andell, J. E. Backvall, and C. Moberg, *Acta Chem. Scand., Ser. B*, **40**, 184 (1986).
109. U. M. Dzhemilev, R. V. Kunakova, R. L.Gaisin, and G. A. Tolstikov, *Izv. Akad. Nauk SSSR, Ser. Khim.*, 2702 (1979); *Chem. Abstr.*, **91**, 128509q (1979).
110. T. Ito and Y. Takami, *Tetrahedron Lett.*, 5049 (1973); *Bull. Chem. Soc. Jpn.*, **51**, 1220 (1978).
111. D. R. Coulson, *J. Org. Chem.*, **37**, 1253 (1972).
112. Reviews: J. Tsuji, *Pure Appl. Chem.*, **51**, 1235 (1979), **53**, 2371(1981).
113. J. Tsuji, I. Shimizu, and Y. Naito, *J. Am. Chem. Soc.*, **101**, 5070 (1979).
114. I. Shimizu, Y. Naito, and J. Tsuji, *Tetrahedron Lett.*, **21**, 487 (1980).
115. J. Tsuji, Y. Kobayashi, and T. Takahashi, *Tetrahedron Lett.*, 483 (1980).
116. J. Tsuji, H. Yasuda, and T. Mandai, *J. Org. Chem.*, **43**, 3606 (1978).
117. J. Tsuji, M. Masaoka, and T. Takahashi, *Tetrahedron Lett.*, 2267 (1977).
118. J. Tsuji, M. Masaoka, T. Takahashi, A. Suzuki, and N. Miyaura, *Bull. Chem. Soc. Jpn.*, **50**, 2507 (1977).
119. J. Tsuji, H. Nagashima, T. Takahashi, and M. Masaoka, *Tetrahedron Lett.*, 1917 (1977).
120. J. Tsuji and T. Mandai, *Tetrahedron Lett.*, 3285 (1977).
121. T. Takahashi, S. Hashiguchi, K. Kasuga, and J. Tsuji, *J. Am. Chem. Soc.*, **100**, 7424 (1978).
122. J. Tsuji and H. Yasuda, *J. Organomet. Chem.*, **131**, 133 (1977).
123. J. Tsuji, M. Kaito, T. Yamada, and T. Mandai, *Bull. Chem. Soc. Jpn.*, **51**, 1915 (1978).

4 Reaction of Allenes Catalyzed by Pd(0)

The stoichiometric reaction of allenes with Pd(II) is treated in Chapter 3, Section 9, and catalytic reactions with organic halides are in this chapter, Section 1.1.1.3. Other catalytic reactions of allenes are surveyed in this section.

Like butadiene, allene undergoes dimerization and addition of nucleophiles to give 1-substituted 3-methyl-2-methylene-3-butenyl compounds. Dimerization–hydration of allene is catalyzed by Pd(0) in the presence of CO_2 to give 3-methyl-2-methylene-3-buten-1-ol (**1**). An addition reaction with MeOH proceeds without CO_2 to give 2-methyl-4-methoxy-3-methylene-1-butene (**2**)[1]. Similarly, piperidine reacts with allene to give the dimeric amine **3**, and the reaction of malonate affords **4** in good yields. Pd(0) coordinated by maleic anhydride (MA) is used as a catalyst[2].

Intramolecular reaction of the allenyl carbamate **5** in the presence of a large excess of allylic chloride catalyzed by $Pd_2(dba)_3$ or $PdCl_2(PhCN)_2$ affords the substituted oxazolidin-2-one **6**. Since the reaction is catalyzed by both Pd(II) and Pd(0), its mechanism is not clear[3].

In the coupling of the allenyl ester **7** with a terminal alkyne, an electron-deficient phosphine (Ph_3P) gave the enyne-conjugated ester **8** as the major product, while an electron-rich phosphine (TDMPP or TTMPP) yielded the non-conjugated enyne esters (*E*)- and (*Z*)-**9**[4].

	8 : 9E : 9Z
$Pd(OAc)_2$ / Ph_3P, 43%	81 : 19 : 0
TCPC / TTMPP, 64%	9 : 47 : 44

Cycloaddition of norbornadiene with allene takes place to yield the cyclobutene derivative **10**[5]. Cyclodimerization of 1,2-cyclononadiene (**11**) affords a mixture of stereoisomers of the cyclobutane derivatives **12**[6,7].

Metallation of allene with various organometallic compounds proceeds smoothly. Chlorodisilane adds to allene to give the organosilicon compound **13** containing vinylsilane and allylsilane units[8]. The regiochemistry of distannation is different depending on temperature, and **14** is obtained at higher temperature[9]. The functionalized allylstannane **15** is obtained by hydrostannation of alkoxyallene[10]. Cyanosilylation of a terminal allene with trimethylsilyl cyanide proceeds regioselectively to give the *E*-isomer **16** as the main product[11].

$$\text{Bu}\!\!-\!\!=\!\!=\ +\ Me_3SnSnMe_3\ \xrightarrow[92\%,]{Pd(Ph_3P)_4}\ \text{Bu}\diagup\!=\!\diagdown\!\!-\!SnMe_3,\ -SnMe_3$$

14   E / Z = 3 / 1

$$\text{MeO}\!\!-\!\!=\!\!=\ +\ HSnBu_3\ \xrightarrow[72\%]{Pd(Ph_3P)_4}\ \text{MeO}\diagup\!=\!\diagdown H,\ -SnBu_3$$

15   E / Z = 1 / 3

$$C_6H_{13}\,/\!\!=\!\!=\ +\ Me_3SiCN\ \xrightarrow[70\%]{PdCl_2}\ C_6H_{13}\,\diagup\!\!=\!\!\diagdown\,CN,\,-SiMe_3\ +\ C_6H_{13}\,\diagup\!\!=\!\!\diagdown\,-SiMe_3,\,CN$$

16

E/Z = 89 : 11

The reaction of CO_2 with allene using bis(dicyclohexylphosphino)ethane as a ligand affords a mixture of esters of methacrylic acid 17 and crotonic acid 18, and the lactone 19[12].

$$=\!\!=\!\!=\ +\ CO_2\ \xrightarrow[\text{dchpe, 40\%}]{p\text{-}C_3H_5PdCl}$$

17   18   19

4.1 References

1. Y. Inoue, Y. Ohtsuka, and H. Hashimoto, *Bull. Chem. Soc. Jpn.*, **57**, 3345 (1984).
2. D. R. Coulson, *J. Org. Chem.*, **38**, 1483 (1973).
3. M. Kimura, K. Fugami, S. Tanaka, and Y. Tamaru, *J. Org. Chem.*, **57**, 6377 (1992).
4. B. M. Trost and G. Kottirsch, *J. Am. Chem. Soc.*, **112**, 2816 (1990).
5. D. R. Coulson, *J. Org. Chem.*, **37**, 1253 (1972).
6. P. Binger and U. Schuchardt, *Chem. Ber.*, **113**, 1063 (1980).
7. Review, U. M. Dzhemilev, R. I. Khusnutdinov, and G. A. Tolstikov, *J. Organomet. Chem.*, **409**, 15 (1991).
8. H. Watanabe, M. Saito, N. Sutou, K. Kishimoto, J. Inose, and Y. Nagai, *Chem. Commun.*, 617 (1981); *J. Organomet. Chem.*, **225**, 343 (1982).
9. H. Killing and T. N. Mitchell, *Organometallics*, **3**, 1318 (1984); T. N. Mitchell and U. Schneider, *J. Organomet. Chem.*, **407**, 319 (1991).
10. K. Koerber, J. Gore, and J. M. Vatele, *Tetrahedron Lett.*, **32**, 1187 (1991).
11. N. Chatani, T. Takeyasu, and T. Hanafusa, *Tetrahedron Lett.*, **27**, 1841 (1986).
12. A. Dohring and P. W. Joly, *Tetrahedron Lett.*, **21**, 3021 (1980).

5 Reactions of Propargylic Compounds Catalyzed by Pd(0)

5.1 Reaction Patterns

Propargylic (or 2-alkynyl) compounds are derivatives of alkynes. However, Pd-catalyzed reactions of propargylic derivatives, particularly esters and halides, are very different mechanistically from those of simple alkynes, except in a few cases. Therefore, the reactions of propargylic esters and halides are treated in this section separately from those of other alkynes. However, some reactions of propargylic alcohols, which behave similarly to simple alkynes, are treated in Section 6.

It has been found that derivatives of propargyl halides add oxidatively to Pd(Ph₃P)₄ to give the σ-allenylpalladium complex 1 and the (σ-prop-2-ynyl)-palladium complex 2 (or propargylpalladium complex), depending on the substituents[1]. The latter is formed when there is a bulky group at C-3. Most of the Pd-catalyzed reactions of propargylic compounds are explained by the formation of the σ-allenylpalladium complex 1. Only a few reactions via the propargylpalladium intermediate 2 are known. Mainly the reactions of propargylic compounds via the complexes 1 and 2 are treated in this section.

Among several propargylic derivatives, the propargylic carbonates 3 were found to be the most reactive and they have been used most extensively because of their high reactivity[2,2a]. The allenylpalladium methoxide 4, formed as an intermediate in catalytic reactions of the methyl propargylic carbonate 3, undergoes two types of transformations. One is substitution of σ-bonded Pd, which proceeds by either insertion or transmetallation. The insertion of an alkene, for example, into the Pd—C σ-bond and elimination of β-hydrogen affords the allenyl compound 5 (1,2,4-triene). Alkene and CO insertions are typical. The substitution of Pd methoxide with hard carbon nucleophiles or terminal alkynes in the presence of CuI takes place via transmetallation to yield the allenyl compound 6. By these reactions, various allenyl derivatives can be prepared.

Another reaction occurs by the attack of a soft nucleophile at the central carbon to form the π-allylpalladium complex **7**, which undergoes further reaction with the nucleophile typical of π-allylpalladium complexes to form the alkene **8**.

Carbonates are the most versatile among derivatives of propargylic compounds and their reactions proceed under neutral conditions. Propargylic alcohols are poor substrates. Propargylic acetates can be used in the presence of a base, but they are less reactive than carbonates[1]. Propargyl halides react with hard carbon nucleophiles. The competitive reaction of a mixture of the allyl carbamate **9** and the propargyl carbamate **10** with a carbon nucleophile (dimedone) gave only the allylated product **11**, and the propargyl carbamate **10** was recovered, showing that the allyl carbamate **9** is more reactive than the propargyl carbamate **10**[3].

5.2 Reactions with Alkenes and Terminal Alkynes

The alka-1,2,4-trienes (allenylalkenes) **12** are prepared by the reaction of methyl propargyl carbonates with alkenes. Alkene insertion takes place into the Pd—C bond of the allenylpalladium methoxide **4** as an intermediate and subsequent elimination of β-hydrogen affords the 1,2,4-triene **12**. The reaction proceeds rapidly under mild conditions in the presence of KBr. No reaction takes place in the absence of an alkali metal salt[4].

Terminal alkynes react with propargylic carbonates at room temperature to afford the alka-1,2-dien-4-yne **14** (allenylalkyne) in good yield with catalysis by Pd(0) and CuI[5]. The reaction can be explained by the transmetallation of the σ-allenylpalladium methoxide **4** with copper acetylides to form the allenyl(alkynyl)palladium **13**, which undergoes reductive elimination to form the allenyl alkyne **14**. In addition to propargylic carbonates, propargylic chlorides and acetates (in the presence of $ZnCl_2$) also react with terminal alkynes to afford allenylalkynes[6]. Allenylalkynes are prepared by the reaction of the alkynyl-oxiranes **15** with zinc acetylides[7].

The unsaturated *exo*-enol lactone **17** is obtained by the coupling of propargylic acetate with 4-pentynoic acid in the presence of KBr using tri(2-furyl)-phosphine (TFP) as a ligand. The reaction is explained by the oxypalladation of the triple bond of 4-pentynoic acid with the allenylpalladium and the carboxylate as shown by **16**, followed by reductive elimination to afford the lactone **17**. The (*E*)-alkene bond is formed because the oxypalladation is *trans* addition[8].

5.3 Carbonylation

Carbonylation of propargylic carbonates proceeds under mild neutral conditions (50 °C, 1–10 atm) using Pd(OAc)$_2$ and Ph$_3$P as a catalyst, yielding the 2,3-alkadienoates **18** in good yields[9,10]. The 2,3-alkadienoates isomerize to 2,4-dienoates during the reaction depending on the solvents and reaction time. 2-Decynyl methyl carbonate is converted into methyl 2-heptyl-2,3-butadienoate (**19**) in 82% yield.

Introduction of an ester group into propargylic carbonates enhances their reactivity. The propargyl-type acetylenic ester **20** (3-methoxycarbonyl-2-alkynyl carbonate), easily prepared from propargylic alcohol, is a highly reactive compound and undergoes vicinal dicarbonylation at room temperature and 1 atm using bidentate phosphines, particularly dppf, as a ligand, yielding the derivative of malonate **21** without stopping at the monocarbonylation step[11]. The diacid **22** (alkylidenesuccinic acid), obtained by the hydrolysis and decarboxylation of **21**, resembles a product of the Stobbe condensation **23**[12].

The carbonylation of propargylic acetates was carried out under mild conditions (55 °C, 1 atm) in a two-phase system (aqueous NaOH and 4-methyl-2-pentanone) in the presence or absence of Bu$_4$NBr to give the allenic acid **24** (2,3-alkadienic acid)[13]. Propargylic acetates and phosphates are also carbonylated in the presence of bases under a higher pressure. Carbonylation of the optically active propargylic phosphate **25** gave the optically active 2,3-dienyl ester **26** with retention of chirality. The use of 1,6-bis(diphenylphosphino)hexane as a ligand is important[14]. The carbonylation of both propargyl bromide (**27**) and the 1-bromo-1,2-butadiene **28** under pressure in the presence of Et_3N at room temperature affords 2,3-alkadienoate[15].

In addition to alcohols, some other nucleophiles such as amines and carbon nucleophiles can be used to trap the acylpalladium intermediates. The α-vinylidene-β-lactam **30** is prepared by the carbonylation of the 4-benzylamino-2-alkynyl methyl carbonate derivative **29**[16]. The reaction proceeds using TMPP, a cyclic phosphite, as a ligand. When the amino group is protected as the *p*-toluenesulfonamide, the reaction proceeds in the presence of potassium carbonate, and the α-alkynyl-β-lactam **31** is obtained by the isomerization of the allenyl (vinylidene) group to the less strained alkyne.

When active methylene compounds are used as nucleophiles in carbonylation at 50 °C and 1 atm, ketones are obtained. As an example, the reaction of
1,3-cyclohexanedione affords the trione **32**[17].

The allenyl moiety (2,3-alkadienyl system) in the carbonylation products is a
reactive system and further reactions such as intramolecular Diels–Alder and
ene reactions are possible by introducing another double bond at suitable
positions of the starting 2-alkynyl carbonates. For example, the propargylic
carbonate **33** which has 1,8(or 1,9)-diene-3-yne system undergoes tandem carbonylation and intramolecular Diels–Alder reaction to afford the polycyclic
compound **34** under mild conditions (60 °C, 1 atm). The use of dppp as ligand
is important. One of the double bonds of the allenyl ester behaves as part of the
diene[18].

The carbonylation of dehydrolinalyl methyl carbonate (**35**) at room temperature affords the cyclopentene derivative **37** formed by the ene reaction of

the primary product **36**, showing that the allenic ester moiety is very reactive[10]. Introduction of an ester group into the triple bond enhances the reactivity further, and the 3-methoxycarbonyl-2-alkynyl carbonate **38** having a 1,6-enyne system undergoes tandem carbonylation and intramolecular ene-type cyclization to afford the five-membered ring compound **39**. The ene-type cyclization is faster than the second carbonylation at the central carbon of the allene system to give the triester **21**. The spiro-type cyclohexadiene **40** is prepared by this reaction and double bond isomerization. Either a five- or a six-membered ring is formed from the 1,6-enyne system, depending on its structure. As the ligand, dppp gives the best results[19].

The Diels–Alder reaction to give **34** is possible only intramolecularly. No intermolecular Diels–Alder reaction of the trienyl ester system takes place in the presence of an alkene as a dienophile. Instead, the carbonylation of the propargylic carbonate **41** (4-alken-2-ynyl carbonate), which has a conjugated enyne structure, at room temperature using bidentate ligands such as dppe and dppp produces the 4-oxo-5-alkylidene-2-cyclopentenecarboxylate **45** in a good yield by vicinal dicarbonylation[20]. The reaction is explained by the following mechanism. The attack of CO at the central carbon of the allenic ester **42** (or hydrocarbonylation of one of the double bonds) takes place to give the acyl-palladium species **43**, which undergoes intramolecular alkene insertion to form the cyclopentanonepalladium. Finally, the elimination of β-hydrogen affords the cyclopentenone **45**. Another possibility is via the formation of a pallada-cyclopentene **44**, which undergoes CO insertion, reductive elimination, and double bond isomerization to give **45**. The *exo*-double bond in the product **45** or **46** is the E form.

Propargylic alcohols are less reactive and their carbonylation proceeds under severe conditions. The Pd-catalyzed carbonylation of propargyl alcohol in the presence of HCl affords the itaconate **47** as the main product and the aconitate **48** as a minor product. In addition, a small amount of methyl 2-methoxy-methacrylate (**49**) is formed. 2,3-Butadienoate is an intermediate of itaconate **47**, because carbonylation of 2,3-butadienoate, prepared independently, affords itaconate **47**. The formation of the aconitate **48** by tricarbonylation is explained by the oxidative carbonylation of the triple bond, followed by allylic carbonylation[21]. As supporting evidence, the aconitate **48** was obtained selectively in 70% overall yield by the Pd-catalyzed two-step carbonylation of propargyl alcohol as described in Chapter 3, Section 8[22].

The carbonylation of 2-methyl-3-butyn-2-ol (**50**) in benzene gives teraconic anhydride (**51**). Fulgide (**53**) (a dimethylenesuccinic anhydride derivative), which is a photochromic compound, can be prepared by the carbonylation of 2,5-dimethyl-3-hexyne-2,5-diol (**52**)[21]. The reaction proceeds under milder conditions when Pd(OAc)₂ is used as a catalyst in the presence of iodine [23].

5.4 Hydrogenolysis

Pd-catalyzed hydrogenolysis of propargylic compounds, particularly carbonates, offers a good synthetic method for allenes and internal alkynes. Propargylic carbonates which have a terminal triple bond are hydrogenolyzed with ammonium formate to give allenes[24]. The hydrogenolysis proceeds more smoothly under mild conditions by use of the propargylic formate **54**, and alkyne **56** and allene derivatives **58** are obtained. Their ratios depend on the substituents[25,26]. As summarized in the general scheme, alkyne **56** is obtained when both R^1 and R^2 are not H by decarboxylation and reductive elimination of the σ-propargylpalladium formate **55**. The allene **58** is obtained when either R^1 or R^2 is H via the allenylpalladium formate **57**. Hydrogenolysis of the propargylic carbonate **59** (the formate of tertiary propargyl alcohol cannot be prepared) possessing a terminal acetylenic bond using ammonium formate affords the allene **60** exclusively. With an excess of ammonium formate, further reduction proceeds regioselectively to give the terminal alkene **61**[24]. Allenes are obtained cleanly by the treatment of propargylic formates with a Pd catalyst in the absence of ammonium formate. On the other hand, the reaction of the propargylic formate **62** which has an internal alkyne bond gives the internal alkyne **63** selectively[25,26]. This is one of the rare examples of Pd-catalyzed reactions of propargyl compounds via the propargylpalladium intermediate **55**. This reaction is useful for the preparation of disubstituted alkynes with functional groups, because disubstituted alkynes are usually prepared by alkylation of terminal alkynes, but the alkylation of terminal alkynes is not easy, particularly in the presence of labile functional groups. On the other hand, the preparation of propargylic alcohols by the reaction of alkynes with aldehydes or ketones is much easier, and the hydroxy group can be removed by the Pd-catalyzed hydrogenolysis mentioned here. The primary propargylic formate **64** exceptionally affords the allene **65** selectively, although it is an internal alkyne.

The propargylic mesylates **66**, bromides, and phosphates are converted into a mixture of the allene **67** and alkyne **68** with metal hydrides. LiBHEt$_3$ gives the allene **67** with high selectivity[27]. The propargylic acetate **69** is converted mainly into allene by the Pd-catalyzed reaction with SmI$_2$ in the presence of a proton source. Alkenylpalladium is reduced with Sm(II) to give allenyl anion[28]. The reaction was applied to the generation of 2,3-naphthoquinodi-methane (**70**), which undergoes facile Diels–Alder reaction with fumarate to give a tetrahydroanthracene derivative[29]. Reaction of propargylic acetate with SmI$_2$ generates a synthetic equivalent of propargylic anion, which reacts intramolecularly with a ketone or aldehyde to give the homopropargyl cycloalkanol **71**[30].

5.5 Reactions with Carbon and Oxygen Nucleophiles, and Other Reactions

The reactions of propargylic compounds with hard and soft carbon nucleophiles give different products. The reactions of propargylic halides, acetates, and phosphates with hard carbon nucleophiles such as Grignard reagents[31] or organozinc reagents[32,33] give the allenyl compounds **73**. The reaction can be explained by transmetallation to generate the allenylphenylpalladium intermediate **72** and its reductive elimination. The reaction of the 17α-ethynyl steroid **74** with phenylzinc chloride gives mainly the *anti* product **75**[34]. *anti*-Stereoselectivity was observed in the reaction of (*R*)-(−)-1-acetoxy-1-phenyl-2-propyne(**76**) with phenylzinc chloride to give the levorotatory allene (*R*)-**77**[34].

The 2-(1-alkynyl)oxirane **78** reacts with an organozinc reagent yielding the β-allenylic alcohol **79**[35].

Organoboranes react with propargylic carbonates. Usually, addition of a base is essential for the Pd-catalyzed reactions of organoboranes, but the reaction with propargylic carbonates proceeds without addition of the base, because methoxide is generated *in situ* from carbonates. For example, the 1,2,4-triene **80** is prepared by the reaction of alkenylborane under neutral conditions[36].

No reaction of soft carbon nucleophiles takes place with propargylic acetates[37], but soft carbon nucleophiles, such as β-keto esters and malonates, react with propargylic carbonates under neutral conditions using dppe as a ligand. The carbon nucleophile attacks the central carbon of the σ-allenylpalladium complex **81** to form the π-allylpalladium complex **82**, which reacts further with the carbon nucleophile to give the alkene **83**. Thus two molecules of the α-monosubstituted β-keto ester **84**, which has one active proton, are introduced[2]. The reaction of propargyl carbonate with 2,5-hexanedione-3,4-dicarboxylate (**85**) affords the methylenecyclobutane **86** by *C*-dialkylation and the 3-methylenedihydropyran derivative **87** by *C*- and *O*-alkylations. The former is converted into the latter by the prolonged Pd-catalyzed reaction[38,39]. Two molecules of malonate, even if it is unsubstituted, react with propargyl carbonate to give the unsaturated tetraesters **88a** and **88b**.

β-Keto esters such as acetoacetate without a substituent at the α-carbon, namely with two acidic protons, first attack the central carbon of the allenylpalladium to form the π-allyl complex **89**. Then intramolecular attack of the enolate oxygen of the β-keto ester at the π-allyl system takes place to form the methylenedihydrofuran **90** as a primary product, which is easily isomerized to the furan **91**. The β-diketone **92** reacts similarly to give the furan **93**[40]. The reaction can be applied to the synthesis of the phenylthiomethyl-substituted furan **94**, which is useful for the synthesis of natural products such as neoliacine. [41]

2-(1-Alkynyl)oxiranes also react with carbon nucleophiles to afford furan derivatives. Furanes of different types are obtained depending on the structure of the substrates. 2-Methyl-2-ethynyloxirane (**95**) reacts with acetoacetate to give the furan **97** by the elimination of formaldehyde from the cyclized product **96**. The hydroxy ester of the alkylidenefuran **98** and the corresponding lactone **99** are obtained by the reaction of 1-methyl-2-(2-propynyl)oxirane[40, 42].

98   99

5.6 Preparation of Enynes by Elimination

When propargylic carbonates are treated with a Pd catalyst in refluxing dioxane in the absence of other reactants, elimination of β-hydrogen from the propargylpalladium intermediate takes place to afford the conjugated enyne **100**[26,43]. Formation of the 1,2,3-triene **101** by the elimination of the allenylpalladium intermediate is not observed. The enyne **102** is obtained in good yield using bidentate phosphines such as dppf as a ligand. Alternatively, the propargylic alcohol **103** was converted into the conjugated enyne **104** by a Pd-catalyzed elimination reaction in the presence of a stoichiometric amount of $SnCl_2$.[44].

5.7 Miscellaneous Reactions Catalyzed by Pd(0) and Pd(II)

Unlike propargyl carbonates, which react easily with Pd(0) to form allenylpalladium complexes with cleavage of the C—O bond, propargyl acetates do not undergo such complex formation with cleavage of the carbon–acetoxy bond. A typical example is given which shows the difference between proargyl carbonates and acetates in the presence of Pd(0) catalyst. Addition of acetic acid to the propargylic acetate **105** takes place to form the allylic geminal diacetate **106** using Pd$_2$(dba)$_3$ and Ph$_3$P as a catalyst. A 14-membered lactone **107** was prepared in 52% yield by an intramolecular version of this reaction[45].

PdCl$_2$(MeCN)$_2$ catalyzes interesting reactions of propargylic acetates involving transposition of the OAc group, although the mechanism is not clear. For example, the PdCl$_2$-catalyzed reaction of the propargylic acetate **108** gives a mixture of products and the conjugated triene **109**, a homo-coupling product, is the major product. Addition of **108** to 1-decene gives the diene **110** [46].

The (1-ethynyl)-2-propenyl acetate derivative **111** undergoes an interesting PdCl$_2$(PhCN)$_2$-catalyzed cyclization to form the 2-cyclopentenone **112**[47]. A Pd–carbene complex is assumed to be an intermediate of the formation of **112**.

5.8 References

1. C. J. Elsevier, H. Kleijn, J. Boersma, and P. Vermeer, *Organometallics*, **5**, 716 (1986); *Chem. Commun.*, 1529 (1983).
2. J. Tsuji, H. Watanabe, I. Minami, and I. Shimizu, *J. Am. Chem. Soc.*, **107**, 2196 (1985).
2a. Review: J. Tsuji and T. Mandai, *Angew. Chem., Int. Ed. Engl.*, (1996).
3. H. X. Zhang, F. Guibe, and G. Balavoine, *Tetrahedron Lett.*, **29**, 623 (1988).
4. T. Mandai, M. Ogawa, H. Yamaoki, T. Nakata, H. Murayama, M. Kawada, and J. Tsuji, *Tetrahedron Lett.*, **32**, 3397 (1991).
5. T. Mandai, T. Nakata, H. Murayama, H. Yamaoki, M. Ogawa, M. Kawada, and J. Tsuji, *Tetrahedron Lett.*, **31**, 7179 (1990); *J. Organomet. Chem.*, **417**, 305 (1991).
6. S. Gueugnot and G. Linstrumelle, *Tetrahedron Lett.*, **34**, 3853 (1993).
7. H. Kleijn, J. Meijer, G. C. Overbeek, and P. Vermeer, *Recl. Trav. Chim. Pays-Bas,* **101**, 97 (1982).
8. D. Bouyssi, J. Gore, G. Balme, D. Louis, and J. Wallach, *Tetrahedron Lett.*, **34**, 3129 (1993).
9. Review: J. Tsuji and T. Mandai, *J. Organomet. Chem.*, **451**, 15 (1993).
10. J. Tsuji, T. Sugiura, and I. Minami, *Tetrahedron Lett.*, **27**, 731 (1986).
11. T. Mandai, Y. Tsujiguchi, S. Matsuoka, S. Saito, and J. Tsuji, *Tetrahedron Lett.*, **35**, 5697 (1994).
12. W. S. Johnson and G. H. Daub, *Org. React.*, **6**, 1 (1951).
13. H. Arzoumanian, M. Choukrad, and D. Nuel, *J. Mol. Catal.*, **85**, 287 (1993).
14. S. Murahashi, Y. Imada, T. Mori and T. Kitamura, *Abstr. Jpn. Chem. Soc. Annu. Meeting II*, 345 (1993).
15. N. D. Trieu, C. J. Elsevier, and K. Vrieze, *J. Organomet. Chem.*, **325**, C23 (1987).
16. T. Mandai, K. Ryoden, M. Kawada, and J. Tsuji, *Tetrahedron Lett.*, **32**, 7683 (1991).
17. T. Mandai, H. Kunitomi, K. Higashi, and M. Kawada, J. Tsuji, *Synlett*, 697 (1991).
18. T. Mandai, S. Suzuki, A. Ikawa, T. Murakami, M. Kawada, and J. Tsuji, *Tetrahedron Lett.*, **32**, 7687 (1991).
19. T. Mandai, Y. Tsujiguchi, S. Saito, and J. Tsuji, *Tetrahedron Lett.*, **35**, 5701 (1994).
20. T. Mandai, Y. Tsujiguchi, J. Tsuji, and S. Saito, *J. Am. Chem. Soc.*, **115**, 5865 (1993).
21. T. Nogi and J. Tsuji, *Tetrahedron*, **25**, 4099 (1969).
22. B. Gabriele, M. Costa, G. Salerno, and G. P. Chiusoli, *Chem. Commun.*, 1007 (1992).
23. J. Kiji, H. Konishi, T. Okano, S. Kometani, and A. Iwasa, *Chem. Lett.*, 313 (1987).
24. J. Tsuji, T. Sugiura, M. Yuhara, and I. Minami, *Chem. Commun.*, 922 (1986); J. Tsuji, T. Sugiura, and I. Minami, *Synthesis*, 603 (1987).
25. T. Mandai, T. Matsumoto, M. Kawada, and J. Tsuji, *Tetrahedron Lett.*, **34**, 2161 (1993).
26. T. Mandai, Y. Tsujiguchi, S. Matsuoka, and J. Tsuji, *J. Organomet. Chem.*, **473**, 343 (1994).
27. Y. Colas, B. Cazes, and J. Gore, *Bull. Soc. Chim. Fr.*, 165 (1987); *Tetrahedron Lett.*, **25**, 845 (1984).
28. T. Tabushi, J. Inanaga, and M. Yamaguchi, *Tetrahedron Lett.*, **27**, 5237 (1986).
29. J. Inanaga, Y. Sugimoto, and T. Hanamoto, *Tetrahedron Lett.*, **33**, 7035 (1992).
30. J. M. Aurrecoechea and R. F. Anton, *J. Org. Chem.*, **59**, 702 (1994).

31. T. Jeffery-Luong and G. Linstrumelle, *Tetrahedron Lett.*, **21**, 5019 (1980).
32. K. Ruitenberg, H. Kleijn, C. J. Elsevier, J. Meier, and P. Vermeer, *Tetrahedron Lett.*, **22**, 1451 (1981).
33. K. Ruitenberg, H. Kleijn, H. Westmijze, J. Meier, and P. Vermeer, *Recl. Trav. Chim. Pays-Bas*, **101**, 405 (1982).
34. C. J. Elsevier, P. M. Stehouwer, H. Westmijze, and P. Vermeer, *J. Org.Chem.*, **48**, 1103 (1983).
35. H. Kleijn, I. Meijer, G. C. Overbeek, and P. Vermeer, *Recl. Trav. Chim. Pays-Bas*, **101**, 97 (1982).
36. T. Moriya, N. Miyaura, and A. Suzuki, *Synlett*, 149 (1994).
37. E. Keinan and E.Bosch, *J. Org. Chem.*, **51**, 4006 (1986).
38. L. Geng and X. Lu, *Tetrahedron Lett.*, **31**, 111 (1990).
39. L. Geng and X. Lu, *J. Chem. Soc., Perkin Trans. 1*, 17 (1992).
40. I. Minami, M. Yuhara, H. Watanabe, and J. Tsuji, *J. Organomet. Chem.*, **334**, 225 (1987).
41. N. Greeves and J. S. Torode, *Synthesis*, 1109 (1993).
42. I. Minami, M. Yuhara, and J. Tsuji, *Tetrahedron Lett.*, **28**, 629 (1987).
43. T. Mandai, Y. Tsujiguchi, S. Matsuoka, and J. Tsuji, *Tetrahedron Lett.*, **34**, 7615 (1993).
44. Y. Masuyama, J. P. Takahara, K. Hashimoto, and Y. Kurusu, *Chem. Commun.*, 1219 (1993).
45. B. M. Trost, W. Brieden, and K. H. Baringhaus, *Angew. Chem., Int. Ed. Engl.*, **31**, 1335 (1992).
46. V. Rautenstrauch, *Tetrahedron Lett.*, **25**, 3845 (1984).
47. V. Rautenstrauch, *J. Org. Chem.*, **49**, 950 (1984).

6 Reactions of Alkynes Catalyzed by Pd(0) and Pd(II)

Alkynes undergo a variety of reactions using either Pd(II) or Pd(0), and they are treated separately: oxidative reactions of alkynes with Pd(II) are treated in Chapter 3, Section 8, Pd(0)-catalyzed reactions of alkynes with halides in Section 1.1.2 in this chapter, and other reactions in this section.

6.1 Carbonylation

Alkynes are reactive compounds for Pd(0)-catalyzed carbonylation, and mono- and dicarbonylations take place depending on the reaction conditions. Monohydroesterification gives α, β-unsaturated esters, and two regioisomers are obtained depending on the reaction conditions[1,2]. The hydroesterification of alkynes is accelerated by the addition of HCl or HI, and explained by the insertion of alkynes into the Pd—H bond, followed by CO insertion. Methyl methacrylate (**1**) is obtained in high yields and high regioselectivity by the carbonylation of propyne using Pd black in the presence of hydrogen iodide[3,4]. This regioselective reaction of terminal alkynes has been applied to the synthesis of the carbapenem derivative **2**[5] and methyl 2-(6-methoxy-2-naphthyl)propenoate[6]. Pd(OAc)$_2$, coordinated by 2-pyridyldiphenylphosphine, is a very selective (99% selective) and active catalyst (40 000 turnover) for the carbonylation of terminal alkynes to give branched unsaturated esters in the presence of *p*-toluenesulfonic acid[7]. Carbonylation of a terminal alkyne to give the branched unsaturated amide **3** in good yield proceeds smoothly at 120 °C under CO pressure (5 atm) in diethylamine by using $PdCl_2(Ph_3P)_2$ and a catalytic amount of methyl iodide as an iodide source[8]. Carbonylation of terminal alkynes with substituted phenols can be carried out under 1 atm of CO using $Pd(OAc)_2$ and dppf to give 2-substituted 2-propenoate aryl esters. Also $Pd(dba)_2$ + 4 Ph_3P is a good catalyst for the carbonylation of terminal and internal alkynes in alcohol in the presence of TsOH[9].

$$Me\!-\!\!\!\equiv\ +\ CO\ +\ MeOH \quad \xrightarrow[\text{80°C, 15 atm}]{\text{Pd, HI}} \quad \overset{Me}{\underset{\mathbf{1}\ \ CO_2Me}{=\!\!<}}$$

$$+\ CO\ +\ MeOH \quad \xrightarrow[\text{65°, 16hr., 70%}]{\text{Pd, HI, 20 atm}}$$

$$+\ CO\ +\ Et_2NH \quad \xrightarrow[\text{120°, 5 atm., 92%}]{\text{PdCl}_2(\text{Ph}_3\text{P})_2,\ \text{MeI}}$$

On the other hand, linear unsaturated esters are formed from terminal alkynes depending on the substrates and reaction conditions. The ratio of linear to branched α, β-unsaturated esters from higher terminal alkynes changes depending on the ligands. The linear ester **4** was obtained with 81% selectivity (65% total yield) from 1-heptyne when $PdCl_2$, coordinated by dimethylphenylphosphine and $SnCl_2$, was used at 80 °C and 240 atm. With $PdCl_2(Ph_3P)_2$ in the absence of $SnCl_2$, the selectivity was 19% (84% yield)[4].

$$C_5H_{11}\!-\!\!\!\equiv\ +\ CO\ +\ MeOH \quad \xrightarrow[\text{80°, 240 atm., 65%}]{(Me_2PhP)_2PdCl_2\text{-SnCl}_2}$$

$$\overset{C_5H_{11}}{\underset{\mathbf{4}\quad CO_2Me}{\diagdown\!=\!\diagup}} \quad + \quad \overset{C_5H_{11}}{\underset{CO_2Me}{=\!\!<}}$$

81%selectivity

The presence of formates, oxalates, formic acid, and oxalic acid in the carbonylation of alkynes affects the regioselectivity. Also, the regioselectivity can be controlled to some extent by the proper selection of the ligands. A linear α,

β-unsaturated ester is formed from a terminal alkyne by the reaction of alkyl formate and oxalate. The linear α, β-unsaturated ester **5** is obtained from the terminal alkyne using dppb as a ligand by the reaction of alkyl formate under CO pressure. On the other hand, a branched ester, *t*-butyl atropate (**6**), is obtained exclusively by the carbonylation of phenylacetylene in *t*-BuOH even by using dppb[10]. Reaction of alkynes and oxalate under CO pressure also gives linear α, β-unsaturated esters **7** and dialkynes. The use of dppb is essential[11]. Carbonylation of 1-octyne in the presence of oxalic acid or formic acid using Ph_3P–dppb (2 : 1) and Pd on carbon affords the branched α, β-unsaturated acid **8** as the main product. Formic acid is regarded as a source of H and OH in the carboxylic acids[12].

Ph—≡ + HCO_2t-Bu + CO → (Pd2(dba)3, dppb, 100°, 80 atm., 50%) Ph—=—CO_2t-Bu **5**

Ph—≡ + CO + t-BuOH → (Pd2(dba)3, dppb, DME, 80 atm., 150°, 56%) Ph / =CO_2t-Bu **6**

Ph—≡ + CO_2Me / CO_2Me + CO → (Pd(OAc)2, dppb, 100°, 80 atm.) Ph—=—CO_2Me **7** (64%) + Ph—≡—≡—Ph (48%)

C_6H_{13}—≡ + CO + HCO_2H → (Pd/C, dppb, Ph_3P, DME, 6.8 atm., 110°, 75%) =C_6H_{13}/CO_2H **8** + C_6H_{13}—=—CO_2H 96 : 4

Monocarbonylation of alkyn-4-ols in the presence of thiourea as an additive gives α-methylene-γ-lactones, a structure widely distributed in certain natural products[13]. A derivative of vernolepine (**9**) was prepared by this method[14]. This carbonylation is explained by the following mechanism. Initially Pd-carboxylate complex **10** is formed by insertion of CO into Pd alkoxide containing an alkynic moiety. Then intramolecular insertion of the alkyne is followed by protonolysis with the alkynol to afford the lactone **11** and to regenerate the Pd alkoxide **12**[15]. Carbonylation of 4-pentyn-1-ol gives a six-membered lactone in a trace amount. On the other hand, the rigid system of 2-*exo*-ethynyl-7-*syn*-

norbornanol (**13**) was carbonylated with PdCl$_2$ in the presence of thiourea at 50 °C to give the α-methylene-δ-lactone **14** in 47% yield[16]. However, the carbonylation of the propargylic alcohol **15** catalyzed by Pd$_2$(dba)$_3$ and dppb affords the unsaturated γ-lactone **16**. In this reaction, CO attacks the terminal carbon[17].

The acetylenedicarboxylate **17** is a reactive compound and is carbonylated smoothly at room temperature to give the ethanetetracarboxylate **18** as the main product and ethene- and ethanetricarboxylates as minor products. Acetylenemonocarboxylate is converted into the ethanetricarboxylate **19** as the main product with several other products[18].

Using a catalyst system of PdCl$_2$, CuCl$_2$, HCl, and O$_2$, the internal alkyne **20** is carbonylated at room temperature and 1 atm to give unsaturated esters[19]. This apparently oxidizing system leads to non-oxidative *cis*-hydroesterification. With terminal alkynes, however, oxidative carbonylation is observed.

EtO_2C—≡—CO_2Et + CO + EtOH $\xrightarrow{PdCl_2}$

17

EtO₂C, CO₂Et / H, CO₂Et

trace

+

EtO₂C, CO₂Et / CO₂Et

24%

+

EtO₂C, CO₂Et / EtO₂C, CO₂Et

18 51%

≡—CO_2Et + CO + EtOH $\xrightarrow[\text{rt.}]{PdCl_2, HCl}$

EtO₂C, CO₂Et / CO₂Et

19

32%

+

EtO₂C, / CO₂Et

20%

+

EtO₂C, CO₂Et / H, CO₂Et

8%

+

EtO₂C, EtO₂C, / CO₂Et, CO₂Et

10%

The (E)-α-alkyl-β-silylacrylate **22** is prepared by regio- and stereoselective carbonylation of the trimethylsilylalkyne **21** using a Pd catalyst coordinated by SnCl₂ and dppf[20].

Bu—≡—Me + CO + MeOH $\xrightarrow[\text{O}_2\text{, rt, 1 atm., 72\%}]{PdCl_2, CuCl_2, HCl}$ Bu, Me / H, CO₂Me

20

C_8H_{17}—≡—TMS + CO + EtOH $\xrightarrow[\text{90°, 20 atm. 91\%}]{PdCl_2(dppf), SnCl_2}$ C₈H₁₇, TMS / MeO₂C **22**

21

Carbonylation of the tetrasubstituted bispropargylic amine **23** using PdCl₂ and thiourea under mild conditions affords the carboxylated pyrrolidine derivatives **24a** and **b** in good yields. Thiourea is regarded as effective for the oxidative carbonylation of alkynes, but no oxidative carbonylation was observed in this case[21].

HN (structure 23) + CO + MeOH $\xrightarrow[\text{thiourea, 65\%}]{PdCl_2}$ HN (structure 24a) CO₂Me + HN (structure 24b) CO₂Me

23   **24a**   **24b**

6.2 Coupling Reactions

Oxidative coupling of terminal alkynes to form 1,3-diynes with Pd(II) is treated in Chapter 3, Section 8. The Pd-catalyzed head-to-tail homocoupling of a terminal alkyne to form the 1-en-3-yne **25** proceeds by using a sterically crowded ligand such as TDMPP. Cross-coupling of the terminal alkyne **27** with the electron-deficient internal alkyne **26** gives the 1-en-3-yne **28** using the same ligand[22]. In an intramolecular version of the cross-coupling, the presence of an EWG in one alkyne gives better results, and the unsaturated lactone **30** was obtained by the intramolecular coupling of the diyne **29**. The reaction is explained by the oxidative addition of a terminal alkyne to form an alkynyl-palladium hydride, the insertion of another alkyne, and reductive coupling[23]. The Pd-catalyzed head-to-head homocoupling of the silylacetylene **31** takes place to give the conjugated enyne **32** using Pd(Ph₃P)₄ as a catalyst[24].

6.3 Cyclization of Nonconjugated Enynes, Diynes, and Triynes

Interesting cyclization reactions of 1,6- and 1,7-enynes and diynes are possible to give a variety of cyclized compounds by treatment with Pd catalysts, offering useful synthetic methods for five- and six-membered rings[25]. Different Pd catalysts give different products. The 1,6-enyne **33** undergoes Pd-catalyzed cyclization to form mainly the five-membered ring compounds **34** (ene-type reaction) and **35**. The 1,7-enyne **36** can be cyclized to give the six-membered ring compound **37**. In addition, the reductive cyclization of 1,6-diynes and 1,6-enynes takes place in the presence of hydrosilane as a hydrogen source. Metathesis cyclization is also possible. Various Pd(II) and Pd(0) species have been used as catalysts in these reactions.

The 1,6-enyne **38** and 1,7-enyne undergo Pd-catalyzed formal ene reactions smoothly at room temperature to give the 1,4-diene **39**. In this reaction, N,N'-bis(benzylidene)ethylenediamine (BBEDA) is used as a ligand[26]. Formation of the Pd–hydride species by reaction of Pd(0) with AcOH is assumed to take place[27] and the insertion of a terminal alkyne into H—PdOAc starts the reaction as a key step. This reaction is explained by the preferential insertion of an alkyne into H—PdOAc to generate the vinylpalladium **41**, followed by intramolecular alkene insertion to give **42**. Insertion of the triple bond is faster than that of the double bond. The final step is the elimination of β-hydrogen to yield either the 1,4-diene **39** or 1,3-diene **40**, depending on which hydrogen is eliminated, and the H—Pd—OAc species is regenerated.

The cyclization is general. The Pd-catalyzed cyclization of the 1,6-enyne **43** gives **44** as a major product as the ene reaction-type product and **45** as a minor product. On the other hand, a thermal ene reaction (625 °C) gives **45** exclusively[28]. Petiodial has been synthesized by this cyclization method[29]. A bicyclic picrotoxane skeleton (**47**) has been constructed by this Pd-catalyzed ene reaction-type cyclization of **46**[30]. The 1,6-enyne system of the alkynyl *N*-acylenamine **48** undergoes a similar cyclization using BBEDA (5%) to give **49**[31]. Asymmetric inductions up to 71% have been observed using tartaric acid derivatives as chiral auxiliaries[32].

Pd(OAc)₂(Ph₃P)₂	73%	2.9 : 1
Pd(OAc)₂	80%	16.9 : 1
no catalyst, 625°,	83%	0 : 1

Depending on the substituents of 1,6-enynes, their cyclization leads to 1,2-dialkylidene derivatives (or a 1,3-diene system). For example, cyclization of the 1,6-enyne **50** affords the 1,3-diene system **51**[33–35]. Furthermore, the 1,6-enyne **53**, which has a terminal alkene, undergoes cyclization with a shift of vinylic hydrogen to generate the 1,3-diene system **54**. The carbapenem skeleton **56** has been synthesized based on the cyclization of the functionalized 1,6-enyne **55**[36]. Similarly, the cyclization of the 1,7-enyne **57** gives a six-membered ring **58** with the 1,3-diene system.

The 1,3-diene system formed by cyclization is useful for further modifications, typically Diels–Alder reactions. The 1,3-diene **51** reacts with maleimide to give **52**[35]. Similarly, the 1,3-diene **59** undergoes a Diels–Alder reaction, and this sequence was used for the syntheses of sterepolide (**60**)[37] and merulidial[38].

[4 + 3] Cycloaddition involving the Pd-catalyzed trimethylenemethane (TMM) fragment **63** and the 1,3-diene **61** with an EWG offers a good synthetic method for the hydroazulene skeleton **65**. The cycloaddition of trimethylenemethane generated from 2-(trimethylsilylmethyl)allyl acetate (**62**) (see Section 2. 2. 2.1) with the 1,3-diene ester **61** via Michael addition to generate **64** and subsequent trapping of the carbanion with π-allylpalladium in **64** afford exclusively the seven-membered ring compound **65** in a good yield, rather than a five-membered ring compound[39].

Cyclization and addition reactions of 1,6-enynes with carbon nucleophiles take place via the 1,3-dienes, giving different products depending on the ligands. The five-membered ring compound **66** is obtained with dppp and its reaction with a carbon nucleophile gives **67**. On the other hand, the six-membered ring product **68** is formed by the use of bis(di-2-methoxyphenylphosphino)propane (dmppp) as a ligand and converted into **69**[40]. (For the 1,4-addition reaction of carbon nucleophiles to 1,3-dienes, see Section 3.1.6.)

Another interesting transformation is the intramolecular metathesis reaction of 1,6-enynes. Depending on the substrates and catalytic species, very different products are formed by the intramolecular enyne metathesis reaction of 1,6-enynes[41]. The cyclic 1,3-diene **71** is formed from a linear 1,6-enyne. The bridged tricyclic compound **73** with a bridgehead alkene can be prepared by the enyne metathesis of the cyclic enyne **72**. The first step of these enyne metatheses is the cycloaddition of an enyne to form the highly strained cyclobutenes **70** and **73**, and their ring opening gives the metathesis products **71** and **74**[42,43]. For the metathesis reaction, 2,3,4,5-tetrakis-(methoxycarbonyl)palladacyclopentadiene (TCPC) (**75**), combined with tri(*o*-tolyl) phosphite (TOTPO), is used as the catalyst. TCPC (**75**) has been synthesized as a tetramer by the reaction of $Pd_2(dba)_3$ with DMAD[44]. Particularly, the esters with EWG groups such as trifluoroethyl ($TCPC^{TFE}$) and heptafluorobutyl ($TCPC^{HFB}$) esters in **75** give better results, and the metathesis is the main reaction path, rather than the ene-type reaction. Terminal substitutions on both the alkenes and the alkynes, particularly an EWG on the alkynes, promote the metathesis. The reaction is carried out in the presence of 1 equiv. of dimethyl acetylenedicarboxylate (DMAD).

From the cyclopentene derivative of 1,6-enyne **76**, the normal metathesis product **77** formed via the very highly strained cyclobutene intermediate was a minor product, and the head-to-tail type coupling product **78** of the double bond with the triple bond was obtained as the major product. The cyclohexene analogue **79** affords three products, **82**, **83**, and **84**, in a ratio of 2 : 1 : 1. The previously unknown bicyclo[5.2.1]deca-1(10),2-diene **82** is the metathesis product formed via the highly strained tricyclic compound **81**. Further isomerization of **82** to **83** takes place to relieve strain energy. The product **84** is formed by double bond isomerization of **81** without opening the cyclobutene ring. Isolation of the cyclobutene **84** strongly supports the formation of the cyclobutene **81** as an intermediate of the enyne metathesis[45]. Formation of the cyclobutene **81** via the palladacyclopentene **80** [Pd(IV) species] has been proposed. A detailed mechanism of the reaction has been presented[41,46].

76 → 77 + 78

TCPC^TFE
DMAD, 65%

1 : 3.7

79 → 80 → 81 → 82 + 83 + 84

TCPC^TFE
DMAD, 59%

82 : 83 : 84 = 2 : 1 : 1

When a 1,6-enyne is substituted by a vinyl group, such as in **85**, an interesting dimer is obtained. The reaction is explained by the formation of the cyclopropylalkylidene complex **87**, which is trapped with the vinyl group of **85** to give the cycloaddition product **88**. In order to trap this intermediate **87** with an activated alkene, an intermolecular reaction of **85** with methyl (*E*)-2,4- pentadienoate (**89**) was carried out using TCPC^HFB as a catalyst in the absence of a ligand, and the adduct **90** was obtained in 80–66% yield. The reaction involves the formation of the cyclopropylalkylidene complex **87** from the palladacycle **86**, which undergoes further cycloaddition with **89** to afford **90**[47].

Another reaction is reductive cyclization. 1,6-Diynes and 1,6-enynes undergo reductive cyclization using hydrosilanes as a hydrogen source in AcOH. The 1,6-diynes **91** and **95** are converted into the 1,2-dialkylidenecyclopentane derivatives (1,3-dienes) **94** and **96**. Triethylsilane is used as a hydrogen donor for the reaction[48]. The reaction involves the formation of a vinylpalladium bond in **92** via the insertion of an alkyne into the Pd—H bond, followed by the alkyne insertion to give **93**, which is hydrogenolyzed with Si—H to give the 1,3-dienes **94** and **96**.

Also, the 1,6-enynes **97** and **99** undergo reductive cyclization to give the methylenecyclopentanes **98** and **100** by using triethylsilane or polymethylhydrosiloxane (PMHS) as a hydrogen donor and tri(*o*-tolyl)phosphine or BBDEA as a ligand in AcOH in the presence of allyl acetate[49]. The reductive cyclization starts by the insertion of an alkyne into the H—Pd bond, formed by the oxidative addition of AcOH to Pd(0). The reaction is terminated by trapping the alkylpalladium with SiH. This reaction has been applied to the stereoselective construction of the five-membered ring moiety **102** from **101** in the total synthesis of phorbol[50] and phyllanthocin[51]. Similarly, the reductive cyclization of the 1,7-enyne **103** yields the six-membered ring compound **104**.

The above-mentioned reductive cyclization starts by the insertion of alkynes into the Pd—H bond to generate alkenylpalladium species. Cyclization of polyenynes can also be initiated by Pd—H species formed by the oxidative addition of AcOH to Pd(0). The pentaenyne **105** undergoes an efficient penta-cyclization to give **106** in 86% yield by use of Ph₃Sb as a ligand in AcOH[52]. The reaction is called a Pd-catalyzed zipper reaction. The first step is the hydropalladation of the triple bond to generate vinylpalladium, which under-goes tandem alkene insertions. For this tandem insertion, 2,2-disubstituted alkenes must be used to generate the neopentylpalladium, which has no β-hydrogen to be eliminated, and the insertion continues without termination. The [3.3.3]propellane skeleton **109** can be constructed from the dienyne **107** via the neopentylpalladium **108**.

The cyclization of the enediynes **110** in AcOH gives the cyclohexadiene derivative **114**. The reaction starts by the insertion of the triple bond into Pd—H to give **111**, followed by tandem insertion of the triple bond and two double bonds to yield the triene system **113**, which is cyclized to give the cyclohexadiene system **114**. Another possibility is the direct formation of **114** from **112** by *endo*-type insertion of an *exo*-methylene double bond[53]. The appropriately structured triyne **115** undergoes Pd-catalyzed cyclization to form an aromatic ring **116** in boiling MeCN, by repeating the intramolecular insertion three times. In this cyclization too, addition of AcOH (5 mol%) is essential to start the reaction[54].

The intramolecular [$2\pi + 2\sigma$] cycloaddition of methylenecyclopropane with the alkyne in **117** using isopropyl phosphite as a ligand affords the methylene-cyclopentene **118**[55].

Intermolecular trimerization of alkynes to give benzene derivatives is catalyzed by Pd(II). The hexasubstituted benzenes **119, 120,** and **121** are obtained by the intermolecular coupling reaction of unsymmetrically disubstituted alkynes such as methylphenylacetylene with PdCl$_2$[56]. The cyclization proceeds catalytically with PdCl$_2$(PhCN)$_2$ in methylene chloride or chloroform[57]. The mechanism of the cyclization is explained by the insertion of an alkyne into the Pd—Cl bond, followed by tandem insertion twice to give the complex **122**. The formation of the intermediate complex **122** and its conversion into substituted benzene have been confirmed. The formation of the cyclobutadiene complex of PdCl$_2$ is also explained by this mechanism[58]. This is the first example of the insertion of an alkyne into a vinyl–Pd bond in Pd chemistry. This cyclotrimerization is a good synthetic method for hexasubstituted benzenes. A very efficient catalyst for the cyclization of internal alkynes is prepared by the treatment of Pd on carbon with TMSCl, and internal alkynes cyclize in refluxing THF nearly quantitatively[59].

Dimethyl acetylenedicarboxylate (DMAD) (**125**) is a very special alkyne and undergoes interesting cyclotrimerization and co-cyclization reactions of its own using the poorly soluble polymeric palladacyclopentadiene complex (TCPC) **75** and its diazadiene stabilized complex **123** as precursors of Pd(0) catalysts. Cyclotrimerization of DMAD is catalyzed by **123**[60]. In addition to the hexasubstituted benzene **126**, the cyclooctatetraene derivative **127** was obtained by the co-cyclization of trimethylsilylpropargyl alcohol with an excess of DMAD (**125**)[61]. Co-cyclization is possible with various alkenes. The naphthalenetetracarboxylate **129** was obtained by the reaction of methoxyallene (**128**) with an excess of DMAD using the catalyst **123**[62].

The cyclohexadiene derivative **130** was obtained by the co-cyclization of DMAD with strained alkenes such as norbornene catalyzed by **75**[63]. However, the linear 2 : 1 adduct **131** of an alkene and DMAD was obtained selectively using bis(maleic anhydride)(norbornene)palladium (**124**)[64] as a catalyst[65]. A similar reaction of allyl alcohol with DMAD is catalyzed by the catalyst **123** to give the linear adducts **132** and **133**[66]. Reaction of a vinyl ether with DMAD gives the cyclopentene derivatives **134** and **135** as 2 : 1 adducts, and a cyclooctadiene derivative, although the selectivity is not high[67].

6.4 1,2-Dimetallation, Hydrometallation, and Hydrogenation Reactions

Organometallic compounds which have main group metal–metal bonds, such as S—B, Si—Mg, Si—Al, Si—Zn, Si—Sn, Si—Si, Sn—Al, and Sn—Sn bonds, undergo 1,2-dimetallation of alkynes. Pd complexes are good catalysts for the addition of these compounds to alkynes. The 1,2-dimetallation products still have reactive metal–carbon bonds and are used for further transformations.

The *cis* thioboration of terminal alkynes with 9-(arylthio)-9-BBN is catalyzed by Pd(Ph$_3$P)$_4$ in the presence of styrene. The product **136** is converted into the vinyl sulfides **137** and **138** by the treatment with MeOH or by Pd-catalyzed cross-coupling with aryl or alkenyl halides using K$_3$PO$_4$ in DMF[68]. No thioboration takes place with internal alkynes.

Silyl metal compounds are prepared *in situ* by the reaction of R$_3$SiLi with RMgCl, Et$_2$AlCl, or ZnBr$_2$, and react with alkynes in the presence of Pd catalysts. The reaction is not regioselective. CuI is a better catalyst for higher regioselectivity. The vinylsilanes **139** and **140** are obtained by the addition of Ph$_3$Si-MgMe prepared *in situ*, followed by protonolysis[69]. The vinylaluminum, formed by the addition of PhMe$_2$SiAlEt$_2$, can be converted into the vinyl iodides **141** and **142** by treatment with iodine. The main product **139** is useful

$$C_8H_{17} \longrightarrow \equiv \quad + \quad PhS-9\text{-}BBN \quad \xrightarrow[\text{styrene, 81\%}]{Pd(Ph_3P)_4} \quad \begin{array}{c} C_8H_{17} \\ PhS \end{array}\diagdown\!\!=\!\!\diagup\, B\diagdown \quad \mathbf{136}$$

$$\xrightarrow{\text{MeOH}} \quad \begin{array}{c} C_8H_{17} \\ PhS \end{array}\diagdown\!\!=\!\!\diagup\, H \quad \mathbf{137}$$

$$\xrightarrow[K_3PO_4,\ 95\%]{PhI,\ Pd(Ph_3P)_4} \quad \begin{array}{c} C_8H_{17} \\ PhS \end{array}\diagdown\!\!=\!\!\diagup\, Ph \quad \mathbf{138}$$

for the stereoselective synthesis of a steroidal side-chain. The Zn—C bond in the products of silylzincation of 1-alkyne with **143** is easily protonated to give the vinylsilanes **139** and **140**[70].

$$C_{10}H_{21} \longrightarrow \equiv \quad + \quad PhMe_2SiLi \quad + \quad MeMgI \quad \xrightarrow[76\%]{PdCl_2(PPh_3)_2} \quad \xrightarrow{H_3O^+}$$

$$\begin{array}{c} C_{10}H_{21} \\ \end{array}\diagdown\!\!=\!\!\diagup\, SiMe_2Ph \quad + \quad \begin{array}{c} C_{10}H_{21} \\ PhMe_2Si \end{array}\diagdown\!\!=\!\!$$
$$\mathbf{139} \qquad\qquad \mathbf{140}$$
$$60 : 40$$

$$Me \longrightarrow \equiv \quad + \quad PhMe_2SiLi \quad + \quad Et_2AlCl \quad \xrightarrow{PdCl_2[(tol)_3P]_2} \quad \xrightarrow{I_2}$$

$$\begin{array}{c} I \\ \end{array}\diagdown\!\!=\!\!\diagup\, SiMe_2Ph \quad + \quad \begin{array}{c} I \\ \end{array}\diagdown\!\!=\!\!\diagup\, SiMe_2Ph$$
$$\mathbf{141} \qquad\qquad \mathbf{142}$$
$$12 : 88$$

$$C_{10}H_{21} \longrightarrow \equiv \quad + \quad \underset{\mathbf{143}}{PhMe_2SiZnEt_2Li} \quad \xrightarrow[2.\ H_2O,\ 80\%]{1.\ Pd(Ph_3P)_4}$$

$$\begin{array}{c} C_{10}H_{21} \\ \end{array}\diagdown\!\!=\!\!\diagup\, SiMe_2Ph \quad + \quad \begin{array}{c} C_{10}H_{21} \\ PhMe_2Si \end{array}\diagdown\!\!=\!\!$$
$$\mathbf{139} \qquad\qquad\qquad \mathbf{140}$$
$$3 \quad : \quad 1$$

Stannylalumination of 1-alkynes proceeds smoothly[71]. Silylstannation of 1-alkynes with **144** proceeds to give **145** with high regio- and stereoselectivity. The reaction is *cis* addition and Sn always adds to the internal position. The addition products are converted into either a vinylsilane or a 2-stannylalk-ene[72–76]. The 2-stannylalkene **146** is prepared by silylstannation of alkynes, followed by treatment with tributylammonium fluoride[77]. Rapamycin has been synthesized applying the silylstannation as a key step. The *cis*-adduct

147 was converted into **148** by the Pd-catalyzed displacement of the stannyl group with acyl chloride[78]. The stannylzincation product was converted into the 2-stannylalkene **149**[76].

R—≡ + Bu₃Sn-SiMe₂Ph →(Pd(Ph₃P)₄, 62 ~ 95%)→ **145** (Bu₃Sn / SiMe₂Ph, R)

144

→(Bu₄NF)→ **146** (Bu₃Sn, R)

Furyl-SiMe₂SnBu₃ + ≡—Me →(Pd(Ph₃P)₄, 49%)→ **147** (SnBu₃, SiMe₂, furyl)

→(i-PrCOCl, PdCl₂(MeCN)₂, 40%)→ **148**

C₁₀H₂₁—≡ + (Bu₃Sn)₂Zn →(Pd(Ph₃P)₄, 70%)→ **149** (C₁₀H₂₁, Bu₃Sn)

The (*E*)-vinylsilane **151** was prepared by treatment of the silylstannation product **150** with hydrogen iodide[75] and the silylzincation product with water[70]. The silylstannylation of 1-ethoxyacetylene proceeds at room temperature using Pd(OAc)₂ and 1,1,3,3-tetramethylbutyl isocyanide regioselectively and an Si group is introduced at the ethoxy-bearing carbon. Subsequent CuI- and Pd-catalyzed displacement of the stannyl group in the product **152** with allyl halide, followed by hydrolysis, affords the acylsilane **153**[79].

(CO₂Et cyclopentanone alkyne) + Bu₃SnSiMe₃ →(Pd(Ph₃P)₄, THF, 78%)→ **150** (CO₂Et, SnBu₃, SiMe₃) →(HI, Bu₄NI, 99%)→ **151** (CO₂Et, SiMe₃)

However, Bu$_3$SnSiMe$_2$H (**154**) reacts with terminal alkynes to give the 3,4-disubstituted silacyclopenta-2,4-diene **156** at room temperature. The dimethylsilylene **155** is an intermediate[80]. (Me$_3$Sn)$_2$ undergoes facile addition to alkynes to give the 1,2-distannylalkene **157**[81–83].

The 1,2-disilacyclopentane **158** adds to alkynes with cleavage of the Si—Si bond to give the seven- and fourteen-membered ring compounds **159** and **160**[84,85]. Bis-silylation of alkynes proceeds more smoothly with the disilane **161** bearing electron-donating groups[85–87]. However, efficient bis-silylation of alkynes with hexalkyldisilane can be carried out by use of the sterically crowded isocyanide **162** and Pd(OAc)$_2$[88] or Pd$_2$(dba)$_3$ and TMPP as a ligand[89]. The non-twisted tetrakis(organosilyl)ethene **164** was prepared by the intramolecular reaction of **163** under pressure using **162**[90]. Addition of a bifunctional polyphenyldisilane to alkynes has been applied to polymer synthesis[91]. The 3,4-bis(alkylidene)-1,2-disilacyclobutane **165** was converted into the 3,4,7,8-tetrakis(alkylidene)-1,2,5,6-tetrasilacyclooctane **166** by homocoupling when R = methyl, and a 1 : 1 adduct **167** was formed by the Pd-catalyzed reaction with alkynes[92].

Alkynes insert into the silacyclobutane **168** to form the silacyclohexene **169**[93]. Also, the silacyclopropene **170** is expanded to the silacyclopentadiene **171** by the insertion of an alkyne[94]. The insertion product **173** was obtained by the Pd-catalyzed reaction of the neopentylidenesilirane **172** with acetylene[95].

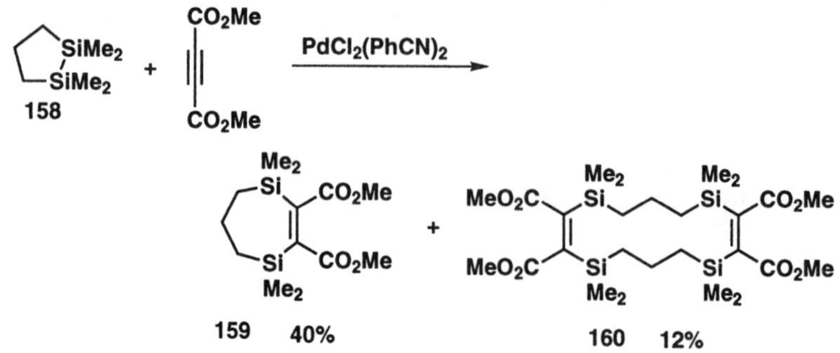
SiMe2
SiMe2
158
CO2Me
CO2Me
PdCl2(PhCN)2
Me2
Si
CO2Me
CO2Me
Si
Me2
159 40%
MeO2C
MeO2C
Me2
Si
Si
Me2
Me2
Si
Si
Me2
CO2Me
CO2Me
160 12%

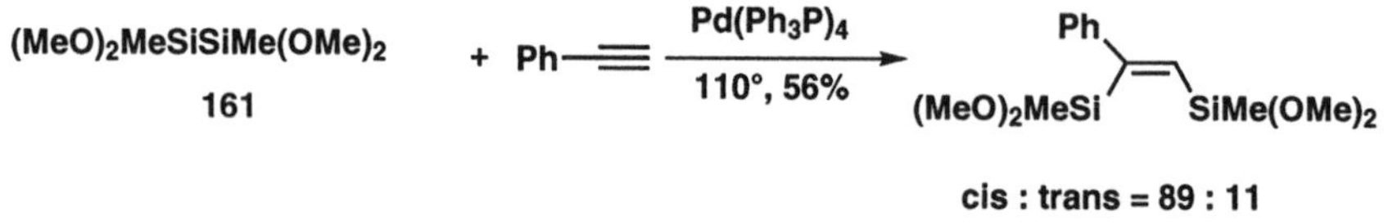
(MeO)2MeSiSiMe(OMe)2
161
Ph
Pd(Ph3P)4
110°, 56%
Ph
(MeO)2MeSi
SiMe(OMe)2
cis : trans = 89 : 11

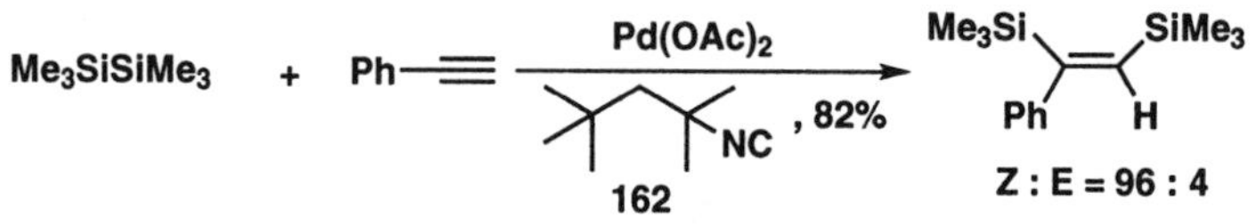
Me3SiSiMe3
Ph
Pd(OAc)2
NC
, 82%
162
Me3Si
SiMe3
Ph
H
Z : E = 96 : 4

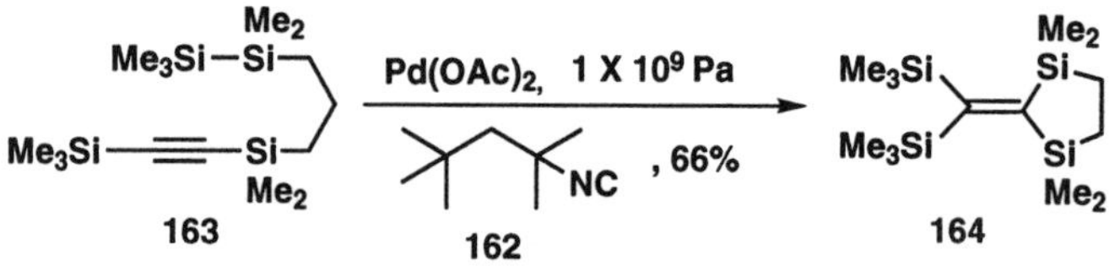
Me2
Me3Si
Si
Me3Si
Si
Me2
163
Pd(OAc)2, 1 X 10^9 Pa
NC
, 66%
162
Me3Si
Me3Si
Me2
Si
Si
Me2
164

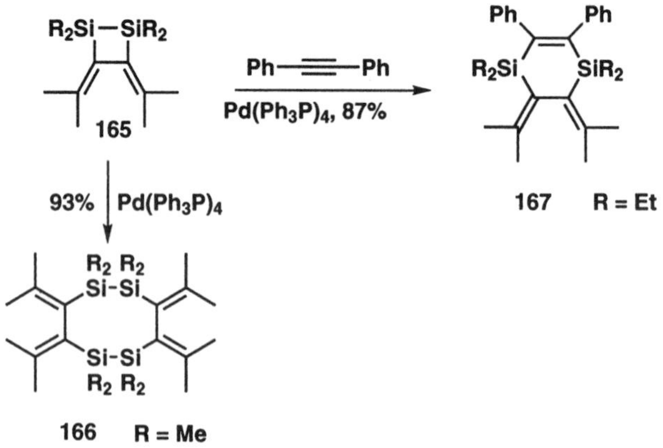
R2Si
SiR2
165
Ph
Ph
Pd(Ph3P)4, 87%
Ph
Ph
R2Si
SiR2
167 R = Et
93%
Pd(Ph3P)4
R2 R2
Si-Si
Si-Si
R2 R2
166 R = Me

The reaction of 1,4-bis(trimethylsilyl)-1,3-butadiyne (**174**) with disilanes, followed by treatment with methylmagnesium bromide, produces 1,1,4,4-tetra(-trimethylsilyl)-1,2,3-butatriene (**175**) as a major product[96]. The reaction of octaethyltetrasilylane (**176**) with DMAD proceeds by ring insertion to give the six-membered ring compounds **177** and **178**[97]. The 1-sila-4-stannacyclohexa-2,5-diene **181** was obtained by a two-step reaction of two alkynes with the disilanylstannane **179** via the 1-sila-2-stannacyclobutane **180**[98].

Me₃Sn—SiMe₂—SiMe₃ + ≡—CO₂Et →[Pd(Ph₃P)₄][63%] Me₃Sn   SiMe₂SiMe₃ / H   CO₂Et

179

⟶ Me₂Sn—SiMe₂ / EtO₂C →[Ph—≡][Pd(Ph₃P)₄, 100%] Me₂Sn   SiMe₂ / EtO₂C **181**

180

The cyanosilylation of alkynes with trimethylsilyl cyanide proceeds by either Pd or Ni catalysis to give **182**[99]. When an excess of trimethylsilyl cyanide is used, the 5-aminopyrrole-2-carbonitrile **183** is obtained[100,101].

Ph—≡ + Me₃SiCN →[PdCl₂][py, 96%] Ph   H / NC   SiMe₃

182

Ph—≡ + Me₃SiCN →[NiCl₂][AlEt₃, 66%] NC   N   N(SiMe₃)₂ / H

183

The Pd(0)-catalyzed addition of trimethylsilyl iodide to an alkyne, followed by capture with alkynylstannane, affords the stereo-defined enyne **186**. The reaction is explained by the oxidative addition of iodosilane, the insertion of an alkyne to generate the vinylpalladium **185**, and the capture of **185** with the alkynylstannane **184**[102].

Ph—≡ + Me₃SiI + Ph—≡—SnBu₃ →[Pd(Ph₃P)₄, dioxane][60°, 71%]

184

[Me₃Si—Pd-I ⟶ Ph   H / IPd   SiMe₃] ⟶ Ph   H / SiMe₃

185   **186**

The double germylation of alkynes takes place by the reaction of the digermane **187** and trigermane. Pd₂(dba)₃ combined with TMPP is an active catalyst[103]. The ring expansion product **189** was obtained by the Pd-catalyzed insertion of an alkyne into the digermirane **188**[104]. The *cis* cyanogermylation of terminal alkynes with trimethylgermyl cyanide yields **190**[105]. Dimethylgermylene (**191**), generated from 7-germanorbornadiene, reacts with alkynes to form the 1-germacyclopenta-2,4-diene **192**[106].

Ph—≡ + ClMe₂GeGeMe₂Cl →[Pd₂(dba)₃ / TMPP, 72%] (product: Ph / ClMe₂Ge / GeMe₂Cl)

187

Ar₂Ge–GeAr₂ + MeO₂C—≡—CO₂Me →[Pd(Ph₃P)₄ / 51%] (product: Ar₂Ge / GeAr₂ / MeO₂C / CO₂Me)

188   **189**

Ar = (2,6-diethylphenyl)

Ph—≡ + Me₃GeCN →[PdCl₂, PhMe / reflux, 99%] (product: Ph / H / NC / GeMe₃)

190

≡—Bu + Me₂Ge →[Pd(Ph₃P)₄] (product: Bu / Bu / Ge / Me Me)

191   **192**

Diaryl disulfides and diselenides add to alkynes to afford the (Z)-1,2-bis(arylthio)alkenes **193** and (Z)-1,2-bis(arylseleno)alkenes **194**. Under CO pressure, carbonylative addition takes place to give thio esters and the selenoketones **195**[107]. The selenoketones are converted into the β-seleno-α, β-unsaturated aldehydes **196** by Pd-catalyzed hydrogenolysis with HSnBu₃[108,109].

(alkyne) + PhSSPh →[Pd(Ph₃P)₄ / 91%] (product: PhS / SPh)

193   91%

(alkyne) + PhSeSePh →[Pd(Ph₃P)₄ / 82%] (product: PhSe / SePh)

194

(alkyne) + CO + (PhSe)₂ →[Pd(Ph₃P)₄ / 80%] (product: PhSe / SePh / O)

195

→[Bu₃SnH / Pd(Ph₃P)₄, 91%] (product: PhSe / CHO)

196

Hydrometallation is catalyzed by Pd. Hydroboration of 1-buten-2-methyl-3-yne (**197**) with catecholborane (**198**) gives the 1,4-adduct **199** with 84% selectivity. The ratio of Pd to phosphine (1:1.5) is important[110]. The vinyl sulfide **201** is prepared by a one-pot reaction of the thioalkyne **200** via a Pd-catalyzed hydroboration–coupling sequence using dppf as a ligand[111].

Hydrostannation of alkynes offers a good synthetic method for vinylstannanes, which can be converted further into other functionalized alkenes. The free radical-type hydrostannation of alkynes with HSnBu₃ proceeds to give the *trans* adducts, but metal catalysts accelerate the addition. Pd-catalyzed hydrostannation proceeds rapidly at room temperature to give the *cis* adducts 202–206. However, the reaction is said to be not completely regioselective and a mixture of regioisomers such as 205 and 206 is obtained[112–114]. In some cases, highly regioselective *cis* addition is observed. The formation of the (*E*)-1-phenylthiovinylstannane 207 is an example[115].

Stereoselective and chemoselective semihydrogenation of the internal alkyne **208** to the *cis*-alkene **210** is achieved by the Pd-catalyzed reaction of some hydride sources. Tetramethyldihydrosiloxane (TMDHS) (**209**) is used in the presence of AcOH[116]. (EtO)$_3$SiH in aqueous THF is also effective for the reduction of alkynes to *cis*-alkenes[117]. Semihydrogenation to the *cis*-alkene **211** is possible also with triethylammonium formate with Pd on carbon[118]. Good yields and high *cis* selectivity are obtained by catalysis with Pd$_2$(dba)$_3$–Bu$_3$P[119].

Arylthiols (but not alkylthiols) add to terminal alkynes regioselectively to afford a Markovnikov-type adduct **212** in good yield using Pd(OAc)$_2$ as a catalyst[120]. This result is clearly different from the *anti*-Markovnikov addition induced by a radical initiator. The hydroselenation of terminal alkynes with benzeneselenol catalyzed by Pd(OAc)$_2$ affords the terminal alkene **213**, which undergoes partial isomerization to the internal alkene **214**[121].

6.5 Addition of Carbon, Oxygen, Nitrogen, and Sulfur Nucleophiles

The intramolecular carbopalladation (or insertion) of the triple bond in dimethyl 4-pentynylmalonate (**215**) with Pd—H species and malonate anion as shown by **216** proceeds in the presence of *t*-BuOK and 18-crown ether, affording the methylenecyclopentane derivatives **217** and **218**, the amounts of which depend on the reaction conditions. The Pd—H species may be formed by the oxidative addition of BuOH to Pd(0), and the triple bond inserts into this bond[122].

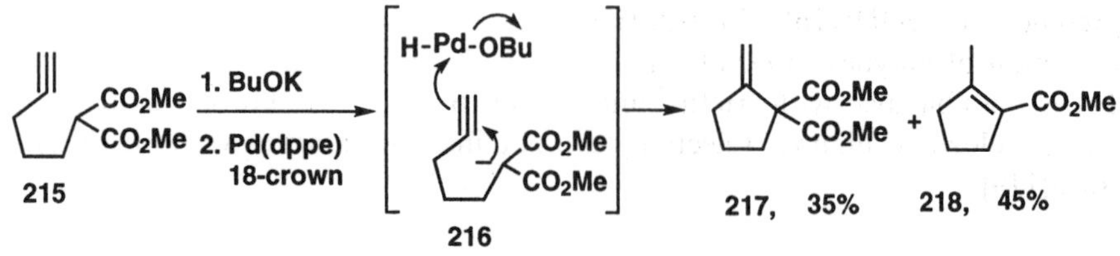

The reactions of alkynes treated in the preceding sections are mostly catalyzed by Pd(0). On the other hand, some addition reactions to alkynes proceed using Pd(II) catalysts by a different mechanism. These examples are shown here. Typically the addition of carboxylic acids, alcohols, and amines to alkynes proceeds by catalysis by Pd(II) salts. The intermolecular stereospecific *trans* hydroacetoxylation of the 2-butynoate **219** with Li acetate to give **220** is catalyzed by Pd(OAc)$_2$[123]. Particularly facile addition takes place intramolecularly[124]. Acetylenecarboxylic acids are converted into unsaturated lactones by using PdCl$_2$(PhCN)$_2$ in THF in the presence of Et$_3$N. The unsaturated lactones **222**, **224**, and **226** are obtained in good yields from 3-pentynoic acid (**221**), 4-pentynoic acid (**223**), and 5-hexynoic acid (**225**)[125].

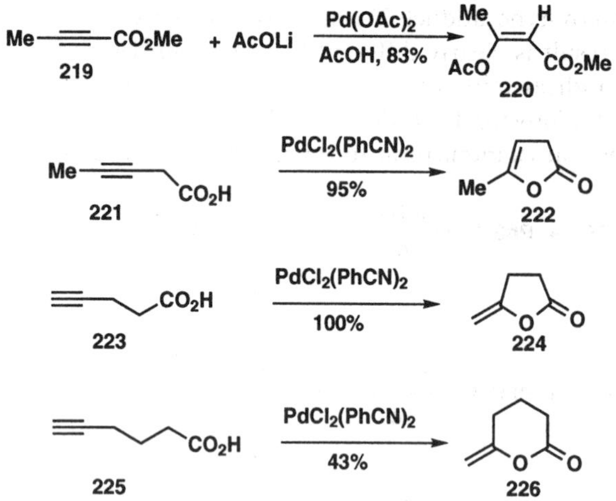

The first step of the reaction is the oxypalladation of the triple bond with PdCl$_2$ as shown by **228** to form the alkenylpalladium species **229**, and the Pd is displaced with proton to regenerate Pd(II) species and the lactone **224**. The alkenylpalladium species **229** can be utilized for further reaction. When allyl chloride (**230**) is added, double bond insertion is followed by elimination of

PdCl$_2$, and the allylated lactone **232** is formed. Regeneration of PdCl$_2$ as shown by **231** makes the reaction catalytic. In this reaction, use of the Li salt **227** of 4-pentynoic acid (**223**) is recommended. Reaction of lithium 3-octynoate (**233**) with allyl chloride affords the unsaturated lactone **234**, which is converted into the γ-keto acid **235** by hydrolysis[126].

The Pd(0)-catalyzed cyclization and allylation of allyl 4-pentynoate (**236**) give the γ-alkylidene-γ-lactone **239**. The reaction is explained by oxypalladation of the triple bond with π-allylpalladium as shown by **237** to generate **238**, followed by reductive coupling to give **239**[127]. The γ-methylene-γ-butyrolactone **242** is prepared similarly in good yields by the Pd(0)-catalyzed reaction of the diallyl 2-propynylmalonate **240** with formic acid[128]. The reaction is explained by the oxypalladation with one of the allyl esters to generate **241** and its hydrogenolysis with formic acid. The other allyl ester undergoes decarboxylation–hydrogenolysis with formic acid.

Propargylic alcohol, after lithiation, reacts with CO_2 to generate the lithium carbonate **243,** which undergoes oxypalladation. The reaction of allyl chloride yields the cyclic carbonate **244** and $PdCl_2$. By this reaction hydroxy and allyl groups are introduced into the triple bond to give the α-allyl ketone **245**[129]. Also the formation of **248** from the keto alkyne **246** with CO_2 via *in situ* formation of the carbonate **247** is catalyzed by Pd(0)[130].

Addition of a hydroxy group to alkynes to form enol ethers is possible with Pd(II). Enol ether formation and its hydrolysis mean the hydration of alkynes to ketones. The 5-hydroxyalkyne **249** was converted into the cyclic enol ether **250**[124]. Stereoselective enol ether formation was applied to the synthesis of prostacyclin[131]. Treatment of the 4-alkynol **251** with a stoichiometric amount of $PdCl_2$, followed by hydrogenolysis with formic acid, gives the cyclic enol ether **253.** Alkoxypalladation to give **252** is *trans* addition, because the $Z : E$ ratio of the alkene **253** was 33 : 1.

The cyclic enol ether **255** from the functionalized 3-alkynol **254** was converted into the furans **256** by the reaction of allyl chloride, and **257** by elimination of MeOH[132]. The alkynes **258** and **260**, which have two hydroxy groups at suitable positions, are converted into the cyclic acetals **259** and **261**. Carcogran and frontalin have been prepared by this reaction[124].

The 4-keto group in the alkyne **262** as an enol form adds to the triple bond to give the furan **263**[133]. Even the conjugated keto alkyne **264** was converted into the furan **266** via isomerization to the allenyl ketone **265**[134].

Hydrolysis of enol ethers yields ketones. Simple alkynes cannot be hydrated to give ketones using a Pd catalyst, but properly functionalized alkynes can be hydrated by participation of the functional group. By the participation of a suitably located ketone, the triple bond is converted into a ketone regioselectively. The 1,5-diketone **268** was prepared by the hydration of the 5-keto alkyne of the prostaglandin derivative **267**[135]. In aqueous MeCN, the regioselective hydration of **269** gives the ketol **270**, which is oxidized to 1,4-diketone **271**, the latter being used for the synthesis of dihydrojasmone (**272**)[124,136].

Amines undergo aminopalladation to alkynes. The intramolecular addition of amines to alkynes yields cyclic imines. The 3-alkynylamine **273** was cyclized to the 1-pyrroline **274**, and the 5-alkynylamine **275** was converted into the 2,3,4,5-tetrahydropyridine **276**[137]. Cyclization of *o*-(1-hexynyl)aniline (**277**) affords the palladated indole **278**, which is trapped with proton, or allyl chloride to give the 3-allylindole **279**[138].

The 2-alkylideneindanone **282** is formed by carbopalladation via ring expansion of the alkynylcyclobutenol **280** with palladium trifluoroacetate to yield an intermediate **281** and its protonolysis. 4-Oxygenated 5-alkylidenecyclopentenones react similarly[139].

Diphenylketene (**283**) adds to terminal alkynes to form the diphenylmethylalkyne **285** via the Pd–carbene intermediate **284**[140].

6.6 Isomerization

Isomerization of alkynes to allenes and then to conjugated dienes is catalyzed by Pd(0) species such as Pd(OAc)$_2$ and Ph$_3$P or Pd$_2$(dba)$_3$. The conjugated ketoalkyne **286** was converted into the conjugated dienone **287**[141,142].

However, it was found that the conjugated ketoalkynes are converted into conjugated dienones by catalysis of Ph₃P without using a Pd catalyst. Similarly, the alkadienyl ester **289** was obtained by the Ph₃P-catalyzed isomerization of the conjugated alkynoate **288** in boiling toluene in the presence of AcOH[143]. Propargylic alcohols are converted into α,β-unsaturated ketones using Pd₂(dba)₃ in refluxing MeCN containing ethylene glycol[144]. The 1,4-alkadiyn-3-ol **290** is isomerized to (*E,E,E*)-trienone **291**[145]. A 1,4-diketone is prepared by the isomerization of a 3-alkyne-1,4-diol using Pd(OAc)₂ and *n*-Bu₃P. Ph₃P is not effective. For example, 3-hexyne-2,5-diol (**292**) was converted into 2,5-hexanedione (**293**)[146]. In the presence of Nafion H, isomerization and cyclization of **292** via **294** take place to give 2,5-dimethylfuran (**295**)[147].

6.7 Miscellaneous Reactions

The reaction of allyl halides with terminal alkynes by use of PdCl₂(PhCN)₂ as a catalyst affords the 1-halo-1,4-pentadienes **297**. π-Allylpalladium is not an intermediate in this reaction. The reaction proceeds by chloropalladation of the triple bond by PdCl₂, followed by the insertion of the double bond of the allyl halide to generate **296**. The last step is the regeneration by elimination of PdCl₂, which recycles[148]. The *cis* addition of allyl chloride to alkynes is supported by formation of the cyclopentenone **299** from the addition product **298** by Ni(CO)₄-catalyzed carbonylation[149].

Butyrolactones are prepared by intramolecular reaction of haloallylic 2-alkynoates. The α-chloromethylenebutyrolactone **301** is prepared by the intramolecular reaction of **300**[150,151]. 4'-Hydroxy-2'-alkenyl 2-alkynoates can be used instead of haloallylic 2-alkynoates, and in this reaction, Pd(II) is regenerated by elimination of the hydroxy group[152]. As a related reaction, the α-(chloromethylene)-γ-butyrolactone **304** is obtained from the cinnamyl 2-alkynoate **302** in the presence of LiCl and CuCl$_2$[153]. Isohinokinin (**305**) has been synthesized by this reaction[154]. The reaction is explained by chloropalladation of the triple bond, followed by intramolecular alkene insertion to generate the alkylpalladium chloride **303**. Then PdCl$_2$ is regenerated by attack of CuCl$_2$ on the alkylpalladium bond as a key step in the catalytic reaction.

6.8 References

1. Review: D. J. Thompson, in *Comprehensive Organic Synthesis*, Vol. 3, Pergamon Press, Oxford, 1991, p. 1015.
2. H. M. Colquhoun, D. J. Thompson, and M. V. Twigg, *Carbonylation*, Plenum Press, New York, 1991.
3. K. Mori, T. Mizoroki, and A. Ozaki, *Chem. Lett.*, 39 (1975).
4. J. F. Knifton, *J. Mol. Catal.*, **2**, 293 (1977).
5. T. Iimori and M. Shibasaki, *Tetrahedron Lett.*, **27**, 2149 (1986).
6. T. Hiyama, N. Wakasa, T. Ueda, and T. Kusumoto, *Bull. Chem. Soc. Jpn.*, **63**, 640 (1990).
7. E. Drent, P. Arnold, and P. H. M. Budzelaar, *J. Organomet. Chem.*, **455**, 247 (1993).
8. S. Torii, H. Okumoto, M. Sadakane, and L. H. Xu, *Chem. Lett.*, 1673 (1991).
9. K. Itoh, M. Miura, and M. Nomura, *Tetrahedron Lett.*, **33**, 5369 (1992); Y. Kushino, K. Itoh, M. Miura, and M. Nomura, *J. Mol. Catal.*, **89**, 151 (1994).
10. B. El Ali and H. Alper, *J. Mol. Catal.*, **67**, 29 (1991); H. Alper, M. Saldana-Maldonado, and I. J. B. Lin, *J. Mol. Catal.*, **49**, L27 (1988).
11. H. Alper and M. Saldana-Maldonado, *Organometallics*, **8**, 1124 (1989)
12. B. El Ali, G. Vasapollo, and H. Alper, *J. Org. Chem.*, **58**, 4739 (1993); D. Zargarian and H. Alper, *Organometallics*, **12**, 712 (1993).
13. J. R. Norton, K. E. Shenton, and J. Schwartz, *Tetrahedron Lett.*, 51 (1975); T. F. Murray, V. Varma, and J. R. Norton, *Chem. Commun.*, 907 (1976); *J. Am. Chem. Soc.*, **101**, 4107 (1979).
14. C. G. Chardarian, S. L. Wov, R. C. Clark, and C. H. Heathcock, *Tetrahedron Lett.*, 1769 (1976).
15. T. F. Murray, E. G. Samsel, V. Varma, and J. R. Norton, *J. Am. Chem. Soc.*, **193**, 7520 (1981).
16. T. F. Murray, V. Varma, and J. R. Norton, *J. Org. Chem.*, **43**, 353 (1978); *J. Am. Chem. Soc.*, **99**, 8085 (1977).
17. B. El Ali and H. Alper, *J. Org. Chem.*, **56**, 5357 (1991).
18. J. Tsuji and T. Nogi, *Tetrahedron Lett.*, 1801 (1966).
19. H. Alper, B. Despeyroux, and J. B. Woell, *Tetrahedron Lett.*, **24**, 5691 (1983).
20. R. Takeuchi, M. Sugiura, N. Ishii, and N. Sato, *Chem. Commun.*, 1358 (1992); *J. Chem. Soc., Perkin Trans. 1*, 1031 (1993).
21. G. P. Chiusoli, M. Costa, E. Masarati, and G. Salerno, *J. Organomet. Chem.*, **255**, C35 (1983).
22. B. M. Trost, C. Chan, and G. Ruhter, *J. Am. Chem. Soc.*, **109**, 3486 (1987).
23. B. M. Trost, S. Marsaubara, and J. J. Caringi, *J. Am. Chem. Soc.*, **111**, 8745 (1989).
24. M. Ishikawa, J. Ohshita, Y. Ito, and A. Minato, *J. Organomet. Chem.*, **346**, C58 (1988).
25. Review: B. M. Trost, *Acc. Chem. Res.*, **23**, 34 (1990).
26. B. M. Trost and M. Lautens, *J. Am. Chem. Soc.*, **107**, 1781 (1985); B. M. Trost, M. Lautens, C. Chan, D. J. Jebaratnam, and T. Mueller, *J. Am. Chem. Soc.*, **113**, 636 (1991); G. J. Engelbrecht and C. W. Holzapfel, *Tetrahedron Lett.*, **32**, 2161 (1991).
27. B. M. Trost, D. C. Lee, and F. Rise, *Tetrahedron Lett.*, **30**, 651 (1989).
28. B. M. Trost and M. Lautens, *Tetrahedron Lett.*, **26**, 4887 (1985).
29. B. M. Trost and K. Matsuda, *J. Am. Chem. Soc.*, **110**, 5233 (1988).
30. B. M. Trost and D. J. Jebaratnam, *Tetrahedron Lett.*, **28**, 1611 (1987).
31. B. M. Trost and C. Pedregal, *J. Am. Chem. Soc.*, **114**, 7292 (1992).

32. B. M. Trost and B. A. Czeskis, *Tetrahedron Lett.*, **35**, 211 (1994).
33. B. M. Trost and J. Y. L. Chung, *J. Am. Chem. Soc.*, **107**, 4586 (1985).
34. B. M. Trost and G. J. Tanoury, *J. Am. Chem. Soc.*, **109**, 4753 (1987).
35. B. M. Trost and D. C. Lee, *J. Org. Chem.*, **54**, 2271 (1989).
36. B. M. Trost and S. F. Chen, *J. Am. Chem. Soc.*, **108**, 6053 (1986).
37. B. M. Trost, P. A. Hipkind, J. Y. L. Chung, and C. Chan, *Angew. Chem., Int. Ed. Engl.*, **28**, 1502 (1989).
38. B. M. Trost and P. A. Hipkind, *Tetrahedron Lett.*, **33**, 4541 (1992).
39. B. M. Trost and D. T. MacPherson, *J. Am. Chem. Soc.*, **109**, 3483 (1987).
40. B. M. Trost, L. Zhi, and K. Imi, *Tetrahedron Lett.*, **35**, 1361 (1994).
41. B. M. Trost and V. K. Chang, *Synthesis*, 824 (1993).
42. B. M. Trost and G. J. Tanoury, *J. Am. Chem. Soc.*, **110**, 1636 (1988).
43. B. M. Trost and M. K. Trost, *J. Am. Chem. Soc.*, **113**, 1850 (1991).
44. K. Moseley and P. M. Maitlis, *Chem. Commun.*, 1604 (1971); *J. Chem. Soc., Dalton Trans.*, 169 (1974).
45. B. M. Trost and M. K. Trost, *Tetrahedron Lett.*, **32**, 3647 (1991).
46. B. M. Trost, M. Yanai, and K. Hoogsteen, *J. Am. Chem. Soc.*, **115**, 5294 (1993).
47. B. M. Trost and A. S. K. Hashmi, *Angew. Chem., Int. Ed. Engl.*, **32**, 1085 (1993).
48. B. M. Trost and D. C. Lee, *J. Am. Chem. Soc.*, **110**, 7255 (1988).
49. B. M. Trost and F. Rise, *J. Am. Chem. Soc.*, **109**, 3161 (1987); B. M. Trost and R. Braslau, *Tetrahedron Lett.*, **29**, 1231 (1988).
50. P. A. Wender and F. E. McDonald, *J. Am. Chem. Soc.*, **112**, 4956 (1990).
51. B. M. Trost and E. D. Edstrom, *Angew. Chem., Int. Ed. Engl.*, **29**, 520 (1990); B. M. Trost and Y. Kondo, *Tetrahedron Lett.*, **32**, 1613 (1991).
52. B. M. Trost and Y. Shi, *J. Am. Chem. Soc.*, **113**, 701 (1991); **115**, 9421 (1993).
53. B. M.Trost and Y. Shi, *J. Am. Chem. Soc.*, **114**, 791 (1992).
54. E. Negishi, L. S. Harring, Z. Owczarczyk, M. M. Mohamud, and A. Ay, *Tetrahedron Lett.*, **33**, 3253 (1992).
55. S. A. Bapuji, W. B. Motherwell, and M. Shipman, *Tetrahedron Lett.*, **30**, 7107 (1989); R. T. Lewis, W. B. Motherwell, and M. Shipman, *Chem. Commun.*, 948 (1988).
56. A. T. Blomquist and P. M. Maitlis, *J. Am. Chem. Soc.*, **84**, 2329 (1962).
57. H. Dietl and P. M. Maitlis, *Chem. Commun.*, 481 (1968).
58. M. Dietl, H. Reinheimer, J. Moffat, and P. M. Maitlis, *J. Am. Chem. Soc.*, **92**, 2276, 2285 (1970).
59. A. K. Jhingan and W. F. Maier, *J. Org. Chem.*, **52**, 1161 (1987).
60. M. tom Dieck, C. Munz, and C. Muller, *J. Organomet. Chem.*, **384**, 243 (1990).
61. M. tom Dieck, C. Munz, and C. Muller, *J. Organomet. Chem.*, **326**, C1 (1987).
62. C. Munz, C. Stephan, and M. tom Dieck, *J. Organomet. Chem.*, **395**, C42 (1990).
63. H. Suzuki, K. Itoh, Y. Ishii, K. Simon, and J. A. Ibers, *J. Am. Chem. Soc.*, **98**, 8494 (1976); L. D. Brown, K. Itoh, H. Suzuki, K. Hirai, and J. A. Ibers, *J. Am. Chem. Soc.*, **100**, 8232 (1978).
64. K. Itoh, F. Ueda, K. Hirai, and Y. Ishii, *Chem. Lett.*, 877 (1977).
65. K. Itoh, K. Hirai, and M. Sasaki, *Chem. Lett.*, 865 (1981).
66. C. Munz, C. Stephan, and M. tom Dieck, *J. Organomet. Chem.*, **407**, 413 (1991).
67. C. Stephan, C. Munz, and M. tom Dieck, *J. Organomet. Chem.*, **452**, 223 (1993).
68. T. Ishiyama, K. Nishijima, N. Miyaura, and A. Suzuki, *J. Am. Chem. Soc.*, **115**, 7219 (1993).
69. H. Hayami, M. Sato, S. Kanemoto, Y. Morizawa, K. Oshima, and H. Nozaki, *J. Am. Chem. Soc.*, **105**, 4491 (1983).

70. K. Wakamatsu, T. Nonaka, Y. Okuda, W. Tuckmantel, K. Oshima, K. Utimoto, and H. Nozaki, *Tetrahedron*, **42**, 4427 (1986); Y. Okuda, K. Wakamatsu, W. Tuckmantel, K. Oshima, and H. Nozaki, *Tetrahedron Lett.*, **26**, 4629 (1985).
71. S. Sharma and A. C. Oehlschlager, *J. Org. Chem.*, **54**, 5064 (1989).
72. S. Matsubara, J. Hibino, Y. Morizawa, K. Oshima, and H. Nozaki, *J. Organomet. Chem.*, **285**, 163 (1985).
73. M. Kosugi, T. Ohta, and T. Migita, *Bull. Chem. Soc. Jpn.*, **56**, 3539 (1983).
74. T. N. Mitchell, H. Killing, R. Dicke, and R. Wickenkamp, *Chem. Commun.*, 354 (1985); T. N. Mitchell, R. Wickenkamp, A. Amamria, R. Dicke, and U. Schneider, *J. Org. Chem.*, **52**, 4868 (1987).
75. M. Mori, N. Watanabe, N. Kaneta, and M. Shibasaki, *Chem. Lett.*, 1615 (1991).
76. B. L. Chenard and C. M. van Zyl, *J. Org. Chem.*, **51**, 3561 (1986); B. L. Chenard, E. D. Laganis, F. Davidson, and T. V. RajanBabu, *J. Org. Chem.*, **50**, 3666 (1985).
77. K. Ritter, *Synthesis*, 218 (1989).
78. M. C. Norley, P. J. Kocienski, and A. Faller, *Synthesis*, 77 (1994).
79. M. Murakami, H. Amii, N. Takizawa, and Y, Ito, *Organometallics*, **12**, 4223 (1993).
80. K. Ikenaga, K. Hiramatsu, N. Nasaka, and S. Matsumoto, *J. Org. Chem.*, **58**, 5045 (1993).
81. H. Watanabe, M. Kobayashi, M. Saito, and Y. Nagai, *J. Organomet. Chem.*, **216**, 149 (1981).
82. E. Piers and R. T. Skerlj, *Chem. Commun.*, 626 (1986).
83. T. M. Mitchell, A. Amamria, H. Killing, and D. Rutschow, *J. Organomet. Chem.*, **241**, C45 (1983); **304**, 257 (1986).
84. H. Sakurai, Y. Kamiyama, and Y. Nakadaira, *J. Am. Chem. Soc.*, **97**, 931 (1975).
85. H. Okinoshimna, K. Yamamoto, and M. Kumada, *J. Organomet. Chem.*, **86**, C27 (1975); K. Tamao, T. Hayashi, and M. Kumada, *J. Organomet. Chem.*, **114**, C19 (1976).
86. H. Sakurai, Y. Kamiyama, and Y. Nakadaira, *J. Am. Chem. Soc.*, **97**, 931 (1975).
87. H. Watanabe, M. Kobayashi, K. Higuchi, and Y. Nagai, *J. Organomet. Chem.*, **186**, 51 (1980); H. Matsumoto, I. Matsubara, T. Kato. K. Shono, H. Watanabe, and Y. Nagai, *J. Organomet. Chem.*, **199**, 43 (1980); **216**, 149 (1981).
88. Y. Ito, M. Suginome, and M. Murakami, *J. Org. Chem.*, **56**, 1948 (1991).
89. H. Yamashita, M. Catellani, and M. Tanaka, *Chem. Lett.*, 241 (1991).
90. M. Murakami, M. Suginome, K. Fujimoto, and Y. Ito, *Angew. Chem., Int. Ed. Engl.*, **32**, 1473 (1993).
91. H. Yamashita and M. Tanaka, *Chem. Lett.*, 1547 (1992).
92. T. Kusukawa, Y. Kabe, and W. Ando, *Chem. Lett.*, 985 (1993).
93. H. Sakurai and T. Imai, *Chem. Lett.*, 891 (1975).
94. H. Sakurai, Y. Kamiyama, and Y. Nakadaira, *J. Am. Chem. Soc.*, **99**, 3879 (1977).
95. H. Sano and W. Ando, *Chem. Lett.*, 1567 (1988).
96. T. Kusumoto and T. Hiyama, *Tetrahedron Lett.*, **28**, 1807 (1987).
97. C. W. Carlson and R. West, *Organometallics*, **2**, 1801 (1983).
98. M. Murakami, Y. Morita, and Y. Ito, *Chem. Commun.*, 428 (1990).
99. N. Chatani and T. Hanafusa, *Chem. Commun.*, 838 (1985); N. Chatani, T. Takeyasu, N. Horiuchi, and T. Hanafusa, *J. Org. Chem.*, **53**, 3539 (1988).
100. N. Chatani and T. Hanafusa, *Tetrahedron Lett.*, **27**, 4201 (1986).
101. T. Kusumoto, T. Hiyama, and K. Ogata, *Tetrahedron Lett.*, **27**, 4197 (1986).
102. N. Chatani, N. Amishiro, and S. Murai, *J. Am. Chem. Soc.*, **113**, 7778 (1991).

103. T. Hayashi, H. Yamashita, T. Sakakura, Y. Uchimaru, and M. Tanaka, *Chem. Lett.*, 245 (1991); K. Mochida, C. Hodota, H. Yamashita and M. Tanaka, *Chem. Lett.*, 1635 (1992).
104. T. Tsumuraya and W. Ando, *Organometallics*, **8**, 2286 (1989).
105. N. Chatani, N. Horiuchi, and T. Hanafusa, *J. Org. Chem.*, **55**, 3393 (1990).
106. W. P. Brauer and W. P. Neumann, *Synlett*, 431 (1991).
107. H. Kuniyasu, A. Ogawa, S. Miyazaki, I. Ryu, N. Kambe, and N. Sonoda, *J. Am. Chem. Soc.*, **114**, 9796 (1992).
108. H. Kuniyasu, A. Ogawa, K. Higaki, and N. Sonoda, *Organometallics*, **11**, 3937 (1992).
109. H. Kuniyasu, A. Ogawa, and N. Sonoda, *Tetrahedron Lett.*, **34**, 2491 (1993).
110. Y. Matsumoto, M. Naito, and T. Hayashi, *Organometallics*, **11**, 2732 (1992).
111. I. D. Gridnev, N. Miyaura, and A. Suzuki, *J. Org. Chem.*, **58**, 5351 (1993).
112. H. Miyake and K. Yamamura, *Chem. Lett.*, 981 (1989).
113. H. X. Zhang, F. Guibe, and G. Balavoine, *J. Org. Chem.*, **55**, 1857 (1990).
114. Y. Ichinose, H. Oda, K. Oshima, and K. Utimoto, *Bull. Chem. Soc. Jpn.*, **60**, 3468 (1987).
115. P. A. Magriotis, J. T. Brown, and M. E. Scott, *Tetrahedron Lett.*, **32**, 5047 (1991).
116. B. M. Trost and R. Braslau, *Tetrahedron Lett.*, **30**, 4657 (1989).
117. J. M. Tour, J. P. Cooper, and S. L. Pendalwar, *J. Org. Chem.*, **55**, 3452 (1990).
118. J. R. Weir, B. A. Patel, and R. F. Heck, *J. Org. Chem.*, **45**, 4926 (1980).
119. K. Tani, N. Ono, S. Okamoto, and F. Sato, *Chem. Commun.*, 386 (1993).
120. H. Kuniyasu, A. Ogawa, K. Sato, I. Ryu, N. Kambe, and N. Sonoda, *J. Am. Chem. Soc.*, **114**, 5902 (1992).
121. H. Kuniyasu, A. Ogawa, K. Sato, I. Ryu, and N. Sonoda, *Tetrahedron Lett.*, **33**, 5525 (1992).
122. N. Monteiro, J. Gore, and G. Balme, *Tetrahedron Lett.*, **32**, 1645 (1991); *Tetrahedron*, **48**, 10103 (1992).
123. X. Lu, G. Zhu, and S. Ma, *Tetrahedron Lett.*, **33**, 7205 (1992).
124. K. Utimoto, *Pure Appl. Chem.*, **55**, 1845 (1983).
125. C. Lambert, K. Utimoto, and H. Nozaki, *Tetrahedron Lett.*, **25**, 5323 (1984).
126. N. Yanagihara, C. Lambert, K. Iritani, K. Utimoto, and H. Nozaki, *J. Am. Chem. Soc.*, **108**, 2753 (1986).
127. T. Tsuda, Y. Ohashi, N. Nagahama, S. Sumiya, and T. Saegusa, *J. Org. Chem.*, **53**, 2650 (1988).
128. T. Mandai, K. Ohta, N. Baba, M. Kawada, and J. Tsuji, *Synlett*, 671 (1992).
129. K. Iritani, N. Yanagihara, and K. Utimoto, *J. Org. Chem.*, **51**, 5499 (1986).
130. Y. Inoue, K. Ohuchi, S. Imaizumi, H. Hagiwara, and H. Uda, *Synth. Commun.*, **20**, 3063 (1990).
131. M. Suzuki, A. Yanagisawa, and R. Noyori, *J. Am. Chem. Soc.*, **110**, 4718 (1988).
132. Y. Wakabayashi, Y. Fukuda, H. Shiragami, K. Utimoto, and H. Nozaki, *Tetrahedron*, **41**, 3655 (1985)
133. Y. Fukuda, H. Shiragami, K. Utimoto, and H. Nozaki, *J. Org. Chem.*, **56**, 5816 (1991).
134. H. Sheng, S. Lin, and Y. Huang, *Tetrahedron Lett.*, **27**, 4893 (1986).
135. K. Imi, K. Imai, and K. Utimoto, *Tetrahedron Lett.*, **28**, 3127 (1987).
136. A. Arcadi, S. Cacchi, and F. Marinelli, *Tetrahedron*, **49**, 4955 (1993).
137. Y. Fukuda, S. Matsubara, and K. Utimoto, *J. Org. Chem.*, **56**, 5812 (1991).
138. K. Iritani, S. Matsubara, and K. Utimoto, *Tetrahedron Lett.*, **29**, 1799 (1988).
139. L. S. Liebeskind, D. Mitchell, and B. S. Foster, *J. Am. Chem. Soc.*, **109**, 7908 (1987); *J. Org. Chem.*, **109**, 7908 (1990).

140. T. Mitsudo, M. Kadokura, and Y. Watanabe, *Tetrahedron Lett.*, **26**, 3697 (1985).
141. C. Guo and X. Lu, *Tetrahedron Lett.*, **33**, 3659 (1992).
142. B. M. Trost and T. Schmidt, *J. Am. Chem. Soc.*, **110**, 2301 (1988).
143. B. M. Trost and U. Kazmaier, *J. Am. Chem. Soc.*, **114**, 7933 (1992).
144. C. Guo and X. Lu, *J. Organomet. Chem.*, **428**, 259 (1992).
145. C. Guo and X. Lu, *Synlett*, 405 (1992).
146. X. Lu, J. Ji, D. Ma, and W. Shen, *J. Org. Chem.*, **56**, 5774 (1991).
147. J. Ji and X. Lu, *Chem. Commun.*, 764 (1993).
148. K. Kaneda, T. Uchiyama, Y. Fujiwara, T. Imanaka, and S. Teranishi, *J. Org. Chem.*, **44**, 55 (1979); *Tetrahedron Lett.*, 1067 (1974); 2833 (1975); 2005 (1977).
149. F. Camp, J. Coll, A. Llebaria, and J. M. Moreto, *Tetrahedron Lett.*, **29**, 5811 (1988).
150. S. Ma and X. Lu, *Chem. Commun.*, 733 (1990); *J. Org. Chem.*, **56**, 5120 (1991).
151. S. Ma, G. Zhu, and X. Lu, *J. Org. Chem.*, **58**, 3692 (1993).
152. S. Ma and X. Lu, *J. Organomet. Chem.*, **447**, 305 (1993).
153. S. Ma and X. Lu, *J. Org. Chem.*, **58**, 1245 (1993).
154. X. Lu and G. Zhu, *Synlett*, 68 (1993).

7 Reactions of Alkenes Catalyzed by Pd(0) and Pd(II)

Oxidative reactions of alkenes with Pd(II) are treated in Chapter 3, Section 2, catalytic reactions of alkenes with organic halides are discussed in this chapter, Section 1.1, and other catalytic reactions of alkenes are discussed in this section.

7.1 Carbonylation

The oxidative carbonylation of alkenes, which consumes Pd(II), is described in Chapter 3, Section 2.7. The carbonylation reactions treated in this section proceed with a catalytic amount of Pd(0) complexes, involving addition of hydrogen and CO to alkenes, and they are mechanistically different from oxidative carbonylation. Pd(0) is an active catalyst for the carbonylation of alkenes under mild conditions. Most of the Pd(0)-catalyzed carbonylation can be carried out conveniently in a laboratory without high-pressure apparatus[1,2]. The reaction is called hydrocarboxylation (carboxylic acid formation) or hydroesterification (alkoxycarbonylation, ester formation) of alkenes. The carbonylation is explained by two different mechanisms. In one mechanism, the first step is the insertion of alkenes into H—PdX to give an alkylpalladium bond, followed by CO insertion to give the acylpalladium complex **1**. The last step is the reaction of the acylpalladium complex **1** with nucleophiles such as an alcohol or water to give an ester or an acid with regeneration of H—PdX. As another possibility, the Pd–carboxylate complex **2** is formed initially by insertion of CO into a Pd–alkoxide bond. Insertion of the alkene into this complex **2** to give **3** is followed by protonolysis to give an ester and regeneration of the Pd alkoxide. These two mechanisms seem to be operative depending on the reaction conditions and substrates.

$$\text{Pd + HCl} \quad + \quad \text{H-Pd-Cl} \xrightarrow{\quad R\diagup\!\!=\quad} \text{RCH}_2\text{CH}_2\text{-Pd-Cl} \xrightarrow{\quad CO\quad}$$

$$\text{RCH}_2\text{CH}_2\text{CO-Pd-Cl} \xrightarrow{\quad R'OH\quad} \text{RCH}_2\text{CH}_2\text{CO}_2\text{R}' \quad + \quad \text{H-Pd-Cl}$$
$$\qquad\qquad \mathbf{1}$$

$$\text{X-Pd-OR}' \xrightarrow{\quad CO\quad} \text{X-Pd-CO}_2\text{R}' \xrightarrow{\quad R\diagup\!\!=\quad} \text{X-Pd-CRHCH}_2\text{CO}_2\text{R}'$$
$$\qquad\qquad \mathbf{2} \qquad\qquad\qquad\qquad \mathbf{3}$$

$$\xrightarrow{\quad R'OH\quad} \text{RCH}_2\text{CH}_2\text{CO}_2\text{R}' \ + \ \text{X-Pd-OR}'$$

Extensive studies on the efficient carbonylation of alkenes have been carried out under different conditions using several types of Pd catalysts. Usually a mixture of linear and branched saturated esters is obtained by the carbonylation of 1-alkenes. Saturated esters are formed by the carbonylation of simple alkenes in alcohols using $PdCl_2$ as a catalyst. Pd(0) formed from $PdCl_2$ is an active catalyst, and even Pd on carbon is active[3]. $PdCl_2(Ph_3P)_2$ is a more suitable catalyst, and various alkenes are carbonylated smoothly under pressure in alcohols to give high yields of mixtures of saturated esters[4]. Addition of a small amount of HCl accelerates the reaction. Saturated carboxylic acids are obtained by carbonylation in aqueous AcOH[5]. The ratio of linear and branched esters changes depending on the reaction conditions, such as temperature, CO pressure, and kinds and amount of phosphine ligands, and considerable attention has been paid to the control of the regioselectivity. For example, the carbonylation of styrene affords a linear ester with a bidentate ligand and a branched ester with Ph_3P[6]. 6-Methoxy-2-vinylnaphthalene is regioselectively carbonylated to give the branched ester **4** in a good yield using cyclohexyldiphenylphosphine as a ligand[7].

A catalyst system of PdCl$_2$ and CuCl$_2$ in aqueous HCl under oxygen is used for carbonylation under mild conditions (1 atm, room temperature) to give branched esters **5** with high selectivity. This catalyst system is expected to lead to oxidative carbonylation. However, unexpected simple non-oxidative hydroesterification takes place. Pd-oxo complex has been claimed to be an active catalyst for hydroesterification[8]. Asymmetric carbonylation of prochiral alkenes using various chiral ligands has been attempted by several groups, giving rather low *ee*. But suprisingly high regioselectivity and enantioselectivity (91% *ee*) in the carbonylation at room temperature under atmospheric pressure to give the branched ester **4** using (*S*)-(+)-1,1′-binaphthyl-2,2′-diyl hydrogen phosphate as a chiral ligand were reported. The reaction was applied to preparation of optically active Ibuprofen and Naproxen. Selection of the ratio (the substrate/the ligand/PdCl$_2$ = 7.7/0.38/1.0) has been claimed to be critical[9]. The carbonylation of alkenes to give branched esters as the main products proceeds at room temperature and 1 atm using Pd on carbon and an excess of CuCl$_2$, although the reaction is very slow (100% yield after 8 days)[10]. The carbonylation of alkenes in the presence of a formate ester with a similar PdCl$_2$–CuCl$_2$ catalyst system also proceeds under very mild conditions (1 atm and room temperature) to give the branched ester as the main product[11].

On the other hand, the linear ester **7** can be prepared as the major product by the carbonylation of a 1-alkene in the presence of a formate ester using Pd(0) with dppb as a catalyst[12]. The linear acid **8** is obtained as the main product by using Pd(OAc)$_2$ or even Pd on carbon as a catalyst and dppb as a ligand in DME in the presence of formic acid or oxalic acid under CO pressure[13]. The linear ester **9** is obtained from a 1-alkene as the main product using PdCl$_2$(Ph$_3$P)$_2$ coordinated by SnCl$_2$[14].

In the carbonylation of *trans,trans,cis*-CDT, the *trans* double bond is attacked preferentially to give the monoester **10**, and then the diester **11**. Attack of the *cis* double bond to give the triester is slow[15]. Only the C-16 alkene was carbonylated regio- and stereoselectively to give the 16α-carboxylate **12** by carbonylation of the C-5 and C-16 unsaturated steroid[16].

Carbonylation of alkenes bearing suitable functional groups proceeds regioselectively. Carbonylation of vinylsilane and β-substituted vinylsilanes proceeds with high β-regioselectivity to give the β-silyl ester **13**[17]. Carbonylation of acrylate affords dimethyl methylmalonate (**14**) with high regioselectivity[18]. On the other hand, carbonylation of methyl vinyl ketone proceeds in EtOH containing HCl regioselectively to give ethyl levulinate (**15**) by attack at the terminal carbon[19]. This regioselectivity may be explained by the formation of 4-chloro-2-butanone as an intermediate, rather than direct carbonylation of the olefinic bond.

Substituted allylic alcohols are carbonylated using the oxidizing system of PdCl$_2$ and CuCl$_2$ in the presence of HCl and oxygen at room temperature and 1 atm of CO to give the γ-lactone **16** in moderate yields[20]. Carbonylation of secondary and tertiary allylic alcohols catalyzed by Pd$_2$(dba)$_3$ and dppb affords the γ-lactone **17** by selective attack of CO at the terminal carbon under fairly severe conditions[21].

Sometimes, keto esters are formed by successive carbonylation. Carbonylation of ethylene in an alcohol affords propionate as the main product, accompanied by a small amount of 4-oxohexanoate (**18**)[3]. The homo-angelica lactones **19** and **20** are obtained by the carbonylation of ethylene in an aprotic solvent[22]. The 1,3-dioxolan-4-one **21** is obtained by the carbonylation of isopropenyl acetate[23].

Keto esters are obtained by the carbonylation of alkadienes via insertion of the alkene into an acylpalladium intermediate. The five-membered ring keto ester **22** is formed from 1,5-hexadiene[24]. Carbonylation of 1,5-COD in alcohols affords the mono- and diesters **23** and **24**[25]. On the other hand, bicyclo[3.3.1]-2-nonen-9-one (**25**) is formed in 40% yield in THF[26]. 1,5-Diphenyl-3-oxopentane (**26**) and 1,5-diphenylpent-1-en-3-one (**27**) are obtained by the carbonylation of styrene. A cationic Pd–diphosphine complex is used as the catalyst[27].

Copolymerization to form polyketones proceeds by the carbonylation of some alkenes in the absence of nucleophiles. Copolymerization of CO and norbornadiene takes place to give the polyketone **28**[28]. Reaction of ethylene and other alkenes with CO affords the polyketones **29**. The use of cationic Pd catalysts and bipyridyl or 1,10-phenanthroline is important for the polymerization [29–31].

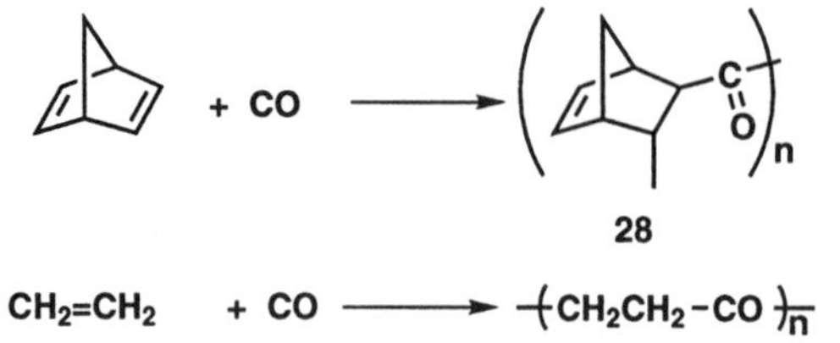

28

$$CH_2=CH_2 \quad + \quad CO \longrightarrow -(CH_2CH_2-CO)_n^-$$

29

Although the yields are not high, $PdCl_2$ or Pd black catalyzes oxo reactions of alkenes with CO and H_2 to give aldehydes[32].

7.2 Addition Reactions (Oligomerization and Co-oligomerization)

A tail-to-tail dimerization of methyl acrylate to give dimethyl 2-hexenedioate (**30**) is catalyzed by Pd(II) salts[33,34]. $Pd(MeCN)_2X_2$ combined with Lewis acids is a good catalyst. High catalytic activity was observed when X = BF_4, and addition of $LiBF_4$ increased the rate of reaction and prolonged the catalytic life. Addition of a vinylcuprate to **30** is followed by Dieckmann condensation to afford the 3-vinylcyclopentanonecarboxylate **31**[35,36]. Cyclization of the 4,4-disubstituted 1,6-diene **32** affords the cyclopentene derivative **33** with Pd(II) catalyst in $CHCl_3$ containing HCl[37]

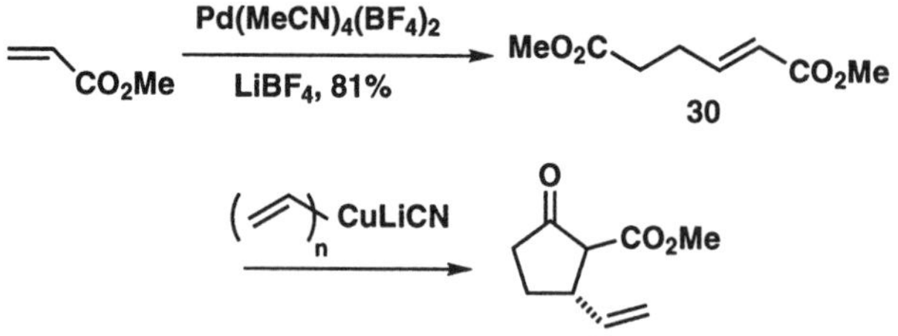

30

31

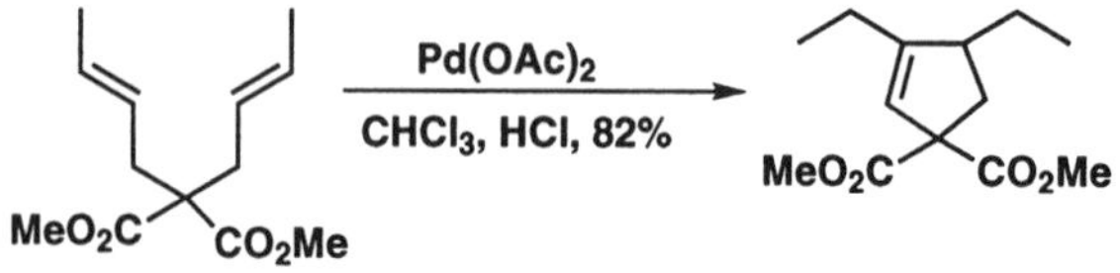

32   **33**

Pd(II)-catalyzed cyclization of the siloxyhexatriene **34** offers a cyclohexenone annulation method. The Pd enolate **35**, formed by transmetallation of the silyl enol ether with Pd(II), is an intermediate which undergoes intramolecular *endo*-alkene insertion. Then Pd(II) is regenerated to give **36**, and finally cyclohexenone is formed[38].

7.3 Hydrosilylation, Bis-silylation, and Reduction

Various Pd–phosphine complexes catalyze the hydrosilylation of alkenes to give the 1-silylalkanes **37** almost exclusively[39]. Their catalytic activity is lower than that of the widely used Pt catalysts. Mainly trichloro- or dichlorosilane is used for the reaction. The reactivity of trialkylsilanes is low. It is difficult to hydrosilylate internal alkenes with Pd catalysts. Hydrosilylation of conjugated dienes is treated in Section 3.1.7. Usually, a silyl group is introduced selectively at the terminal carbon by the palladium-catalyzed hydrosilylation of 1-alkenes. Trifluoropropene afforded 2-silylated fluoroalkane as an exception[40]. However, by use of an optically active monodentate phosphine, namely 2-(diphenylphosphino)-2'-methoxy-1,1'-binaphthyl (MOP) (**38**), as a ligand, surprisingly a trichlorosilyl group is introduced mainly at the internal carbon of terminal alkenes. Moreover, this remarkable ligand accelerates the reaction, which is highly enantioselective. The trichlorosilylalkane **39** is converted into the alcohol **40** in overall 94% *ee*[41]. Norbornanol was obtained in 96% *ee* from norbornene. Monofunctionalization of norbornadiene was achieved with high chemo- and enantioselectivity to give *exo*-2-trichlorosilyl-5-norbornene (**41**) with 1 equiv. of trichlorosilane and converted into the alcohol **42** in 95% *ee*. With 2.5 equiv., the chiral disilylnorbornane **43** was obtained rather than the *meso* isomer **44** (18:1)[42]. Hydrosilylation of 2,5-dihydrofuran (**45**) gave 3-(trichlorosilyl)tetrahydrofuran (**46**) in 65% yield, which was converted into the optically active alcohol **47** (95% *ee*)[43].

Reaction of triethylsilane with α, β-unsaturated aldehydes catalyzed by Pd on carbon gives a *trans*-1,4-adduct as the main product. Reaction of acrolein gave the adduct in 86% yield, in which the 1,4-adduct **48** was 97% and the 1,2-adduct was 3%[44].

1,4-Addition of dihydrosilane to α, β-unsaturated carbonyl compounds such as citral (**49**), followed by hydrolysis, affords saturated citronellal (**50**) directly. The reaction is used for the selective reduction of conjugated double bonds[45,46]. In addition to Pd catalyst, the use of a catalytic amount of

ZnCl$_2$ is essential. The reaction was utilized in the synthesis of strophanthidin. Only the α, β-alkene in the α, β- and γ, δ-unsaturated ketone **51** is reduced selectively[47]. Triethoxysilane is another reducing agent of the enone **52** and simple alkenes[48].

1,4-Addition of difluorotetramethyldisilane or 1,1-dichloro-1-phenyl-2,2,2-trimethyldisilane to the α,β-unsaturated ketone **53** takes place to give the silyl enolate **54** with catalysis by Pd(Ph$_3$P)$_4$ or PdCl$_2$(Ph$_3$P)$_2$[49,50]. Asymmetric 1,4-disilylation of the enone **55** using BINAP as a chiral ligand, followed by oxidation, afforded the 3-hydroxyketone **56** in 92% *ee*[51]. It is important to use unsymmetrically substituted disilanes. Intramolecular bis-silylation of the double bond **57**, followed by oxidative cleavage of C—Si bonds in the product **58**, offers a stereoselective synthetic method for 1,2,4-triol **59**[52]. Similar selective reduction of the double bond in an α, β-unsaturated aldehyde such as citral (**49**) to give citronellal (**50**) is possible by hydrostannation in the presence of AcOH[53,54].

Triethylammonium formate is another reducing agent for α, β-unsaturated carbonyl compounds. Pd on carbon is better catalyst than Pd–phosphine complex, and citral (**49**) is reduced to citronellal (**50**) smoothly[55]. However, the trisubstituted butenolide **60** is reduced to the saturated lactone with potassium formate using Pd(OAc)$_2$. Triethylammonium formate is not effective. Enones are also reduced with potassium formate[56]. Sodium hypophosphite (**61**) is used for the reduction of double bonds catalyzed by Pd on charcoal[57].

7.4 Reactions of Methylenecyclopropanes

Methylenecyclopropane (**62**) and its substituted derivatives function as tri-methylenemethane (TMM) equivalents in the presence of Pd catalyst (for other preparative methods for the TMM complex, see Sections 2.2.2.1 and 2.2.2.2). They can be regarded as reactive alkenes. They undergo Pd-catalyzed [3 + 2] cycloaddition (codimerization) with various alkenes through the Pd–TMM complex **63**, generated by distal ring opening[58]. Methylenecyclopropane reacts with acrylate and norbornene stereoselectively using *i*-Pr$_3$P at 100–140 °C to give the five-membered ring compounds **64** and **65**[58a]. The

isomeric diphenyl-substituted methylenecyclopropanes **66** and **67** react with cyclopentene and butadiene, giving products **68** and **69** of a similar type. Electron-poor alkenes of cyclopentenone and cycloheptenone react with **66** and **67** regioselectively to give the bicyclic ketones **70** and **71** in high yields[59].

An interesting synthetic method for the [3.3.3]propellane **74** by intramole-cular cycloaddition of a disubstituted methylenecyclopropane with an unsaturated ester in the cyclooctane derivative **72** via the trimethylenemethane-like intermediate **73** in nearly quantitive yield has been reported[60].

Cycloaddition of CO_2 with the dimethyl-substituted methylenecyclopropane **75** proceeds smoothly above 100 °C under pressure, yielding the five-membered ring lactone **76**. The regiochemistry of this reaction is different from that of above-mentioned diphenyl-substituted methylenecyclopropanes **66** and **67**[61]. This allylic lactone **76** is another source of trimethylenemethane when it is treated with Pd(0) catalyst coordinated by dppe in refluxing toluene to generate **77**, and its reaction with aldehydes or ketones affords the 3-methylenetetrahydrofuran derivative **78** as expected for this intermediate. Also, the lactone **76** reacts with α, β-unsaturated carbonyl compounds. The reaction of coumarin (**79**) with **76** to give the chroman-2-one derivative **80** is an example[62].

The reaction of phenylmethylenecyclopropane with trimethylsilyl cyanide catalyzed by $PdCl_2$ affords the allylsilane **81** in 71% yield[63].

7.5 Miscellaneous Reactions

Addition of nucleophiles to both activated and unactivated alkenes is catalyzed by Pd(II). Addition of alcohols or AcOH to alkenes bearing EWGs is catalyzed by $PdCl_2(PhCN)_2$ to give the corresponding ethers and esters. The addition of an alcohol to the cyclic acetal of acrolein **82** to give the ether **83** is also possible with the same catalyst[64]. Amines add to the vinylic ether **84** to give **85**, but not to simple alkenes[65].

Ethyl *trans*-2-butenyl sulfone (**86**) together with some ethyl vinyl sulfone are obtained by the reaction of ethylene and SO_2 in wet benzene using $PdCl_2$. SO_2 behaves mechanistically similarly to CO in this reaction[66]. Hydrosulfination of alkenes with SO_2 and H_2 is catalyzed by the Pd(dppp) complex. The sulfinic acid **87** is a primary product, which reacts further to give the S-alkyl alkanethiosulfonates **88** as the major product, and **89** and the sulfonic acid **90** as minor products[67].

Conjugate addition of a phenylmercury reagent to an enone with catalysis by tributylammonium palladium trichloride affords the β-phenyl ketone **91**[68]. Arylboronic acids undergo a Heck-type reaction with α, β-unsaturated ketones, aldehydes, and esters in AcOH by using Pd(OAc)$_2$ as a catalyst. In this reaction an arylpalladium species is formed by the oxidative addition of phenylboronic acid to Pd(0) and undergoes a Heck-type reaction. Methyl cinnamate (**93**) is prepared by the coupling of phenylboronic acid (**92**) with methyl acrylate[69]. However, 3-phenylcyclohexanone (**94**) is obtained by the hydro-phenylation of cyclohexenone with sodium tetraphenylborate when a catalytic amount of SbCl$_3$ is added[70]. Pd-catalyzed hydroarylation of α, β-unsaturated ketones and aldehydes with Ar$_3$Sb proceeds in the presence of silver acetate to give **95**. Styrene, acrylate, and other alkenes undergo a Heck-type reaction with Ar$_3$Sb under the same conditions[71,72]. It should be noted that a Heck-type reaction is possible even with Ph$_3$P.

PdCl$_2$–CuCl$_2$ catalyzes the condensation of branched-chain alkenes with formaldehyde to give the 1,3-dioxanes **96a** and **96b** (Prins reaction)[73]. The yields are much higher than in the conventional acid-catalyzed Prins reaction.

7.6 References

1. Review: D. J. Thompson, in *Comprehensive Organic Synthesis*, Vol. 3, Pergamon Press, Oxford, 1991, p. 1015.
2. H. M. Colquhoun, D. J. Thompson, and M. V. Twigg, *Carbonylation*, Plenum Press, New York, 1991.
3. J. Tsuji, M. Morikawa, and J, Kiji, *Tetrahedron Lett.*, 1437 (1963).
4. K. Bittler, N. V. Kutepow, K. Neubauer, and H. Reis, *Angew. Chem.*, **80**, 352 (1968).
5. D. M. Fenton, *J. Org. Chem.*, **38**, 3192 (1973).
6. Y. Sugi, K. Bando, and S. Shin, *Chem. Ind. (London)*, 397 (1975).
7. T. Hiyama, N. Wakasa, and T. Kusumoto, *Synlett*, 569 (1991).
8. H. Alper, J. B. Woell, B. Despeyroux, and D. J. H. Smith, *Chem. Commun.*, 1270 (1983); H.Alper and D. Leonard, *Tetrahedron Lett.*, **26**, 5639 (1985).
9. H. Alper and N. Hamel, *J. Am. Chem. Soc.*, **112**, 2803 (1990).
10. K. Inomata, S. Toda, and H. Kinoshita, *Chem. Lett.*, 1567 (1990).
11. M. Mlekuz, F. Joo, and H. Alper, *Organometallics*, **6**, 1591 (1987).
12. I. J. B. Lin and H. Alper, *Chem. Commun.*, 248 (1989).
13. B. El Ali and H. Alper, *J. Mol. Catal.*, **77**, 7 (1992); *J. Org. Chem.*, **58**, 3595 (1993); B. El Ali, G. Vasapollo, and H. Alper, *J. Org. Chem.*, **58**, 4739 (1993).
14. J. F. Knifton, *J. Org. Chem.*, **41**, 2885 (1976); *J. Am. Oil Chem. Soc.*, **55**, 496 (1978).
15. J. Tsuji and T. Nogi, *Bull. Chem. Soc. Jpn.*, **39**, 146 (1966).
16. S. Toros and B. Heil, *Tetrahedron Lett.*, **33**, 3667 (1992).
17. R. Takeuchi, N. Ishii, Sugiura, and N. Sato, *Chem. Commun.*, 1247 (1991); *J. Org. Chem.*, **57**, 4189 (1992).
18. G. Consiglio, S. C. A. Nefkens, C. Pisano, and F. Wenzinger, *Helv. Chim. Acta*, **74**, 323 (1991).
19. G. Gavinato and L. Toniolo, *J. Mol. Catal.*, **58**, 251 (1990).
20. H. Alper and D. Leonard, *Tetrahedron Lett.*, **26**, 5639 (1985).
21. B. El Ali and H. Alper, *J. Org. Chem.*, **56**, 5357 (1991).
22. ICI, *Br. Pat.*, 1 148 043; *Neth. Pat. Appl.*, 6 511 995; *Chem. Abstr.*, **65**, 7064 (1966).
23. K. Kudo, K. Mitsuhashi, S. Mori, K. Komatsu, and N. Sugita, *Chem. Lett.*, 1615 (1993).
24. S. Brewis and P. R. Hughes, *Chem. Commun.*, 489 (1965); 71 (1967).
25. J. Tsuji, S. Hosaka, J. Kiji, and T. Susuki, *Bull. Chem. Soc. Jpn.*, **39**, 141 (1966).
26. S. Brewis and P. R. Hughes, *Chem. Commun.*, 6 (1966).
27. C. Pisano, G. Consiglio, A. Sironi, and M. Moret, *Chem. Commun.*, 421 (1991); C. Pisano, A. Mezzetti, and G. Consiglio, *Organometallics*, **11**, 20 (1992); *Angew. Chem., Int. Ed. Engl.*, **30**, 989 (1991).
28. J. Tsuji and S. Hosaka, *Polym. Lett.*, **3**, 703 (1965).
29. E. Dent, J. A. M. van Broekhoven, and M. J. Doyle, *J. Organomet. Chem.*, **417**, 235 (1991).
30. M. Brookhart, F. C. Rix, J. M. DeSimone, and J. C. Barborak, *J. Am. Chem. Soc.*, **114**, 5894 (1992).
31. A. Sen and T. W. Lai, *J. Am. Chem. Soc.*, **104**, 3520 (1982); A. Sen, *Acc. Chem. Res.*, **26**, 303 (1993).
32. J. Tsuji, N. Iwamoto, and M. Morikawa, *Bull. Chem. Soc. Jpn.*, **38**, 2213 (1965).
33. M. G. Barlow, M. J. Bryant, R. N. Haszeldine, and A. G. Mackie, *J. Organomet. Chem.*, **21**, 215 (1970).
34. D. Grenouillet, D. Neibecker, and I. Tkatchenko, *Organometallics*, **3**, 1130 (1984).

35. G. Oehme and H. Pracejus, *Tetrahedron Lett.*, 343 (1979); *J. Organomet. Chem.*, **320**, C56 (1987).
36. W. A. Nugent and R. J. McKinney, *J. Mol. Catal.*, **29**, 65 (1985); W. A. Nugent and F. W. Hobbs, Jr, *J. Org. Chem.*, **48**, 5364 (1983); *Org. Synth.*, **66**, 52 (1987).
37. R. Grigg, T. R. B. Mitchell, and A. Ramasubbu, *Chem. Commun.*, 669 (1979).
38. C. M. Hettrick and W. J. Scott, *J. Am. Chem. Soc.*, **113**, 4903 (1991).
39. M. Hara, K. Ohno, and J. Tsuji, *Chem. Commun.*, 247 (1971); J. Tsuji, M. Hara, and K. Ohno, *Tetrahedron*, **30**, 2143 (1974).
40. I. Ojima, M. Yatabe, and T. Fuchikami, *J. Organomet. Chem.*, **260**, 335 (1984).
41. Y. Uozumi and T. Hayashi, *J. Am. Chem. Soc.*, **113**, 9887 (1991); *Pure Appl. Chem.*, **64**, 1911 (1992).
42. Y. Uozumi, S. Y. Lee, and T. Hayashi, *Tetrahedron Lett.*, **33**, 7185 (1992).
43. Y. Uozumi and T. Hayashi, *Tetrahedron Lett.*, **34**, 2335 (1993).
44. R. Bourhis, E. Frainnet, and F. Moulines, *J. Organomet. Chem.*, **141**, 157 (1977).
45. E. Keinan and N. Greenspoon, *Tetrahedron Lett.*, **26**, 1353 (1985); *J. Am. Chem. Soc.*, **108**, 7314 (1986).
46. J. M. Tour and S. L. Pendalwar, *Tetrahedron Lett.*, **31**, 4719 (1990).
47. P. Kocovsky and I. Stieborova, *Tetrahedron Lett.*, **30**, 4295 (1989).
48. J. M. Tour, J. P. Cooper, and S. L. Pendalwar, *J. Org. Chem.*, **55**, 3452 (1990).
49. K. Tamao, S. Okazaki, and M. Kumada, *J. Organomet. Chem.*, **146**, 87 (1978).
50. T. Hayashi, Y. Matsumoto, and Y. Ito, *Tetrahedron Lett.*, **29**, 4147 (1988).
51. T. Hayashi, Y. Matsumoto, and Y. Ito, *J. Am. Chem. Soc.*, **110**, 5579 (1988); *Tetrahedron*, 50, 335 (1994).
52. M. Murakami, P. G. Andersson, M. Suginome, and Y. Ito, *J. Am. Chem. Soc.*, **113**, 3987 (1991); M. Murakami, M. Suginome, K. Fujimoto, H. Nakamura, P. G. Andersson, and Y. Ito, *J. Am. Chem. Soc.*, **115**, 6487 (1993).
53. P. Four and F. Guibe, *Tetrahedron Lett.*, **23**, 1825 (1982).
54. E. Keinan and P. A. Gleize, *Tetrahedron Lett.*, **23**, 477 (1982).
55. N. A. Cortese and R. F. Heck, *J. Org. Chem.*, **43**, 3985 (1978)
56. A. Arcadi, E. Bernocchi, S. Cacchi, and F. Marinelli, *Synlett*, 27 (1991).
57. R. Sala, G. Doria, and C. Passarotti, *Tetrahedron Lett.*, **25**, 4565 (1984).
58. Reviews: D. M. T. Chan, in *Comprehensive Organic Synthesis*, Vol. 5, Pergamon Press, Oxford, 1991, p. 271; T. Ohta and H. Takaya, p. 1185.
58a.P. Binger and U. Schuchardt, *Chem. Ber.*, **114**, 3313 (1981); *Angew. Chem., Int. Ed. Engl.*, **16**, 255 (1977).
59. P. Binger and P. Bentz, *Angew. Chem., Int. Ed. Engl.*, **21**, 622 (1982).
60. S. Yamago and E. Nakamura, *Chem. Commun.*, 1112 (1988); *Tetrahedron*, **45**, 3081 (1989).
61. Y. Inoue, T. Hibi, M. Satake, and H. Hashimoto, *Chem. Commun.*, 982 (1979).
62. Y. Inoue, S. Ajika, M. Toyofuku, A. Mori, T. Fukui, Y. Kawashima, S. Miyano, and H. Hashimoto, *J. Mol. Catal.*, **32**, 91 (1985).
63. N. Chatani, T. Takeyasu, and T. Hanafusa, *Tetrahedron Lett.*, **29**, 3979 (1988).
64. T. Hosokawa, T. Shinohara, Y. Ooka, and S. Murahashi, *Chem. Lett.*, 2001 (1989).
65. T. Baig, J. Jenck, and P. Kalck, *Chem. Commun.*, 1552 (1992).
66. H. S. Klein, *Chem. Commun.*, 377 (1968).
67. W. Keim and J. Herwig, *Chem. Commun.*, 1592 (1993).
68. S. Cacchi, D. Misiti, and G. Palmieri, *Tetrahedron*, **37**, 2941 (1981); S. Cacchi, F. La Torre, and D. Misiti, *Tetrahedron Lett.*, 4591 (1979).
69. C. S. Cho and S. Uemura, *J. Organomet. Chem.*, **465**, 85 (1994).
70. C. S. Cho, S. Motofusa, and S. Uemura, *Tetrahedron Lett.*, **35**, 1739 (1994).

71. C. S. Cho, K. Tanabe, and S. Uemura, *Tetrahedron Lett.*, **35**, 1275 (1994).
72. T. Kawamura, K. Kikukawa, M. Takagi, and T. Matsuda, *Bull. Chem. Soc. Jpn.*, **50**, 2021 (1977).
73. S. Sakai, Y. Kawashima, Y. Takahashi, and Y. Ishii, *Chem. Commun.*, 1065 (1970).

Chapter 5

Various Reactions Catalyzed by Pd(II) and Pd(0)

In this chapter, reactions catalyzed by either Pd(0) or Pd(II), which do not belong to the reactions classified in Chapters 3 and 4, are treated. The mechanisms of most of them are not clear, and sometimes it is difficult to tell which is the real catalytic species, Pd(0) or Pd(II). Therefore, it is inevitable that they are treated rather arbitrarily.

1 Exchange Reactions of Alkenyl Ethers and Esters Catalyzed by Pd(II)

Vinyl ethers and vinyl esters are activated by the coordination of Pd(II) catalyst. An alcoholic component of vinyl ethers can be exchanged with other alcohols by use of PdCl$_2$ via reversible alkoxypalladation and dealkoxypalladation[1]. The exchange reaction of ethyl vinyl ether (1) to give butyl vinyl ether (2) is catalyzed by PdCl$_2$(PhCN)$_2$ at $-40\,^\circ$C, whereas at higher temperature a mixture of acetals 3a, 3b, and 3c is formed. Actually the acetal formation is catalyzed by HCl generated from PdCl$_2$. Therefore, only ethyl vinyl ether (1) is obtained cleanly from butyl vinyl ether (2) in the presence of an excess of ethanol when Pd(OAc)$_2$ coordinated by 1,10-phenanthroline (phen) or 2,2$'$-bipyridyl (bpy) is used as the catalyst[2].

528

The key step in the total synthesis of rhizobitoxine is the Pd-catalyzed exchange reaction of the methyl alkenyl ether moiety in **4** with the functionalized alcohol, although the yield is low[3]. The enol pyruvate **6** (α-ethoxyacrylic acid) is prepared by the reaction of methyl α-methoxyacrylate or α-methoxyacrylic acid (**5**) with ethanol catalyzed by PdCl$_2$(PhCN)$_2$ at room temperature in the presence of CuCl$_2$ and NaH$_2$PO$_4$[4].

The PdCl$_2$-catalyzed formation of enol ethers and acetals and their exchange reactions are applied to the protection of carbonyl groups as acetals and their deprotection. In these reactions, PdCl$_2$ can be regarded as a Lewis acid. It should be noted that the reaction catalyzed by Pd(II) compounds can be carried out under almost neutral or slightly acidic conditions. Alcohols add smoothly to 2-benzyloxy-1-propene (**7**) to give the acetal **8** at room temperature by catalysis of PdCl$_2$(COD), and this reaction is used as a protection method for alcohols. Deprotection can be effected by catalytic hydrogenolysis of benzyloxy group[5]. PdCl$_2$(MeCN)$_2$ is an efficient catalyst for the hydrolysis of acetals in aqueous MeCN. The best method of deprotection is the Pd(II)-catalyzed exchange reaction of acetals and ketals, e.g. **9**, with an excess of acetone at room temperature to regenerate the carbonyl group[6].

The chemoselective desilylation of one of the two different silyl enol ethers in **10** to give the monosilyl enol ether **11** is realized by the Pd-catalyzed reaction of Bu$_3$SnF. The chemoselectivity is controlled by steric congestion and the relative amount of the reagent[7,8]. An interesting transformation of the 6-alkoxy-2,3-dihydro-6H-pyran-3-one **12** into the cyclopentenone derivative **13** proceeds smoothly with catalysis by Pd(OAc)$_2$ (10 mol%)[9].

Exchange reaction of the acid component of vinyl acetate (**14**) with other acids is catalyzed by PdCl$_2$[10–12]. Pd(OAc)$_2$ coordinated by 1,10-phenanthroline or 2,2′-bipyridyl is a good catalyst for the exchange reaction[2]. A trans-vinylation mechanism has been proposed based on the fact that the labeled oxygen is entirely incorporated in vinyl acetate in the exchange reaction of vinyl propionate with ^{18}O-labeled AcOH[11]. Methyl vinyl itaconate (**16**) is prepared by the reaction of the half-ester of itaconic acid (**15**) with vinyl acetate using Li$_2$PdCl$_4$ as a catalyst[13]. Vinyl esters of various carboxylic acids are produced commercially by the exchange reaction of cheaply available vinyl acetate with carboxylic acids. Vinylation of an amide group to form an en-amide is possible by the reaction of vinyl acetate. The N-vinyllactam **18** and a cyclic imide are prepared by the exchange reaction of the lactam **17** and imide with vinyl acetate[14].

2 Palladium-Catalyzed Decomposition of Azo Compounds, Azides, and Peroxides

Various Pd(II) and Pd(0) compounds are good catalysts for the decomposition of azo compounds. Cyclopropanation of alkenes by the addition of carbenes generated from diazomethane and diazoacetate is carried out using π-allylpalladium chloride[15], Pd(OAc)$_2$, and PdCl$_2$(Ph$_3$P)$_2$ as the catalysts[16–20]. The method is particularly useful for the conversion of α,β-unsaturated carbonyl compounds into the cyclopropyl ester **19** and ketone **20**[21]. The reaction has been applied to steroids, prostaglandin derivatives[22], and norbornadiene[23]. The highly chemoselective Pd(OAc)$_2$-catalyzed reaction of diazomethane on the terminal alkene without attacking the ketone in FK 506 has been reported. In the absence of the Pd catalyst, the ketone is attacked[24].

α-Diazo-β-keto esters add to alkenes[25]. 3-Methylcyclopentenone-2-carboxylate (**22**) is prepared by intramolecular insertion of the carbene generated from the α-diazo-β-keto ester **21**[26]. The enynyl-α-diazo keto ester **23** undergoes intramolecular insertion of the alkyne and alkene to the Pd–carbene to give **24**. In the reaction of the diazo keto ester **23**, the Pd catalyst behaves differently from the Rh catalyst commonly used for the decomposition[27].

The reaction of cyclohexene with the diazopyruvate **25** gives unexpectedly ethyl 3-cyclohexenyl malonate (**26**), involving Wolff rearrangement. No cyclopropanation takes place[28]. 1,3-Dipolar cycloaddition takes place by the reaction of acrylonitrile with diazoacetate to afford the oxazole derivative **27**[29]. Bis(trimethylstannyl)diazomethane (**28**) undergoes Pd(0)-catalyzed rearrangement to give the *N*-stannylcarbodiimide **29** under mild conditions[30].

The thiepin system **31** is formed quantitatively by ring expansion of the diazoacetate derivative **30** via carbene formation catalyzed by π-allylpalladium chloride and its intramolecular insertion[31]. The 4-diazomethyl-4*H*-pyrane **32** is expanded to the oxepine **33** in quantitative yield with the same catalyst[32].

Metallic Pd is a good catalyst for the conversion of the primary azide **34** into the nitrile **35** in the presence of a hydrogen acceptor such as diphenylacetylene[33]. By this method, organic halides can be converted into nitriles without increasing the carbon number. Reaction of the azidoformate **36** with an allylic ether affords the *N*-alkoxycarbonyl imine **37** using PdCl$_2$(PhCN)$_2$ as a catalyst[34].

PhCH$_2$CH$_2$Cl + NaN$_3$ ⟶ PhCH$_2$CH$_2$N$_3$ $\xrightarrow[\text{Pd black, 68%}]{\text{Ph}\equiv\equiv\text{Ph}}$ PhCH$_2$CN

34 **35**

$$\text{36} + N_3CO_2Me \xrightarrow[95\%]{PdCl_2(PhCN)_2} \text{37}$$

The α,β-unsaturated linear carbonyl compound **39** is obtained by the decomposition of the cyclic hydroperoxide **38** with PdCl$_2$[35]. The α, β-epoxy ketone **40** is isomerized to the β-diketone **41** with Pd(0) catalyst[36]. The 1,4-epiperoxide **42** is converted into the β-hydroxy ketone **43** and other products[37].

38 $\xrightarrow[80\%]{PdCl_2}$ **39**

40 $\xrightarrow[\text{dppe, 62%}]{Pd(Ph_3P)_4}$ **41**

42 ⟶ **43** HO 40% + OHC 25% + HO 20%

3 Cope Rearrangement and Related Reactions

The first report on the Pd(II)-promoted Cope rearrangement is the conversion of *cis,trans*-1,5-cyclodecadiene (**44**) into *cis*-1,2-divinylcyclohexane–PdCl$_2$ complex (**45**) with a stoichiometric amount of PdCl$_2$(PhCN)$_2$ at room temperature. The complex formation is the driving force of this unusual rearrangement [38,39]. A similar transformation of germacrane (1,5-dimethyl-8-isopropylidene-*trans,trans*-1,5-cyclodecadiene) takes place[40].

Cope rearrangement of 1,5-dienes such as **46** is accelerated considerably with a catalytic amount of PdCl$_2$(PhCN)$_2$ and proceeds at room temperature in benzene or CH$_2$Cl$_2$ without forming the PdCl$_2$ complex of the rearranged 1,5-diene **47**[41–43]. Pd(OAc)$_2$ as a precursor of Pd(0) and Pd(Ph$_3$P)$_4$ are inactive. The rearrangement is catalyzed by Pd(II), but not by Pd(0). Optically active (3R,5E)-2,3-dimethyl-3-phenyl-1,5-heptadiene (**48**) (99% *ee*) rearranges to (2E,5R)-3,5-dimethyl-2-phenyl-2,6-heptadiene (**49**) and (2E,5S)-3,5-dimethyl-2-phenyl-2,6-heptadiene (**50**) in a ratio of 7 : 3 with virtually complete 1,4-transfer of the chirality[44]. This is the same sense of asymmetric induction as observed in the thermal Cope rearrangement of **48**. This result demonstrates that the PdCl$_2$-catalyzed Cope rearrangement of **48** occurs preferentially with a chair topology. However, the failure of this catalytic method with some substituted 1,5-hexadienes has been reported. Successful Pd catalysis appears to require that the C-2 and C-5 atoms of the substituted 1,5-hexadienes have one H and one 'non hydrogen' substituent[42]. 1,5-Hexadienes which are substituted with an EWG at C-3 cyclize to cyclohexenes. For example, the 1,5-hexadiene **51** does not rearrange under the standard conditions at room temperature, and instead, the cyclohexene derivative **52** is formed in 53% yield[45].

The mechanism of the PdCl$_2$-catalyzed Cope rearrangement has been studied by use of the partially deuterated 1,5-diene **53**[46]. The coordination of Pd(II) activates the alkene, and cyclization (carbopalladation) takes place to give the 4-palladacyclohexyl cation **54** as a rate-determining step. Then the rearranged 1,5-diene **55** is formed by the cleavage of the ring **54**.

Oxy-Cope rearrangement of **56** to form the cyclic ketone **57** can be carried out at room temperature with catalysis by $PdCl_2(PhCN)_2$[47].

Bicyclo[6.1.0]non-4-ene (**58**) is converted into *cis,cis*-1,5-cyclononadiene (**59**) coordinated by $PdCl_2$ in a high yield[48]. The cyclopropane ring of bicyclo[5.1.0]oct-3-ene is also opened[49].

The $PdCl_2$-catalyzed instantaneous rearrangement of *N*-carbethoxy-8-azabicyclo[5.1.0]oct-3-ene (**60**) takes place at room temperature to give *N*-carbethoxy-8-azabicyclo[3.2.1]oct-2-ene (**61**)[50]. The azepine **62** undergoes a smooth skeletal rearrangement to give **63**, and the diazepine **64** is converted into the open-chain product[51]. Beckmann fission of the oxime **65** of ketones and aldehydes to give the nitrile **66** is induced by a Pd(0) complex and oxygen [52,53].

The skeletal rearrangement of various strained cyclic compounds is carried out with a catalytic amount of soluble complexes of $PdCl_2$. Namely, the rearrangements of bulvalene (**67**) to bicyclo[4.2.2]deca-2,4,7,9-tetraene (**68**)[54], cubane (**69**) to cuneane (**70**)[55], hexamethyl Dewar benzene (**71**) to hexamethylbenzene (**72**)[56], and 3-oxaquadricyclanes[57] and quadricyclane (**73**) to norbornadiene[58–60] take place mostly at room temperature. Reaction of iodocubane (**74**) with a terminal alkyne catalyzed by Pd(0) and CuBr unexpectedly affords an alkynylcyclooctatetraene **75**, without giving the desired cubylalkyne **76**. Probably the rearrangement is a Pd-catalyzed reaction[61].

4 Decarbonylation of Aldehydes and Related Compounds

Pd is an efficient catalyst for the decarbonylation of aldehydes above 180 °C to form alkenes and the corresponding alkanes[62,63]. Metallic Pd is an active catalyst. In the synthesis of irone from α-pinene (77), *cis*-pinonic aldehyde (78) was decarbonylated at 220 °C to give pinonone (79) and pinonenone (80) in 80% yield[64,65]. α-Pinene (77) is converted into myrtenal (81), which is decarbonylated with Pd on BaSO₄ at 195 °C to give apopinene (82)[66]. Low-boiling aldehydes are decarbonylated in the gas phase using a supported Pd catalyst. Isobutyraldehyde (83), the less useful product of the oxo reaction of propylene, can be converted into CO, propylene, and H₂ by the decarbonylation[67]. Aldehydes are decarbonylated with free radical generators accompanied by skeletal rearrangement. On the other hand, the Pd-catalyzed decarbonylation of branched aldehydes such as tetralin-2-carboxaldehyde, β-phenylisovaleraldehyde (84), and 3,3-dimethyl-4-phenylbutyraldehyde proceeds without undergoing such a rearrangement[68–70].

Decarbonylation of aromatic aldehydes proceeds smoothly[71]. Terephthalic acid (86), commercially produced by the oxidation of *p*-xylene (85), contains *p*-formylbenzoic acid (87) as an impurity, which is removed as benzoic acid (88) by Pd-catalyzed decarbonylation at a high temperature. The benzoic acid produced by the decarbonylation can be separated from terephthalic acid (86) based on the solubility difference in water[72].

The cyano ketone **89** is converted into the nitrile **90** by heating at 140 °C with Pd(Ph₃P)₄[73,74]. The α-ketophosphonate **91** is decarbonylated with PdMe₂(PMePh₂)₂ complex to give the phosphonate **92**[75].

5 Reactions of Carbon Monoxide and Carbon Dioxide

Reductive carbonylation of nitro compounds is catalyzed by various Pd catalysts. Phenyl isocyanate (**93**) is produced by the PdCl₂-catalyzed reductive carbonylation (deoxygenation) of nitrobenzene with CO, probably via nitrene formation. Extensive studies have been carried out to develop the phosgene-free commercial process for phenyl isocyanate production from nitrobenzene[76]. Effects of various additives such as phenanthroline have been studied[77–79]. The co-catalysts of montmorillonite–bipyridylpalladium acetate and Ru₃(CO)₁₂ are used for the reductive carbonylation of nitroarenes[80,81]. Extensive studies on the reaction in alcohol to form the *N*-phenylurethane **94** have also been carried out[82–87]. Reaction of nitrobenzene with CO in the presence of aniline affords diphenylurea (**95**)[88].

$$94$$

$$95$$

As another example of nitrene formation, the reaction of *o*-nitrostilbene (**96**) with CO in the presence of $SnCl_2$ affords 2-phenylindole (**97**). The reaction is explained by nitrene formation by deoxygenation of the nitro group with CO, followed by the addition of the nitrene to alkene. Similarly, the 2*H*-indazole derivative **99** was prepared by reductive cyclization of the *N*-(2-nitrobenzyli-dene)amine **98**[89].

$$96 \qquad\qquad\qquad\qquad\qquad\qquad\qquad 97$$

$$98 \qquad\qquad\qquad\qquad\qquad\qquad 99$$

On the other hand, the carbonylation of the nitroalkane **100** at 190 °C using Pd on carbon and $FeCl_3$ afforded the trialkylpyridine **101**[90].

$$100 \qquad\qquad\qquad\qquad\qquad\qquad 101$$

The α-methylene-β-lactam **103** is obtained by the carbonylation of the methyleneaziridine **102** under mild conditions[91]. The azirine **104** undergoes an interesting dimerization–carbonylation to form the fused β-lactam **105**[92].

$$102 \qquad\qquad\qquad\qquad\qquad 103$$

$$104 \qquad\qquad\qquad\qquad\qquad 105$$

The chiral siloxycyclopropane **106** undergoes carbonylative homocoupling to form the 4-ketopimelate derivative **108** via the palladium homoenolate **107** without racemization. The reaction is catalytic in $CHCl_3$, but stoichiometric in benzene[93].

The Pd-catalyzed reductive carbonylation of methyl acetate with CO and H_2 affords acetaldehyde. The net reaction is the formation of acetaldehyde from MeOH, CO, and H_2[94]. Methyl formate (**109**) is converted into AcOH under CO pressure in the presence of LiI and $Pd(OAc)_2$[95].

$$MeCO_2Me + CO + H_2 \longrightarrow MeCO_2H + MeCHO$$

$$MeCO_2H + MeOH \longrightarrow MeCO_2Me + H_2O$$

$$MeOH + CO + H_2 \longrightarrow MeCHO + H_2O$$

$$\underset{\textbf{109}}{HCO_2Me} + CO \xrightarrow[\text{LiI, NMP}]{Pd(OAc)_2} CH_3CO_2H$$

Formic acid is formed by the reaction of H_2 and CO_2 catalyzed by the dppe complex of Pd[96]. In alcohol, alkyl formates are obtained[97]. DMF is obtained by the reaction of CO_2 (40 atm) and Me_2NH under a high pressure of H_2 (80 atm) in the presence of a base in methyl Cellosolve[98]. The formate formation is explained by the insertion of CO_2 into a Pd—H bond to form Pd–formate species. Tetraethylurea (**110**) and diethylformamide (**111**) are obtained by the reaction of Et_2NH and CO_2[99].

$$CO_2 + H_2 + H_2O \xrightarrow[\text{amine}]{Pd / dppe} HCO_2H$$

$$CO_2 + H_2 + ROH \longrightarrow HCO_2R$$

$$CO_2 + H_2 + Me_2NH \xrightarrow[150°, 99\%]{PdCl_2, K_2CO_3} Me_2NHCO$$

$$Et_2NH + CO_2 \xrightarrow[\text{CCl}_4, HCO_2Na]{PdCl_2(MeCN)_2, Ph_3P} \underset{\textbf{110}}{Et_2NCONEt_2} + \underset{\textbf{111}}{Et_2NCHO}$$

6 Reduction

Formic acid is a good reducing agent in the presence of Pd on carbon as a catalyst. Aromatic nitro compounds are reduced to aniline with formic acid[100]. Selective reduction of one nitro group in 2,4-dinitrotoluene (**112**) with triethylammonium formate is possible[101]. *o*-Nitroacetophenone (**113**) is first reduced to *o*-aminoacetophenone, then to *o*-ethylaniline when an excess of formate is used[102]. Ammonium and potassium formate are also used for the reduction of aliphatic and aromatic nitro compounds. Pd on carbon is a good catalyst[103,104]. NaBH$_4$ is also used for the Pd-catalyzed reduction of nitro compounds[105]. However, the α,β-unsaturated nitroalkene **114** is partially reduced to the oxime **115** with ammonium formate[106].

The heterocyclic rings in quinoline (**116**) and isoquinoline are selectively reduced by Pd on carbon-catalyzed reaction of ammonium formate[107]. Some benzene rings are also reduced. For example, nitrobenzene is reduced to cyclohexylamine (**117**) with formic acid. It is important to use a sevenfold excess of formic acid[108].

Aromatic nitriles are converted into a methyl group with ammonium formate[109]. Aldehydes and ketones are reduced to alcohols[110].

The alkyl azide **118** is reduced to a primary amine by the Pd on carbon-catalyzed reaction of ammonium formate in MeOH at room temperature. No racemization takes place with chiral azides[111,112].

Debenzylation of benzylamines and benzyl ethers is carried out with ammonium formate[113,114]. Hydrosilanes are also used for debenzylation[115].

The α, β-epoxy ketone **119** and esters are hydrogenolyzed with triethylammonium formate or H_2 chemoselectively to aldols[116].

7 Miscellaneous Reactions

Double silylation of the α-diketone **120** with $(Me_3Si)_2$ is catalyzed by $PdCl_2(Me_3P)_2$ to give the 1,2-bis(siloxy)ethylene **121** in a good yield[117]. Hydrosilylation of phenyl isocyanate (**122**) catalyzed by $PdCl_2$ affords the *N*-

silylformamide **123**, and the carbodiimide is converted into the *N*-silylforma-
midine **124**[118].

Silylation of alcohols, amines and carboxylic acids with hydrosilanes is catalyzed by Pd catalysts[119]. Based on this reaction, silyl protection of alcohols,
amines, and carboxylic acids can be carried out with *t*-butyldimethylsilane
using Pd on carbon as a catalyst. This method is simpler and more convenient
than the silylation with *t*-butyldimethylsilyl chloride, which is used commonly
for the protection. Protection of β-hydroxymethyl-β-lactam (**125**) is an example[120].

Isocyanide is isoelectronic with CO and a reactive compound in the presence
of Pd catalysts. The heterobicyclic compound **127** is obtained by the successive
insertion of 2,6-xylyl isocyanide (**126**) into the Pd–hydride bond formed from
the hydrosilane[121]. Aryl isocyanide inserts into the Si—Si bond in oligo-
silanes. For example, 3 mol of 2,6-xylyl isocyanide insert into the tetrasilane
128 to give **129**[122].

Me-Si(Me)(Me)–Si(Me)(Me)–Si(Me)(Me)–Si(Me)(Me)-Me + ArNC **128**

$\xrightarrow[\text{DMF, 55\%}]{\text{Pd(OAc)}_2}$

Me–Si(Me)(Me)–C(=NAr)–Si(Me)(Me)–C(=NAr)–Si(Me)(Me)–C(=NAr)–Si(Me)(Me)-Me **129**

Ar = 2,6-xylyl

1,2-Diisocyanobenzene (**130**) undergoes living polymerization to form the poly(quinoxaline-2,3-diyl)s **131**, and the optically active helical poly(quinoxaline-2,3-diyl) **132** is prepared from **131**[123].

130 $\xrightarrow[\text{100\%}]{\text{MePdBrL}_2}$ **131** $\xrightarrow{\text{MeMgBr}}$ **132**

n = 1 ~ 8

The selective monochlorination of the methyl group in toluene to give benzyl chloride with SO$_2$Cl$_2$ is possible with catalysis by Pd(Ph$_3$P)$_4$[124].

Ph–CH$_3$ + SO$_2$Cl$_2$ $\xrightarrow[\text{PhH}]{\text{Pd(Ph}_3\text{P)}_4}$ Ph–CH$_2$Cl + Ph–CHCl$_2$

98 : 2

Malononitrile (**133**) undergoes Pd(0)-catalyzed dimerization to give **134**[125,126]. The trimerization of malononitrile in boiling benzene catalyzed by potassium tetrakis(pentafluorophenyl)palladate gives 4,6-diamino-3,5-dicyano-2-cyanomethylpyridine (**135**)[127].

CH$_2$(CN)$_2$ **133** $\xrightarrow{\text{Pd(Ph}_3\text{P)}_4}$ **134** (NC)(NC)C=C(NH$_2$)(CH$_2$CN)

CH$_2$(CN)$_2$ **133** $\xrightarrow[\text{65\%}]{\text{K}_2[\text{Pd(C}_6\text{F}_5)_4]}$ **135**

The imidazolidenimine **138** is formed by the addition of the aziridine **136** to the carbodiimide **137**[128].

Pd(II) salts are active catalysts for the oxidation of some organic compounds with peroxides. Pd(II) salt-catalyzed oxidative ring cleavage of the five-, six-, and seven-membered cyclic acetals **139** with t-BuO$_2$H affords the monoesters **140** of diols in good yields[129]. A catalytic amount of Pd(OCOCF$_3$)(t-BuO$_2$) or Pd(OAc)$_2$ is used. Deprotection of the 4,6-O-benzylidene ring in the carbohydrate **141** to give the monoesters **142** and **143** can be carried out by this method[130,131]. The ene-lactam **144** is oxidatively cleaved with H$_2$O$_2$ to form the imide **145** and N-fused azabicyclic compounds. The reaction was applied to the synthesis of macrocyclic keto imides[132].

8 References

1. J. Smidt, W. Hafner, R. Jira, R. Sieber, J. Sedlmeier, and A. Sabel, *Angew. Chem., Int. Ed. Engl.*, **1**, 80 (1962); J. Smidt, *Chem. Ind.*, 54 (1962).
2. J. E. McKeon, P. Fitton, and A. A. Griswold, *Tetrahedron*, **28**, 227, 233 (1972).
3. D. D. Keith, J. A. Tortora, K. Ineichen, and W. Leimgruber, *Tetrahedron*, **31**, 2633 (1975).
4. G. A. Divers and G. A. Berchtold, *Synth. Commun.*, **7**, 43 (1977).
5. T. Mukaiyama, M. Oshima, and M. Murakami, *Chem. Lett.*, 265, 615 (1984).

6. B. H. Lipshutz, D. Pollart, J. Monforte, and H. Kotsuki, *Tetrahedron Lett.*, **26**, 705 (1985).
7. H. Urabe, Y. Takano, and I. Kuwajima, *J. Am. Chem. Soc.*, **105**, 5703 (1983).
8. L. A. Paquette, P. E. Wiedeman, and P. C. Bulman-Page, *Tetrahedron Lett.*, **26**, 1611 (1985).
9. B. Mucha and H. M. R. Hoffmann, *Tetrahedron Lett.*, **30**, 4489 (1989).
10. P. M. Henry, *J. Am. Chem. Soc.*, **94**, 7316 (1972); **93**, 3853 (1971).
11. P. M. Henry, *Acc. Chem. Res.*, **6**, 16 (1973).
12. A. Sabel, J. Smidt, R. Jira, and H. Prigge, *Chem. Ber.*, **102**, 2939 (1969).
13. D. W. Bjorkquist, R. D. Bush, F. S. Ezra, and T. Keough, *J. Org. Chem.*, **51**, 3192 (1986).
14. E. Bayer and K. Geckeler, *Angew. Chem.*, **18**, 533 (1979).
15. R. K. Armstrong, *J. Org. Chem.*, **31**, 618 (1966).
16. R. Paulissen, A. J. Hubert, and Ph. Teyssie, *Tetrahedron Lett.*, 1465 (1972); A. J. Anciaux, A. J. Hubert, A. F. Noels, N. Petiniot, and P. Teyssie, *J. Org. Chem.*, **45**, 695 (1980).
17. A. Nakamura, T. Koyama, and S. Otsuka, *Bull. Chem. Soc. Jpn.*, **51**, 593 (1978).
18. I. G. Dinulescu, L. N. Enescu, A. Ghenculescu, and M. Avram, *J. Chem. Res.*, 456 (1978).
19. M. W. Majchrzak, A. Kotelko, and J. B. Lambert, *Synthesis*, 469 (1983).
20. M. Suda, Synthesis, 714 (1981).
21. U. Mende, B. Raduchel, W. Skuballa, and H. Vorbruggen, *Tetrahedron Lett.*, 629 (1975).
22. B. Raduchel, U. Mende, G. Cleve, G. A. Hoyer, and H. Vorbruggen, *Tetrahedron Lett.*, 633 (1975).
23. J. Kottwitz and H. Vorbruggen, *Synthesis*, 636 (1975).
24. A. J. F. Edmunds, K. Baumann, M. Grassberger, and G. Schulz, *Tetrahedron Lett.*, **32**, 7039 (1991).
25. F. L. M. Smeets, L. Thijs, and B. Zwanenberg, *Tetrahedron*, **36**, 3269 (1980).
26. D. F. Taber, J. C. Ameddio, and R. G. Sherrill, *J. Org. Chem.*, **51**, 3382 (1986).
27. T. R. Hoye, C. J. Dinsmore, D. S. Johnson, and P. F. Korkowski, *J. Org. Chem.*, **55**, 4518 (1990).
28. S. Bien and Y. Segal, *J. Org. Chem.*, **42**, 1685 (1977).
29. R. Paulissen, Ph. Moniotte, A. J. Hubert, and Ph. Teyssie, *Tetrahedron Lett.*, 3311 (1974).
30. G. Veneziani, R. Reau, and G. Bertrand, *Organometallics*, **12**, 4289 (1993).
31. K. Nakasuji, K. Kawawamura, T. Ishihara, and I. Murata, *Angew. Chem.*, **88**, 650 (1976).
32. K. L. Hoffmann and M. Regitz, *Tetrahedron Lett.*, **24**, 5355 (1983).
33. H. Hayashi, A. Ohno, and S. Oka, *Bull. Chem. Soc. Jpn.*, **49**, 506 (1976).
34. T. Migita, M. Chiba, K. Takahashi, N. Saitoh, S. Nakaido, and M. Kosugi, *Bull. Chem. Soc. Jpn.*, **55**, 3943 (1982); *Chem. Lett.*, 1403 (1978); T. Migita, N. Saitoh, H. Iizuka, and C. Ogyu, *Chem. Lett.*, 1015 (1982).
35. K. Formanek, J. P. Aune, M. Jouffret, and J. Metzger, *Nouv. J. Chim.*, **1**, 13 (1977).
36. M. Suzuki, A. Watanabe, and R. Noyori, *J. Am. Chem. Soc.*, **102**, 2095 (1980).
37. M. Suzuki, R. Noyori, and N. Hamanaka, *J. Am. Chem. Soc.*, **103**, 5606 (1981).
38. J. C. Trebellas, J. R. Olechowski, and H. B. Jonassen, *J. Organomet. Chem.*, **6**, 412 (1966).
39. P. Heimbach and M. Molin, *J. Organomet. Chem.*, **49**, 477, 483 (1973).

40. E. D. Brown, T. M. Sam, J. K. Sutherland, and A. Torre, *J. Chem. Soc., Perkin Trans. 1*, 2326 (1975).
41. Review: R. P. Lutz, *Chem. Rev.*, **84**, 205 (1984).
42. Review: L. E. Overman, *Angew. Chem., Int. Ed. Engl.*, **23**, 579 (1984).
43. L. E. Overman and F. M. Knoll, *J. Am. Chem. Soc.*, **102**, 865 (1980); L. E. Overman and J. Jacobsen, *J. Am. Chem. Soc.*, **104**, 7225 (1982); L. E. Overman and A. F. Renaldo, *Tetrahedron Lett.*, **24**, 3757 (1983).
44. L. E. Overman and E. J. Jacobsen, *J. Am. Chem. Soc.*, **104**, 7225 (1982).
45. L. E. Overman and A. F. Renaldo, *Tetrahedron Lett.*, **24**, 2235 (1983).
46. L. E. Overman and A. F. Renaldo, *J. Am. Chem. Soc.*, **112**, 3945 (1990).
47. N. Bluthe, M. Malacria, and J. Gore, *Tetrahedron Lett.*, **24**, 1157 (1983).
48. G. Albelo and M. F. Rettig, *J. Organomet. Chem.*, **42**, 183 (1972).
49. G. Albelo, G. R. Wiger, and M. F. Rettig, *J. Am. Chem. Soc.*, **97**, 4510 (1975).
50. G. R. Wiger and M. F. Rettig, *J. Am. Chem. Soc.*, **98**, 4168 (1976).
51. K. Sato, Y. Horie, and K. Takahashi, *J. Organomet. Chem.*, **363**, 231 (1989).
52. K. Maeda, I. Moritani, T. Hosokawa, and S. Murahashi, *Chem. Commun.*, 689 (1975).
53. A. J. Leusink, T. G. Meerbeek, and J. G. Noltes, *Recl. Trav. Chim. Pays-Bas*, **95**, 123 (1976).
54. E. Vedejs, *J. Am. Chem. Soc.*, **90**, 4751(1968).
55. L. Cassar, P. E. Eaton, and J. Halpern, *J. Am. Chem. Soc.*, **92**, 6366 (1970).
56. H. Dietl and P. M. Maitlis, *Chem. Commun.*, 759 (1967).
57. H. Hogeveen and B. J. Nusse, *J. Am. Chem. Soc.*, **100**, 3110 (1978).
58. H. Hogeveen and H. C. Volger, *J. Am. Chem. Soc.*, **89**, 2486 (1967).
59. R. B. King and R. M. Hanes, *J. Org. Chem.*, **44**, 1092 (1979).
60. S. Miki, T. Ohno, H. Iwasaki, Y. Maeda, and Z. Yoshida, *Tetrahedron*, **44**, 55 (1988).
61. P. E. Eaton and D. Stossel, *J. Org. Chem.*, **56**, 5138 (1991).
62. Review: J. Tsuji and K. Ohno, *Synthesis*, 157 (1969).
63. J. Tsuji and K. Ohno, *J. Am. Chem. Soc.*, **90**, 94 (1968).
64. H. E. Eschinazi, *J. Am. Chem. Soc.*, **81**, 2905 (1959).
65. J. M. Conia and C. Faget, *Bull. Soc. Chim. Fr.*, 1963 (1964).
66. H. E. Eschinazi and H. Pines, *J. Org. Chem.*, **24**, 1369 (1959).
67. J. Falbe, H. Tumes, and H. Hahn, *Angew. Chem.*, **82**, 181 (1970).
68. M. S. Newman and H. V. Zahm, *J. Am. Chem. Soc.*, **65**, 1097 (1966); M. S. Newman and N. Gill, *J.Org. Chem.*, **31**, 3860 (1966).
69. J. W. Wilt and V. P. Abegg, *J. Org. Chem.*, **33**, 923 (1968).
70. N. E. Hoffman and T. Puthenpurackal, *J. Org. Chem.*, **30**, 420 (1965).
71. H. E. Eschinazi, *Bull. Soc. Chim. Fr.*, 967 (1952).
72. K. Matsuzawa, T. Kimura, Y. Murao, and H. Hashizume, *Ger. Offen.*, 2 232 252; *Chem. Abstr.*, **78**, 84043, 98244 (1973).
73. S. Murahashi, T. Naota, and N. Nakajima, *J. Org. Chem.*, **51**, 898 (1986).
74. R. F. C. Brown, F. W. Eastwood, and B. E. Kissler, *Tetrahedron Lett.*, **29**, 6861 (1988).
75. H. Nakazawa, Y. Matsuoka, I. Nakagawa, and K. Miyoshi, *Organometallics*, **11**, 1385 (1992).
76. W. B. Hardy and R. P. Bennett, *Tetrahedron Lett.*, 961 (1967).
77. P. Braunstein, R. Bender, and J. Kervennal, *Organometallics*, **1**, 1286 (1982).
78. R. Ugo, R. P. Sato, M. Pizzotti, P. Nardi, C. Dossi, A. Andretta, and G. Capparella, *J. Organomet. Chem.*, **417**, 211 (1991).

79. S. B. Halligudi, R. V. Chaudhari, and L. K. Doraiswamy, *Ind. Eng. Chem. Process Res. Dev.*, **23**, 794 (1984).
80. B. M. Choudary, K. K. Rao, S. B. Pirdzkov, and A. L. Lapidus, *Synth. Commun.*, **21**, 1923 (1991).
81. V. L. K. Valli and H. Alper, *J. Am. Chem. Soc.*, **115**, 3778 (1993).
82. J. Zajacek and J. J. McCoy, *US Pat.*, 3 993 685 (1976); *Chem. Abstr.*, **86**, 89440 (1977).
83. E. Alessio and G. Mestroni, *J. Organomet. Chem.*, **291**, 117 (1985).
84. H. Alper and G. Vasapollo, *Tetrahedron Lett.*, **28**, 6411 (1987).
85. A. Boontempi, E. Alessio, G. Chanos, and G. Mestroni, *J. Mol. Catal.*, **42**, 67 (1987).
86. P. Leconte, F. Metz, A. Mortreux, J. Osborn, and F. Paul, *Chem. Commun.*, 1616 (1990).
87. Y. Izumi, Y. Satoh, and K. Urabe, *Chem. Lett.*, 795 (1990).
88. H. A. Dieck, R. M. Laine, and R. F. Heck, *J. Org. Chem.*, **40**, 2819 (1975).
89. M. Akazome, T. Kondo, and Y. Watanabe, *Chem. Lett.*, 769 (1992); M. Akazome, T. Kondo, and Y. Watanabe, *Chem. Commun.*, 1466 (1991).
90. A. G. Mohan, *J. Org. Chem.*, **35**, 3982 (1970).
91. H. Alper and N. Hamel, *Tetrahedron Lett.*, **28**, 3237 (1987).
92. H. Alper and C. P. Mahatantila, *Organometallics*, **1**, 70 (1982).
93. S. Aoki, E. Nakamura, and I. Kuwajima, *Tetrahedron Lett.*, **29**, 1541 (1988).
94. J. L. Graff and M. G. Romanelli, *Chem. Commun.*, 337 (1987).
95. G. Jenner, *Tetrahedron Lett.*, **31**, 3887 (1990).
96. Y. Inoue, H. Izumida, Y. Sasaki, and H. Hashimoto, *Chem. Lett.*, 863 (1976).
97. Y. Inoue, Y. Sasaki, and H. Hashimoto, *Chem. Commun.*, 718 (1975).
98. K. Kubo, H. Phala, N. Sugita, and Y. Takezaki, *Chem. Lett.*, 1495 (1977).
99. Y. Morimoto, Y. Fujiwara, H. Taniguchi, Y. Hori, and Y. Nagano, *Tetrahedron Lett.*, **27**, 1809 (1986).
100. I. D. Entwistle, A. E. Jackson, R. A. W. Johnstone, and R. P. Telford, *J. Chem. Soc., Perkin Trans. 1*, 443 (1977).
101. M. O. Terpko and R. F. Heck, *J. Org. Chem.*, **45**, 4992 (1980).
102. J. R. Weir, B. A. Patel, and R. F. Heck, *J. Org. Chem.*, **45**, 4926 (1980).
103. S. Ram and R. F. Ehrenkaufer, *Tetrahedron Lett.*, **25**, 3415 (1984).
104. H. Wiener, J. Blum, and Y. Sasson, *J. Org. Chem.*, **56**, 4481 (1991).
105. M. Petrini, R. Ballini, and G. Rosini, *Synthesis*, 713 (1987).
106. G. W. Kabalka, R. D. Pace, and P. P. Wadgaonkar, *Synth. Commun.*, **20**, 2453 (1990).
107. P. Bakczewski and J. A. Joule, *Synth. Commun.*, **20**, 2815 (1990).
108. H. Alper and G. Vasapollo, *Tetrahedron Lett.*, **33**, 7477 (1992).
109. G. R. Brown and A. J. Foubuster, *Synthesis*, 1036 (1982).
110. S. Ram and L. D. Spicer, *Tetrahedron Lett.*, **29**, 3741 (1988).
111. T. Gartiser, C. Selve, and J. J. Delpuleh, *Tetrahedron Lett.*, **24**, 1609 (1983).
112. R. Dharanipragada, E. Nicolas, G. Toth, and V. J. Hruby, *Tetrahedron Lett.*, **30**, 6841 (1989).
113. S. Ram and L. D. Spicer, *Tetrahedron Lett.*, **28**, 515 (1987); *Synth. Commun.*, **17**, 415 (1987).
114. T. Bieg and W. Szeja, *Synthesis*, 76 (1985).
115. Y. Watanabe, Y. Maki, K. Kikuchi, and H. Sugiyama, *Chem. Ind. (London)*, 272 (1984).
116. S. Torii, H. Okumoto, S. Nakayasu, and T. Kotani, *Chem. Lett.*, 1975 (1989).
117. H. Yamashita, N. P. Reddy, and M. Tanaka, *Chem. Lett.*, 315 (1993).

118. I. Ojima and S. Inaba, *J. Organomet. Chem.*, **140**, 97 (1977).
119. L. H. Sommer and J. E. Lion, *J. Am. Chem. Soc.*, **91**, 7061 (1967).
120. K. Yamamoto and M. Teramae, *Bull. Chem. Soc. Jpn.*, **62**, 2111 (1989).
121. T. Tanase, T. Ohizumi, K. Kobayashi, and Y. Yamamoto, *Chem. Commun.*, 707 (1992).
122. Y. Ito, M. Suginome, T. Matsuura, and M. Murakami, *J. Am. Chem. Soc.*, **113**, 8899 (1991); **110**, 3692 (1988).
123. Y. Ito, E. Ihara, M. Murakami, and M. Shiro, *J. Am. Chem. Soc.*, **112**, 6446 (1990); Y. Ito, E. Ihara, and M. Murakami, *Angew. Chem., Int. Ed. Engl.*, **31**, 1509 (1992).
124. H. Matsumoto, T. Nakano, M. Kato, and Y. Nagai, *Chem. Lett.*, 223 (1978).
125. K. Takahashi, A. Miyake, and G. Hata, *Bull. Chem. Soc. Jpn.*, **44**, 3484 (1971).
126. K. Schorpp, P. Kreutzer, and W. Beck, *J. Organomet. Chem.*, **37**, 397 (1972).
127. G. Lopez, G. Sanchez, G. Garcia, J. Galvez, and J. Ruiz, *J. Organomet. Chem.*, **321**, 273 (1987).
128. J. O. Baeg and H. Alper, *J. Org. Chem.*, **57**, 157 (1992).
129. T. Hosokawa, Y. Yamada, and S. Murahashi, *Chem. Commun.*, 1245 (1983).
130. K. Sato, T. Igarashi, Y. Yanagisawa, N. Kawaguchi, H. Hashimoto, and J. Yoshimura, *Chem. Lett.*, 1699 (1988).
131. F. E. Ziegler and J. S. Tung, *J. Org. Chem.*, **56**, 6530 (1991).
132. T. Naota, S. Sasao, K. Tanaka, H. Yamamoto, and S. Murahashi, *Tetrahedron Lett.*, **34**, 4843 (1993).

Index